**Aboriginal Man**
**and Environme**
**in Australia**



*The Continent of Australia is so vast, and the dialects, customs, and ceremonies of its inhabitants so varied in detail, though so similar in general outline and character, that it will require the lapse of years, and the labours of many individuals, to detect and exhibit the links which form the chain of connection in the habits and history of tribes so remotely separated; and it will be long before any one can attempt to give to the world a complete and well-drawn outline of the whole.*

*Edward John Eyre, 1845*

# Aboriginal Man and Environment in Australia

Editors
D. J. Mulvaney
and J. Golson

Australian National University Press Canberra 1971

Registered in Australia for transmission by post as a book.
National Library of Australia Card no. and ISBN 0 7081 0452 5.
Library of Congress Catalog Card no. 75-135160

# Preface

It is appropriate that this wide ranging series of studies directed to greater human understanding of the Australian Aboriginal past is published so close to the two hundredth anniversary of Captain James Cook's arrival on the Australian coast. Cook expressed sympathetic interest in many aspects of Aboriginal life and was one of the earliest observers to advance objective comment on such issues as racial origins, language, material culture, economy, and foreign cultural influences. Almost 125 years ago another distinguished explorer, Edward John Eyre, hoped for 'a complete and well-drawn outline' of Aboriginal prehistory; yet the subject was sadly neglected until around the present decade.

All the papers included in this volume were presented originally at seminars at the Australian National University between October and December 1968 and reflect the research situation at that time. They represent the most comprehensive attempt yet made in Australian prehistoric studies to correlate research data in different disciplines, to assess the current state of knowledge, and to explore avenues for future inter-disciplinary research.

As organisers of the seminar series, we naturally invited participants for their expertise in their particular fields, but in addition we had to consider their availability and proximity to Canberra. Consequently, Canberra and Sydney institutions contributed all but two of the speakers. This bias is offset, to some extent, by the fact that those two centres contain the only two academic Prehistory Departments in Australia, and that relevant Divisions of the Commonwealth Scientific and Industrial Research Organization are also based in Canberra. A few topics were neglected because the appropriate person was unavailable at that time; some other potential thematic or regional assessments are unrepresented simply because directed research in that field is lacking. If the southeastern regions of the continent appear to be emphasised, this is a fair reflection of present knowledge.

Therefore, inevitable gaps in coverage exist, and this is not the comprehensive survey which may be possible within a few years. Yet it constitutes a landmark in Australian prehistoric research, in its attempt to re-evaluate evidence critically, in its interdisciplinary approach, and in its expression of some contemporary trends in research.

It is worth observing that only about one-fifth of the contributors worked in the Australian field more than a decade ago, at which time neither the Australian National University nor the University of Sydney offered any training in prehistoric archaeology. It is relevant that during 1969 the first Sydney-trained undergraduate prehistorians completed their postgraduate research theses in the Prehistory Department at the ANU. During this past decade also, the Commonwealth Government established the Australian Institute of Aboriginal Studies, which has sponsored much of the field work described here. One major theme omitted from this seminar—the protection and conservation of Aboriginal antiquities through means of legislation, public education, and laboratory research and treatment—was the subject of a conference organised by the A.I.A.S. during 1968.

Although the editors are both prehistorians, they have attempted to maintain a catholic balance between the human and natural aspects of prehistory, and many contributors concerned themselves with problems involving both aspects. Nine prehistorians presented papers, together with three anthropologists, a historical geographer, a linguist, a geneticist, and a physical anthropologist, all of whom provided insights into aspects of man's colonisation of Australia and his regional and temporal adaptations to its challenges. Environmental and ecological conditions were assessed by six geologists or geomorphologists, a plant ecologist, and a biologist. The chronological coverage begins with the earliest evidence for human occupation, and the volume title was chosen to emphasise that it was Aboriginal man, and not merely his prehistory, which was the chief concern. Several contributions therefore utilise evidence from the recent historical past, while others are set in the context of the 'ethnographic present'. All contributors have presented bibliographies which direct readers to the relevant literature for pursuing further research. All papers were revised in the light of symposium comment.

Several articles contain specific illustration, but in order to minimise the number of text figures, some general purpose illustrations serve the balance. In the case of these latter chapters, the endpaper map contains references to the most important topographic features and place names referred to in these contributions. Stone artifact typology is represented at appropriate places in the text. We realise that 'ideal' implement types and universal implement nomenclatures are repugnant to some workers. It was considered, however, that some visual aids were necessary for the non-specialist reader, in order to conceptualise several frequently used terms. For this reason, line drawings of 'typical' selected stone artifacts have been included.

Australian prehistorians, and those scientists whose evidence is related to Aboriginal prehistory, feel a sense of purpose and immediacy, and this conviction was evident during symposium discussion. Australia, the only continent whose prehistory ended with the Industrial Revolution, possesses unique advantages which make it a laboratory for the study of both ecological and cultural relationships between man and nature and man and man, in the stone using, hunter-gatherer state of human social organisation.

These papers attempt the first comprehensive specialist sketch of prehistoric environment, land use and human adaptation in Australia. Man the hunter is envisaged as a dynamic agent in the environment, whose role was not that of the tradition-bound and parasitic nomad of so many conventional accounts.

On the evidence assembled here, it is possible that Australia was settled before the New World was penetrated. Even 20,000 years ago Aborigines possessed artistic and technological skills which earlier workers would not have credited. In both adapting to and shaping the Australian landscape, the Aborigines left traces which systematic inter-disciplinary research can transform into prehistory.

Australia has been ignored in many works dealing with society on a global scale. Participants at this symposium urged that the time has arrived when the status of the Australian evidence has achieved world significance. This volume is offered as a preliminary statement of this proposition, and a focus from which further research may be directed.

D.J.M.
J.G.
CANBERRA 1969

# Notes on Contributors

**J. N. Jennings,** Professorial Fellow in Geomorphology in the Research School of Pacific Studies, Australian National University, graduated in Geography at the University of Cambridge. After army service during World War II, he lectured at the University of Leicester until assuming his Australian appointment in 1952. Since then he has worked extensively in the karst areas and on sea level changes around the Australian coast. With J. A. Mabbutt he edited *Landform Studies from Australia and New Guinea* for the ANU Press.

**R. W. Galloway,** Senior Research Scientist, Division of Land Research, Commonwealth Scientific and Industrial Research Organization, graduated in Geography at the University of Edinburgh. After some years of research on glacial and periglacial features of Europe and the Arctic, including a doctorate from the University of Edinburgh on periglacial phenomena in Scotland, he joined CSIRO in 1958. In Australia he has worked on regional surveys in Queensland and New South Wales and on Quaternary environments.

**A. B. Costin,** Senior Principal Research Scientist, Division of Plant Industry, Commonwealth Scientific and Industrial Research Organization, joined CSIRO in 1955. After graduating in Agricultural Science at the University of Sydney, he worked with the Soil Conservation Service of New South Wales and the Soil Conservation Authority of Victoria. He is an ecologist whose special studies on the soils, vegetation, and climatic history of southeastern Australia have gained him a D.Sc.Agr. from the University of Sydney.

**Simon Pels** is Regional Research Officer at Deniliquin with the Water Conservation and Irrigation Commission of New South Wales. Since graduating in Geology at the University of Melbourne, he has spent twelve years working on hydrological and geomorphological subjects, particularly in relation to drainage and salinity problems in irrigation districts.

**J. M. Bowler** is a Fellow in the Department of Biogeography and Geomorphology, Research School of Pacific Studies, Australian National University. He graduated in Geology at the University of Melbourne, where he subsequently held a lectureship. After initial work on problems of Tertiary statigraphy, he has concentrated on the Quarternary and his doctoral thesis, presented at the Australian National University, is in this field. He has co-operated in archaeological projects in the Australian field.

**J. A. Mabbutt,** Professor of Geography, University of New South Wales, formerly Principal Research Scientist, Division of Land Research, Commonwealth Scientific and Industrial Research Organization, is a graduate in Geography of the University of Cambridge. Joining the staff of the University of Capetown, he studied desert landforms in southern Africa, and since coming to Australia he has continued his arid zone research. With J. N. Jennings he edited *Landform Studies from Australia and New Guinea* for the ANU Press.

**J. H. Calaby** is a Principal Research Scientist, Division of Wildlife Research, Commonwealth Scientific and Industrial Research Organization. He is especially interested in the biology, taxonomy, and distribution of native mammals and the effects of prehistoric and historical factors on their distribution and abundance. He has advised archaeologists on problems of identifying excavated remains of marsupial fauna.

**R. (Bud) Frank** graduated in Geology at the University of Texas. As a Research Scholar in the Department of Biogeography and Geomorphology, Research School of Pacific Studies, Australian National University, he completed research on cave sediments in New South Wales, particularly as these relate to karst development and to surface and underground palaeoenvironments. He was invited to participate in the Koonalda project by the Australian Institute of Aboriginal Studies. He is now lecturing in the Department of Geology and Geography, University of the West Indies.

**R. V. S. Wright** is Senior Lecturer in the Department of Anthropology, University of Sydney. He graduated in Prehistoric Archaeology at the University of Cambridge and spent some time working in East Africa. Since coming to Australia in 1961, he has carried out field work in New South Wales and Cape York, and in 1967 he directed the Australian Institute of Aboriginal Studies project at Koonalda Cave. He is a Council Member of the AIAS.

**R. J. Lampert** is Field Officer in the Department of Prehistory, Research School of Pacific Studies, Australian National University. Before taking up this post in 1964, he excavated English archaeological sites for the British Ministry of Works. He has done intensive work on Aboriginal sites on the south coast of New South Wales as well as a number of projects in New Guinea.

**Carmel White** is a graduate in Archaeology and Anthropology of the Universities of Capetown and Cambridge. Her postgraduate research in Arnhem Land was completed in the Department of Prehistory, Research School of Pacific Studies, Australian National University. Subsequently she was a Lecturer in Prehistory at the University of Sydney but has recently moved to North America, where she is teaching at the Binghampton campus of the State University of New York.

**I. C. Glover** graduated in Anthropology at the University of Sydney. As a Research Scholar in the Department of Prehistory, Research School of Pacific Studies, Australian National University, he excavated in Portuguese Timor. He now holds a lectureship at the Institute of Archaeology, University of London.

**J. Peter White** is a graduate of the Universities of Melbourne (History) and Cambridge (Prehistoric

Archaeology). His doctoral thesis in the Department of Prehistory, Research School of Pacific Studies, Australian National University, was concerned with New Guinea Highlands prehistory. As Assistant Curator of Anthropology, Australian Museum, Sydney, he has continued to work primarily in this field, with special emphasis on technological ethnoarchaeology.

**Jack Golson** is the first Professor of Prehistory, Research School of Pacific Studies, Australian National University. A graduate in History and Archaeology of the University of Cambridge, he spent several years at the University of Auckland, during which period he carried out extensive field work in Polynesia. He moved to Canberra in 1961 and since then has excavated in Australia and New Guinea and directed several other research projects in those areas.

**Nicolas Peterson**, who graduated in Anthropology at the University of Cambridge, spent three years in the Northern Territory as Research Officer of the Australian Institute of Aboriginal Studies. In Arnhem Land he lived with one of the last Aboriginal groups still subsisting largely 'off the land'. This field work formed the subject of his doctoral dissertation at the University of Sydney.

**Roger Lawrence** graduated in Geography at the University of Sydney and completed a research project on Aboriginal habitat and economy while an M.A. student in the Department of Geography, School of General Studies, Australian National University. He is now a Lecturer in Geography at the Victoria University of Wellington, New Zealand.

**J. P. M. Long** is a graduate in History of the University of Sydney. He served in the Welfare Branch of the Northern Territory Administration from 1955 to 1968, during which time he carried out prolonged field surveys in central Australia. He is now an Assistant Director in the Office of Aboriginal Affairs, Canberra.

**Rhys Jones** joined the staff of the Department of Anthropology, University of Sydney, after graduating in Prehistoric Archaeology at the University of Cambridge. Since 1963 he has undertaken field work in northwestern Tasmania, a project sponsored by the Australian Institute of Aboriginal Studies. He is now a Research Fellow in the Department of Prehistory, Research School of Pacific Studies, Australian National University.

**Diane E. Barwick** is an Anthropology graduate of the University of British Columbia, Vancouver. Her postgraduate research on the Aboriginal population of Victoria was completed at the Australian National University, where she is now a Research Fellow in the Department of Anthropology and Sociology, Research School of Pacific Studies.

**A. G. Thorne** is a Lecturer in Anatomy at the University of Sydney. In 1967 he gained his Master's degree for research into the osteology and affinities of the Tasmanian Aborigines, based upon material excavated by Rhys Jones. He is currently engaged on a major field program to recover Pleistocene human skeletal remains in a lunette complex in northern Victoria.

**R. L. Kirk** is Senior Fellow in Human Genetics, John Curtin School of Medical Research, Australian National University. A graduate of the University of Birmingham, he holds the degree of D.Sc. from the University of Western Australia, where he was a staff member from 1949 to 1965. Before taking up his Canberra post, he served as Chief, Human Genetics Unit, World Health Organization, Geneva.

**D. T. Tryon** graduated from the University of Canterbury, New Zealand, and obtained his doctorate in the Department of Linguistics, Research School of Pacific Studies, Australian National University, where he is now a Research Fellow. His linguistic interests are in Australian Aboriginal languages, on a descriptive and comparative level, and also in Austronesian, particularly Melanesian, linguistics.

**Robert Edwards** is Curator of Anthropology at the South Australian Museum, Adelaide. For some years he has been making a systematic photographic coverage of Aboriginal art sites in South Australia and the Northern Territory. He is a member of the Australian Institute of Aboriginal Studies Advisory Committee on Film Production.

**D. J. Mulvaney** was Senior Fellow in Prehistory, Research School of Pacific Studies, Australian National University, until his appointment to the first Chair of Prehistory in the School of General Studies. He graduated in History at the University of Melbourne and in Prehistoric Archaeology at the University of Cambridge. After lecturing in History at the University of Melbourne, he transferred to Canberra in 1965. He has published both on Australian archaeology and Aboriginal history and is an Executive Member of the Australian Institute of Aboriginal Studies.

# Acknowledgments

We are grateful to all participants of the symposium for their co-operation in submitting revised papers for publication. In a volume involving such a complexity of authors, delays and difficulties during its assembly are inevitable; we are grateful to the staff of the Australian National University Press for their advice and patience.

Individual authors make their own appropriate acknowledgments, but we are indebted particularly for specialist assistance from the following: Lois White of our Prehistory Department typed numerous chapter drafts and the entire final copy; most of the maps and other line illustrations are the work of Winifred Mumford, of the same Department. Robin Dowie checked all bibliographical references, a painstaking task which saved the editors many worries. Josephine Flood and Rosemary Groube compiled the index and Rosemary Groube read the proofs.

Unless stated otherwise, all photographs in this volume come from the camera of Robert Edwards, Curator of Anthropology, South Australian Museum.

Plate VII has been reproduced from the Nan Kivell Collection with the permission of the National Librarian, National Library of Australia, Canberra.

We particularly wish to thank Sir Keith Hancock for the interest which he has taken both in the symposium itself and in this publication.

# Contents

Preface v
Notes on Contributors ix
Acknowledgments xiii

1 Sea Level Changes and Land Links
J. N. Jennings 1
2 Evidence for Late Quaternary Climates
R. W. Galloway 14
3 Vegetation, Soils, and Climate in Late Quaternary Southeastern Australia
A. B. Costin 26
4 River Systems and Climatic Changes in Southeastern Australia
Simon Pels 38
5 Pleistocene Salinities and Climatic Change: Evidence from Lakes and Lunettes in Southeastern Australia
J. M. Bowler 47
6 The Australian Arid Zone as a Prehistoric Environment
J. A. Mabbutt 66
7 Man, Fauna, and Climate in Aboriginal Australia
J. H. Calaby 80
8 Cave Sediments as Palaeoenvironmental Indicators, and the Sedimentary Sequence in Koonalda Cave
R. Frank 94
9 The Archaeology of Koonalda Cave
R. V. S. Wright 105
10 Coastal Aborigines of Southeastern Australia
R. J. Lampert 114
11 Prehistory in the Cape York Peninsula
R. V. S. Wright 133
12 Man and Environment in Northwest Arnhem Land
Carmel White 141
13 Prehistoric Research in Timor
I. C. Glover 158
14 New Guinea and Australian Prehistory: the 'Neolithic Problem'
J. Peter White 182
15 Australian Aboriginal Food Plants: Some Ecological and Culture-Historical Implications
Jack Golson 196

16 Open Sites and the Ethnographic Approach to the Archaeology of Hunter-Gatherers
Nicolas Peterson 239

17 Habitat and Economy: A Historical Perspective
Roger Lawrence 249

18 Arid Region Aborigines: The Pintubi
J. P. M. Long 262

✓19 The Demography of Hunters and Farmers in Tasmania
Rhys Jones 271

20 Changes in the Aboriginal Population of Victoria, 1863-1966
Diane E. Barwick 288

21 The Racial Affinities and Origins of the Australian Aborigines
A. G. Thorne 316

22 Genetic Evidence and its Implications for Aboriginal Prehistory
R. L. Kirk 326

23 Linguistic Evidence and Aboriginal Origins
D. T. Tryon 344

24 Art and Aboriginal Prehistory
Robert Edwards 356

25 Aboriginal Social Evolution: A Retrospective View
D. J. Mulvaney 368

Appendix 15:1
Species List of the Food Plants of Arnhem Land, Cape York, and Central Australia 210

Appendix 15:2
Generic List of the Food Plants of Arnhem Land, Cape York, and Central Australia 233

Appendix 15:3
Genera and Species Recorded as Food Sources in Malaysia, Present but Apparently Unused in Australia 236

Index 381

# Plates

Between pp. 362 and 363


I Upland and sand desert in the Rumbalara Hills, N.T.
II Riverine desert, southwestern Queensland
III Koonalda Cave, S.A., from the groundsurface
IV The Gallus site, Koonalda Cave, S.A.
V Wright's deep pit, Koonalda Cave, S.A., with upper chalky rubble and lower red sediments
VI Detail of the lower red unit, Koonalda Cave, S.A.
VII Group of Tasmanian Aborigines around their camp fires on a shell midden
VIII Circular depression on the surface of a shell midden at West Point, Tasmania
IX Large engraved pavement fractured by natural weathering, Ooraminna, central Australia
X Human figures at Mootwingee, western N.S.W.
XI Surviving fragment of a former extensive engraved pavement, Panaramitee, S.A.
XII Engraved designs, Eucolo Creek, near Woomera, S.A.
XIII Representative designs on a rock face at Panaramitee, S.A.
XIV Engravings at Amulda Springs, central Australia
XV Simple portrayal of an adult emu, Florina, S.A.
XVI Emu tracks at Tukulnga rockhole, western central Australia
XVII Unusual male and female figures at N'Dahla Gorge, central Australia
XVIII Large fern-like designs at N'Dahla Gorge, central Australia
XIX Abraded grooves and pecked engravings at Ingaladdi, N.T.
XX Recent motifs at Ingaladdi, N.T.
XXI Abraded grooves at Nackara Springs, S.A.
XXII Abraded grooves at Cleland Hills, central Australia
XXIII Finger markings in Koonalda Cave, S.A.
XXIV Abraded grooves, Koonalda Cave, S.A.
XXV Surface fretting at Ooraminna, central Australia
XXVI Heart-shaped faces, Cleland Hills, central Australia
XXVII Cleland Hills art, central Australia
XXVIII Excavated hearth at Panaramitee, S.A.
XXIX The large gallery at Mt Cameron West, Tasmania
XXX Grinding the edge of a diorite axe blank, near Yuendumu, central Australia
XXXI Stages and techniques of hafting

# Figures


1:1 Continental shelves around Australia 5
1:2 Bass Strait 6
1:3 Torres Strait 8
3:1 Diagrammatic presentation of $K_1$, $K_2$, and $K_3$ groundsurfaces 28
3:2 Diagrammatic representation of the history of climatic conditions, soil-forming intervals, and depositional systems 30
3:3 Phases of landscape development on the Canberra-Queanbeyan Plain 31
3:4 Southeastern Australia 34
4:1 The Riverine Plain 39
4:2 Cross-section of a prior stream 40
4:3 Cross-section of an ancestral river 43
4:4 Possible interpretation of geological events 45
5:1 Relationship between distribution of linear dunes and lunette lakes 50
5:2 Regional climatic map of southeastern Australia 54
5:3 Geomorphic map of northwestern Victoria and southwestern N.S.W. 56
5:4 Detailed geomorphic map of the Willandra lakes 57
5:5 Generalised stratigraphic sections through lunettes of Willandra lake system 58
5:6 Stratigraphic sections through southern end of the Walls of China 60
6:1 The Australian arid zone: organisation of drainage 68
6:2 The Australian arid zone: desert habitats and dune ridges 75
9:1 Selected specimens of tula adze flakes 108
9:2 Section through Koonalda Cave 109
10:1 Excavated sites in eastern N.S.W. 115
10:2 Selected flake tools older than 5,000 years 117
10:3 Selected backed blades 118
10:4 Selected eloueras 119
10:5 Vertical section of south face of trench, Burrill Lake shelter 122
10:6 a. Vertical section of west face of trench, Currarong shelter 1 124
b. Distribution of implements in trenches 1 and 2. Currarong shelter 1 124
12:1 Archaeological sites in northwest Arnhem Land 142
12:2 Diagrammatic presentation of contents of excavated sites 145
12:3 Pleistocene edge-ground axes from Arnhem Land 148
12:4 Selected examples of one thick, steeply retouched flake and three core-scrapers 150

**12:5** Selected examples of bifacial and unifacial points 151
**13:1** Timor, showing the main geological formations and archaeological sites 162
**13:2** Uai Bobo 1 (TO1), plan and sections 165
**13:3** Uai Bobo 2 (TO2), plan and sections 166
**13:4** Provisional depth-age calculations for TO2, correlated with TO1 167
**13:5** Numbers of waste flakes per $m^3$ 168
**13:6** Numbers of sherds per $m^3$ 170
**13:7** Percentages of the major classes of flaked stone tools 173
**14:1** Australia and New Guinea at Torres Strait 186
**19:1** Tasmanian localities 272
**19:2** Tribal and linguistic map of Tasmania 275
**19:3** Distribution of Aboriginal man in Tasmania 276
**19:4** Presumed decline and collapse of Tasmanian Aboriginal population 283
**19:5** European population growth in Tasmania compared with Aboriginal population decline 285
**20:1** Locations of Victorian Stations and Reserves, 1860-1966 290
**20:2** Age structure of nine tribal remnants, 1863 302
**20:3** Age at death, Victorian Aboriginal Station residents, 1876-1912 304
**20:4** Age at death, classified by sex and caste, 1876-1912 305
**20:5** Known causes of death, Victorian Aboriginal Station residents, 1876-1912 306
**20:6** Mean family size, Victorian Aboriginal women, 1863-1966 311
**20:7** Infant and child mortality, issue of women, 1863-1966 312
**22:1** Area distribution in Australia of Rh-E positive persons 329
**22:2** Localities mentioned in Table 22:5 333
**22:3** Frequency ranges for various alleles and phenotypes in Aboriginal populations in Australia and New Guinea. 336
**22:4** Frequency ranges for various alleles and phenotypes in Aboriginal populations in Australia and the Ainu of Japan 337
**22:5** Frequency ranges for various alleles and phenotypes in Aboriginal populations in Australia and Bushmen in Africa 338
**22:6** Frequency ranges for various alleles and phenotypes in Aboriginal populations in Australia and the Veddoids in Ceylon and south India 340
**24:1** Art sites mentioned in text 357

# Tables


3:1 Climates indicated by some vegetation disjunctions in southeastern Australia 27
3:2 C-14 determinations of the K-cycle terraces, south coast, N.S.W. 29
3:3 Some C-14 dates from southeastern Australia and their probable significance 32
5:1 Summary of stratigraphic sequence, radiocarbon chronology, and evidence of human occupation from the dry lakes of the Willandra Creek, N.S.W. 62
6:1 Expected number of rainfalls per year at selected stations in the arid zone 67
6:2 Extent of Australian desert habitats 73
10:1 Chronology and cultural sequence on excavated sites in eastern Australia 129
11:1 Changes in the size of implements over time 137
11:2 Characteristics of Cape York archaeological sites 139
13:1 Flaked stone: numbers in each horizon of TO1 and TO2 169
13:2 Numbers of incised sherds 172
13:3 Numbers of shell ornaments 174
13:4 Preliminary analysis of faunal remains 175
15:1 Uses in Malaysia of shared species and genera used for food in Arnhem Land 199
15:2 Distribution by Specht's ecological categories of 65 Arnhem Land food plants 200
15:3 Produce of 65 Arnhem Land food plants by season and ecological zone 200
15:4 Botanical affiliations of food plants in three Australian regions 200
15:5 Food plants shared between three Australian regions 202
15:6 Numbers of species used in one area, present but unused in another 202
15:7 Absence and presence in central Australia of Arnhem Land food genera by use 202
15:8 Utilised species in three Australian regions by food use 204
15:9 Absence and presence in Arnhem Land of central Australian food genera by use 204
15:10 Genera and species supplying seed foods in central Australia, present in Arnhem Land 204
15:11 Absence and presence in Arnhem Land of central Australian food genera by botanical affiliation 205

**15:12** Uses in Malaysia of shared genera used for food in central Australia 206
**15:13** Uses in Malaysia of shared genera used for food in Arnhem Land and Cape York 206
**15:14** Uses in Malaysia of plants used for food in three Australian regions and to some degree domesticated in Malaysia 208
**15:15** Occurrence in other parts of Australia of Arnhem Land food plants 209
**19:1** Calculations of Tasmanian Aboriginal population 280
**19:2** Decline of Tasmanian Aboriginal population 282
**19:3** European population of Tasmania 284
**20:1** Enumeration of Aborigines in Victoria, 1863 292
**20:2** Enumeration of Aborigines in Victoria, 1877 294
**22:1** Human blood group systems 327
**22:2** Polymorphic serum protein and enzyme systems in man 328
**22:3** Range of gene frequencies and phenotype frequencies in Australia 328
**22:4** Range of gene frequencies for serum protein groups in Australia 330
**22:5** Distribution of 6 PGD types in northern Australia 332
**24:1** Frequency of occurrence of designs 362

# 1 Sea Level Changes and Land Links

J. N. Jennings

In 1932 when a block of peat trawled from 36 m down on the floor of the North Sea fell onto the deck of the steam drifter *Colinda*, it split open to reveal a Mesolithic bone harpoon which had been enshrined there for some seven millennia. No such dramatic fusion of environmental and human prehistory has enlivened the pages of Australian archaeology. Yet archaeologists and biologists must inevitably pose to their colleagues concerned with the environment questions about the changing relative level of land and sea and about land links, questions which they are regrettably ill-equipped to answer at the present time (Bulmer and Bulmer 1964; Jones 1966; Macintosh and Barker 1965; Ridpath and Moreau 1966). Though there are universal elements in the answers to such questions, even moderately precise responses can only issue from detailed regional investigations and nowhere round the coast of this continent has a knowledge comparable with that achieved for the North Sea yet been approached.

### Universal Elements in Sea Level Change

The universal elements are the eustatic ones, the variations in the general level of the sea itself. There are many causes of such variations—alterations in the water volume through temperature and salinity changes and through the addition of new water by volcanoes, changes in the capacity of the ocean basins through sedimentation, vulcanicity, and movement of their floors. Nevertheless, the significant cause during the period with which the archaeologist is concerned—several orders greater in magnitude than any other (Fairbridge 1962)—is the withdrawal from the oceans and the return to them of water locked temporarily in the fluctuating masses of land ice. These glacioeustatic shifts affect all sea coasts equally and represent the net effects of all the separate changes, positive and negative, in the budgets of the glaciers and icesheets of the world. Detailed observations over the last century or so make it evident that though some glaciers advance whilst others retreat, there are preponderant tendencies. Increasingly better founded reconstruction of their earlier histories supports the idea that there has been broadly synchronous change in glacier mass throughout the world in the late Pleistocene and Holocene. However, rates of change may not have been the same even when the direction of change was. For example, the final Wisconsin Laurentide icesheet in North America seems to have retained substantially greater area some millennia after the equivalent Weichsel icesheet of north Europe (Manley 1955). Such differences do not affect the universality of the consequent sea level change which reflects the algebraic sum of the different rates of decline but they are a warning against over-simplified glacioeustatic inferences from gross glacial chronologies.

**Differential Elements in Sea Level Change**

The regional elements in the land and sea level equation are differential changes in sea level and movements of the solid crust. All known causes of the former can be shown to have been negligible in the period of time under consideration. As regards crustal movement, it is customary to distinguish between tectonic and isostatic elements, though in practice they may not readily be differentiated.

Tectonic movements comprise both localised folding and faulting, the orogenic or mountain-building movements, and gentler, extensive warpings, the epeirogenic or continent-building movements. The latter may vary so modestly from place to place that their expression at the coast may not easily be disentangled from eustatic effects. Some parts of the crust are tectonically much more unstable than others. Extreme instability is found in northern New Guinea, where Hossfeld (1964) has determined an uplift of 60 m in 5,000 years at Aitape, and Chappell (in Bird 1968) 250 m in 30,000 years on the Finsch Coast. By contrast Australia has been thought to be generally very stable. However, more detailed work is beginning to show that here, as in other continents, neotectonism is more widespread than has been thought in the past. Persistence of movement into the Quaternary is likely in the Cape Range of Western Australia where Lower Miocene limestones are folded anticlinally to an altitude of over 300 m. Twidale (1966) has argued for late Pleistocene-Holocene movement in the Selwyn Upwarp south of the Gulf of Carpenteria. But it is in southeastern Australia that the evidence for neotectonism is surest, with the Pleistocene shorelines of the southeast of South Australia warped up (Blackburn *et al.* 1965) and the Pleistocene sediments of the Latrobe Valley warped down (Boutakoff 1957). Bowler (1967) has been able to show that the latest movement of the Cadell Fault in the Riverina occurred between events with C-14 dates of 20,000 and 13,000 years BP. Assumption of complete tectonic stability in the Upper Quaternary thus becomes less justifiable anywhere in Australia.

The imposition of fresh loads on the crust results in downwarping as part of a process of restoration of balance in crustal loading, i.e. of isostasy. Conversely upwarping occurs when loads are removed. Thus when a river delta reaches a critical mass, it bows down the crust beneath it. An extreme instance is that of the Mississippi delta where Pleistocene deltaic beds have been carried down 5,000 m. Australia has few deltas of appreciable size and isostatic adjustment has not yet been demonstrated in any of them. The most familiar kind of isostatic adjustment is that associated with the north European and Laurentide final icesheet, where rebound following their disappearance has reached as much as 250 m. Recent studies have reduced the probable dimensions of Pleistocene glaciers in Tasmania and the southeastern mainland and there is little likelihood that isostatic movements of this origin affected Australian coastlands.

Nevertheless there remains another kind of isostasy which the Australian littoral cannot have escaped. This is the effect of changing sea level on the continental shelves (Higgins 1965). When the oceans flood the shelves in interglacial times, the latter are over-loaded and sink; when sea level is low in glacial periods, the converse applies. Bloom (1967) has attempted to measure this element of isostatic sinking due to the Flandrian or Postglacial glacioeustatic transgression at various points along the Atlantic coast of the U.S.A. It varies in amount because of different shelf widths, depths, and conformation. More empirical testing is needed but, for the present discussion, the theoretical basis seems unimpugnable for differential shifts of a significant order (up to 45 m). It is relevant that the main ocean floors will also adjust isostatically to the glacioeustatic fluctuations of load and this will modify the glacioeustatic changes of level themselves, i.e. there will be a kind of negative feedback.

**Inferences about Former Shorelines from Glacioeustatic Depth-Time Curves**

Enough has been said of the variables involved to make it clear that the changing relative level of land and sea is a regional thing to be established independently along each sector of coastline. The interpretation of the regional evidence bearing witness to this changing level is furthermore complicated by many other factors. The changing conformation of coastal water bodies will have affected tidal range; different sediments will have been subject to varying amounts of compaction; there will have been advance and retreat of the shorelines as a result of depositional and erosional factors independent of sea level change, to mention only the more outstanding of these complexities. Couple with them the fact that much of the relevant evidence lies on the sea floor and it is not surprising that scarcely anywhere round the coasts of Australia can the pertinent questions of archaeologist and biologist be answered on the only sound basis there is, the local basis.

There has consequently been pressure to take short cuts, on the axiom that, tectonically unstable areas apart, the fastest and largest shifts in relative level of land and sea in the late Pleistocene and Holocene are due to glacioeustasy. The procedure is to apply C-14 dated glacioeustatic depth-time curves established elsewhere to Australian contexts. Within certain limits of accuracy, estimates of the times at which shorelines lay at given levels and the positions of shorelines at given times can be made in this way. There are very real dangers, however, of not keeping well aware of these limits and of using this procedure in unjustifiably fine ways. The differences between published glacioeustatic curves are salutary in this respect; each one attempts to average different, if overlapping groups of data, which, even after careful selection, inevitably incorporate the effects of factors other than the universal eustatic ones.

For the period before the last glacial maximum about 20,000 years ago, direct dated evidence of sea level is very slight at the moment and so this inferential approach cannot be employed except in the broadest way. For the period since, the glacioeustatic curves of different authorities have certain attributes in common. All show rapid, substantial rise in sea level from the last glacial maximum till about 6-7,000 years ago and much less change since then. It is the long rapid rise which permits certain estimates to be made about landlinks of possible biological and prehistoric significance.

**Geography of Glacial Maxima Low Sea Levels**

Glacial low sea levels were first estimated by calculating the volume of ice in Pleistocene icesheets. Direct evidence has come subsequently from the depths of drowned and buried river valleys cut when base level was much lower and from coastal features, sediments, and fossils found deeply submerged. Present estimates for the glacial minima vary between 100 m and 150 m (Cotton 1962). Curray (1965) defends −145 m for sea levels at the early Wisconsin maximum glaciation at some time between 40,000 and 100,000 BP and −125 m for the final Wisconsin maximum about 18-20,000 BP. Between these times, sea level rose substantially towards its present level once or possibly twice during interstadials.

The outer margins of the continental shelves slope rather more steeply than the inner parts so that, despite the uncertainty attaching to these glacial sea level minima, it is sure that the shelves were very largely exposed at these times. For the same reason the use in Fig. 1:1 of the 200 m submarine contour because of its availability on published maps is not significantly misleading on a small scale map about the extent of land exposed at sea level minima in the Upper Pleistocene.

At these minima there were not only broad land bridges linking Australia to Tasmania and to New Guinea in which interest is naturally greatest, but there were also wide coastal plains elsewhere onto which the Aborigines were able to spread at these times from the present continental area. The Gulf of Carpentaria, most of the Arafura Sea and of the Timor Sea made up a tremendous additional land area on the north. From study of the bottom contours and deposits van Andel *et al.* (1967) have been able to determine some of the terrestrial characteristics of the Timor shelf and to locate certain shorelines. The lowest sea level stand is registered in a small cliff between −110 m and −130 m, and littoral shells trawled from −132 m have yielded a C-14 date of 16,910 ± 500 BP. In Queensland the land reached to the outer margin of the Barrier Reefs. The South Australian gulfs disappeared as such and calcareous dune systems have been located at the mouth of St Vincent Gulf at a depth of 50 m by diving geologists (Sprigg 1965).

Two other coastal plains of Pleistocene low sea level times, those south of the Nullarbor Plain and northwest of the Great Sandy Desert, are of particular interest, for it is in these sectors that the desert heart of Australia approaches the present coastline most closely. The question arises therefore whether these coastal plains, as well as the obvious land bridges of Torres Strait and Bass Strait, may not have acted as links or isthmuses for plants, animals, and men. There is no direct evidence as yet to indicate how long these shelves were exposed for migration by terrestrial organisms but on general grounds it can be argued that they were available in this way for many millennia during the climax of each major Wisconsin ice advance. However, the climate of these shelves at these times is equally pertinent to the question. Climatic gradients are perpendicular to the Nullarbor coast (Fig. 1:1). If the glacials were also pluvials in the sense of increased

*Fig. 1:1 Continental shelves around Australia. Isobath based on Geological Map of the World 1/5,000,000 Series, Sheets 6 and 7, Australia and Oceania. Climate boundaries from Troll (1964).*

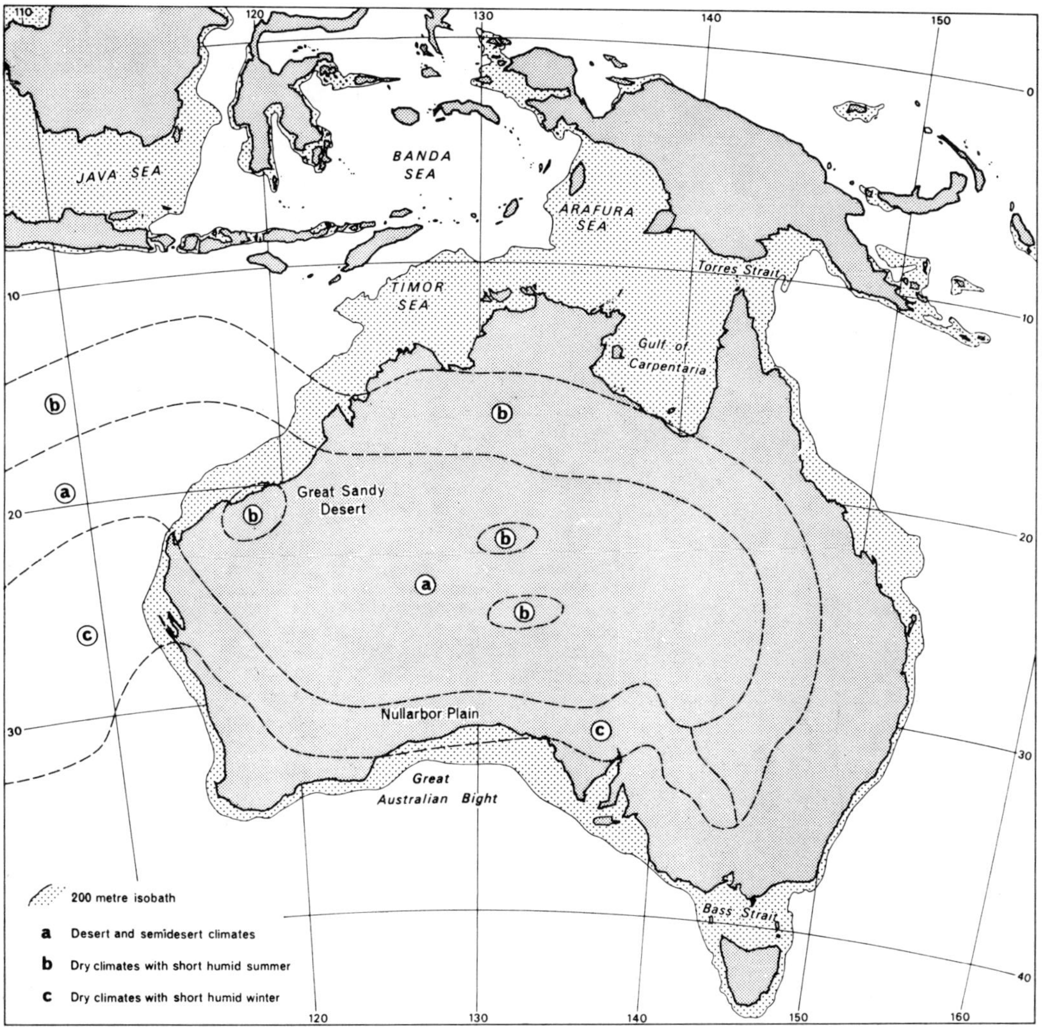


absolute precipitation, the coastal plain south of the Nullarbor must have provided the kind of migration corridor which biologists (e.g. Crocker and Wood 1947) have required to explain plant and animal distributions and which they have sought in a wetter Nullarbor Plain itself. Even if the glacials were drier, the effective precipitation of the coastal plain, lying as it does south of the Nullarbor Plain, might still have been more favourable as an environment than the latter is today.

But the northwest coastal plain is different in that it lies athwart the climatic belts. If in the glacials the mid-latitude and the equatorial humid belts expanded simultaneously,

*Fig. 1:2 Bass Strait. Based on Jennings (1959) and later R.A.N. charts.*

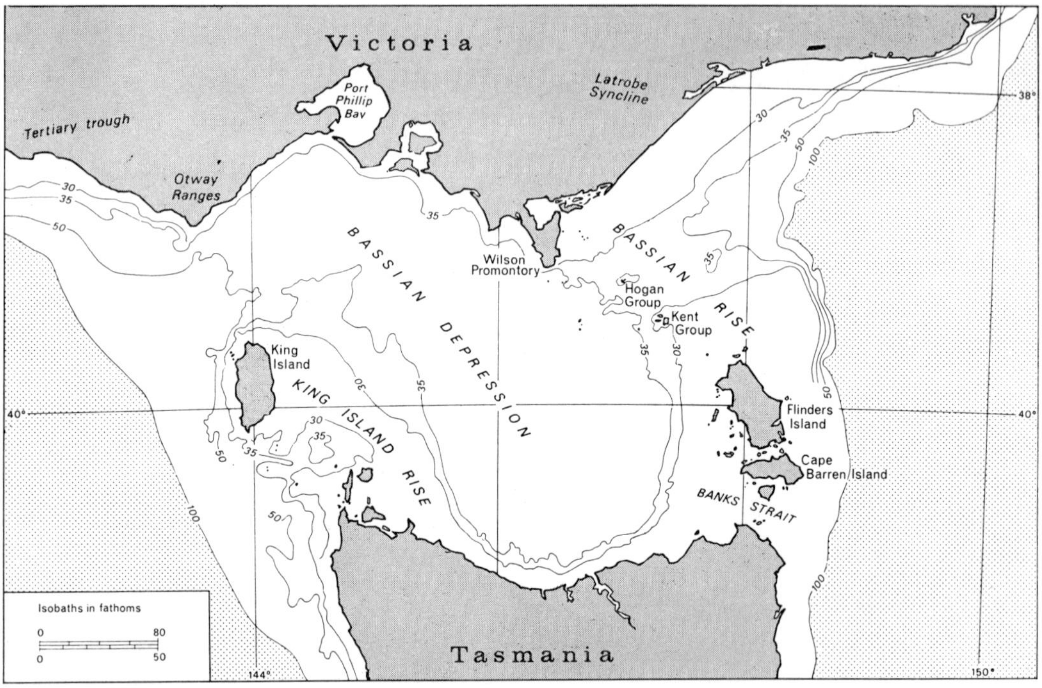


the dry tropical high pressure belt would have contracted, not disappeared, both on the continent and over the ocean on its western side so that this coastal plain would still have remained arid in part (contrast Gill 1965a). However, Mayr (1944) has argued that the equatorial humid belt retreated northwards at this time and van Andel *et al.* (1967) have found evidence in support of this, notably in the finding of kunkar concretions on the Timor shelf, which are interpreted as fossil pedocal relicts. The burial of fixed desert dunes beneath Flandrian transgression deposits in the Fitzroy estuary have also been cited as evidence of glacial aridity there (Fairbridge 1964), though at the moment these dunes cannot be proven to be older than 7,500 BP (Jennings, unpublished data). It seems much less likely therefore that this coastal plain was ever a favourable biological corridor in the Upper Pleistocene.

Although New Guinea was linked by glacial low sea level with Australia, this was not the case between New Guinea and south-east Asia. The western part of the Indonesian archipelago formed a great extension of the Asiatic continent centred on the Java Sea—Sundaland—but between the two continental masses intervenes a belt of active orogenic arcs and troughs in the Celebes, Moluccas, and Banda Sea area. Even if future research takes glacial low levels lower still, this would not alter significantly the width of the deep straits which must have persisted between New Guinea, the Banda island arc, and the Java-Flores arc. Channels of the order of 50 to 100 km would need to be crossed in the surviving half of the archipelago. If the sighting of smudges of land on the horizon was a necessary prompter for primitive men to venture forth on logs or rafts, then the necessary stimulus of a chain of intervisibility would have existed through the Celebes and Moluccas to New Guinea. The strait between Timor and the extended Australia would not have been very much greater, though Macintosh (1965) and van Andel *et al.* (1967) have underestimated the width of the Timor Trough at a depth of −130 m and lack of intervisibility of its shores would have made it an additional degree less favourable for migration.

## Final Drowning of Bass Strait Land Links

Whereas the full extent of Bass Strait (Fig. 1:2) may have been land for only a comparatively few millennia around the maximum of the final Wisconsin, the broad Bassian Rise from Wilson Promontory to Flinders Island maintained connection between the mainland and Tasmania for much longer. It cannot be assumed, however, as Ridpath and Moreau (1966) do, that this link was maintained throughout the Wisconsinian, since in one or more interstadials sea level rose considerably.

The critical sill depth on the Bassian Rise lies between 55 and 64 m (30 and 35 fathoms) on either side of the Kent and Hogan groups and between them (Jennings 1959), whereas the lowest sill depths between Flinders Island and Tasmania are between 27 and 37 m (15 and 20 fathoms). The lowest parts of the King Island Rise between King Island and Tasmania lie between 46 and 55 m (25 and 30 fathoms). Before these sill depths are applied to eustatic curves, consideration must be given to what may have happened to the sills after they were drowned in the Postglacial or Flandrian transgression. They may have been raised or lowered tectonically, and they may have been built up by bottom deposits or cut down by submarine erosion, introducing errors in determining the dates of overspilling by the sea.

It has already been pointed out that the Latrobe Syncline in East Gippsland is a neotectonic area. This is only the eastern end of a zone of instability through the Port Phillip Sunkland to the deep Tertiary sedimentary trough beneath the basalts of western Victoria. Moreover Foster, in South Gippsland, is a focus of seismicity. However, wood dated 8,780 ± 200 BP (Gill 1956) from a buried Yarra River channel at −19 m (−63 ft) sits well with other data from many parts of the world used to construct the curve of the Flandrian transgression (e.g. Godwin *et al.* 1958). Gill (1967) has also argued from the agreement between later sea level evidence at five sites stretching through southern Victoria, that there has been no significant tectonic movement there in the last 5,000 years. Moreover, there are granite basement exposures at frequent intervals from Wilson Promontory to northeastern Tasmania, and though there are submarine features suggestive of fault scarps, there is nothing about them which requires an age younger than Tertiary. It is likely therefore that any tectonic shifts in Bass Strait since 20,000 BP have been modest.

As regards sedimentation, there are shoals in Bass Strait which may have been built up substantially in the relevant period. Such are Beagle Spit off the northeastern corner of Flinders Island and the East Bank and Moriarty Shoals on the northern side of Banks Strait. But these are very different in size, form, and disposition from the broad sills between Flinders Island and Wilson Promontory, King Island and Tasmania, Cape Barren Island and Tasmania. Even though Dannevig (1915) attributed the King Island Rise to accumulation of sediment in

*Fig. 1:3 Torres Strait. Based in part on R.A.N. charts. The 'high islands' are chiefly of granite such as is found in Cape York Peninsula and at Mabaduan in Papua. The 'low islands' are of recent geological construction.*

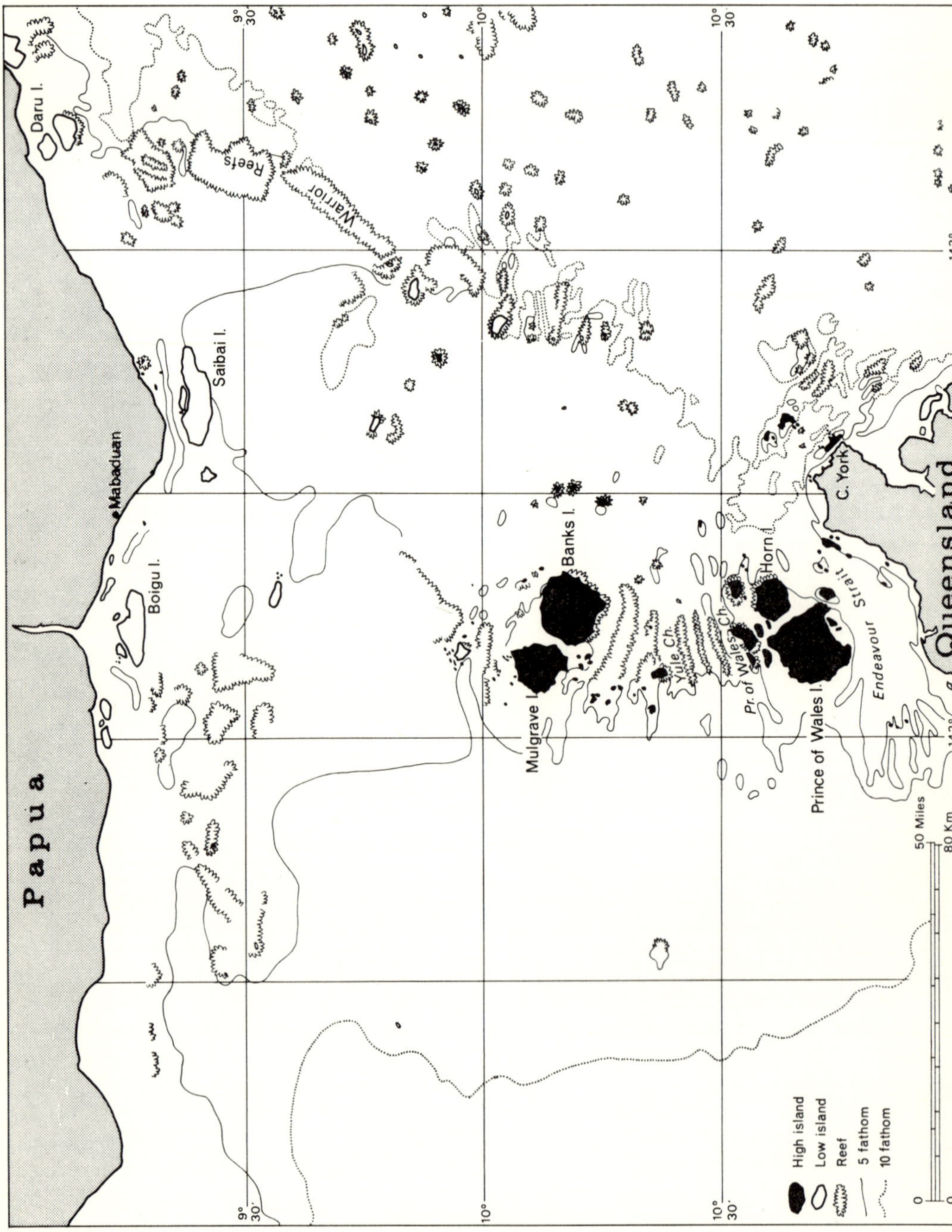

less disturbed waters than those to the windward directly between King Island and Tasmania, sedimentation on this and the other broad protuberances would only have occurred very slowly.

On the side of erosion, there are deep holes and channels in Bass Strait most probably due to tidal action, though their development may have been lengthy and as much the result of inhibition of sedimentation as of actual excavation. However, all three sills under discussion lie outside these centres of tidal scour and this factor can be disregarded in the present connection. Widespread erosion over the broad rises constituting the sills must have been very small over the period concerned.

The rising Postglacial sea first of all cut the direct route from the Otways to King Island and then spread southeastwards over the Bassian Depression. Then came the most biologically significant break between Victoria and Flinders Island. If we take the range of 55 to 64 m (30 to 35 fathoms) to allow for modest amounts of subsequent crustal movement, sedimentation, and erosion, and apply this to the glacioeustatic curves of a number of authorities (Godwin *et al.* 1958; Shepard 1960, 1964; Fisk and McFarlan 1955; Fairbridge 1961; Jelgersma 1966; Curray 1961, 1965; Schofield 1964), the dates for the overspilling of this land link range from 10,000 to 15,000 BP. However, the highest values (Fisk and McFarlan 1955; Shepard 1964) may be unduly influenced by dates from isostatically subsiding shelves and the weight of the estimates lies between 12,000 and 13,500 BP. There is the complication that this event lies within the period of quite sharp late glacial oscillations in climate and ice limits in the Northern Hemisphere, registered paleobotanically in the Bølling and Allerød interstadials. Sea level must have been affected and empirical evidence is beginning to be found in support of this theoretical inference. It may be, then, that the main Bass Strait sill was flooded before or in the Bølling, temporarily re-exposed in the Older Dryas period and finally submerged in the Allerød.

For the King Island Rise, a vertical range of 46 to 55 m (25 to 30 fathoms), again allowing latitude for some subsequent modification of the sill's depth, gives a similar range of dates as for the Bassian Rise but with a preponderance between 10,000 and 12,500 BP. A time lag of the order of a millennium between the elimination of this link and that of the major link with the mainland is probably not significant, biologically at least.

The Banks Strait link, lying between 27 and 37 m (15 and 20 fathoms), may on the same procedure have been lost between 8,500 and 10,000 BP. The significance of this 3,000-year time lag after the isolation of the whole from the mainland would be much reduced if there were an intervening re-establishment, even for a short time, of the northern link as suggested above.

After this essay was written and presented at the seminar series in October 1968, my attention was drawn to a paper I had overlooked, by Littlejohn and Martin (1965: 247). In this, estimates are given for the dates of the drowning of the Wilson Promontory-Flinders Island Link (*c.* 12,000), the King Island-Tasmanian Link (*c.* 11,000) and the Banks Strait Link (*c.* 10,000). Since these are based on my bathymetric map and on the Godwin, Suggate, and Willis glacioeustatic curve, both of which I employ also, it is natural that there should be agreement between my estimates and theirs. My purpose, however, has been to give an idea of the degree of uncertainty attaching to such estimates and the reasons therefor.

**Final Drowning of the Torres Strait Land Link**

Much of New Guinea is violently unstable tectonically, as has been indicated, but fortunately for the present purpose the Oriomo Plateau on the northern side of Torres Strait is the most stable part of that island; with the granite basement exposed at Mabaduan, it is structurally part of Australia. Moreover, the deepest channels (Fig. 1:3) lie between the granite high islands of the Strait, which belong structurally to Cape York Peninsula where the geomorphological work done so far has given no hint of neotectonism (Valentin 1959, 1961). On the other hand, there is much coral reef growth here and the deepest channels with sills at 11 to 13 m (6 to 7 fathoms) lie between them. Reef growth shallows the sea, but at the same time constricts tidal forces, so that currents scour the intervening bottom more vigorously. Furthermore, these conflicting processes have been at work in the later part of the Flandrian transgression when sea level was changing more slowly, so that errors in estimating the former level of the sill have more effect on the inferred date of the establishment of the Strait. A precautionary range of depths of 9 to 18 m (5 to 10 fathoms) may be adopted therefore and applied to published depth-time curves (Shepard and Suess 1956; Godwin *et al.* 1958; Shepard 1960; Fairbridge 1961; Curray 1961, 1965; Jelgersma 1966; Schofield 1964; Coleman and Smith 1964). The resulting estimates for the final drowning of Torres Strait lie between 5,000 and 11,000 BP, though most fall in the span of 6,500 and 8,000.

Since this estimate was made at the 1968 seminar, Phipps (1970) has presented results of marine geological work in the Gulf of Carpentaria. A core from −60 m in the Gulf showed an alternation of shallow marine and non-marine sediments ranging in age between 6,500 and 19,500 BP. On the basis of the world glacioeustatic curve these sediments should be non-marine when they are in fact shallow marine, and deeper water when they are in fact shallow water. As a result Phipps infers that the bottom of the Gulf rose tectonically to keep close to sea level commensurately with the Flandrian transgression until about 6,500 BP. Subsequently it has subsided to bring the shallow water sediments into deep water, as the further rise of sea level itself is inadequate to have done this.

These conclusions cast some doubt on the assumption made above that the Strait has been more or less stable in the Holocene. However, there is a structural boundary between the Gulf of Carpentaria and Cape York Peninsula, which belong to quite distinct tectonic provinces. Phipps (1967) indicates that the Gulf is a fault trough running north-south which reaches to the latitude of Torres Strait. The assumption of stability for the Strait may therefore continue to be made but what is plainly needed is the extension of such submarine investigations to the Strait itself.

**The Last 5,000 Years**

Some 5,000 years ago sea level had risen practically to its present stand, so no major changes in coastal geography can have been brought about by changing sea level since then. However, at the local scale, there may have been geomorphological changes large enough to have an influence on the prehistory of particular communities, and so have a bearing on the interpretation of individual archaeological sites. There are controversial views as to what has happened to sea level in this time:

**1** that sea level has been stationary;

**2** that there has been a small but persistent rise;

**3** that sea level rose to about $+2$ m, since which time there has been net emergence, with smaller oscillations superimposed and ending in a rise in the last few centuries.

Oscillations in glacier extent and volume are known in this period (e.g. Goldthwait 1966) so that sea level movements in sympathy must have occurred, though there is doubt whether they would be of the magnitude which has been claimed for them in this time (Mercer 1968). The detailed relative sea level curves for Scandinavia, representing the interplay of eustatic and isostatic factors, have been dissected into these two separate effects by Schofield (1964) in favour of Holocene emergence. Nevertheless, other well documented histories from the Atlantic and Gulf Coasts of the United States (Coleman and Smith 1964; Bloom 1967) and from Holland (Jelgersma 1966) support the first or the second view.

Within Australia, Fairbridge (1961) and Gill (1961) have argued strongly in favour of the third view involving one or more emergences. However, much of the evidence for emergence has not been dated and there is confusion with features inherited from interstadial or even interglacial times, which lie a little above present sea level. The conflict of views persists to this day, with Russell (1963), Logan (1967), Hails (1965), Langford-Smith and Thom (1969) against Holocene eustatic emergence, and Gill (1965b), Kriewaldt (1966), Hopley (1968), Logan (1968), and Jessup (1968) favouring it.

Not only is there no consensus about changing sea level in this period but also any eustatic shifts then have been so small and slow that other processes—tectonic movement, varying storminess, changing supply of sediment—may have produced effects of the same order of magnitude. There is, therefore, no alternative within this period to relating local prehistory to palaeogeography locally reconstructed. Intensification of oceanographic research around Australia and the beginning of search for and exploitation of submarine economic minerals hold promise for this, and even more, in rendering increasingly certain and precise our knowledge of the earlier times, when the shallow sea floors around the continent were alternately made land and reconquered by the sea.

## References


Bird, E. C. F. 1968. *Coasts*. Canberra: Australian National University Press.

Blackburn, G., Bond, R. D., and Clarke, A. R. P. 1965. *Soil development associated with stranded beach ridges in south-east South Australia*. Soil Publ. CSIRO Aust. No. 22.

Bloom, A. L. 1967. Pleistocene shorelines: a new test of isostasy. *Bull.geol.Soc.Am.* **78**: 1477-94.

Boutakoff, N. 1957. Lake Wellington Formation, Gippsland. *Min.geol.J.* **6**:46-9.

Bowler, J. M. 1967. Quaternary chronology of Goulburn Valley sediments and their correlation in southeastern Australia. *J.geol. Soc.Aust.* **14**:287-92.

Bulmer, Susan and Bulmer, R. 1964. The prehistory of the Australian New Guinea Highlands. *Am.Anthrop.* (Special Publ.) **66** (4):39-76.

Coleman, J. M. and Smith, W. G. 1964. Late Recent rise of sea level. *Bull.geol.Soc.Am.* **75**:833-40.

Cotton, C. A. 1962. Low sea levels in the late Pleistocene. *Trans.R.Soc.N.Z.—Geology* **1**:249-52.

Crocker, R. L. and Wood, J. G. 1947. Some historical influences on the development of the South Australian vegetation communities and their bearing on concepts and classification in ecology. *Trans.R.Soc.S.Aust.* **71**:91-136.

Curray, J. R. 1961. Late Quaternary sea level: a discussion. *Bull.geol.Soc.Am.* **72**: 1707-12.

——1965. Late Quaternary history, continental shelves of the United States. In *The Quaternary of the United States* (eds. H. E. Wright and D. G. Frey), pp. 723-35. Princeton: Princeton University Press.

Dannevig, H. C. 1915. The continental shelf of the east coast of Australia. *Fisheries, Sydney* **3**: 339-44.

Fairbridge, R. W. 1961. Eustatic changes in sea level. *Physics Chem.Earth* **4**: 99-185.

——1962. World sea-level and climatic changes. *Quaternaria* **6**: 111-34.

——1964. African ice-age aridity. In *Problems in palaeoclimatology* (ed. A. E. M. Nairn), pp. 356-63. London: Interscience.

Fisk, H. N. and McFarlan, E. 1955. Late Quaternary deltaic deposits of the Mississippi River. In *Crust of the earth (a symposium)* (ed. A. Poldervaart), pp. 279-302. Geological Society of America Special Paper No. 62.

Gill, E. D. 1956. Radiocarbon dating of late Quaternary shorelines in Australia. *Quaternaria* **3**: 133-8.

——1961. Changes in the level of the sea relative to the land in Australia during the Quaternary Era. *Z.Geomorph.* Suppbd **3**: 73-9.

——1965a. The paleogeography of Australia in relation to the migrations of marsupials and men. *Trans.N.Y.Acad.Sci.* **28**: 5-14.

——1965b. Radiocarbon dating of past sea levels in S.E. Australia. In *Abstracts International Association for Quaternary Research VII International Congress general sessions.* 1965, Boulder and Denver, Colorado. pp. 167-8.

——1967. Description of Quaternary shorelines with special reference to the tectonic factor. *J.Inst.Polytech.Osaka Cy Univ.Geoscience* **10**: 131-4.

Godwin, H., Suggate, R. P., and Willis, E. H. 1958. Radiocarbon dating of the eustatic rise in ocean-level. *Nature, Lond.* **181**: 1518-19.

Goldthwait, R. P. 1966. Evidence from Alaskan glaciers of major climatic changes. In *Proceedings of the International Symposium on world climate 8,000 to 0 B.C.* (ed. J. S. Sawyer), pp. 40-53. London: Royal Meteorological Society.

Hails, J. R. 1965. A critical review of sea-level changes in eastern Australia since the Last Glacial. *Aust.geogrl Stud.* **3**: 63-78.

Higgins, C. G. 1965. Causes of relative sea-level changes. *Am.Scient.* **53**: 464-76.

Hopley, D. 1968. Morphology of Curacao Island Spit, north Queensland. *Aust.J.Sci.* **31**: 122-3.

Hossfeld, P. 1964. The Aitape calvarium. *Aust.J.Sci.* **27**: 179.

Jelgersma, S. 1961. Holocene sea level changes in the Netherlands. *Meded.geol. Sticht.* Ser. C, **6** (7).

——1966. Sea-level changes during the last 10,000 years. In *Proceedings of the International Symposium on world climate 8,000 to 0 B.C.* (ed. J. S. Sawyer), pp. 54-71. London: Royal Meteorological Society.

Jennings, J. N. 1959. The submarine topography of Bass Strait. *Proc.R.Soc.Vict.* **71**: 49-72.

Jessup, R. W. 1968. Soil development in coastal South Australia in relation to glacio-eustatic changes of sea-level. In *Transactions 9th International Congress of Soil Science, Adelaide, Australia, 1968.* Vol. 4, pp. 641-9. Sydney: International Society of Soil Science.

Jones, R. 1966. A speculative archaeological sequence for north-west Tasmania. *Rec. Queen Vict.Mus.* No. 25.

Kriewaldt, M. 1966. Shell beds and a change in sea level along the northwest coast of W.A. *Rep.geol.Surv.West.Aust.* 1965: 57-62.

Langford-Smith, T. and Thom, B. G. 1969. New South Wales coastal morphology. *J.geol.Soc.Aust.* **16**:572-80.

Littlejohn, M. J. and Martin, A. A. 1965. The vertebrate fauna of the Bass Strait islands: I, The amphibia of Flinders and King Islands. *Proc.R.Soc.Vict.* **79**:247-56.

Logan, B. 1967. Western Australia. In Further Australasian research in Quaternary shorelines (ed. E. D. Gill) *Aust.J.Sci.* **30**:15-16.

——1968. Western Australia. In Quaternary shorelines research in Australia and New Zealand (ed. E. D. Gill) *Aust.J.Sci.* **31**:110.

Macintosh, N. W. G. 1965. The physical aspect of man in Australia. In *Aboriginal man in Australia* (eds. R. M. Berndt and Catherine H. Berndt), pp. 29-70. Sydney: Angus and Robertson.

——and Barker, B. C. W. 1965. The osteology of aboriginal man in Tasmania. *Oceania Monogr.* No. 12.

Manley, G. 1955. A climatological survey of the retreat of the Laurentide ice sheet. *Am.J.Sci.* **253**:256-73.

Mayr, E. 1944. Timor and the colonization of Australia by birds. *Emu* **44**:113-30.

Mercer, J. H. 1968. Antarctic ice and Sangamon sea level. *Int.Ass.Sci.Hydrol.* **79**:217-25.

Phipps, C. V. G. 1967. The character and evolution of the Australian continental shelf. *The A.P.E.A. J.* 1967:44-9.

——1970. Dating of eustatic events from cores taken in the Gulf of Carpentaria and samples taken from the New South Wales continental shelf. *Aust.J.Sci.* **32**:329-30.

Ridpath, M. G. and Moreau, R. E. 1966. The birds of Tasmania: ecology and evolution. *Ibis* **108**:348-93.

Russell, R. J. 1963. Recent recession of tropical cliffy coasts. *Science, N.Y.* **139**:9-15.

Schofield, J. C. 1964. Postglacial sea levels and isostatic uplift. *N.Z.Jl Geol.Geophys.* **7**:359-70.

Shepard, F. P. 1960. Rise of sea level along northwest Gulf of Mexico. In *Recent sediments, northwest Gulf of Mexico, 1951-1958* (eds. F. P. Shepard, F. B. Phleger, and T. H. van Andel), pp. 338-44. Tulsa: American Association of Petroleum Geologists.

——1964. Sea level changes in the past 6000 years: possible archaeological significance. *Science, N.Y.* **143**:574-6.

——and Suess, H. E. 1956. Rates of postglacial rise of sea level. *Science, N.Y.* **123**:1082-3.

Sprigg, R. C. 1965. Sea floor sedimentation in St. Vincent Gulf and its approaches. In *Abstracts Sect. C. Australian and New Zealand Association for the Advancement of Science 38th Congress.* Hobart: Australian and New Zealand Association for the Advancement of Science.

Troll, C. 1964. Karte der Jahreszeitenklimate der Erde. *Erdkunde* **18**:5-28.

Twidale, C. R. 1966. Late Cainozoic activity of the Selwyn Upwarp, northwest Queensland. *J.geol.Soc.Aust.* **13**:491-4.

Valentin, H. 1959. Geomorphological reconnaissance of the north-west coast of Cape York Peninsula (northern Australia). In *Second coastal geography conference* (ed. R. J. Russell), pp. 213-31. Washington D.C.: Office of Naval Research, Geography Branch.

——1961. The central west coast of Cape York Peninsula. *Aust.Geogr.* **8**:65-72.

van Andel, T. H., Heath, G. R., Moore, T. C., and McGeary, D. F. R. 1967. Late Quaternary history, climate, and oceanography of the Timor Sea, northwestern Australia. *Am.J.Sci.* **265**:737-58.

# 2 Evidence for Late Quaternary Climates

R. W. Galloway

*What the soldier said isn't evidence.*
*Army adage*

Great climatic changes have affected Australia, in common with the rest of the world, during the Quaternary Era. Glaciers formerly existed in Tasmania and the Snowy Mountains, closed lakes and salt pans were bigger, great dune systems evolved and are now stable, floral and faunal distributions have changed.

It is hardly necessary to stress the importance of past climates to our understanding of Australia's landscapes and prehistory but there is perhaps room for a brief assessment of how much we actually know. This essay attempts to state where we stand and to point out some directions in which rapid progress towards a better knowledge of past climates should be possible. Attention will be confined mainly to work published since 1960. Gentilli's extensive review (1961) gives an account of earlier work and more recently Merrilees (1968) and Jones (1968) have also reviewed palaeoclimates while discussing problems of faunal extinction.

Current palaeoclimatic reconstructions are mainly concerned with the simpler climatic elements of mean annual or seasonal temperature, precipitation, and winds with some consideration of evaporation, temperature ranges, and rates of precipitation.

We must always bear in mind that palaeoclimates were not constant for long periods but fluctuated widely when viewed in time perspectives greater than a few centuries. Studies of late glacial conditions in western Europe have shown how great and rapid such fluctuations might be. According to van der Hammen, Maarleveld, Vogel, and Zagwijn (1967), within two millennia the vegetation of the Netherlands oscillated from tundra to forest to tundra and back to forest, with concomitant temperature fluctuations of 5°C or more. Quite likely there were not simply one or two last-glacial and post-glacial climates in Australia but rather a whole gamut of climates even in one locality. Furthermore, Australia is a large land area spanning several major climatic belts and may have experienced different climatic changes in various parts of the continent at the same time (Gentilli 1961).

Clearly, if scattered field observations are to contribute to a coherent picture of past climates they must be dated to within a few centuries; absolute dates are essential. Earlier palaeoclimatic work has suffered because a satisfactorily dated framework has not existed and field observations have been interpreted according to unproven assumptions; these interpretations have in turn been cited later as proof of the correctness of the assumptions. 'Evidence' for a 'mid-Holocene arid period' has largely built up in this way. This is not to say that undated palaeoclimatic deductions have no value. On the contrary they help to define the phenomena which palaeoclimatic reconstructions must account for and date. Furthermore, information on the succession of palaeoclimatic events, even when their absolute age is unknown, should help towards eventually understanding their causes.

Attention is confined here to roughly the last 25,000 years covering the last major advance and retreat of the icesheets in the Northern Hemisphere, the associated regression and transgression of the sea, and post-glacial climatic history. The post-glacial or Holocene is taken as beginning 10,000 years ago.

## Lines of Evidence

Climate and weather are transient phenomena which we cannot measure directly at a later date. Our knowledge must depend on deductions from disequilibria between physical features or fossils which developed under past conditions and the present climate (Flint 1963). Some of the bases for such deductions will be briefly reviewed in the Australian context.

### BIOTA

Pollen analysis is of course one of the Great White Hopes of palaeoclimatic research, but at the time of writing this hope is yet to be realised and awaits completion of essential background studies on the properties of Australian pollen and the provision of funds and appropriate personnel. Walker (in Walker, Churchill, and Moar 1966) stresses that we have as yet little definite knowledge from Australia or New Guinea. Churchill (Appendix A in Walker *et al.* 1966) has provided reasonably good evidence for humid conditions in Western Australia about 4,500-6,000 years ago (in the 'mid-Holocene arid period').

When pollen evidence is eventually used to reconstruct past climates, let us hope that where feasible the *range* of possibilities will be given rather than solely that which represents the *minimal* change from present conditions. Much of the conflict between palynological and geological evidence in North America can be resolved when it is realised that estimates of the full-glacial timber line and temperature depression derived from pollen studies are only minima and the possible range of values has not been specified (Galloway 1970).

Plant macrofossils, not being subject to wind transport, can give datable, location-specific information about vegetation from which palaeoclimatic deductions can be made. For example, lake beds and bogs in Tasmania have furnished datable macrofossil plant remains (Peterson 1968). It would be particularly helpful if carefully dated plant macrofossils could be used to determine the altitude of the tree line at various periods, since the tree line both in Australia and in other parts of the world correlates quite well with summer temperatures over a wide range of precipitation (Galloway 1965; Wardle 1965). A mid-Holocene tree line higher than the present would be strong evidence for warmer conditions though it would say little about the precipitation. Knowledge of the lowest position of the tree line in the late Pleistocene would enable full-glacial summer temperatures to be estimated.

Lacustrine and marine shells can give reasonable information about former water temperatures. In Victoria abundant warm water shells in raised beaches indicate warmer seas (Gill 1956). As a rule shells tell us little concerning precipitation but some deductions can be made from the presence of *Coxiella* shells dated at 13,725 ±350 BP (Gill 1955a, 1955b:50) in lacustrine sediments in western Victoria. These shells indicate that the water was brackish and the lakes did not overflow despite the lower evaporation; consequently the climate is unlikely to have been pluvial (Galloway 1965).

Mammalian remains and particularly the implications of extinction have been repeatedly invoked as evidence of former climates. Merrilees (1968) and Jones (1968) have shown how little reliance can be placed on this sort of evidence particularly when dating is as confused as it is, for instance, at Lake Callabonna. Far-reaching conclusions have often been based on very scanty evidence when interpreting mammalian fauna. For example, Tedford's suggestion (1967) that precipitation during the late Pleistocene at Lake Menindee, western New South Wales, 'averaged more than 10 [250 mm] and nearer 20 inches [500 mm] per year' (compared to 8 to 9 inches—200 to 225 mm—today) is apparently based on a single, poorly preserved koala bone. Koalas can feed on the foliage of river redgum (*Eucalyptus camaldulensis*), which grows in the area today (cf. Calaby, see ch. 7). The fossil bone neither proves nor disproves climatic change.

Disjunct distributions of plants and animals such as the palms of central Australia certainly imply climatic changes and they have been freely invoked as evidence in support of various hypotheses. However, since we do not know when these biota reached their present isolated positions, their value as evidence for any given palaeoclimatic scheme is limited. Disjunct distributions may well require palaeoclimatic theories to explain them, but they provide little information towards evolving and testing such theories.

We cannot be sure how plants and animals would have behaved as individuals or as communities in conditions which have no modern counterpart. Consequently it is difficult to test palaeoclimatic hypotheses which postulate such conditions (e.g. Galloway 1965).

### FIRN LIMITS

During glacial stages snow lines (better termed firn limits) were lower than they are today and glaciers occupied high ground in Tasmania and the Snowy Mountains. Sufficient absolute dates are now available to demonstrate that at least the major phases of the last Pleistocene glaciation ran parallel in both Northern and Southern Hemispheres. Davies (1967) and Peterson (1968) have shown that the altitude of the firn limit and the nature of the cirques and other glacial landforms reflect climatic gradients between western and eastern Tasmania and between Tasmania and the Snowy Mountains similar to those existing today. Peterson points out that the form of cirques in the Snowy Mountains implies only moderate snowfall when the glaciers existed. However, without knowing summer temperatures it is difficult to say how far the depression of firn limits in glacial times was due to lower summer temperatures and how far, if at all, to greater winter snowfall.

PERIGLACIAL FEATURES

Evidence of former frost action such as deep shattering of the bedrock, solifluction rubble mantles now stabilised, and patterned ground can be used as guides to past temperature conditions (Galloway 1961a, 1965). Sporadic occurrences of medium or large-scale patterned ground and blockfields in the mountains of southeast Australia and Tasmania (Carr and Costin 1955; Talent 1965; Costin *et al.* 1967; Caine 1968b) point to colder climates but their time of formation and climatic connotations are insufficiently known for more accurate deductions. Furthermore similar forms have been described from non-periglacial environments. For example, sorted stripes exist at low altitudes in the humid tropics of New Guinea (Haantjens 1965).

The absence of fossil tundra polygons and ice wedge casts suggests that really severe conditions did not exist in Australia even at the height of the last glaciation, except perhaps on the very highest mountains. However, there is still room for surprises in this direction as the recent discovery of fossil tundra polygons at low altitudes in southern France has shown (Cailleux and Rousset 1968). Continued search in older valley floor sediments in southern Tasmania might prove rewarding although Caine (1968a) has shown that some of this material is post-glacial.

Periglacial solifluction mantles are primarily related to temperatures and are largely independent of precipitation in all but very dry areas (Galloway 1961b; Davies 1967). Consequently they reflect temperature conditions more accurately than do firn limits, which depend on both snowfall and summer temperature. The difficulty is that very similar slope mantles can develop under very different climates. For instance mudflow deposits, periglacial solifluction mantles, and colluvium formed by soil creep and rainwash are not easily distinguished. Because of their increasing thickness and frequency with increasing altitude, dominant down-slope orientation of included stones, and angular frost-shattered fragments, there can be little doubt that slope mantles in the mountains of southeastern Australia are periglacial; but stony colluvium exists at much lower levels far to the north (e.g. Hawkins and Walker 1960 and recent excavation for the Moonie pipeline in southern Queensland). Here a periglacial origin is hard to accept because it would imply such a catastrophic deterioration of climate. Unless reliable criteria can be developed for distinguishing periglacial solifluction deposits from other slope mantles their value in palaeoclimatic reconstruction will be limited.

LAKES

Evidence of former higher lake levels has been cited as proving former conditions of higher rainfall. However, former higher lake levels could have been due to reduced evaporation rather than to greater rainfall. Galloway (1965, 1967) has presented a case for higher lake levels and increased run-off in southeastern Australia during the last glaciation because of the drastic reduction in evaporation and despite decreased precipitation. However, this case depends on palaeotemperature deductions from periglacial features which, as we have seen, are not entirely unambiguous and on the belief that Lake George (near Canberra) was a closed basin at the time of higher lake levels. It is certainly closed today but may not have been when the water stood higher and seepage to the lower Yass valley only a few kilometres to the west and about 150 m lower might have occurred.

On the basis of two not unequivocal radiocarbon dates, Johns (1962) tentatively suggests that Lake Eyre was dry and subject to deflation at some time between 20,000 and 40,000 BP. This period of time included glacial maxima and an important interstadial and consequently it is of great palaeoclimatic interest to determine more exactly when the deflation occurred. Krinsley, Woo, and Stoertz (1968), on finding aeolian sand a few feet below lacustrine sediments at Lake Frome, South Australia, concluded that here the last glacial period was dry; but their evidence is undated. They also noted the absence of abandoned shoreline features round playas in Western Australia, which suggests that these basins have never contained very much water and that the climate has always been dry ever since their formation.

It can be concluded that many Australian closed lakes and playas show that different climates have prevailed but further work is required before these climates can be defined and dated. Attention should be focused on the wettest and the driest areas where such lakes exist. In the wettest areas a higher proportion of the water supply is derived from direct rainfall on the lake surface. Consequently the value chosen for run-off, which is the most difficult term to estimate when calculating former hydrologic budgets, is less critical than for lakes in dry areas. Potentially suitable lakes exist in Victoria, the Southern Tablelands of New South Wales, eastern Queensland (Lakes Galilee and Buchanan), and southwestern Western Australia. In the driest areas, former arid periods should be clearly revealed by the absence of lacustrine material in the sediments of playa floors, although such lacunae are intrinsically hard to date.

#### AEOLIAN FEATURES

The immense dune systems of Australia, particularly the longitudinal inland dunes, point to drier and/or windier climates in the past. Coastal dunes have little value as palaeoclimatic indicators since they are developing today under a wide range of Australian coastal climates and their formation depends more on sediment supply, marine dynamics, and vegetation than directly on climate. Work in progress on dune trends and dune formation in central Australia and the relation of dune systems to sea level in northern and southern Australia should help to clarify the palaeoclimatic implications, but little definite can be said at the moment. It is certainly naïve and possibly wrong to say that because the inland dunes appear to plunge below present sea level in northwestern Australia (Fairbridge 1961:115) they therefore date from the maximum of the last glaciation when sea level was more than 100 m lower than it is today (Galloway 1965). Nor can we be sure that the inland dunes represent a single system formed contemporaneously across the continent (Jennings 1967).

Lunettes (crescentic dunes of clay, silt, or sand) are commonly found on one side of playas in Australia. Few, if any, are developing today (Bowler 1968) and it therefore seems necessary to postulate climates which have no present counterpart in Australia. Clearly they formed on the lee side of the lakes and the dominant wind direction at the time of their formation can be deduced. Currey (1964) believes lunettes formed when the lakes were full, Stephens and Crocker (1946) when the lakes were dry, and Gill (1966) when the lakes were seasonally wet and dry. Bowler and Harford (1966) postulate full lake conditions at least while the coarser-textured lunettes were forming. Until the mode of origin is definitely determined the value of lunettes as palaeoclimatic indications will be limited.

### STREAM EROSION AND SEDIMENTATION

Much study has been devoted to questions of sedimentation and erosion on the riverine plains of western New South Wales and Victoria (Pels, see ch. 4). That there have been phases of erosion, stream incision, deposition, and aeolian action and that these were related to climatic changes is certainly true. However, the climatic significance and the time of these various phases remain controversial and widely differing opinions have been put forward (Cotton 1963). Part of the argument is based on the work of Langbein and Schumm (1958) and Schumm (1965) who believe from studies in the western United States that sediment yield is at a maximum when precipitation is 25 to 35 cm. Dury (1967) has shown that this belief is based on inadequate data. Much more detailed work with many more absolute dates must be done before any definite conclusions about palaeoclimates can be drawn and all work must bear in mind the possibility of changed run-off as a result of changed temperature rather than changed precipitation.

There is evidence from northeast New South Wales and east Tasmania (Langford-Smith, in Gill *et al.* 1967:17; Goede 1965) for a distinct phase of accelerated fluvial sedimentation about 4,000 to 5,000 years ago, but in the absence of definite information on the sedimentological effects of changes of climate the palaeoclimatic implications remain obscure. One of these authors suggests increased precipitation, the other decreased precipitation.

### CHANNEL FORMS

In many parts of the world former river meanders were much larger than the present generation. The size of meanders may be related to discharge, and Dury, in a series of thought-provoking papers (see Dury 1967 for some references), has calculated that precipitation during glacial times must have been of the order of 1½-2 times greater than at present in order to account for the larger meanders, even when allowance is made for the effect of greater run-off consequent on reduced evaporation. However, some doubt exists regarding the general validity of this conclusion, because it implies extraordinarily high discharges (Geyl 1968); channels formed as recently as 1956 show a spectrum of meander sizes rather than a single dominant size (Speight 1965, 1967), while meander forms are related not only to discharge but also to calibre of stream load (Schumm 1967). Dury (1967) cites a number of American studies as independent evidence for the proposed 'glacial-pluvial' climate but none are diagnostic; all the evidence could equally well be the result of reduced evaporation rather than increased precipitation. It must be concluded that while studies of meander patterns in relation to discharge offer an interesting method of approach they do not yet provide us with reliable data on past rainfall.

### SOILS

Soil-forming and soil-eroding periods have been investigated in southeast Australia and the K-cycle concept has provided a useful framework (Butler 1959). Generally speaking it has been claimed that periods of soil stability represented relatively humid climates perhaps correlated with glacial or cooler phases in higher latitudes, and periods of soil instability represented drier and stormier phases. The concept has been applied to aeolian, colluvial, and alluvial materials. However, serious discrepancies arise when more than strictly local correlations are attempted (Walker 1962, 1963).

Jessup (1968) correlates soil formation on the Eyre Peninsula with humid, glacial low sea level phases, yet one of the soils he cites, which passes below present sea level and is ascribed to *pluvial* conditions during the last glaciation, is a desert loam implying low precipitation when it developed.

Apparently landscape instability and soil erosion can be associated not only with drier conditions and reduced vegetation cover but also with increased rainfall of greater seasonality (Douglas 1967). In this connection it is worth noting that in the drier parts of Australia rainfall intensity is *less* than in the wetter parts (Jennings 1968) and consequently a climatic change to greater aridity does not imply a change to more intense if less abundant rainfall. Past periods of soil instability may even be related to the effects of Aboriginal activities on the vegetation (Jones 1968). The inherent variability of soils, inadequate knowledge of their mode and rate of formation, the difficulty of classifying and dating them, and their considerable dependence on the nature of the parent material lead one to conclude that soils will hardly be of prime importance in palaeoclimatic studies. Soil scientists will be consumers rather than producers of palaeoclimatic evidence.

### CARBONATES

Opinions on the climatic connotations of soil carbonates are far from unanimous and until this question is settled little can be done in the way of satisfactory palaeoclimatic deduction. Isotopic studies of speleothems as carried out in New Zealand offer scope for important advances (Hendy and Wilson 1968).

### MARINE SEDIMENTS

Conolly (1967) has deduced from oxygen isotope studies that the Indian Ocean surface near Australia experienced temperature fluctuations closely paralleling those in comparable oceanic situations in the Northern Hemisphere. Obviously the study of marine sediments offers great potential for development of dated palaeotemperature records. It should eventually be possible to determine how far icebergs drifted north from Antarctica at various stages of the Pleistocene from the mineral matter which they drop onto the sea bottom as they melt. This will give palaeotemperature information which is independent of isotopic studies or conclusions based on foraminifera.

When considering the relationships of marine palaeoclimatic studies to conditions on the continents three points should be borne in mind. Firstly, oxygen isotopic studies involve assumptions about the former isotopic content of the oceans and consequently the deduced palaeotemperatures are not necessarily correct (Emiliani 1966; Shackleton 1967). Secondly, in full-glacial times mean annual temperatures on the continents would have been about 1°C lower than on the oceans because of the effective increase in altitude consequent on the fall in sea level. Thirdly, terrestrial climates would have been somewhat more continental during glacial low sea level phases and this would give relatively higher summer temperatures and reduced precipitation near present day coasts independent of other climatic changes.

Terrestrial sediments preserved on the continental shelf should ultimately tell us a great deal about the climates during low sea level stages of the Quaternary. It will be particularly interesting to discover whether continental type dunes are preserved well below present sea level and if so whether this occurs on both equatorial and polar sides of the continent. Van Andel and Veevers (1967) found kunkar nodules on the Sahul

Shelf which presumably formed in the soil when this area was dry land. Their conclusion that this implied drier conditions during the last glaciation is perhaps premature since kunkar (*sens. lat.*) is more widespread in Australia than they believe (e.g. Rutherford 1964).

CLIMATOLOGICAL THEORY

With increasing knowledge of the mechanisms of present climates and evidence of past climates it may eventually be possible to fill in some of the gaps in our knowledge by reasoning from general climatological theory. It must be stressed, however, that field workers should beware any tendency on the part of theoretical climatologists to inhibit their thinking. The example of continental drift, where theoretical objections held up effective study for many years, should serve as a warning. The very fact that climatic changes have occurred means climatic factors have altered and consequently deductions based on modern climates may be of only limited value in understanding the past.

ANALOGIES WITH OTHER PARTS OF THE WORLD

There is now little doubt that major Quaternary climatic changes were synchronous over the entire world and similar in sense and direction in similar climatic regions (Dury 1967). Thus it should be possible to gain insight into Australian Quaternary climates by looking at the story in similar areas elsewhere. This was a favourite and probably inevitable method in earlier stages of the study of past Australian climates. Data from this country were fitted into a time sequence derived from the European Alps or North America and the interpretation of evidence was strongly influenced by overseas ideas. Unfortunately this method is useful only if truly contemporaneous events are correlated and the overseas sequence is itself soundly based. While correlations will surely be shown to exist, the first condition is only now being met by isotopic dating and the second condition has not been met at all outside glacial and immediately periglacial areas. Palaeoclimatic studies in Africa, Asia, and America are at least as controversial as they are here and insufficient is yet known about the palaeoclimates of low latitudes for satisfactory long-distance correlations. Perhaps in the not too distant future we shall see workers in other continents looking to Australian studies as a guide to interpreting their own data.

## The 'Glacial-Pluvial' and the 'Mid-Holocene Arid Period'

The scheme of a 'glacial-pluvial' corresponding to the last glaciation in higher latitudes, followed by a warming and drying trend leading to the 'mid-Holocene arid period', in turn followed by slightly cooler and moister conditions in recent millennia, can no longer be accepted as a matter of course. When observations were few, scattered, and hard to correlate this scheme was useful but it should now be regarded as only one of several alternative hypotheses which may have limited applicability to parts of the continent and must be tested against field evidence. Two recent papers dealing with late Quaternary deposits at McEachern Cave in southwest Victoria illustrate the need for a more critical approach.

Link (1967:43) states (in German summary) that there was a humid phase 11,000-25,000 BP (misprinted as B.C.). The English text (p. 140) is rather less definite but makes the same claim: '*Sthenurus* sp. in Victoria may indicate a humid climate' and Lundelius (1963) is cited as authority for this belief. Lundelius in turn bases his belief that an undated fossil of *Sthenurus* implies wetter conditions during the last glaciation on earlier papers by Main, Lee, and Littlejohn (1958) and Serventy (1951) whose ideas are derived from previous unconfirmed notions about Quaternary climates. Despite the shaky chain of reasoning it is all too likely that Link's study will in turn be cited as evidence for a glacial-pluvial.

Wakefield (1967) in a preliminary study of the cave sediments and fauna has postulated four successive climatic stages: humid, dry, slightly more humid than today, and modern. A radiocarbon date of 15,200 $\pm$ 32 BP comes from the upper part of sediments correlated with the first stage, which is regarded as reflecting moister conditions during the last glaciation. The second stage is correlated with a presumed mid-Holocene warm-arid phase, but is not actually dated. The third stage is believed to represent a moist post-mid-Holocene phase and the fourth stage corresponds to the modern climate.

Correlation of the second stage with the mid-Holocene is apparently based on the belief that its fauna indicate a drier climate and on the belief that there was a mid-Holocene arid period. Both of these beliefs are questionable. The presence of eleven individuals of *Bettongia lesueur* (a rat-kangaroo) is cited as evidence for the drier climate, because in Victoria this animal is now restricted to mallee country with less than 250 mm annual precipitation compared to 750 mm at McEachern Cave. However, four individuals of *B. lesueur* were recovered from the supposed moist-climate third layer and two from the modern sediments. Furthermore Calaby (see ch. 7) points out that *B. lesueur* lived in wet areas of Western Australia in historic times so it can hardly be regarded as indicating a dry climate. Finally, although a strong case for a mid-Holocene arid period in Victoria has been presented (e.g. Gill 1955a, 1966) the question is by no means settled (Merrilees 1968).

Wakefield's (preliminary) palaeoclimatic conclusions must be regarded as not proven. Since he uses the concept of a mid-Holocene arid period to date and interpret the sediments and fauna, it would be circular reasoning for others to then cite his work as proving the existence and age of such a climatic phase.

It is painfully clear that we really know very little. From the evidence of glacial and periglacial features, sea bed fossils and possibly oxygen isotope studies, we know that lower temperatures existed in southern Australia and Tasmania at a time corresponding to the last great glaciation of the Northern Hemisphere. We do not know how much lower the temperature was nor if there were regional variations in the amount of lowering. Abandoned lake shore features show that the ratio of evaporation to precipitation has been much smaller than it is today but we do not know if actual precipitation was more, about the same, or less. We have no definite evidence for climatic fluctuations within the last glaciation other than some scrappy observations in the very limited glaciated areas of the continent and one or two deep sea cores. Probably a period of higher temperature occurred about the middle of the Holocene though even this is not entirely certain and the concomitant precipitation is unknown.

While this may seem a pessimistic summing-up, we can at least comfort ourselves with the knowledge that things are really little better overseas. It might be felt that this review will discourage people from making palaeoclimatic assessments. No such dampening effect is intended. On the contrary the field is now wide open and we need as much liberty of thought and as wide a spectrum of opinion as possible to encourage the formulation of multiple working hypotheses. Controversy is to be welcomed since it leads to the necessary *testing* of hypotheses against evidence and to clarification of our thinking.

Now that absolute dates are becoming increasingly available and field evidence examined more critically, progress should be more rapid, and we may soon be in a position to give a reasonable account of the major climates under which man has lived in Australia.

**References**

Bowler, J. M. 1968. Australian landform example No. 11: lunette. *Aust.Geogr.* **10**: 402-4.

——and Harford, L. B. 1966. Quaternary tectonics and the evolution of the Riverine Plain near Echuca, Victoria. *J.geol.Soc. Aust.* **13**:339-54.

Butler, B. E. 1959. *Periodic phenomena in landscapes as a basis for soil studies.* Soil Publ. CSIRO Aust. No. 14.

Cailleux, A. and Rousset, C. 1968. Présence de réseaux polygonaux de fentes en coin en Basse-Provence occidentale et leur signification paléoclimatique. *C.r.hebd. Séanc.Acad.Sci., Paris* Sér.D, **266**:669-71.

Caine, N. 1968a. Carbon-14 dates from the Broad River Valley, Mt. Field National Park, Tasmania. *Aust.J.Sci.* **31**:119-20.

——1968b. *The blockfields of northeastern Tasmania.* Canberra: Australian National University, Research School of Pacific Studies, Department of Geography Publications G/6.

Carr, Stella G. M. and Costin, A. B. 1955. Pleistocene glaciation in the Victorian Alps. *Proc.Linn.Soc.N.S.W.* **80**:217-28.

Conolly, J. R. 1967. Postglacial-glacial change in climate in the Indian Ocean. *Nature, Lond.* **214**:873-5.

Costin, A. B., Thom, B. G., Wimbush, D. J., and Stuiver, Minze. 1967. Nonsorted steps in the Mt. Kosciusko area, Australia. *Bull.geol.Soc.Am.* **78**:979-91.

Cotton, C. A. 1963. Did the Murrumbidgee aggradations take place in glacial ages? *Aust.J.Sci.* **26**:54-5.

Currey, D. T. 1964. The former extent of Lake Corangamite. *Proc.R.Soc.Vict.* **77**: 377-86.

Davies, J. L. 1967. Tasmanian landforms and Quaternary climates. In *Landform studies from Australia and New Guinea* (eds. J. N. Jennings and J. A. Mabbutt), pp. 1-25. Canberra: Australian National University Press.

Douglas, I. 1967. Man, vegetation and the sediment yields of rivers. *Nature, Lond.* **215**:925-8.

Dury, G. H. 1967. Climatic change as a geographical backdrop. *Aust.Geogr.* **10**: 231-42.

Emiliani, C. 1966. Isotopic paleotemperatures. *Science, N.Y.* **154**:851-7.

Fairbridge, R. W. 1961. Eustatic changes in sea level. *Physics Chem.Earth* **4**:99-185.

Flint, R. F. 1963. Pleistocene climates in low latitudes. *Geogrl Rev.* **53**:123-9.

Galloway, R. W. 1961a. Periglacial phenomena in Scotland. *Geogr.Annlr* **43**:348-53.

——1961b. Solifluction in Scotland. *Scott. geogr.Mag.* **77**:75-87.

——1965. Late Quaternary climates in Australia. *J.Geol.* **73**:603-18.

——1967. Dating of shore features at Lake George, New South Wales. *Aust.J.Sci.* **29**:477.

——1970. The full-glacial climate in southwestern U.S.A. *Ann.Ass.Am.Geogr.* **60**: in press.

Gentilli, J. 1961. Quaternary climates of the Australian region. *Ann.N.Y.Acad.Sci.* **95**: 465-501.

Geyl, W. F. 1968. Tidal stream action and sea level change as one cause of valley meanders and underfit streams. *Aust. geogrl Stud.* **6**:24-42.

Gill, E. D. 1955a. The Australian 'Arid Period'. *Aust.J.Sci.* **17**:204-6.

——1955b. Radiocarbon dates for Australian archaeological and geological samples. *Aust.J.Sci.* **18**:49-52.

——1956. Radiocarbon dating of Late Quaternary shorelines in Australia. *Quaternaria* **3**:133-8.

——1966. Geochronology of Victorian Mallee ridges. *Proc.R.Soc.Vict.* **79**:555-9.

——*et al.* 1967. Further Australasian research in Quaternary shorelines. *Aust.J.Sci.* **30**: 15-18.

Goede, A. 1965. Geomorphology of the Buckland Basin, Tasmania. *Pap.Proc.R.Soc. Tasm.* **99**:133-54.

Haantjens, H. A. 1965. Morphology and origin of patterned ground in a humid tropical lowland area, New Guinea. *Aust. J.Soil Res.* **3**:111-29.

Hawkins, C. A. and Walker, P. H. 1960. An occurrence of buried soils at Prospect, N.S.W. *J.Proc.R.Soc.N.S.W.* **94**:115-20.

Hendy, C. H. and Wilson, A. T. 1968. Palaeoclimatic data from speleothems. *Nature, Lond.* **219**:48-51.

Jennings, J. N. 1967. Two maps of rainfall intensity in Australia. *Aust.Geogr.* **10**:256-62.

——1968. A revised map of the desert dunes of Australia. *Aust. Geogr.* **10**:408-9.

Jessup, R. W. 1968. Soil development in coastal South Australia in relation to glacio-eustatic changes of sea-level. In *Transactions 9th International Congress of Soil Science, Adelaide, Australia, 1968.* Vol. 4, pp. 641-9. Sydney: International Society of Soil Science.

Johns, R. K. 1962. Investigation of Lake Eyre. Part 1. The sediments of the Lake Eyre Basin, evaporites and brines, their constitution and age. *Rep.geol.Surv.S.Aust.* No. 24: 1-69.

Jones, R. 1968. The geographical background to the arrival of man in Australia and Tasmania. *Archaeol.& phys.Anthrop. Oceania* **3**:186-215.

Krinsley, D. B., Woo, C. C., and Stoertz, G. E. 1968. Geologic characteristics of seven Australian playas. *U.S. Air Force, Office of Aerospace Research, Air Force Cambridge Research Laboratories, Environmental Research Papers* No. 283:59-103.

Langbein, W. B. and Schumm, S. A. 1958. Yield of sediment in relation to mean annual precipitation. *Trans.Am.geophys. Union* **39**:1076-84.

Link, A. G. 1967. Late Pleistocene-Holocene climatic fluctuations; possible solution pipe-foiba relationships; and the evolution of limestone cave morphology. *Z.Geomorph.* **11**:117-45.

Lundelius, E. L. 1963. Vertebrate remains from the Nullarbor caves, Western Australia. *J.Proc.R.Soc.West.Aust.* **46**:75-80.

Main, A. R., Lee, A. K., and Littlejohn, M. J. 1958. Evolution in three genera of Australian frogs. *Evolution, Lancaster, Pa.* **12**:224-33.

Merrilees, D. 1968. Man the destroyer: Late Quaternary changes in the Australian marsupial fauna. *J.Proc.R.Soc.West.Aust.* **51**:1-24.

Peterson, J. A. 1968. Cirque morphology and Pleistocene ice formation conditions in southeastern Australia. *Aust.geogrl Stud.* **6**: 67-83.

Rutherford, G. K. 1964. Soils of the West Kimberley area. In *General report on lands of the West Kimberley area, W.A.* pp. 119-39. Land Res.Ser.CSIRO Aust. No. 9.

Schumm, S. A. 1965. Quaternary paleohydrology. In *The Quaternary of the United States* (eds. H. E. Wright and D. G. Frey), pp. 783-94. Princeton: Princeton University Press.

——1967. Meander wavelength of alluvial rivers. *Science, N.Y.* **157**:1549-50.

Serventy, D. L. 1951. The evolution of the chestnut-shouldered wrens (*Malurus*). *Emu* **51**:113-20.

Shackleton, N. 1967. Oxygen isotope analyses and Pleistocene temperatures re-assessed. *Nature, Lond.* **215**:15-17.

Speight, J. G. 1965. Meander spectra of the Angabunga River. *J.Hydrol.* **3**:1-15.

——1967. Spectral analysis of meanders of some Australasian rivers. In *Landform studies from Australia and New Guinea* (eds. J. N. Jennings and J. A. Mabbutt), pp. 48-63. Canberra: Australian National University Press.

Stephens, C. G. and Crocker, R. L. 1946. Composition and genesis of lunettes. *Trans. R.Soc.S.Aust.* **70**:302-12.

Talent, J. A. 1965. Geomorphic forms and processes in the highlands of eastern Victoria. *Proc.R.Soc.Vict.* **78**:119-35.

Tedford, R. H. 1967. The fossil Macropodidae from Lake Menindee, New South Wales. *Univ.Calif.Publs geol.Sci.* **64**:1-156.

van Andel, T. H. and Veevers, J. J. 1967. *Morphology and sediments of the Timor Sea.* Bur.Min.Res.Aust.Bull. No. 83.

van der Hammen, T., Maarleveld, G. C., Vogel, J. C., and Zagwijn, W. H. 1967. Stratigraphy, climatic succession and radiocarbon dating of the Last Glacial in the Netherlands. *Geologie Mijnb.* **46**:79-95.

Wakefield, N. A. 1967. Preliminary report on McEachern's Cave, S.W. Victoria. *Victorian Nat.* **84**:363-83.

Walker, D., Churchill, D. M., and Moar, N. T. 1966. A commentary on botanical data from New Guinea, Australia and New Zealand. In *Proceedings of the International Symposium on world climate 8,000 to 0 B.C.* (ed. J. S. Sawyer), pp. 149-56. London: Royal Meteorological Society.

Walker, P. H. 1962. Terrace chronology and soil formation on the south coast of N.S.W. *J.Soil Sci.* **13**:178-86.

——1963. Soil history and debris-avalanche deposits along the Illawarra scarpland. *Aust.J.Soil Res.* **1**:223-30.

Wardle, P. 1965. A comparison of Alpine timber lines in New Zealand and North America. *N.Z.Jl Bot.* **3**:113-35.

# 3 Vegetation, Soils, and Climate in Late Quaternary Southeastern Australia

A. B. Costin

Even though I am restricting myself to southeastern Australia, the title of this paper is still ambitious in view of the large gaps in knowledge concerning the Late Quaternary climate, vegetation, soils, and associated landforms of this area. Nevertheless, some important information does exist, and some of this seems to have anthropological relevance. I shall first present a selection of some of this information—this is factual enough; then discuss its climatic significance—some of this is a matter of opinion; and finally suggest some anthropological implications—although this is a field which I shall leave mainly to others.

Southeastern Australia is a varied environment. There is a wide range of climates in close proximity, from mild coastal to cold alpine to almost semi-arid; considerable relief; and contrasting rock types and soils. Wide variations in the environment (but not necessarily the present distribution of these variations) have probably existed for several tens of thousands of years, since there has been little change in overall relief and in the main weather systems for a long time. For example, the inland and coastal mountain systems are likely to have blocked off moisture-bearing winds both from the northwest/southwest and the northeast/southeast, as at present, giving wet coastal and inland mountains with drier tableland areas in between.

Diversified environments such as southeastern Australia are well buffered ecologically, since, except in more extreme situations, plant and animal communities can adjust to considerable climatic changes by relatively local changes in their distribution to other aspects, altitudes, or soil types. Such

Advice and criticism from J. N. Jennings, Department of Biogeography and Geomorphology, Australian National University, Canberra, are gratefully acknowledged.

*Table 3:1 Climates indicated by some vegetation disjunctions in southeastern Australia*

| **Indicated climate** | **Vegetation** | **Typical disjunct occurrences** |
|---|---|---|
| 1 Colder, with more snow and possibly moister | Various subalpine and alpine communities (as in Costin 1954) | Australian Alps, Great Dividing Range north of Nimmitabel |
| | *Eucalyptus delegatensis* wet sclerophyll forest | Mt Delegate, Mt Tingiringi, Mt Clear |
| 2 Windier | *Casuarina nana* heath | Great Dividing Range north of Nimmitabel, and southwards extension to Victoria |
| | Wet mallee scrubs | As above; also Australian Alps |
| 3 Warmer, and possibly moister | Subtropical rainforest | South coast and adjacent ranges; east Gippsland |
| | *E. fastigata* wet sclerophyll forest | Cotter Valley, Goodradigbee Valley, Geehi Valley |
| | *E. macrorhyncha-E. rossii* dry sclerophyll forest | North and south Monaro region with isolated occurrences in between |
| | *E. melliodora-E. blakelyi* savannah woodland | As above |
| 4 Warmer, and possibly drier | ?*E. macrorhyncha-E. rossii* dry sclerophyll forest | As above |
| | ?*E. melliodora-E. blakelyi* savannah woodland | As above |
| | *E. albens-Callitris columellaris* tall woodland | Lower Snowy Valley |
| | *Acacia glaucescens* dry scrub | Lower Snowy Valley |
| | ?Absence of *Nothofagus* in temperate rainforest communities | Geehi Valley |

adaptability is clearly not possible in the more extreme cases of continental glaciation or of widespread aridity over relatively even terrain where the chances of ecological survival within the area are small.

We may now examine some of the plant communities, soils, and associated landforms, in relation to their palaeoclimatic significance.

## Vegetation

Assuming that disjunctions of whole plant communities, as distinct from individual plant species, indicate a formerly more widespread distribution of that community under climatic conditions which have since changed, certain inferences with respect to past climates and their probable sequence can be drawn from disjunct communities in southeastern Australia today.

Some disjunct communities, and the past climates which they probably indicate, are listed in Table 3:1 (after Costin 1954).

Because some of the relatively mesothermal communities (i.e. those under headings 3 and 4, Table 3:1) are at present near their limits with respect to low temperatures, and could hardly be expected to have survived if the last major climatic change was towards much colder conditions, the most likely sequence of climates is for 1 and 2 to have occurred together first, then 3, together with, or grading into, 4. Such a sequence would satisfactorily explain all of the main disjunctions, other than some of those under 4. The

*Fig. 3:1 Diagrammatic presentation of $K_1$, $K_2$, and $K_3$ groundsurfaces on a hillslope to illustrate the relationship of groundsurface zones (after Butler 1959)*

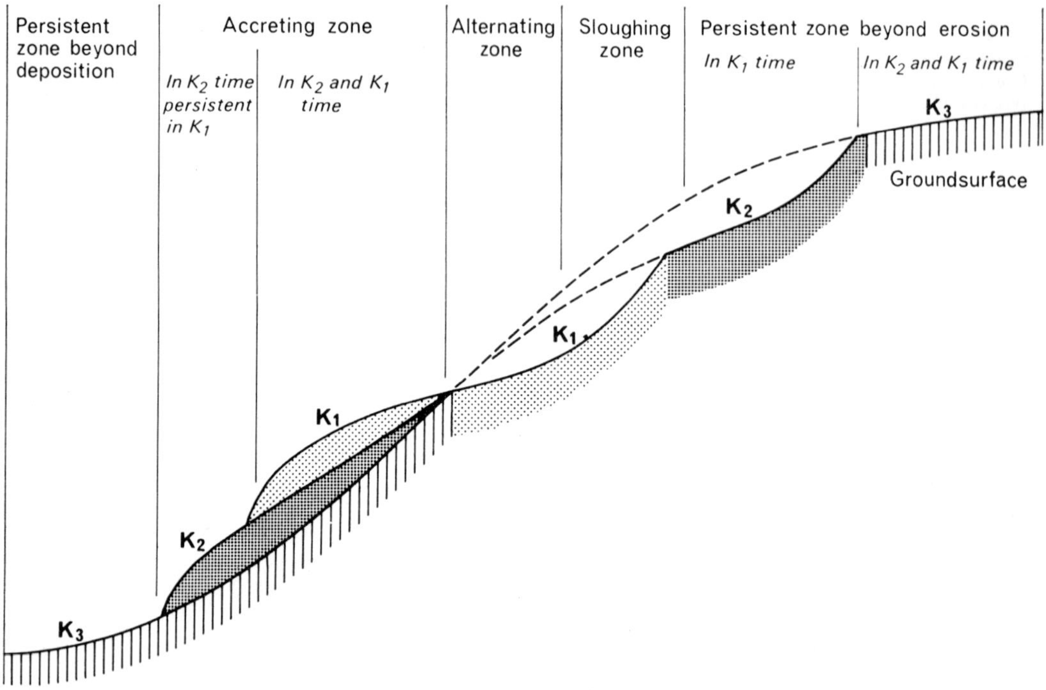


latter would also be accounted for, however, by a further change towards slight cooling-off. Smaller climatic oscillations undoubtedly occurred also, but it is difficult to detect these in the existing vegetation. Unfortunately, there is no way that I can see, using the evidence of vegetation disjunctions alone, of giving absolute dates to the occurrence of the indicated climates. The main point to bring out now is that the 'high mountain' and 'windier' communities would have been much more widespread during the colder period and, if this period coincided with the arrival of Aboriginal man into the general area, he would hardly have been encouraged to occupy such relatively inhospitable country. Furthermore, unless Aboriginal man had already reached the southeast coast, the widespread 'mountain' conditions would have discouraged him from eastward expansion.

The other point to be emphasised is that the existing plant communities are more sensitive to a lowering in temperature than to a lowering in rainfall, so that vegetation instability and associated soil movement (discussed below) may be related more to colder than to warmer periods. The effects of such colder periods would also be more severe if precipitations did not increase materially—i.e. the combination of cold and relatively dry conditions is the type of climate to which the Australian vegetation, including that in the southeast, is least adapted.

### Soils and Associated Landscape Features

There is also evidence from the distribution of fossil soils (i.e. soils with properties not attributable to present conditions) and from buried soils, for climatic change in southeastern Australia.

The work of Butler (1959), van Dijk (1959) and colleagues has produced evidence for several cycles (K-cycles) of groundsurface formation in southeastern Australia, each including an unstable period of erosion and associated deposition (Ku), followed by a more stable period of soil formation (Ks). The Ks phases are considered to represent the more humid periods of denser vegetation and associated soil profile development, and the Ku phases drier, stormier periods. It is assumed that the drier periods were also warmer but, as will be seen from the earlier remarks on the cold-susceptibility of the vegetation, the unstable phases could well have been colder. Not all workers agree with these interpretations: some consider that the Ku phases were wetter, and that the Ks phases were the drier ones. Clearly, therefore, for soils evidence to be most useful, there is need for a better understanding of the conditions associated with slope development and soil profile formation.

Various K-cycles in a region are numbered from $K_0$ (the present cycle of modern erosion) back through $K_1$, $K_2$, $K_3$, etc., as the case may be. In some studies (van Dijk 1959), traces of $K_4$, $K_5$, and older materials have been identified. Because of the different degrees and extents of erosion and deposition on a slope, very complicated soil layering can develop, as shown in Fig. 3:1. The kind of contribution the study of soil layers can make to the understanding of the variety and duration of past climates is indicated in Fig. 3:2. This contribution would, of course, be improved if there were a better understanding of pedogenic processes—for example, whether the Ku features in cold areas are due to colder or warmer conditions—and the inclusion of suitable organic materials in successive K layers would permit the relative chronology to be placed on an absolute basis.

*Table 3:2 C-14 determinations of the K-cycle terraces, south coast, N.S.W.*

| **Terrace cycle** | **Age, years BP** |
|---|---|
| $K_0$ | Modern (0-120) |
| $K_1$ | 390 ± 60 |
| $K_2$ | 3,740 ± 100 |
| $K_3$ | 29,000 ± 800 |

*Source:* Walker 1962.

On the south coast of New South Wales, in the Nowra area, Walker (1962) has identified four main K-cycles—$K_0$, $K_1$, $K_2$, and $K_3$—the climatic sequence and radiometrically-determined dates of which are shown in Table 3:2. In this area, Walker concluded that there are few soils older than $K_3$—that is, most of the soil features now seen could have developed either before or within the probable period of Aboriginal occupation there. It should be pointed out that much of Walker's work was in the lower Shoalhaven Valley, which is likely to reflect changes which occurred in the higher Tableland sections of the catchment as well as on the coast; correlations between events at higher and lower levels might therefore be expected.

In the Southern Tablelands, van Dijk (1959) has studied soil-layering phenomena in considerable detail. Here, as in the coastal sector, climatic rather than topographic changes in the environment are considered to have been responsible for all of the K-cycles except the modern cycle of erosion. Four main cycles can be clearly identified, with remnants of much older ones. Details are set out in Fig. 3:3 (after van Dijk 1959), which shows phases of landscape development on the Canberra-Queanbeyan Plain related to the K-cycles of Butler (1958, 1959). In the Southern Tablelands van Dijk also identified aeolian deposits of apparently $K_2$ or greater age.

*Fig. 3: 2 Diagrammatic representation of the history of climatic conditions, soil-forming intervals, and depositional systems indicated in the Riverine Plain (after Butler 1958)*

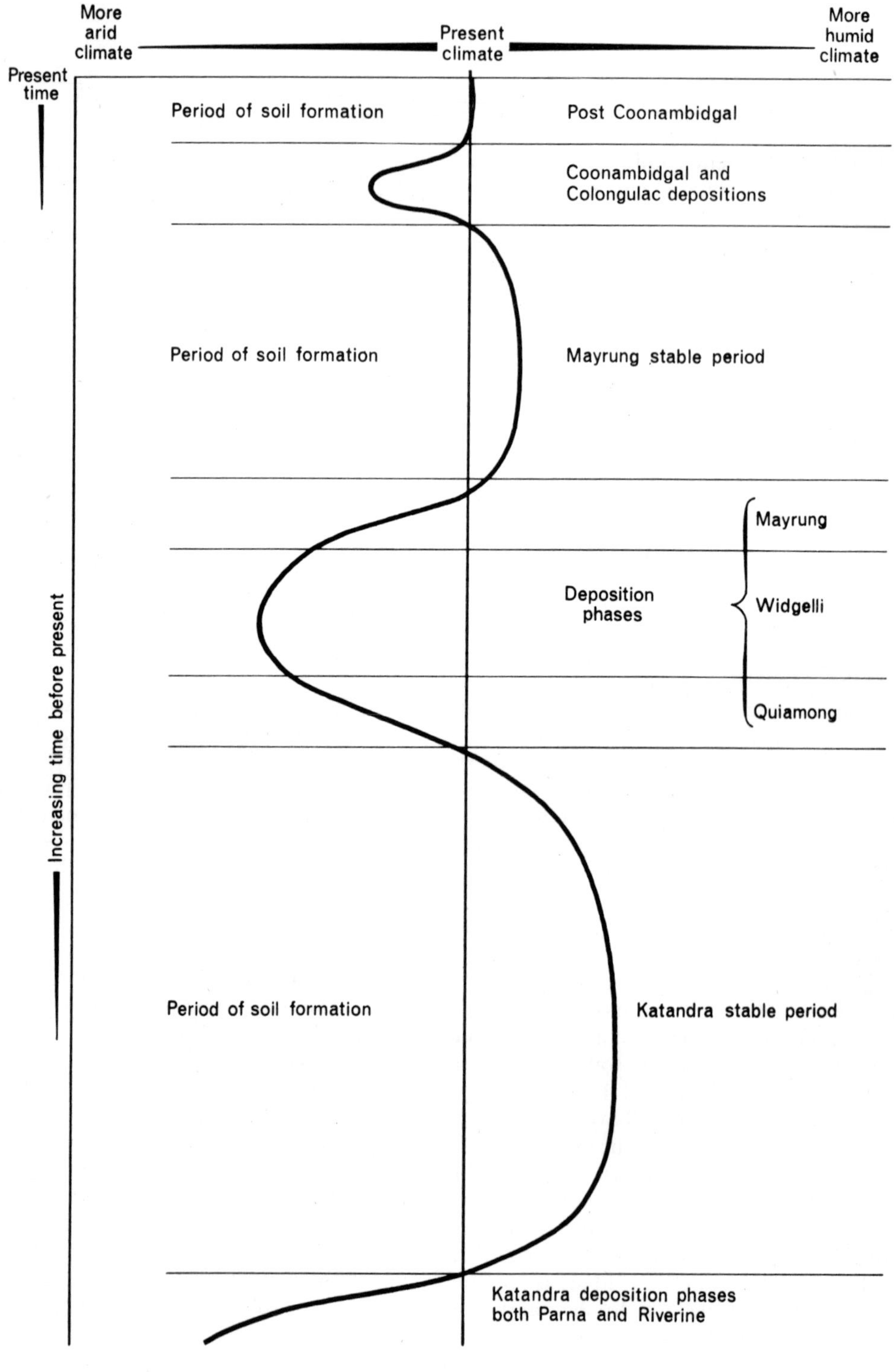

*Fig. 3:3 Phases of landscape development on the Canberra-Queanbeyan Plain related to the K-cycles of Butler (1958, 1959); after van Dijk (1959). s = stable phase of groundsurface development; very restricted hillslope erosion; u = unstable phase; varying intensities of erosion and deposition.*

| K cycles and relative duration | Climatic conditions | | Phases and conditions of groundsurface development |
|---|---|---|---|
| | More arid, stormy (u) | More humid, equable (s) | |
| Present | 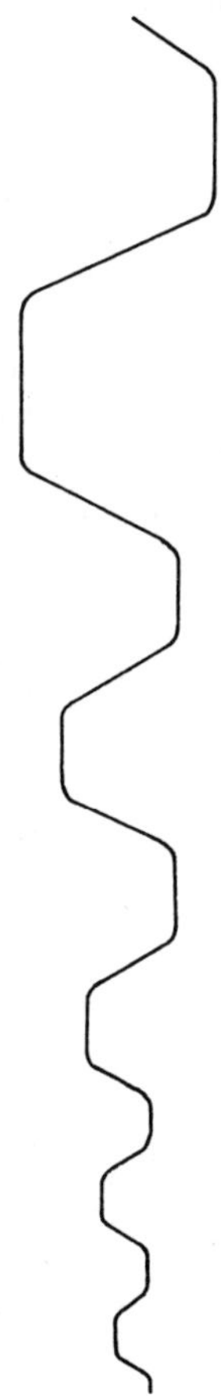 | | Contemporary flooding and river deposits |
| $K_0$—Kambah (very short) | | | s Soil formation (regosol)<br>u Very short aggradational phase |
| $K_1$—Tharwa (short) | | | s Soil formation (prairie soil)<br>u Short aggradational phase |
| $K_2$—Kurrumbene (moderately long) | | | s Soil formation: red earth type; linear entrenchment of streams<br>u Instability of part of the groundsurface on hillsides with sheet erosion and deposition and ravine activity; floodplain aggradation and valley-floor widening by lateral stream erosion |
| $K_3$—Pialligo (long) | | | s Soil formation: podzolic; linear entrenchment of streams<br>u General instability of the groundsurface; vigorous colluvial activity in the whole landscape; floodplain aggradation and valley-floor widening by lateral stream erosion |
| $K_4$—Gundaroo (very long) | | | s Soil formation: planozolic; linear entrenchment of streams<br>u General instability of groundsurfaces, very vigorous colluvial activity; deep deposits of fine and coarse-grained deposits in the lower parts of the landscape |
| $K_5$ (?) | (?) | | Depositional phase followed by soil formation |
| Tertiary (?) | (?) | | High-level gravel deposits and colluvial cones, some of which show features of lateritic profile development |

*Table 3:3 Some C-14 dates from southeastern Australia, and their probable significance*

| **Material** | **Locality and approx. elevation (m)** | **Age (years BP)** | **Interpretation** |
|---|---|---|---|
| Wood fragments at base of periglacial slope deposit | Munyang, Snowy Mts, 1,370 m | 31,300 ± 2,300<br>31,820 ± 2,500<br>30,920 ± 1,800 | Beginning of period of widespread cooling-off, with glacial and periglacial conditions |
| Wood fragments at base of periglacial slope deposit | Geehi, Snowy Mts, 1,370 m | 33,700 ± 2,000 | As above |
| Organic soil and fine wood fragments near base of periglacial slope deposit | Island Bend, Snowy Mts, 1,160 m | 25,700 ± 5,230 | As above |
| Laminated silts and sands on cirque floor | Mt Twynam cirque, Snowy Mts, 1,980 m | 20,200 ± 165 | Cold conditions continuing |
| Basal peats on glacial or periglacial rubble | Upper Snowy Valley, 1,830 m | 15,000 ± 350 | Cold conditions ameliorating enough to permit general development above 1,830 m of alpine vegetation similar to that of today |
| Basal peats on glacial or periglacial rubble | Carruthers Creek, Snowy Mts, 1,950 m | 14,400 ± 250 | As above |
| Basal peat over gravel | Badja, Badja River, 1,070 m | 9,760 ± 170 | Cold conditions ameliorating enough along coastal ranges to permit development of vegetation similar to that of today |
| Basal peats overlying more mineral deposits on cirque floor | Mt Twynam cirque, Snowy Mts, 1,980 m | 8,620 ± 180 | Present-day vegetation also developing in coldest sites, suggesting similar climatic conditions to those at present |
| Wood remains of present-day species over rubble | Perisher Valley, Snowy Mts, 1,675 m | 8,100 ± 250 | As above |
| Wood remains of present-day species in peat over gravel | Wullwye Creek, nr Berridale, 915 m | 7,010 ± 110<br>6,250 ± 90 | Vegetation and climate generally similar to those of today |
| Basal peat and wood remains in sphagnum bog over shallow (?) periglacial rubble | Ginini Bog, Brindabella Range, 1,585 m | 3,240 ± 70<br>3,050 ± 70 | Colder phase prior to about 3,000 BP |
| Wood remains in periglacial solifluction terraces | Mts Lee and Northcote, Snowy Mts, 2,075 m | 2,980 ± 180<br>2,910 ± 130<br>2,860 ± 160<br>2,250 ± 130<br>1,540 ± 160 | Colder conditions in alpine areas 1,500-3,000 years ago |
| Peat over periglacial stone line in deeper 14,400-year-old peat | Carruthers Creek, Snowy Mts, 1,950 m | 1,835 ± 180 | Colder phase, interrupting peat formation, more than 1,800 years ago |
| Basal snowpatch peat | Carruthers Peak, Snowy Mts, 1,980 m | 2,940 ± 110 | Colder phase not ameliorating enough to permit peat formation until about 3,000 BP |
| Basal snowpatch peat | Mt Twynam cirque, Snowy Mts, 1,980 m | 2,520 ± 200 | Colder phase not ameliorating enough to permit peat formation until about 2,500 BP |
| Wood remains in river bank | Brindabella, Goodradigbee River, 760 m | 1,675 ± 60 | Vegetation and climatic conditions generally similar to those of today |

Although soil features of van Dijk's groundsurfaces are similar to those described by Walker on the south coast, and suggest contemporaneity, absolute dates are not available for van Dijk's materials. However, recent soil studies by Coventry (in Polach *et al.* 1968) on the western scarp of Lake George do suggest some correspondence. In particular, there is evidence of instability 2,000 to 3,000 years ago in what appear to be $K_2$ materials.

In the Lake George basin itself, Galloway (1967) has obtained a C-14 date of 15,000 years for charcoal fragments in a sand and gravel spit formed when the lake was about three times deeper than at present. He suggests that this higher lake level, occurring at about the time of the last major phase of the Wisconsin Glaciation in North America, need not necessarily denote higher absolute precipitations; lower temperatures with less evaporation could achieve the same result. It is possible that some of the aeolian sand deposits on the eastern shores of Lake George, and in the broad Shoalhaven Valley, are of this age.

Further south, in the Berridale area (Table 3:3), there are also well-developed terraces with soil features similar to those of van Dijk's K series in the Southern Tablelands. Some of these deposits contain abundant wood remains. In a layer from 6,000 to 7,000 years old there is an abundance of eucalypt wood (affinities with *E. dalrympleana*), epacrids, and *Leptospermum*. The presence of the *E. dalrympleana*, now present in small quantities in the uppermost part of the catchment, suggests that the climate 6,000 to 7,000 years ago was either similar to or only slightly cooler than that of today; it could hardly have been warmer, otherwise *E. dalrympleana* would probably not have occurred in the catchment area. In a terrace of the Goodradigbee River, near Brindabella, wood aged around 1,700 years old and buried beneath 2 m of sediment has been identified as eucalypt having affinities with *E. radiata* (Table 3:3), a common present-day species in the catchment area. In other words, contemporary eucalypt communities appear to have been in the same general area for the last several thousand years.

It is relevant to note that in the Port Phillip area the numerous remains of *E. tereticornis* and *E. camaldulensis* wood discovered in bores and excavations (Gill *et al.* 1968) also suggest that plant communities adapted to similar, or perhaps very slightly warmer, climates compared with today's have been present for at least the last 10,000 years.

Incomplete though the evidence of wood remains is, there is not yet any suggestion that, during the last 10,000 years, plant communities in southeastern Australia were very different from those there today. If Aborigines were in the vicinity at that time, they would probably have found conditions much the same as now.

The Berridale features are also associated with red, wind-blown sands, apparently blown up from the nearby deposits of Wullwye Creek, which here occupies an old floodplain several hundred metres wide. As far as can be ascertained at present, these sands indicate windier (and probably colder) conditions, and are considerably older than the nearby 6,000 to 7,000-years-old deposits containing wood. Similar deposits occur downwind of other floodplain areas in the Monaro and Southern Tablelands Regions.

Also present are extensive slope mantle deposits, apparently the oldest transported materials in the area. These deposits, which are very widespread on the Monaro Tableland, indicate a long period (or periods) of severe instability which is attributed to much colder periglacial conditions, with mean temperatures apparently 8·3°C to 11°C lower

*Fig. 3:4 Southeastern Australia*

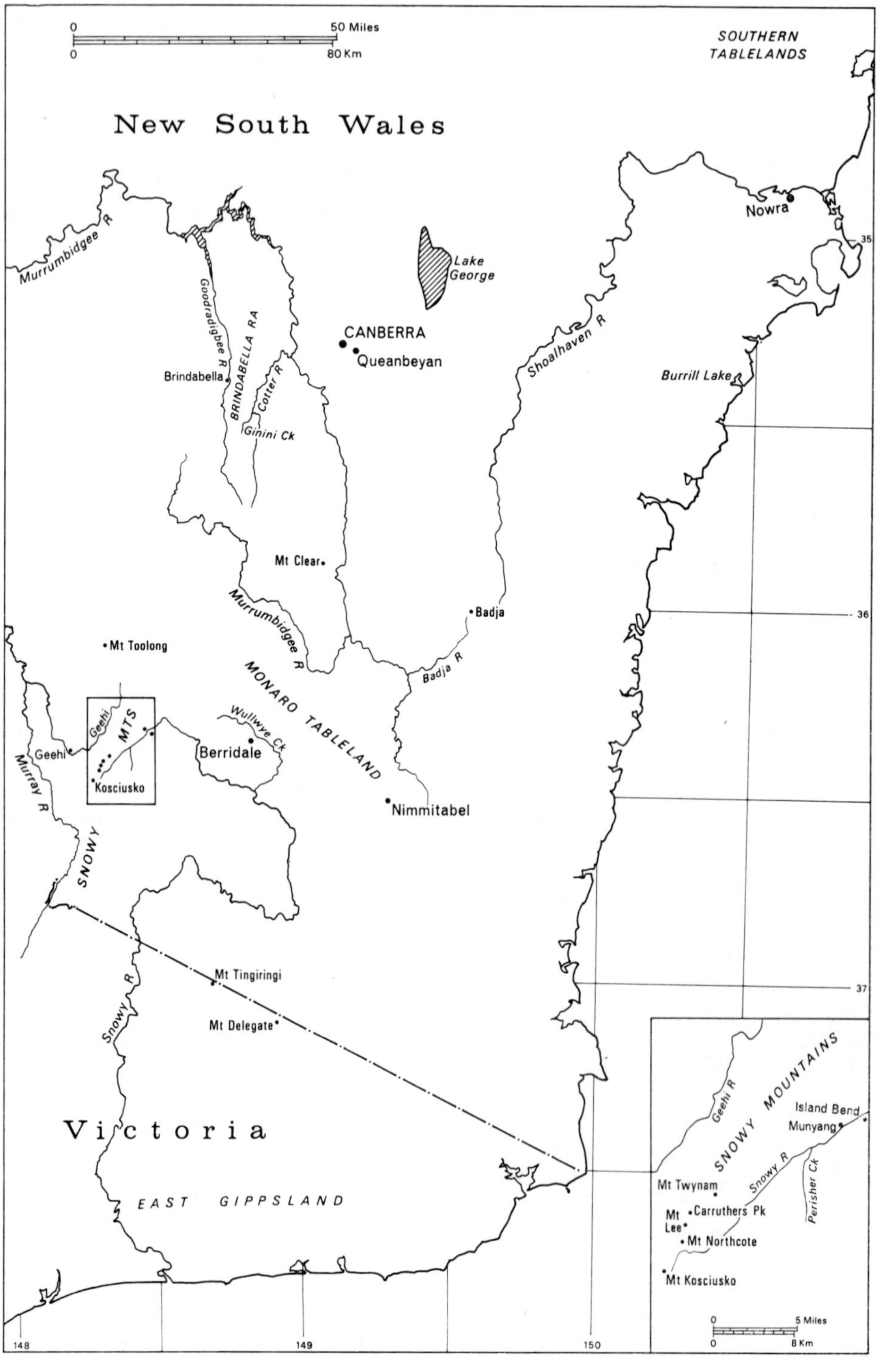

than those of today. An unknown, but considerable age (>30,000 years) for these deposits is indicated. Conditions for plant growth during this long, severe, unstable period must have been very difficult, and it seems likely that only a sparse, alpine type of vegetation could exist. Such conditions would also have been inhospitable for human occupation, assuming that Aboriginal man was in the area at that time.

In the higher, mountain environments more direct information on the nature, extent, and timing of changes in soils, vegetation, and climate is available. The majority of the soils and soil materials studied and sampled here are from pedogenic or stratigraphic discontinuities which, in themselves, indicate changes, apparently climatically induced, in the area under investigation. The main evidence is summarised in Table 3:3. So far, there is a virtual absence of dates between about 8,000 and 3,000 BP; as yet, there is no direct evidence for marked pedogenic or stratigraphic discontinuities (or associated climatic change) within this period.

## Climatic and Ecological Inferences

Reviewing the information presented, it seems clear that the most significant event of the last thirty-odd thousand years was a fairly general cooling-off, with widespread slope instability, at least down to 915 m; the onset of this unstable period seems to have been about 32,000 years ago. The tablelands slope deposits, which in places extend to as low as 610 m, may also belong to this period, although they could be still earlier. Mean temperatures are likely to have been about 8·3°C to 11°C lower than at the present time, i.e. the climate at 610 m—the level of Canberra—would be similar as far as temperatures are concerned to that at Mount Kosciusko at present. Precipitations (cf. the Lake George data, Galloway 1967) were not necessarily greater, however, with the implication that much of the land surface would not be protected by winter/spring snow cover, and would therefore be exposed to frost and wind erosion.

The areas most sensitive to lowering in temperature would be the mountains and adjacent tablelands where fairly major soil and vegetation effects would have been likely, including the disappearance of tree vegetation to lower levels and an increase in shrub and herbaceous cover of the type now found above the tree line (> 1,830 m) and on exposed rocky sites down to 1,220 m. It should be noted, however, that most areas which would have been thus affected are close to lower and varied country where ecological readjustment to climatic change would be possible. In coastal and coastal mountain areas subject to oceanic influences, and in the deep valleys at lower levels on the western side of the Great Dividing Range, the same pronounced temperature changes may not have occurred, and it seems likely that many of the present-day communities could have existed there, available for fairly rapid invasion of the then colder areas when temperatures rose.

On the basis of the wood remains in the Munyang deposits (possibly *Eucalyptus* species) and particularly the 35,000-years-old *Nothofagus* affin. *cunninghamii* stump under the block stream on the Toolong Range, studied by Caine and Jennings (1968), it would seem that, prior to the initiation of general instability, conditions, whilst not greatly different from those of today, were somewhat cooler in summer. This conclusion is based on the observation that the present distribution of *N. cunninghamii* (and *N. moorei* further north) is associated with cool summers and frequent afternoon cloud and fog as in Tasmania, the Baw Baw Plateau, and the Barrington Tops.

In any event, it seems unlikely, at least on the basis of existing evidence on the antiquity of the Aboriginal, that he was well established in southeastern Australia when the 32,000 years BP cooling-off period commenced. Walker's date of 29,000 years for the $K_3$ groundsurface on the south coast, and the 26,480-years-old Linda moraine in Tasmania (Peterson 1968), suggest that this period of instability was prolonged and widespread.

On the evidence of the varve-like laminated silts and sands on the floor of the Twynam cirque, cold conditions with some surviving glacial activity were still in existence 20,000 years ago. (It is noteworthy, in passing, that the main period of excavation of the Twynam cirque must have been much earlier than this.) The basal valley peats over rubble in the Kosciusko Summit area, 14,000 to 15,000 years in age, indicate that, by this time, ice and snow cover had disappeared sufficiently to support stands of vegetation similar to those of today; the peats here are composed partly of species of sedges still important in the area. The date of 8,000 years for peats overlying the varve-like materials in the Twynam cirque indicates that deglaciation was probably incomplete here and in similar snowy situations until within the last 10,000 years; and there is other evidence (e.g. the basal peats of the old Perisher Lake and the Badja Fen) that a new period of soil and vegetation development not very different from that at present also began at this time.

It must be emphasised that this evidence for a long, cold period from about 32,000 to 10,000 years ago is broadly based; further work will most probably show that there were also climatic oscillations within this period.

This evidence suggests, therefore, that when Aboriginal man reached southeastern Australia it was considerably cooler than today. These conditions, with the associated vegetation inhospitable both for man and larger species of game, may have kept Aborigines substantially out of the tableland and mountain areas until conditions improved some 10,000 years ago. Furthermore, if their movement towards the coast was in a general west-east direction, the large area of periglaciated country that would have extended from Victoria towards the New England region of New South Wales would have blocked their way to the southeast coast, even though conditions there may have been quite favourable for human occupation. Alternatively, as now seems more likely in view of the 20,000-years-old dates from the Burrill Lake rock shelter, reported by Lampert (see ch. 10), Aboriginal man may have occupied the coastal zone from the north. Occupation of tableland areas several thousand years later could then have been both from the west and the east.

For the next several thousand years, from about 10,000 BP, the climate appears to have been rather mild, perhaps with slightly warmer temperatures than at present. At this stage of knowledge, there is no definite evidence for arid cycles during this period, although it is fair to say that the unstable phases of earlier K-cycles could be interpreted in this way.

There is a variety of evidence from the sphagnum peats, snowpatch peats, stone lines, and solifluction terraces for a slightly colder period beginning roughly 3,000 to 4,000 years ago and continuing probably intermittently until about 1,500 years BP. Mean annual temperatures were probably 2°C to 3°C lower than at present. Soil data from tableland and coastal areas indicate that the effects of this period were also felt at lower levels, either because of the temperature changes themselves or because of associated changes in effective precipitation. But this change was not comparable in magnitude or in its effects with the colder period of 10,000 to 30,000 years ago.

Summing up, then, the evidence presented indicates that the major climatic changes of the Late Pleistocene in southeastern Australia—the much colder period commencing about 32,000 years ago and continuing probably with gradual amelioration until about 10,000 years BP—either occurred before Aboriginal man reached the southeast, or effectively halted his movement into the tableland areas both from west to east and from the coast. Since about 10,000 years ago the climate and associated soils and vegetation appear to have been generally similar to those at present, except for the colder period 3,000 to 4,000 years ago. However, even this colder 'Little Ice Age', whilst responsible for considerable soil and vegetation instability, does not seem to have had major biological and soil effects as far as Aboriginal man is concerned. Almost certainly, none of the changes in southeastern Australia during the last 10,000 years, except perhaps for some of the 'Little Ice Age' effects at higher elevations, would have caused much inconvenience to the Aborigines.

## References


Butler, B. E. 1958. *Depositional systems of the Riverine Plain of south-eastern Australia in relation to soils*. Soil Publ. CSIRO Aust. No. 10.

——1959. *Periodic phenomena in landscapes as a basis for soil studies*. Soil Publ. CSIRO Aust. No. 14.

Caine, N. and Jennings, J. N. 1968. Some blockstreams of the Toolong Range, Kosciusko State Park, New South Wales. *J.Proc.R.Soc.N.S.W.* **101**:93-103.

Costin, A. B. 1954. *A study of the ecosystems of the Monaro region of New South Wales*. Sydney: Government Printer.

Galloway, R. W. 1967. Dating of shore features at Lake George, New South Wales. *Aust.J.Sci.* **29**:477.

Gill, E. D. *et al.* 1968. Quaternary shorelines research in Australia and New Zealand. *Aust.J.Sci.* **31**:106-11.

Peterson, J. A. 1968. Cirque morphology and Pleistocene ice formation conditions in southeastern Australia. *Aust.geogrl Stud.* **6**:67-83.

Polach, H. A., Golson, J., Lovering, J. F., and Stipp, J. J. 1968. ANU radiocarbon date list II. *Radiocarbon* **10**:179-99.

van Dijk, D. C. 1959. *Soil features in relation to erosional history in the vicinity of Canberra*. Soil Publ. CSIRO Aust. No. 13.

Walker, P. H. 1962. Terrace chronology and soil formation on the south coast of N.S.W. *J.Soil Sci.* **13**:178-86.

# 4 River Systems and Climatic Changes in Southeastern Australia

Simon Pels

The Riverine Plain of southeastern Australia is one of the largest areas in Australia to which a reasonably well understood geological history of relatively short duration can be ascribed. Its extent in New South Wales alone, as shown in Fig. 4:1, is 65,000 km$^2$.

Apart from secondary aeolian features, it consists entirely of water-transported alluvial and lacustrine sediments deposited by inland flowing river systems. These sediments have an apparent affinity with the present rivers and it was initially assumed that the Plain consisted of floodplain deposits of the present river system. Environmental studies, carried out in recent years as an adjunct of irrigation development, have greatly improved the understanding of the region and there is now ample evidence to show that the present rivers are unrelated to the relict alluvial landscape.

The aim of this paper is to outline the present knowledge of this region's geological history, to list the main landscape features and describe how they can be recognised. A further object is to discuss the relative age of the various geomorphic units in terms of stratigraphic evidence.

Although many of the geomorphic units are well preserved and have a youthful appearance, radiocarbon estimations have shown that the bulk of the Plain's sediments are of an age beyond the range of this dating technique, and these sediments are classified as 'Older Alluvium'.

'Younger Alluvium' also features prominently in the landscape, and radiocarbon dating has provided a number of absolute dates back to approximately 30,000 years BP.

By understanding the processes of deposition, or conversely in the case of remnant features, the processes of degradation, insights are gained into climatic conditions of the past which permit inferences contributing

*Fig. 4:1 The Riverine Plain*

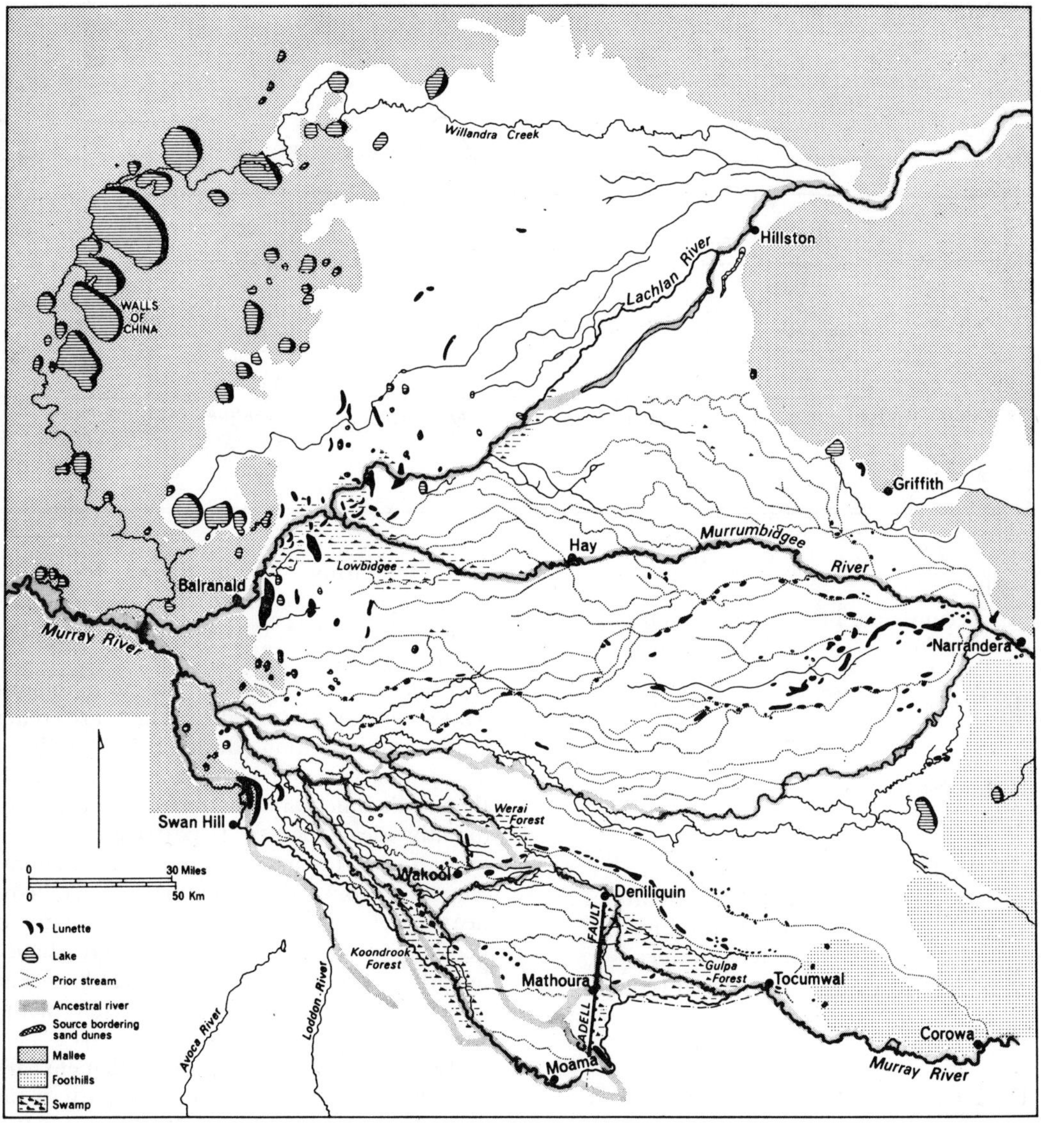


to an evaluation of a regional palaeoclimatology. Such inferences should be regarded as contributions towards an eventual fuller understanding, as the observed processes of deposition and degradation are at times subject to widely differing interpretations.

To facilitate subsequent discussion, the main geomorphic surface features are listed:

*Fig. 4: 2 Cross-section of a prior stream showing geomorphic units, soils, and land use*

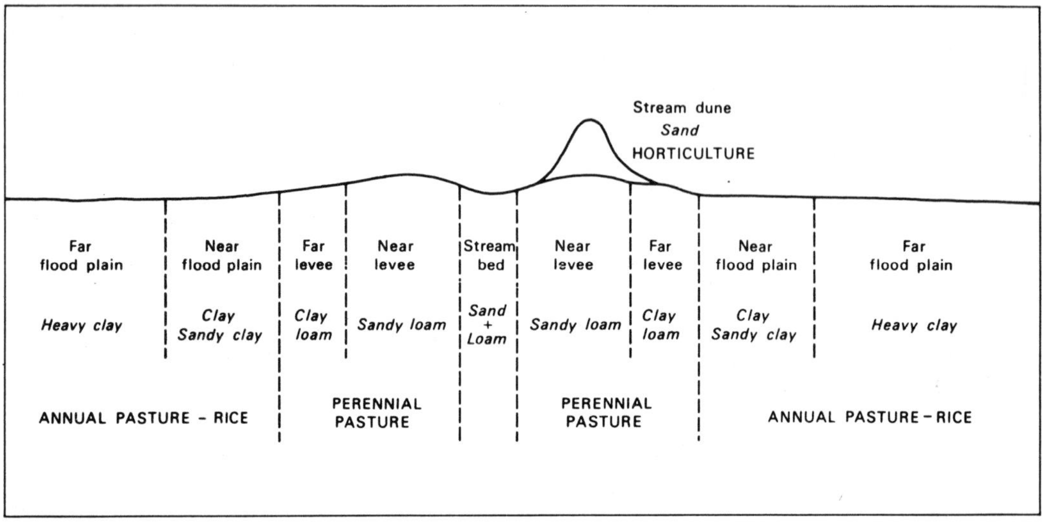


| | |
|---|---|
| I Fluviatile | |
| (a) Terminating on the plain | 1 Prior streams |
| | 2 Box swamps |
| | 3 Deltaic swamps |
| (b) Transitory | 1 Ancestral rivers |
| | 2 Present rivers |
| II Lacustrine | 1 Lake beds |
| | 2 Far floodplains |
| III Aeolian | 1 Lunettes |
| | 2 Crescentic dunes |
| | 3 Stream dunes |
| | 4 River dunes |
| | 5 Deflated levees |
| | 6 Sand sheets |
| | 7 Clay sheets |
| IV Tectonic | 1 Fault scarps |

These geomorphic units occur in a region of low relief with a westerly gradient of 1-2 m per 5 km. The region comprises a number of very gently sloping alluvial fans emanating from gaps in the ranges which form its northeastern, eastern, and southern boundaries. These fans coalesce to form a piedmont plain.

Prior streams (I(a) 1) are a dominant feature of the landscape. They are relict streams of deposition (Butler 1950) and have a characteristic soil pattern associated with them. Sedimentary patterns, and subsequently developed soil patterns, show a gradation of textural differences in a lateral sequence from coarse textured streambeds, near-levees, far-levees, to fine textured floodplains. Source-bordering sand dunes, here described as stream dunes (III 3), occur commonly on the leeward (NE) side of the streambeds.

Land use and mode of settlement are determined by this sedimentary pattern, as shown generally in Fig. 4:2. Although this configuration is fundamentally simple, it is not always evidently so in the field. This is due to the meandering of prior streams and the fact that several stream systems may be superimposed. Frequent diversions also complicate the pattern of soils distribution in the field.

Evidence of prehistoric habitation is largely restricted to levees, streambeds, and sand dunes. It is unlikely that prior streams carried water at the time of human habitation, as C-14 datings of wood samples from the streambed sands indicated ages of over 36,000 years BP (Pels 1964a). Stone fragments, artifacts, and buried fireplaces with baked clay are common on the levees. As the plain does not have any naturally occurring consolidated sediments apart from small gravel, it can be said that all stone found there was carried into the area by man. The only exceptions to this rule are meteorites, which have been found around Barratta Station west of Deniliquin, and small outcrops of laterite in the younger alluvium of river banks.

Mapping of stream traces has shown that prior stream systems fanned out from three gaps in the ranges in New South Wales. These are at Hillston, Narrandera, and Corowa (see Fig. 4:1). Closer to the apices of these fans, i.e. in the eastern half of the plain, stream traces are closely spaced and the streambed-levee components predominate over floodplain components of the landscape. Soils are therefore generally of a lighter texture than in the western part, where wider spacing and less developed levees resulted in a predominance of floodplains.

The lighter textured levees, commonly deflated into elongated tracts of 'scalded country', are common over the plain. Similar deflation of sandhills, or incipient source-bordering sand dune development, have formed sand sheets (III 5), while redeposition of deflated levees (III 6) has given rise to aeolian clay sheet (III 7) deposition.

The far western zone is an area of evaporite accumulation, where soils are more saline and of heavier texture. Lake beds (II 1) and lunettes (III 1) are common in this region. The term lunette, as originally proposed by Hills (1939), describes a crescentic hill of loamy sediments accumulated on the leeward side of a lake bed. True lunettes seem to have formed by deflation and redeposition of lake floor sediments during periods of exposure. Crescentic sand dunes (III 2) are found in similar situations adjacent to lakes, but they appear to have been derived from sandy beach deposits washed up by on-shore waves. The two terms are commonly regarded as synonomous; or the terms loam lunette and sand lunette are used. Many of the lunettes bear evidence of prehistoric habitation in the form of stone artifacts, burial sites, and fireplaces (cf. Bowler, see ch. 5).

The lakes show great variation in water quality. Some are highly saline, while others, often in close proximity, may contain fresh water. Similarly, dry lake beds may consist of non-saline soils, while others contain thick layers of evaporites. It is thought that this phenomenon is closely linked to the hydrologic régime of each lake. Where lake floors are well above the prevailing groundwater level in the region, water is lost from the lake to the water table, and after being filled by flooding from the rivers, water quality remains high although some concentration occurs by evaporation.

Where groundwater levels in the area surrounding the lake are higher than the lake bed, saline groundwater tends to flow into the lakes, and rapid and continuous concentration of salt occurs. This also appears to be the case where, within a chain of hydraulically connected lakes, only certain lakes are held at a higher level by their incorporation into the irrigation supply system or by more frequent flooding from the river.

In the western zone, many former lake beds are now covered with fluviatile sediments deposited from prior streams. Lunettes are still found cropping out of those sediments, and these outcrops indicate the outlines of former lakes to their windward side.

Such situations provide evidence for concluding that lakes in the far western zone were the product of tail flows of prior streams, and point towards their contemporaneous existence. They show that the present lakes are superimposed on former and often much larger lake remains, in the same way as prior stream systems are superimposed all over the plain towards the east.

Prior streams, being superimposed on similar older systems and accompanying frequent diversions of the prior streambeds, have created a further geomorphic unit named Box swamps (I(a) 2). These are linear depressions of a discontinuous nature occurring all over the plain; their origin is unclear. Some are sandy depressions, others are in clay. The name is derived from the frequent but not invariable occurrence of box trees (*Eucalyptus largiflorens*) in these depressions. While the sandy depressions are often the exposed, discontinuous remnants of former prior streambeds, the clayey depressions may be products of local drainage during those periods of higher rainfall which are known to have occurred, from evidence to be discussed subsequently.

The final phase of prior stream deposition has taken place from a well preserved system which has been mapped (Fig. 4:1). It can be seen that it is a distributary system which dissipates in the far western zone.

The geomorphological nature of the landscape described so far suggests that the Riverine Plain at the time was an area of sediment accumulation with little, if any, drainage out of the area. It is possible that some surface drainage took place in a southwesterly direction, i.e. towards the Victorian Wimmera where extensive alluvial sediments also occur.

The prior stream discharge could have been disposed of by evaporation from the vast area of heavy clay soils in the west. There are at least 40,000 $km^2$ of heavy soils in the floodplain environment with a high evaporation of 1,500 mm annually. This would yield a total water loss of 60 $km^3$, which is a great deal more than the annual discharge of the present river system. Lunette remains indicate that northeast of Balranald, lakes existed with surface areas of about 1,000 $km^2$.

It is known that fluviatile deposition transgressed onto the Mallee to the west and that chains of lakes extended into the Mallee landscape in several places. Terminal lakes occur in the present landscape on the Willandra Creek, Loddon River, Avoca River, and the Wimmera River.

That the Riverine Plain was an area of occluded drainage at this time would explain the distribution of salt in the sediments of the region and also the steady deterioration of groundwater quality towards the west. It would further explain the anomalous occurrence of extensive fluviatile deposition along the middle reach of the Murray River system.

The foregoing is a description of the prior stream and lake-lunette landscape which was deposited during the period of older alluvium deposition. It is possibly too old to date by radiocarbon methods, i.e. older than at least 36,000 years BP.

The subsequent period of deposition of younger alluvium appears to have occurred in a drastically changed hydrological environment, with drainage channels having broken out of the area along the course of the present lower Murray River to the sea. Regional surveys have shown that during this period the older alluvium surface was dissected by a system of deep and wide channels which were subsequently filled by coarse sediments. These channels were clearly the ancestors of the

*Fig. 4:3 Cross-section of an ancestral river*

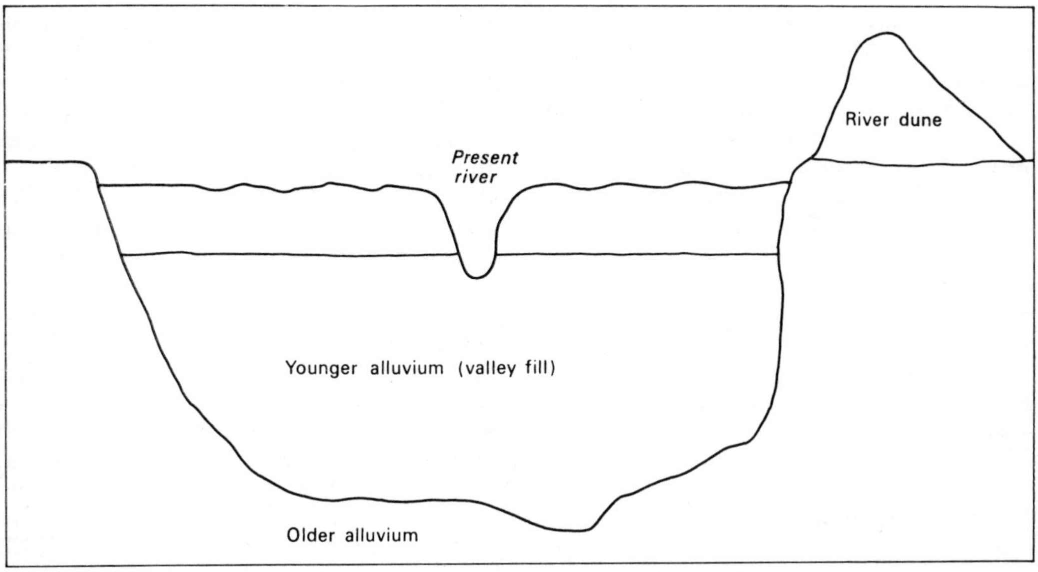


present rivers, and in fact the present rivers represent a continuation of this phase of activity (Fig. 4:3). These channels and the younger sediments they now contain (valley fills) were named ancestral rivers (I (b) 1) (Pels 1964b, 1966). They cut across prior streams and lake beds and clearly postdate the older alluvium surface.

Ancestral rivers enter the plain through the same gaps in the eastern ranges, but combine in the west to form a tributary system in contrast with the prior streams (Fig. 4:1). Drilling across these ancestral rivers has shown that the younger sediments reach depths of up to 21 m. The channel width commonly exceeds 1·5 km, and source-bordering sand dunes occur adjacent to them (river dunes III 4). Ancestral rivers generally form belts of depressed country in the landscape, but in some locations they merge with the older surface or they may be slightly elevated. Generally speaking, however, and with the exception of the dunes, deposition is restricted to the ancestral river channel (Fig. 4:3).

Fig. 4:1 shows a complicated pattern of ancestral rivers in the southwestern portion of the plain in New South Wales. This is caused by the interruption and diversion of the river system by tectonic movement which created a fault scarp (IV 1) between Deniliquin and Moama. This scarp is an interesting and rare example of tectonic breaks in unconsolidated sediments. It is well preserved and reaches a height of 15 m near Mathoura.

The diversion pattern of valley fills around this scarp immediately suggested that the ancestral rivers were not caused simply by one cutting and filling phase. Detailed examination and mapping (Pels 1966) showed that three separate phases of ancestral river activity had occurred, each consisting of a downcutting and backfilling sub-phase. Surface evidence reveals the occurrence of one non-diverted and two diverted channel systems.

Tectonic diversion has caused the present and ancestral Murray River combination to have three different sectors:

**1** Upstream from Tocumwal, where the three valley fills are superimposed and three terrace levels occur adjacent to the present river, which is again superimposed.

**2** The sector between Tocumwal and Wakool Junction where ancestral rivers were laterally separated by diversion around the Cadell Fault. Floodplains occur here which are now deserted by the present river.

**3** The sector downstream from Wakool Junction where the three valley fills are again superimposed and associated with the present river. Three terrace levels are again evident in this sector.

Mapping of the ancestral rivers and investigations into the physical nature of this geomorphic unit has explained many features of the plain previously thought to be anomalous. These include:

(i) soil complexes in irrigation districts;

(ii) the devious paths of floods when river stages exceed present river bank levels and then follow the deserted floodplains;

(iii) the occurrence of belts of waterlogged and salted land;

(iv) the presence of a floodplain on one reach of the river and the absence on another, as at Tocumwal and Swan Hill respectively, is also no longer an anomalous aspect of the river's morphology.

Tectonic movement is furthermore responsible for the formation of deltaic swamps (I(a) 3) which are found upstream from the Cadell Fault, the Lowbidgee Swamp on the lower Murrumbidgee River and the lower Lachlan River 'reed bed' swamp. Tectonics are more indirectly responsible for the regularly inundated areas known as Werai Forest and Koondrook Forest.

The younger alluvium has been dated by radiocarbon samples collected from formations representing the three phases of deposition (Pels 1970), but various interpretations of these dates are possible.

The literature on river morphology shows that it is generally accepted that regional downcutting of river channels occurs under conditions of high discharge, and such conditions are interpreted here as indicating a more pluvial environment. Aggradation in river channels is taken as an indication of more arid conditions.

By applying this reasoning, it follows that the three phases of aggradation are represented by the three valley fills from which carbon samples were obtained for dating. Stratigraphical evidence for three distinct and consecutive phases of river activity was so clear at the time of investigation that this was tentatively put forward as a relative geochronology (Pels 1966). The absolute dates have now confirmed this interpretation.

The datings show that active deposition was taking place at the following points of time:

| | |
|---|---|
| Phase I | Exceeding 28,600 years BP |
| Phase II | 24,050 $\pm$ 835 years BP |
| Phase III | 9,800 $\pm$ 200 years BP |

These findings have been combined with previously published dates of the Goulburn ancestral river system (Bowler 1967). Interpretation of the combined data is shown in diagrammatic form in Fig. 4:4. This also shows the approximate span of time elapsed since the tectonic movement took place, which diverted the ancestral system.

The present river system is shown to have developed during the last 4,000 years and downcutting of the major rivers has occurred during this period. This downcutting phase has been restricted to the present trunk rivers, and this has created a further subdivision of the river system into high level and low level rivers and creeks, features so

*Fig. 4:4 Possible interpretation of geological events, derived from C-14 age estimates*

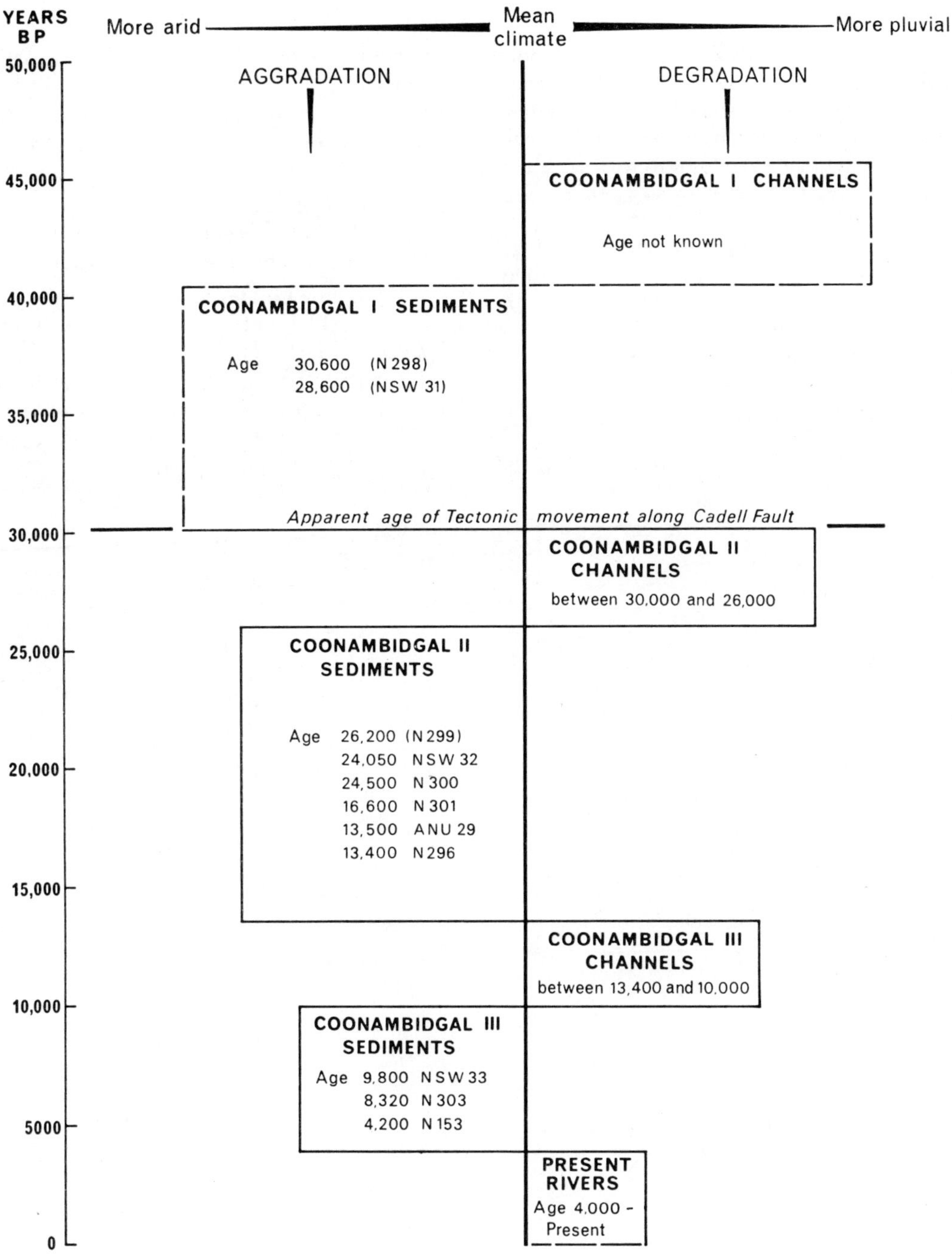

characteristic of the Riverine Plain. It is only during high floods that the relict high level rivers receive water, when the overall system regains the character it held before down-cutting of the trunk rivers began.

**References**

Bowler, J. M. 1967. Quaternary chronology of Goulburn Valley sediments and their correlation in southeastern Australia. *J. geol. Soc.Aust.* **14**:287-92.

Butler, B. E. 1950. A theory of prior streams as a causal factor of soil occurrence in the Riverine Plain of south-eastern Australia. *Aust.J.agric.Res.* **1**:231-52.

Hills, E. S. 1939. The physiography of north-western Victoria. *Proc.R.Soc.Vict.* **51**: 297-323.

Pels, S. 1964a. Quaternary sedimentation by prior streams on the Riverine Plain, south-west of Griffith, N.S.W. *J.Proc.R.Soc. N.S.W.* **97**:107-15.

——1964b. The present and ancestral Murray River system. *Aust.geogrl Stud.* **2**:111-19.

——1966. Late Quaternary chronology of the Riverine Plain of southeastern Australia. *J.geol.Soc.Aust.* **13**:27-40.

——1970. Alluvial sediments of the Riverine Plain of southeastern Australia and their relationship to the Murray River system. *J.Proc.R.Soc.N.S.W.* (in press).

# 5 Pleistocene Salinities and Climatic Change: Evidence from Lakes and Lunettes in Southeastern Australia

J. M. Bowler

The quality and location of inland waters have always exerted a strong influence on both the migration patterns of man and his adaptation to the physical environment. In a country with such limited water resources as Australia, the presence of a large number of lake basins, many of which were filled during the more humid periods of late Quaternary time, would undoubtedly have been an important factor in the ecology of the early inhabitants. Not only did they provide watering places for early man but, depending on their salinities, they also helped sustain the animal populations necessary for his food requirements. In this respect these basins, concentrated mainly in the arid and semi-arid regions, provide focal points for studying the distribution and ecology of Aboriginal populations across vast areas of inland Australia.

In their geomorphic and sedimentary features, these ancient lakes preserve in considerable detail the imprint of past changes in the physical environment to which they responded with particular sensitivity. Such evidence may be derived from the record of strandline fluctuations, the associated terraces and dunes, the salts preserved in their sediments, the aquatic faunas, and the remains of animals and men who lived on their shores. For reconstructing past Quaternary climates, these elements provide data more comprehensive than those of most other physical environments. Moreover, the semi-arid zone with which we are mainly concerned is dominantly an area of deposition and retains a relatively complete sedimentary record. Erosional processes are small in magnitude compared with those of more active geomorphic environments found at higher and colder altitudes or on steep slopes. Thus, by comparison with alpine or glacial environments from which so much previous Pleistocene evidence has been drawn, the non-glaciated semi-arid regions in general, and their lakes in particular, may provide a relatively complete record of the Quaternary stratigraphic sequence. In North America, the evidence from the large Pleistocene lakes Bonneville and Lahontan has already provided a detailed sequence of climatic fluctuations complementing evidence from the glaciated regions (Gilbert 1890; Broecker and Kaufman 1965; Morrison and Frye 1965). In Africa, Lake Chad in the southern Sahara and lakes of the East Africa Rift system are being intensively studied for a similar purpose (Grove and Warren 1968; Richardson 1966; Washbourn 1967). In many cases the Pleistocene stratigraphy is intimately associated with the archaeological record in such a way that the reconstruction of the one is incomplete without an understanding of the other.

It is the purpose of this essay to present some evidence of Pleistocene salinity and hydrologic changes in southeastern Australia, to examine their influence on the physical environment, and in so doing, to define some of the factors affecting the migration and adaptive behaviour of early man in this context. In giving a preliminary account of the evidence, much of which requires further examination, the inadequacies of such a brief statement will be apparent. Of necessity, some claims may be inadequately substantiated, a compromise dictated by limitations of space and definition of the essay. In an attempt to limit the subject to areas of archaeological interest, the discussion is centred on salinity and palaeohydrology in the belief that an understanding of the nature and sequence of these changes will assist in understanding the distribution of early man, the environment in which he lived, and his food gathering economy.

This essay is divided into two major sections. In the first, general features of Pleistocene salinity changes are examined, with emphasis on the significance of lunettes. In the second, using new data from lakes in the Murray Basin, detailed examples of salinity and water level changes in a stratigraphic sequence associated with archaeological evidence are discussed.

## Salinity and Palaeohydrology in the Archaeological Context

The water quality or salt content of Pleistocene lakes exerted both a direct and indirect control on the suitability of the environment for human occupation. Firstly, the quality of the water determined whether it was suitable for human consumption. Secondly, the lakes yielded a variety of aquatic foods, e.g. mollusca and fish, and by providing a suitable watering place they helped sustain a large number and variety of animals sought by the ancient hunter, including kangaroos, emus, and probably at least some representatives of the large extinct Pleistocene marsupials.

The importance of any one lake in the economy of either animals or man was determined by two important factors: the salinity of the lake waters and the salt tolerance of the species involved. Salinity varies according to many complex factors amongst which climate exerts an important influence. Lakes in high rainfall areas from which the salts are frequently flushed by overflow are usually low in salinity and suitable for human consumption. But internal drainage basins in areas of high evaporation progressively accumulate salts while losing water by evaporation. Lakes in such regions are usually highly saline.

The second factor is a biological one, but may indirectly affect the archaeological record. Where lakes are too saline for human consumption, they may sustain only an impoverished aquatic fauna, but may still provide an important watering place for salt-tolerant animals. In the past man may therefore have been drawn to their shores in search of food without basing his economy directly on their environment.

The relevance of salinity can be shown by examining the tolerance of man and some of the animals with which he is commonly associated. Dixey (1966) lists the following as the approximate limits in parts weight of solid per million parts of water (ppm):

| | |
|---|---|
| European man domestic supply | 570 ppm |
| Man in arid lands | 2,500 and rarely to 4,000 ppm |
| Horses | 3,000 and up to 4,000 ppm |
| Sheep | 7,000 ppm |

The salinity of sea water, for comparison, is approximately 35,000 ppm. The upper tolerance of unionids is not accurately known, but they commonly occur in inland fresh and brackish waters with salinities less than 15,000 ppm while the gasteropod *Coxiella* and many ostracods flourish in salinities from 30,000 to more than 60,000 ppm. Moreover, some salts, e.g. dolomite, form under hypersaline conditions in which salinities exceed 50,000 ppm.

The tolerances of Australian desert and semi-arid faunas are not well known. Calaby (pers. comm.) reports that by conservation of body fluids many of the smaller animals can survive on dry food alone. Small wallabies from Rottnest Island, W.A. (*Setonyx brachyurus*) have been found experimentally to drink salt solutions as concentrated as 25,000 ppm sodium chloride (Bentley 1955). Calaby further states:

there seems to be a considerable diversity in the large arid-country kangaroos; there are drinkers and non-drinkers. One could also speculate that there is some diversity in individual ability to make do with brackish water. In a situation in which the available water was becoming increasingly salty, selection would favour those best adapted to deal with brackish water.

During periods of increasing water salinities, the lakes would have become progressively less capable of sustaining suitable faunas, until eventually they would have been abandoned by the hunter-gatherer community.

Since climate is one of the most important factors controlling variations in water salinity, it is important to consider a regional or climatic distribution of lake basins. Fig. 5:1 shows the climatic boundary between the arid and semi-arid zones in Australia based on Thornthwaite's classification.

Within the arid continental interior, the desert or playa lakes usually possess a salt-encrusted surface, and water collecting in them after high intensity rains takes up large quantities of dissolved salts. In the semi-arid zone under present climatic conditions, however, precipitation is usually sufficient to leach salts below the level of the lake floor. Salts concentrated in drying lake waters crystallise in the soil profile, in the underlying sediment, or are carried through to groundwaters, which are notoriously saline below Quaternary lakes.

Within the semi-arid zone lakes are either dry or ephemerally active under the present climate. But the presence of wave-trimmed shorelines, cliffs, terraces, and associated faunas points to a late Quaternary situation in which many remained filled for long periods. Where outlets were developed, as in many lakes associated with major stream systems, salts brought in by solution were removed by overflow, maintaining low salinities throughout the lake-full stage. But where

*Fig. 5:1 The relationship between the distribution of linear dunes in the arid continental interior and the occurrence of lunette lakes. The area discussed in the text lies within the southeastern region as shown. (Data from Gentilli (1948), Jennings (1968), and records of Commonwealth Bureau of Meteorology.)*

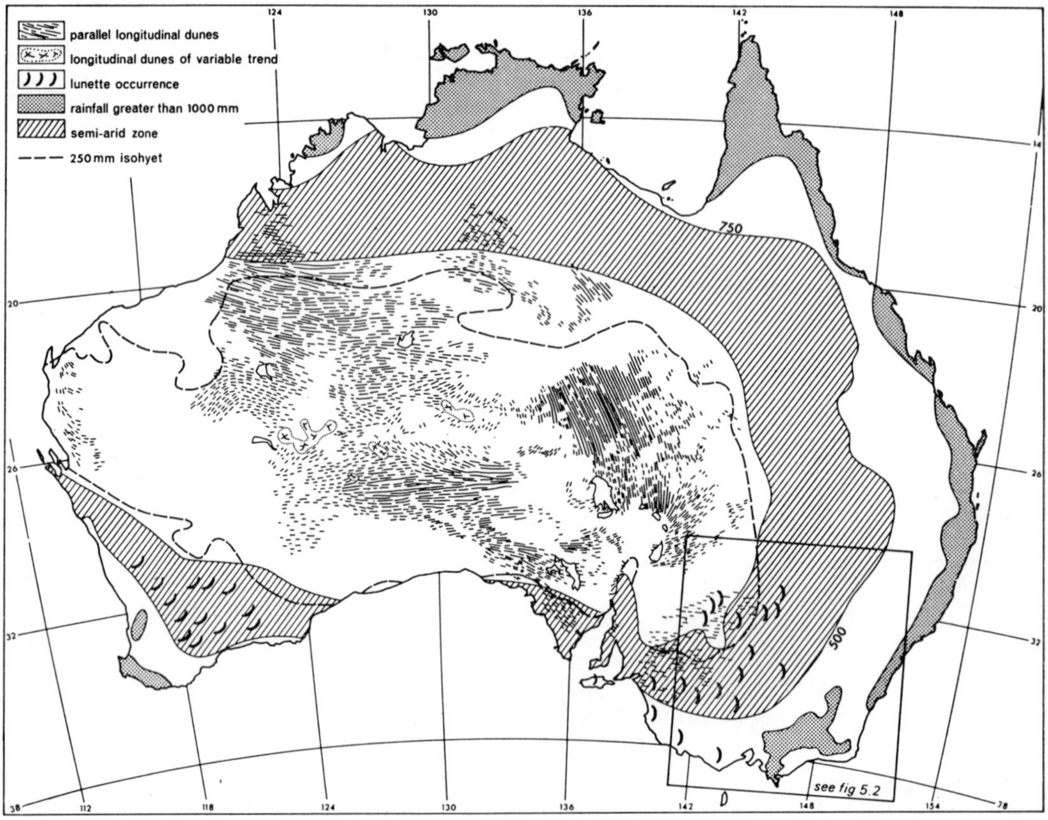


overflow was not possible due to either the configuration of the basin or to a decrease in discharge, the total dissolved salts concentrated by evaporation eventually reached a level where precipitation of the less soluble components began. In examples from modern lakes, e.g. in Lake Keilambete in western Victoria and in the Coorong coastal lagoon in South Australia, aragonite and calcite were first precipitated, resulting in a residual concentration of magnesium. A point is ultimately reached where the concentration of magnesium results in the precipitation of dolomite or high magnesium calcite in the saline waters. Simultaneously, the removal of carbonate from the waters results in a residual concentration of the more soluble chloride ion. The result is a complex sequence of chemical evolution equivalent to a form of 'ageing' in the waters, a process which controls many important geochemical phenomena and one which can be recognised in the sedimentary and geochemical evolution of lakes in the semi-arid zone. Here the occurrence and abundance of various carbonates, sulphates, and chlorides may reflect important palaeosalinity changes. Depending on their type and magnitude, these changes may in turn have influenced both the faunal and human occupational record.

Other factors affecting salinity include the influence of groundwaters, shape and size of the water body, the rate of salt removal in the sediment compared to rate of supply, and many other variables. An example of the complex relationships between two adjacent water bodies is demonstrated in the volcanic crater lakes, Gnotuk and Bullenmerri, near Camperdown, Victoria. Here two adjacent deep lakes in the same geological and groundwater environment contain waters of vastly different quality. Gnotuk, with salinity near 60,000 ppm (Maddocks 1967), supports an impoverished fauna including ostracods and the salt tolerant gasteropod *Coxiella*, while Bullenmerri, with salinity near 7,500 ppm, is capable of maintaining freshwater trout. The difference in this case is due to a combination of factors including size, the volume of water relative to surface area, and different hydrologic histories (Sutcliffe pers. comm.).

Evidence of past salinity variations in lakes of southeastern Australia is provided by the occurrence of sedimentary carbonates consisting predominantly of calcite at one time and dolomite at another. Moreover, freshwater unionid and limnaeid faunas sometimes give way to the salt tolerant gasteropod *Coxiella* at different stratigraphic levels, representing salinity variations through time. Such variations in water quality must be of sufficient duration to permit the growth and reproduction of entire molluscan communities, indicating long-term rather than seasonal or short-term variations.

Some of the best evidence of past hydrologic and geochemical environments is provided by the transverse crescentic dunes or lunettes which occur on the eastern shores of most lake basins in the semi-arid zone of both southeastern and southwestern Australia (Fig. 5:1). These contain a relatively high content of silt and clay and in this important respect they differ from the well-sorted quartz sands typical of most strandline dunes. Although similar features are known from North Africa (Boulaine 1956), and modern clay dunes have been observed to be forming on the Gulf Coast of Texas (Coffey 1909; Price 1963; Fisk 1959), they do not occur elsewhere in such numbers, nor with such widespread and regular distribution. The significance and the age of these features, which were first appreciated and described in Australia by Hills (1939), have been comparatively little understood until recently, but data now available enable us to determine their ages and to reconstruct the circumstances of their formation.

Although lunettes range widely in their latitude and altitude through southern Australia, they are almost entirely relict features formed under the influence of earlier Quaternary environments. In reconstructing those environments, Hills initially postulated that lunette formation was related to a lake-full phase in which dusts in suspension were trapped by spray on the downwind side. Stephens and Crocker (1946) later argued in favour of sediment deflation from dry lake floors, from which silts and clays were blown as pelletal aggregates, requiring for their formation a period of aridity. Bettenay (1962) substantiated the deflation of lake floor aggregates and drew attention to the similarity between the process postulated by Stephens and Crocker and that observed in modern clay dunes on the Gulf Coast of Texas.

Multiple episodes of lunette formation have been recognised on many lakes. At least two aeolian stratigraphic units are often present, separated by periods of soil formation. Episodic aeolian accession has occurred at specific intervals under hydrologic or climatic conditions apparently not represented in the continent today, but which were reproduced more than once during Quaternary time.

The lunette-aridity correlation has been accepted as a basic assumption in many interpretations of Australian Quaternary successions. Gill (1955) used the presence of a lunette at Lake Colungulac in western Victoria, which overlies a horizon dated at about 13,000 BP, to substantiate the claim for post-glacial aridity. Several lines of evidence, however, specifically exclude aridity as the controlling factor in the formation of some lunettes, while its relevance must be qualified in others.

**1** Lunette lakes are mainly restricted to semi-arid regions. Some occur in the cool temperate and humid regions of New South Wales, Victoria, and Tasmania, but they are significantly absent from the arid centre of the continent.

**2** They are associated with lakes which often possess regular symmetrical outlines, often elliptical or near-circular in plan with smooth crescentic eastern margins and with an abrupt slope or cliff developed on the west. These features reflect extensive shoreline modification by wave action, demonstrating the importance of high water stands in the lake-lunette evolution.

**3** The presence of gravel and coarse sand in some lunettes, as at Wagin Lake, W.A. (Stephens and Crocker 1946) and at Echuca (Bowler and Harford 1966), cannot be explained by deflation from a dry lake floor. The presence of lunettes consisting predominantly of quartz sands in close association with those containing high percentages of clay requires special consideration. Bowler and Harford (1966) in describing the occurrence of a large lake in a fault-angle depression near Echuca, specifically ruled out the possibility of lunette formation due to aridity. The sediments in this dune (the Bama Sand Hills) consist of non-calcareous well-sorted micaceous quartz sands with modal diameter near 0·2 mm, in marked contrast to the clay dunes found on lakes further south (Lake Cooper system). Moreover, the sediments of the lake floor adjacent to the quartz sand dune pass out in a regular succession through coarse shoreline sands to fine sands and silts in the deeper part of the basin. Since silts predominate in the lake floor sediments, their deflation could not produce the range of coarse textures found in the dune. These could only be derived by deflation from a sandy beach, requiring continuous replenishment of sand by wave-generated currents under the influence of southwesterly winds. Sands, thrown on to the beach by wave action, dried and were blown into the dune in a manner similar to the formation of coastal foredunes. The presence of a beach implied lake-full conditions synchronous with dune development.

This hypothesis was substantiated by the later discovery of beach sands interdigitating with dune sands in a gravel pit on the Echuca-Barmah road. The sand lunette in this case was therefore formed synchronously with beach formation while the lake was full and required deep water to permit the generation of currents necessary to erode, transport, and maintain the sand supply. The examples of coarse sand and gravels in the Western Australian lunettes (Stephens and Crocker 1946) are probably explained in the same way. Coarse sands in the section shown by Bettenay (1962: Plate 1) provide another example of a lunette development during a lake-full phase; thus it would seem that at least a particular type of lunette, namely sand lunettes, are associated with high water levels rather than with dry lake situations. Currey (1964) had already recorded the deflation of shells into a dune on the present shores of Lake Corangamite contributing in a small way to modern lunette formation.

In the Echuca example clay and silts, components present in most lunettes, are virtually absent. The well-sorted quartz sand dunes formed under high water level conditions are indicative of low salinities. For the calcareous and gypseous silty-clay dunes, however, an alternative explanation must be sought.

In the example of modern clay dune formation in the hypersaline lagoons of the Gulf Coast of Texas (Price 1963), sand-sized clay pellets are blown from drying lake floors under conditions of high salinity with frequent wetting and drying (Fisk 1959). The pellets accumulate on the downwind margin under the influence of an almost uni-directional wind régime. The presence of salts assists in the hygroscopic absorption of moisture, and the pellets become plastic during periods of low temperature and high humidity and quickly stabilise. Thus they rarely form sand-slip faces and high angle bedding characteristic of mobile quartz dunes. The formation of clay dunes in this environment requires a particular combination of salinity, frequency of wetting and drying, and a favourable wind régime.

The sediments of the Australian silt-clay lunettes consist of pelletal aggregates similar to those described from Texas (Bettenay 1962). In thin section, sand-sized pellets of oriented clay sometimes enclosing fine sand, silt, and shell debris are associated with carbonate pellets and gypsum crystals. These were first observed by Sleeman (pers. comm.) and have since been recorded by the author in samples from many localities. Pellet sizes vary, but are usually close to those of saltation sands with modal diameters near 0·2 mm.

The evidence of wave action and high water levels has already been stressed in the lunette context. The clay lunette forming stage was only reached late in the evolution of a lake when falling water levels reached a critical limit due to a deficiency between seasonal additions of water and the evaporation loss in an increasingly saline environment. In the large seasonal fluctuations which followed, vegetation was prevented from colonising the lake floor and the 'fluffy' pelletal clays, formed as a result of desiccation and salt efflorescence, were blown by the dominant winds to the eastern lake shore (cf. Macumber 1969).

These conditions closely circumscribe the hydrologic context necessary for clay dune formation. They correspond to a transitional period from a high water table to a falling water table régime with increased evaporation and high salinities. The conditions necessary to account for lunette occurrence on a continental scale are satisfied only by a major climatic change corresponding to a transition from the terminal phase of a cold or low evaporation régime with high run-off and high water tables, to one of rising temperatures and increased water loss. While the water table is near the basin floors, extensive seasonal efflorescence of salts assists deflation, but when the water table and capillary fringe fall below that surface, as under the present climate, no further lunette formation is possible. If the formation of clay dunes requires that these conditions be satisfied, then the presence of a large number of silty-clay lunettes through extensive areas of southern Australia provides important evidence of hydrologic and climatic changes on a continental scale. Evidence in support of this reconstruction is found in the stratigraphy of lakes through the Murray Basin.

## Evidence from Lakes in the Murray Basin

The Murray-Darling drainage system in which many of the lunette lakes occur forms the largest exoreic drainage system in Australia. Most of the Murray Basin falls within

*Fig. 5:2 Regional climatic map of southeastern Australia showing mean annual distribution of rainfall and evaporation. (Data from Commonwealth Bureau of Meteorology and Water Conservation and Irrigation Commission, N.S.W.)*

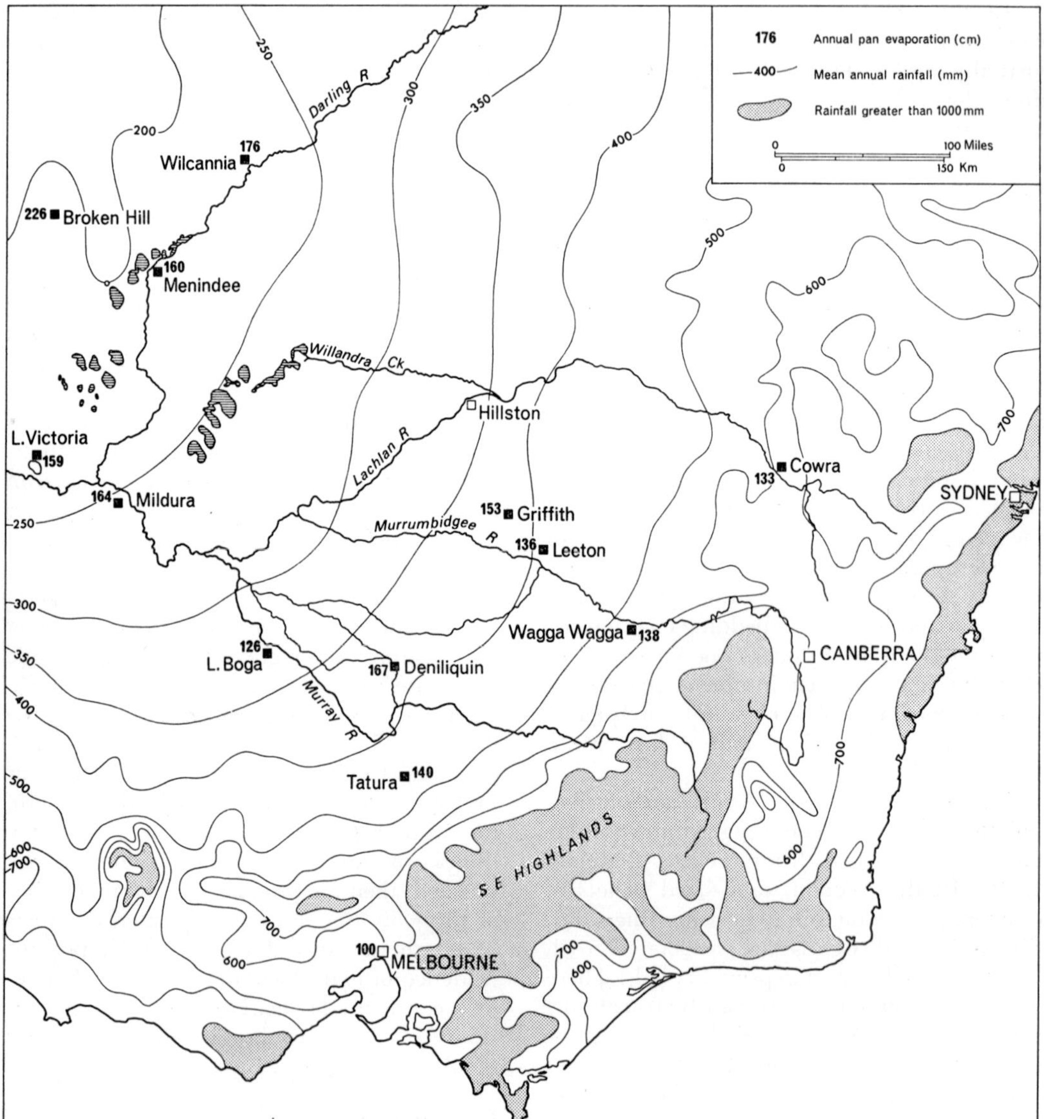

the semi-arid zone of Thornthwaite's climatic classification (Gentilli 1948). Values for mean annual precipitation range from near 200 mm near Broken Hill to 600 mm near Shepparton. Annual pan evaporation across the same region varies from more than 200 cm in the former case to near 100 cm in the latter (Fig. 5:2). This area possesses one of the steepest climatic gradients of inland Australia, and as such is likely to respond sensitively to climatic change. A small shift in temperature or precipitation would result in a significant variation in the climatic pattern across the plain. Moreover, changes in the southeastern highlands, such as the onset of periglaciation, would contribute significantly to variations in run-off and stream discharge with consequent changes in lake levels and the groundwater régime. The zone of climatic sensitivity coincides here with one in which a wide variety of aeolian, fluviatile, and lacustrine processes operated during Pleistocene time, making it one of the best available areas for documenting the Quaternary record in Australia.

The surficial deposits of the Murray Basin fall into two distinct geomorphic and sedimentary provinces. On the east, fluviatile sediments form the Riverine Plain, preserving a record of the complex Quaternary streams which drained the higher alpine regions to the southeast. It passes west into the Mallee, which is characterised by the longitudinal east-west sand ridges. This latter area forms the eastern extremity of the continental dunefield which extends across the length of inland Australia (Fig. 5:1). In the more arid interior, some dunes are still active (Wopfner and Twidale 1967). But in the Mallee, before clearing for agriculture, the dunes were fixed by vegetation and represent relict or fossil landforms from a previous Quaternary climate (Hills 1939; Lawrence 1966). The numerous lakes scattered through the area are concentrated near the Mallee-Riverine boundary (Fig. 5:3), where aeolian activity has interfered with drainage from the south and east, contributing to the formation and development of the lake basins. In places the longitudinal dunes have encroached onto lake floor sediments, providing an opportunity of dating the aeolian activity relative to the lacustrine sequence.

The larger lakes are associated with streams that were responsible for maintaining them. Few lakes carry water under the present climate, except where rain in the catchment provides seasonally high run-off producing an irregular or ephemeral flow in the contributing streams, as in Lakes Hindmarsh or Albacutya on the Wimmera River, or where irrigation has resulted in drastic changes in the water table and groundwater régime sufficient to flood the basins, as in the Kerang Lakes in northern Victoria. Otherwise they remain relatively dry and, like the dunes, bear witness to early Quaternary environments in which they contained large quantities of water for considerable periods.

Although few have been studied in detail, some on the Darling River, like Lake Menindee, have already yielded valuable faunal and archaeological evidence (Tindale 1955; Tedford 1967; Merrilees 1968). All conform to the general description given earlier, with the common characteristics of regular to elliptical or subcircular outlines with transverse dunes developed on their eastern margin. All are shallow, usually less than 15 m below the plain, and possess nearly flat basin floors terminated on the west by an abrupt steep slope, and on the east by the gentle gradient to the lunette crest. The lunette generally rises 7 to 15 m above the lake floor, but on occasions may exceed 30 m on the larger lakes.

Since the lunettes are deflationary in origin, their development implies a break in

*Fig. 5:3 Geomorphic map of northwestern Victoria and southwestern New South Wales showing the distribution of lunette lakes relative to linear dunes and alluvium of the Riverine Plain. (Data compiled from Hills (1939), Rowan and Downes (1963), Bowler and Harford (1966), maps of Geological Survey of New South Wales, and from aerial photographs and mosaics.)*

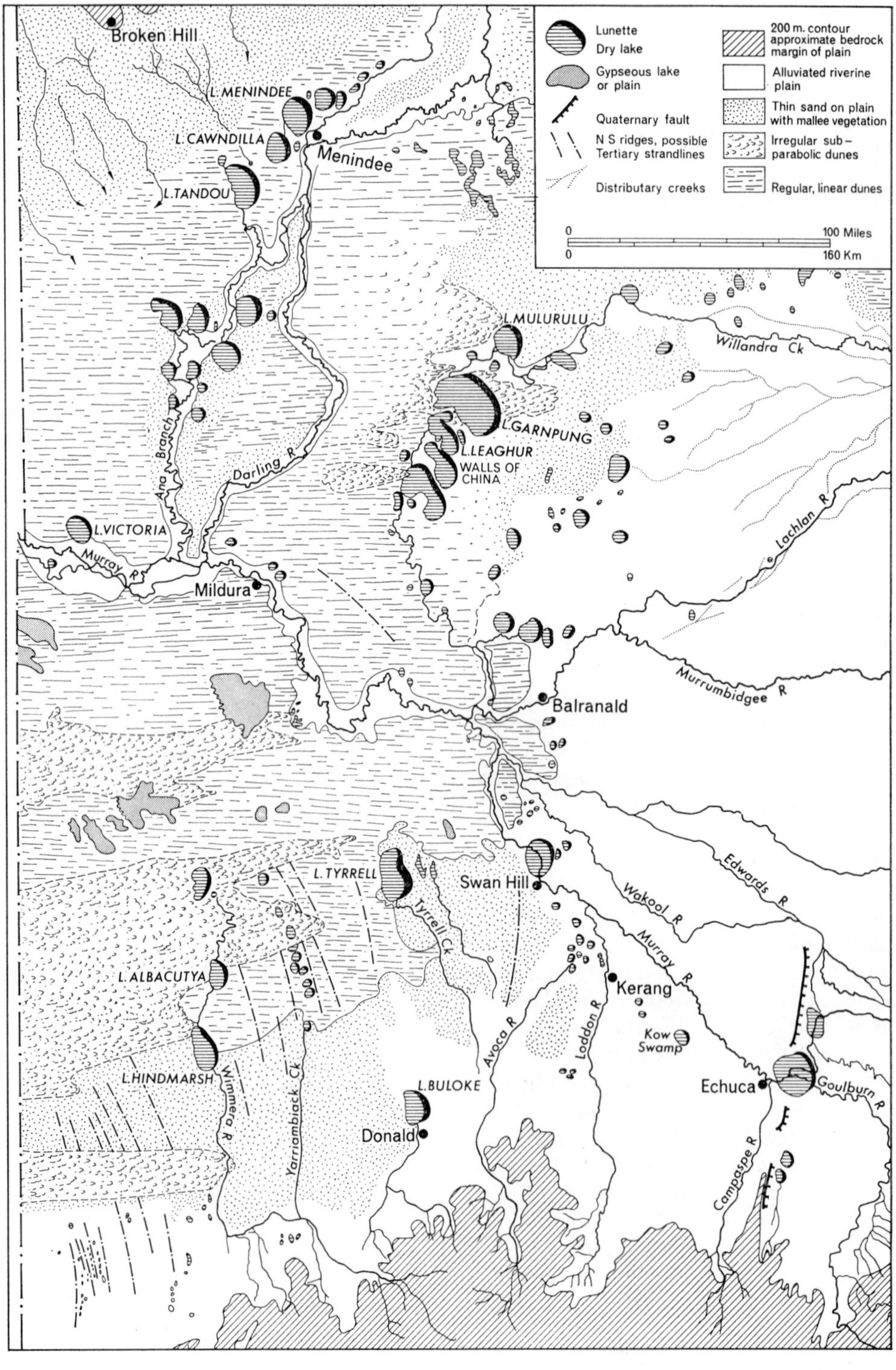

*Fig. 5:4 Detailed geomorphic map of the Willandra lakes. (Compiled from aerial photographs and ground traverses.)*

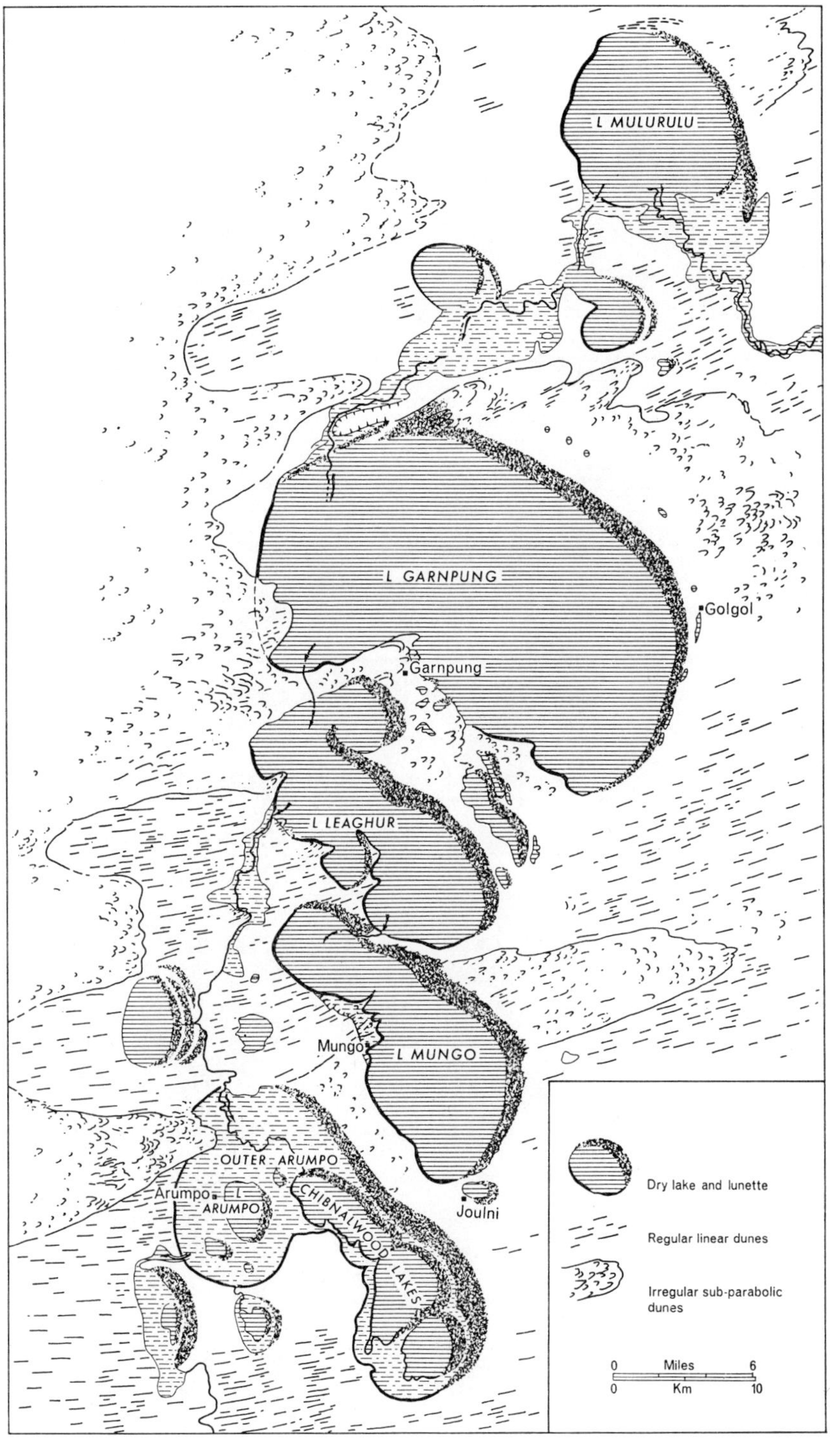

*Fig. 5: 5 Generalised stratigraphic sections through lunettes of the Willandra lake system*

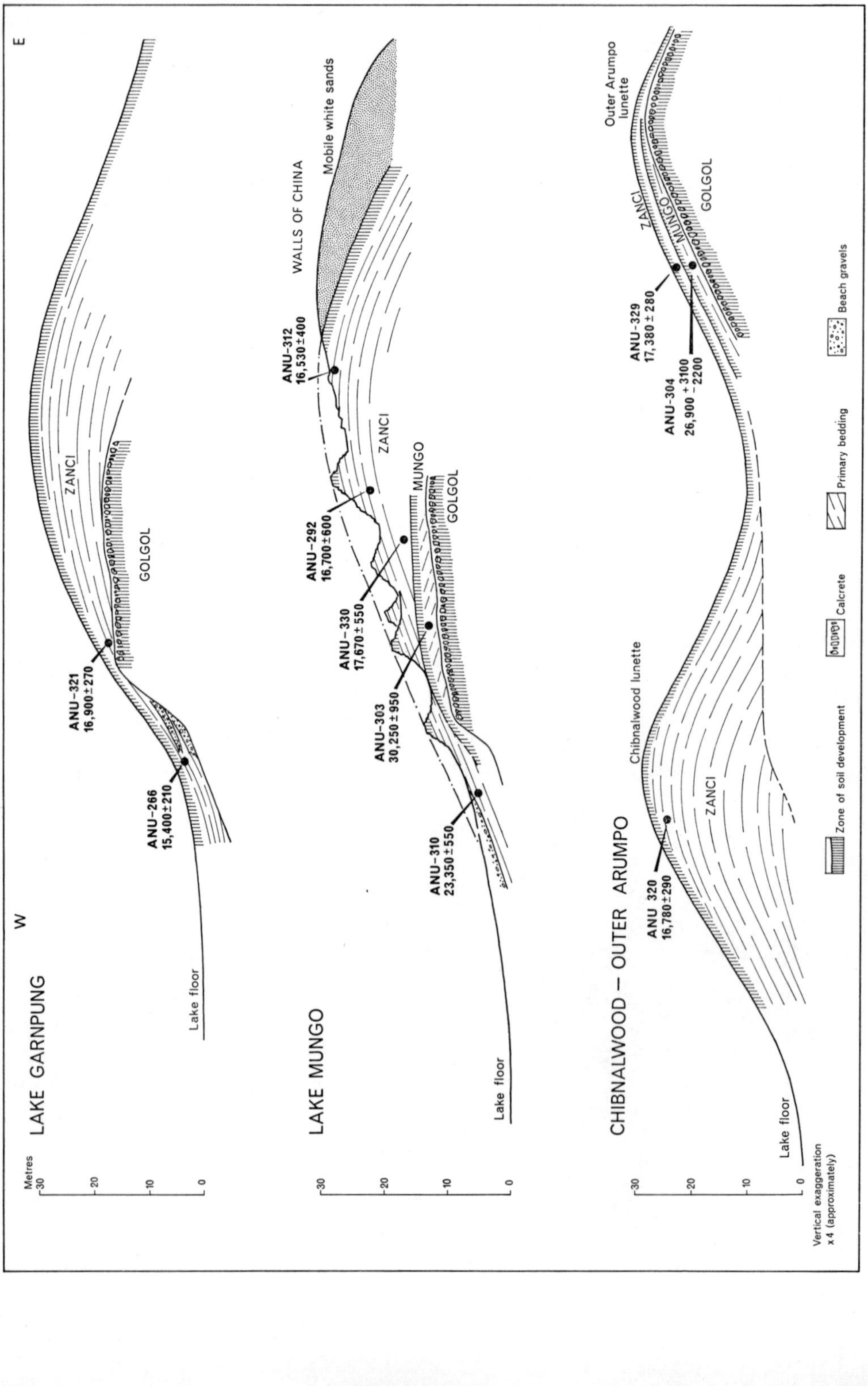

the continuity of the lake floor sedimentary record, but the sequence of faunas and the record of human occupation on the eastern lake shores has often been excellently preserved by the blanket of alkaline aeolian materials deposited there. In this respect, the lunettes provide critical areas for the preservation and examination of such evidence.

### WILLANDRA LAKES

A group of lakes located approximately 100 km northeast from Mildura forms the terminal system of the Willandra Creek, a distributary stream which leaves the main Lachlan channel near Hillston (Figs. 5:2, 5:3). The Willandra Creek now carries water only during peak or flood discharges, but during Pleistocene time it possessed higher discharges than at present and maintained the system at the lake-full stage for long periods. The system overflowed through the lower lake, Outer Arumpo (Fig. 5:4), and a channel carried the surplus drainage south through the linear dunefield to join the Murrumbidgee approximately 25 km west of Balranald. When filled, these lakes covered an area of 1,088 $km^2$ which, under the present evaporation-precipitation régime, would require more than the total Lachlan discharge to maintain them at constant level, leaving no surplus available for overflow. The environment at the time of lake-full activity was therefore one in which evaporation was lower, or discharge much higher, than at present. Alternatively, both factors may have combined to provide the large quantity of water needed to sustain the system.

All lakes in this system possess regular outlines, with high lunettes on eastern margins (Fig. 5:4). On the western shores linear dunes have been trimmed by Quaternary high water levels, but irregular parabolic dunes, representing the final phase of dune instability, have encroached onto the floors of Lakes Garnpung and Outer Arumpo. There is a northeasterly trend to the irregular dunefield margins.

### LAKE DUNE STRATIGRAPHY

The stratigraphic sequences within the lunettes are often exposed in gully sections and in the large areas uncovered by modern deflation. Representative stratigraphic cross-sections through the lunettes of three lakes, Garnpung, Mungo, and Outer Arumpo, are shown in Fig. 5:5. At each locality several sedimentary units are recognised, in which periods of dune formation alternated with periods of stability when soils were formed. These may be briefly described in terms of three soil-sedimentary units:

**1** The basal Golgol unit, consisting of a red calcareous soil with hard nodular to massive laminated calcrete in the B horizon, developed on quartz sands. In it all original sedimentary structures and textural characteristics have been extensively modified due to the development of the mature soil profile. In a lunette section at Lake Garnpung (Fig. 5:5) grey calcareous and argillaceous sands of the Zanci unit rest on an eroded remnant of the heavily calcreted Bca Golgol horizon. Beach sand and gravels representing the last lake-full stage disconformably overlie a buried cliff eroded into the Golgol unit.

**2** The Mungo unit, consisting of red to reddish yellow sandy A horizon passing to a calcareous B horizon. Carbonate occurs both as fine earth and soft concretions; sometimes it cements a zone of sands into a laterally extensive soft calcrete band which hardens on exposure to the atmosphere. Basal sediments of well-sorted quartz sands are overlain by calcareous and argillaceous silty sands. The unit is best exposed on the southern end of the Walls of China (Fig. 5:5).

**3** The Zanci unit, comprising the last major aeolian deposit in the development of the

*Fig. 5:6 Stratigraphic sections through the southern end of the Walls of China lunette, Lake Mungo.*
*A. Section showing facies variations within the Mungo and Zanci units and the relationships of the radiocarbon samples on which age estimates are based.*

*B. General soil—sedimentary relationships of the Mungo and Zanci units at the site of the ancient human burial in the Mungo unit, approximately 1·5 km west of Section A.*

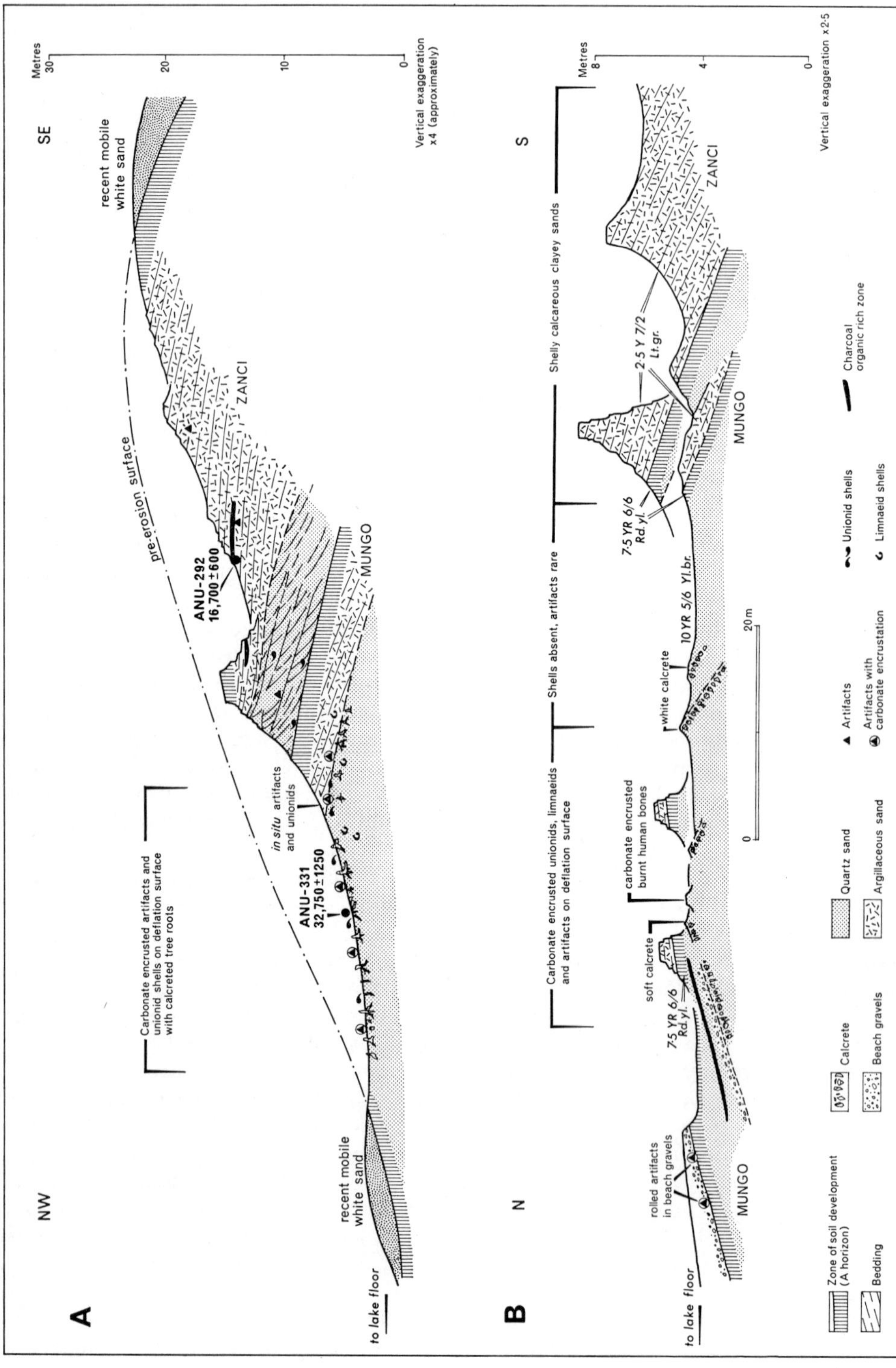

lunette stratigraphy. It consists of up to 32 m of calcareous and argillaceous silts and sands with well-preserved bedding on which a calcareous brown to red-brown earth soil has developed. It often directly overlies the Golgol or is separated from it by the thin Mungo unit, as on the Walls of China and in the gully section through the Outer Chibnalwood lunette (Figs. 5:5 and 5:6). The consistent values of all radiocarbon analyses from the Zanci dune indicate that its formation began about 17,500 BP and continued for approximately 1,500 years. Preservation of the fine bedding reflects the rapid and probably seasonal nature of deposition. After lunette formation ceased, about 15,500 BP, the entire system downstream from Lake Garnpung became inactive.

In each of the two younger units, pedogenesis has been insufficient to destroy original sedimentary features (Fig. 5:6A). Deposition was initiated in both cases by aeolian accumulation of quartz sands corresponding to high water level, beach-derived material, similar to that described from Echuca. In the southern part of the Walls of China, these phases of aeolian activity were associated with shoreline or beach gravels. The freshwater nature of these environments, moreover, is confirmed by the presence of unionid shells (*Velesunio ambiguus*) in the quartzose sediments. On Lake Mungo, the early quartz sands of both the Mungo and Zanci units pass upwards into grey calcareous argillaceous sands. The association of these two facies represents the transition from a high-level freshwater environment to one with fluctuating water levels and increasing salinities.

At Lake Chibnalwood, instead of vertical superposition of the stratigraphic units, the last phase of dune building occurred inside the outer ridge, providing another expression of sequential dune building. In the final stage, represented here by the high Chibnalwood lunette (Fig. 5:6), pelletal sandy clays with carbonate and gypsum were blown from the adjacent lake floor. The presence of *Coxiella* sp. confirms the transition from the freshwater facies (with unionids) to one of high salinity. The carbonates, moreover, show an important transition from calcite in the earlier freshwater deposits to dolomite in the final lunette, representing the late stage of water evolution. Thus the sedimentary, mineralogical, and faunal data present independent evidence of major salinity changes within the sediments of the same lake, changes which coincided with and which controlled the aeolian deposition of the clay dune.

### CHRONOLOGY AND HUMAN OCCUPATION

Evidence of ancient human occupation is found within high shoreline gravels on Outer Arumpo and on the western side of all major lakes. But some of the best exposures occur within modern deflation zones in the lunettes on Lakes Mungo and Garnpung. Here accumulations of unionid shells and stone artifacts are often associated with charcoal concentrations both in undisturbed sediment and as eroded or lag accumulations on deflation surfaces. The cultural assemblage consists almost exclusively of a core and scraper industry. Although the lunette stratigraphy represents a long period in time, no trace of the point or microlithic industry has yet been found within lunette sediments. Since the lakes had the effect of concentrating the Aboriginal populations on the water's edge, as shown by the concentrations of food remains and stone, the absence of a particular industry of known age can serve as corroborative evidence for the age of final drying of the system. When the point or microlithic industry made its impact in western New South

*Table 5:1 Summary of stratigraphic sequence, radiocarbon chronology, and evidence of human occupation from the dry lakes of the Willandra Creek, N.S.W.*

| **Lunette stratigraphic units** | **Soil-sedimentary environments** | **Radiocarbon chronology (years BP)** | **Hydrologic régime** | **Human occupation on Walls of China** |
|---|---|---|---|---|
| | Soil formation | No deposition | Lakes dry | No further concentration on shorelines—ovens on lake floors |
| B | Clay dune formation | 16,530 ± 400 ANU-312[a] | Decreasing discharge, rising temperatures, increased evaporation and drop in water table; high salinity, fluctuating lake levels | Core and scraper industry, unionids and food bones rare except at base of Zanci B |
| ZANCI ........................ | | 17,670 ± 550 ANU-330[a] | | .................... |
| A | Quartz sands deflated from high level beach | 23,350 ± 550 ANU-310[b] | High discharge, low salinity, high water levels, unionid faunas | Core and scraper industry, unionid shell middens common with charcoal and bones |
| | Soil formation | | Outer Arumpo and Mungo dry; perhaps some water in Garnpung | |
| B | Clay dune formation | | Decreasing discharge, fluctuating water levels, increasing salinity | Core and scraper industry associated with fires, burnt bones and unionid shells localised near beach deposits |
| MUNGO ...................... | | 30,250 ± 950 ANU-303[a] | | |
| A | Quartz sands deflated from high level beach | 32,750 ± 1250 ANU-331[c] | High lake levels, low salinity | |
| | Strong soil formation | >38,500 $^{+2950}_{-2150}$ ANU-306[c] | | |
| | Sedimentary features destroyed by pedogenesis | | | No evidence observed |
| GOLGOL ...................... | | | | |
| | Quartz sands | | | |

[a] Charcoal samples sealed within dune sediment.
[b] Calcareous concretions formed from lake waters.
[c] Unionid shells.

*Note:* It is not possible within the limits of this essay to give a complete account of samples dated, their stratigraphic context and limitations of reliability. Additional radiocarbon dates are available consistent with the above interpretation, full details of which will be presented elsewhere.

Wales we may conclude that the lakes were no longer present in the form in which they are shown in Fig. 5:4. Mulvaney (1969) places the age of these cultural innovations at about 5,000 to 7,000 BP.

Radiocarbon dates have been obtained from representative horizons through the present sequence to determine precisely the ages of water level fluctuations and their associated ecological changes. A summary of the dates and their relationship to the stratigraphic units is shown in Table 5:1.

The evidence presently available suggests that the age of Golgol deposition is beyond the range of radiocarbon. It contains only rare traces of bone, but these are associated with neither artifacts nor charcoal. The deep soil profile, strong colours, textural differentiation, and massive calcrete formation suggest a long period of pedogenesis before deposition of the younger sedimentary units.

The oldest datable sediments (ANU-306) exposed in the Outer Arumpo terrace represent a high level, freshwater environment equivalent to the freshwater facies of the Mungo unit. An association of carbonate-encrusted unionid shells with artifacts on the deflation surface truncating the Mungo quartz sands on the southern end of the Walls of China (Fig. 5:6A), suggests that the area was already occupied at least during the final stage of this early freshwater environment. The C-14 age determined on encrusted unionid shells (ANU-331), the presence of which in the dune is difficult to explain by any agency other than by human transport, would place this amongst the earliest dated evidence of human occupation in Australia. Moreover, a core and scraper industry occurs on deflation surfaces with some artifacts *in situ* within the Mungo unit approximately 1·5 km further south. These are associated with burnt human bones cemented by secondary calcrete within the soil of the Mungo unit (Fig. 5:6B). Details of these remains, which from their stratigraphic context are considerably more than 20,000 years old, are being analysed for publication in collaboration with Rhys Jones and Harry Allen, Department of Prehistory, Australian National University, and A. G. Thorne, Department of Anatomy, University of Sydney (Bowler *et al.* 1970).

Dune sediments associated with the onset of the later Zanci high water phase contain artifacts, fire remains, and unionid midden shells. Occupation continued on the lake shores until final drying of the system and rapid deposition of the Zanci calcareous and gypseous pelletal aggregates.

The age of final drying can be determined from the age of clay dune formation because, as stated earlier, the rapid fluctuations necessary to permit formation and deflation of pellets occur only during the falling water level stage accompanying the lead into drier conditions with lower water tables. The radiocarbon evidence shows that the Zanci aeolian silty clays were deposited in a relatively short time between 15,500 and 17,500 BP, corresponding to the transition from a régime of low evaporation and high stream discharge to one in which evaporation or water loss exceeded that supplied to the system. In a trench section at Lake Garnpung (Fig 5:5), unionid shells from bedded beach deposits on the inner edge of the dune were dated at 15,400 $\pm$ 210 BP (ANU-266, Polach *et al.* 1970). These, the youngest sediments on the lake floor in which weak soil formation had occurred, represent the last phase of lacustrine activity in that lake. If this interpretation is correct, all lakes downstream from Garnpung would have been dry for the last 15,000 years. Before 15,000 BP the lakes supported a considerable human and animal population, but in the post-15,000 period there is no evidence of strandline occupation. Aboriginal ovens on the lake floor overlie the 15,000-year-old sediment, indicating later occupation after the main drying of the system. The ovens are not concentrated along strandlines and the associated human economy was probably no longer based on the lacustrine environment.

The important transition from a high water table to a low water table and water deficient régime, with its attendant effects on landscape development, salinity, vegetation, and occupational patterns, began about 17,000 BP, heralding the eventual drying of the lakes. The Arumpo-Chibnalwood lakes, being the last in the chain, were the first to feel the effects of salinity change. As discharge declined relative to evaporation, the level of Outer Arumpo fell below the outlet, preventing flushing by fresh water and permitting salts to accumulate in the system. The large freshwater molluscs were replaced by small salt-tolerant species of no economic value. Salinity increased until a point was reached when the lacustrine environment ceased to have any real economic importance to Aboriginal occupants of the area. The remains of marsupials, artifacts, fires, and other evidence of human occupation are abundant around the shorelines associated with the freshwater sedimentary facies, but are absent from the saline facies represented by the younger Chibnalwood lunette. Within a few hundred years a similar transition progressively affected all lakes upstream from Arumpo, terminating in the drying of Lake Garnpung approximately 2,000 years later.

The evidence presented here for one lake system is believed to be representative of others through the Murray Basin. A preliminary survey of the Darling Lakes in 1968 revealed, in Lake Cawndilla and Lake Tandou, a stratigraphic sequence which bears a strong resemblance to that of the Willandra lakes. Farther south at Lake Albacutya on the Wimmera River, identical salinity or facies changes are associated with lunette sediments which in many ways resemble those described above. From the degree of soil developed in them, they appear to be closely related in time with those of the Willandra system. Additional data will establish these relationships more precisely.

The hydrologic changes demonstrated here, which coincided with the end of the last major cold phase of the Pleistocene, produced widespread and lasting changes in the physical environment throughout an extensive region of southeastern Australia. The implications of these changes are far-reaching and help explain features of present soils, salinities, vegetation distribution and, in fact, bear on almost every aspect of the natural system. But for the archaeologist, an understanding of this sequence and the processes controlling it provides a basis on which to assess the impact of climatic and associated changes on early Aboriginal cultures and to evaluate man's capability of adapting to them.

## References

Bentley, P. J. 1955. Some aspects of the water metabolism of an Australian marsupial *Setonyx brachyurus*. *J.Physiol.,Lond.* **127**:1-10.

Bettenay, E. 1962. The salt lake systems and their associated aeolian features in the semi-arid regions of Western Australia. *J.Soil Sci.* **13**:10-17.

Boulaine, J. 1956. Les lunettes des basses plaines oranaises. Formations éoliennes argileuses liées à l'extension de sols salins La Sebkha de Ben Ziane, la dépression de Chantrit. In *Actes du IV Congrès International du Quaternaire*. Rome-Pisa, 1953 (ed. G. A. Blanc), Vol. 1, pp. 143-50. Rome: International Association for Quaternary Research.

Bowler, J. M. and Harford, L. B. 1966. Quaternary tectonics and the evolution of the Riverine Plain near Echuca, Victoria. *J.geol.Soc.Aust.* **13**:339-54.

——, Jones, R., Allen, H., and Thorne, A. G. 1970. Pleistocene human remains from Australia: a living site and human cremation from Lake Mungo, western New South Wales. *World Archaeology* **2**: 39-60.

Broecker, W. S. and Kaufman, A. 1965. Radiocarbon chronology of Lake Lahontan and Lake Bonneville II, Great Basin.

*Bull.geol.Soc.Am.* **76**: 537-66.
Coffey, G. N. 1909. Clay dunes. *J.Geol.* **17**: 754-5.
Currey, D. T. 1964. The former extent of Lake Corangamite. *Proc.R.Soc.Vict.* **77**: 377-86.
Dixey, F. 1966. Water supply, use and management. In *Arid lands* (ed. E. S. Hills), pp. 77-102. London: Methuen, Paris: UNESCO.
Fisk, H. N. 1959. Padre Island and the Laguna Madre flats, coastal South Texas. In *Second coastal geography conference* (ed. R. J. Russell), pp. 103-52. Washington D.C.: Office of Naval Research, Geog. Br.
Galloway, R. W. 1965. Late Quaternary climates in Australia. *J.Geol.* **73**: 603-18.
Gentilli, J. 1948. Two climatic systems applied to Australia. *Aust.J.Sci.* **11**: 13-16.
Gilbert, G. K. 1890. *Lake Bonneville.* Monogr. U.S. geol. Surv. No. 1.
Gill, E. D. 1955. The Australian 'Arid Period'. *Aust.J.Sci.* **17**: 204-6.
Grove, A. T. and Warren, A. 1968. Quaternary landforms and climate on the south side of the Sahara. *Geogr.J.* **134**: 194-208.
Hills, E. S. 1939. The physiography of north-western Victoria. *Proc.R.Soc.Vict.* **51**: 297-323.
——1940. The lunette, a new land form of aeolian origin. *Aust.Geogr.* **3** (7): 15-21.
Jennings, J. N. 1968. A revised map of the desert dunes of Australia. *Aust.Geogr.* **10**: 408-9.
Jutson, J. T. 1934. *The physiography (geomorphology) of Western Australia.* Bull. geol. Surv.West.Aust. No. 95.
Lawrence, C. R. 1966. Cainozoic stratigraphy and structure of the Mallee Region, Victoria. *Proc.R.Soc.Vict.* **79**: 517-53.
Macumber, P. G. 1969. Interrelationship between physiography, hydrology, sedimentation, and salinization of the Loddon River Plains, Australia. *J.Hydrol.* **7**: 39-57.
Maddocks, G. E. 1967. The geochemistry of surface waters of the Western District of Victoria. *Aust.J.mar.freshwat.Res.* **18**: 35-52.
Merrilees, D. 1968. Man the destroyer: Late Quaternary changes in the Australian marsupial fauna. *J.Proc.R.Soc.West.Aust.* **51**: 1-24.
Morrison, R. B. and Frye, J. C. 1965. *Correlation of the Middle and Late Quaternary successions of the Lake Lahontan, Lake Bonneville, Rocky Mountains (Wasatch Range), southern Great Plains, and eastern midwest areas.* Rep.Nev.Bur.Mines No. 9.
Mulvaney, D. J. 1969. *The prehistory of Australia.* London: Thames and Hudson.
Polach, H. A., Lovering, J. F., and Bowler, J. M. 1970. ANU radiocarbon date list IV. *Radiocarbon* **12**: 1-18.
Price, W. A. 1963. Physicochemical and environmental factors in clay dune genesis. *J.sedim.Petrol.* **33**: 766-78.
Richardson, J. L. 1966. Changes in level of Lake Naivasha, Kenya, during postglacial times. *Nature, Lond.* **209**: 290-1.
Rowan, J. N. and Downes, R. G. 1963. *A study of the land in north-western Victoria.* Soil Conservation Authority, Vic.
Stephens, C. G. and Crocker, R. L. 1946. Composition and genesis of lunettes. *Trans.R.Soc.S.Aust.* **70**: 302-12.
Tedford, R. H. 1967. The fossil Macropodidae from Lake Menindee, New South Wales. *Univ.Calif.Publs geol.Sci.* 64.
Tindale, N. B. 1955. Archaeological site at Lake Menindee, New South Wales. *Rec. S.Aust.Mus.* **11**: 269-98.
Washbourn, Celia K. 1967. Lake levels and Quaternary climates in the Eastern Rift Valley of Kenya. *Nature, Lond.* **216**: 672-3.
Wopfner, H. and Twidale, C. R. 1967. Geomorphological history of the Lake Eyre Basin. In *Landform studies from Australia and New Guinea* (eds. J. N. Jennings and J. A. Mabbutt), pp. 118-43. Canberra: Australian National University Press.

# 6 The Australian Arid Zone as a Prehistoric Environment

J. A. Mabbutt

A widely-used world map of arid zones employs the pragmatic definition of aridity, that the climate is too dry for successful growth of crops in average years (Meigs 1953: 210). This is hardly appropriate to a consideration of the habitat of hunters and gatherers, but the map limits of aridity are in fact drawn on parameters expressive of an excess of potential evaporation over precipitation (the Thornthwaite Index of —40). On this criterion about 70 per cent of Australia is arid (Fig. 6:1). This arid zone, reflecting airmass stability within the subtropical high-pressure belt, continentality, a predominantly easterly airflow, and the rainshadow effect of the eastern uplands, occupies the centre and west of the continent between latitudes 20° and 33°S. Within it there are important contrasts in the amount, seasonal incidence, intensity, and reliability of rainfall.

The area of least rainfall is in the east-central, lowest-lying part, which averages below 12 cm annually. In general, rainfall increases outwards from this dry core near Lake Eyre, but most markedly northwards and eastwards, so that the arid zone is bounded approximately by the 50 cm annual isohyet in the north and by the 25 cm isohyet in the south. No part of arid Australia experiences the extreme desert conditions of the central Sahara or of the cold-water coastal deserts such as the Namib of South West Africa.

The zone includes a southern belt of predominantly winter rainfall, but northwards of latitude 30°S the proportion of summer rainfall increases progressively. At Oodnadatta (26°S) 60 per cent of a mean annual rainfall of 12 cm falls during the summer half-year; at Alice Springs (23°30′S) the summer increment accounts for 75 per cent of a mean total of 26·5 cm; and at Tennant Creek (20°S) summer rainfall constitutes 85 per cent of a total of 35 cm. With this northward

*Table 6:1 Expected number of rainfalls per year at selected stations in the arid zone*

| Station | Mean annual rainfall (cm) | Exceeding 1·25 cm Summer | Winter | Exceeding 5 cm Summer | Winter |
|---|---|---|---|---|---|
| Charlotte Waters (27°S) | 12·9 | 4·3 | 1·9 | 0·8 | — |
| Alice Springs (23°30′S) | 25·2 | 9·8 | 4·1 | 3·4 | 0·4 |
| Tennant Creek (20°S) | 35·2 | 13·2 | 2·5 | 4·0 | — |

*Source:* Adapted from Slatyer 1962:109.

increase in summer rainfall is associated an increase in the average intensity of rainfall as reflected in amount of rain per rain day, from less than 5 mm per rain day in the south to more than 13 mm in the north and northeast of the arid zone, a matter of considerable importance to run-off (Jennings 1967:258).

There is also a northward increase in the reliability of rainfall, following the general rule that variability decreases as the amount received rises. Nevertheless, throughout the arid zone it is more realistic to consider rainfall as a phenomenon to which a probability of occurrence can be attached, rather than in terms of average totals. Slatyer (1962:109) has expressed this for central Australia (Table 6:1), showing the improvement in rainfall prospects northwards across the arid zone, associated with an increasing likelihood of invasion by humid tropical air-masses.

Notwithstanding these regional variations in rainfall régime, the arid zone is dominated throughout by deficiency of rainfall, for over much of it evaporation from a free water surface attains 3 m annually. Lack of water affects not only plant cover but a range of phenomena linked with the hydrologic cycle, including weathering and soil formation, the nature and balance of erosion and deposition, and the resulting landforms. Geomorphologists have recognised a desert morphogenetic system; one in which, as a result of inadequate water supply, the vegetal cover and soils are too scanty to ensure effective protection of rock surfaces against atmospheric agents (Tricart and Cailleux 1960:3).

## Desert Water Supplies

The overriding limitation of a desert habitat for prehistoric hunters and gatherers is water supply for man and for the animals and plants that provide his food. Accordingly, this review first considers some physical factors controlling the availability of water in deserts, and second the types of Australian desert environment defined by those factors.

### PHYSICAL FACTORS IN WATER YIELD

When rain falls on the desert, losses tend to be high, initially due to absorption by hot dry soils and subsequently through intense evaporation. Under such conditions the intensity of rainfall, ground slope, and surface permeability are quite critical to water yield.

Records of run-off from sloping non-sandy surfaces in the Sahara (Dubief 1953:423) indicate that an initial loss of 5-10 mm is normal following a dry spell, that yields thereafter are closely related to intensity and duration of rainfall, and that more than half the balance of rainfall will run off with an intensity of about 0·5 mm per minute. Pluviograph records are not available from the Australian arid zone, but in practice the frequency of run-off on sloping terrain will approximate that of rainfalls exceeding 1·25

*Fig. 6:1 The Australian arid zone: organisation of drainage*

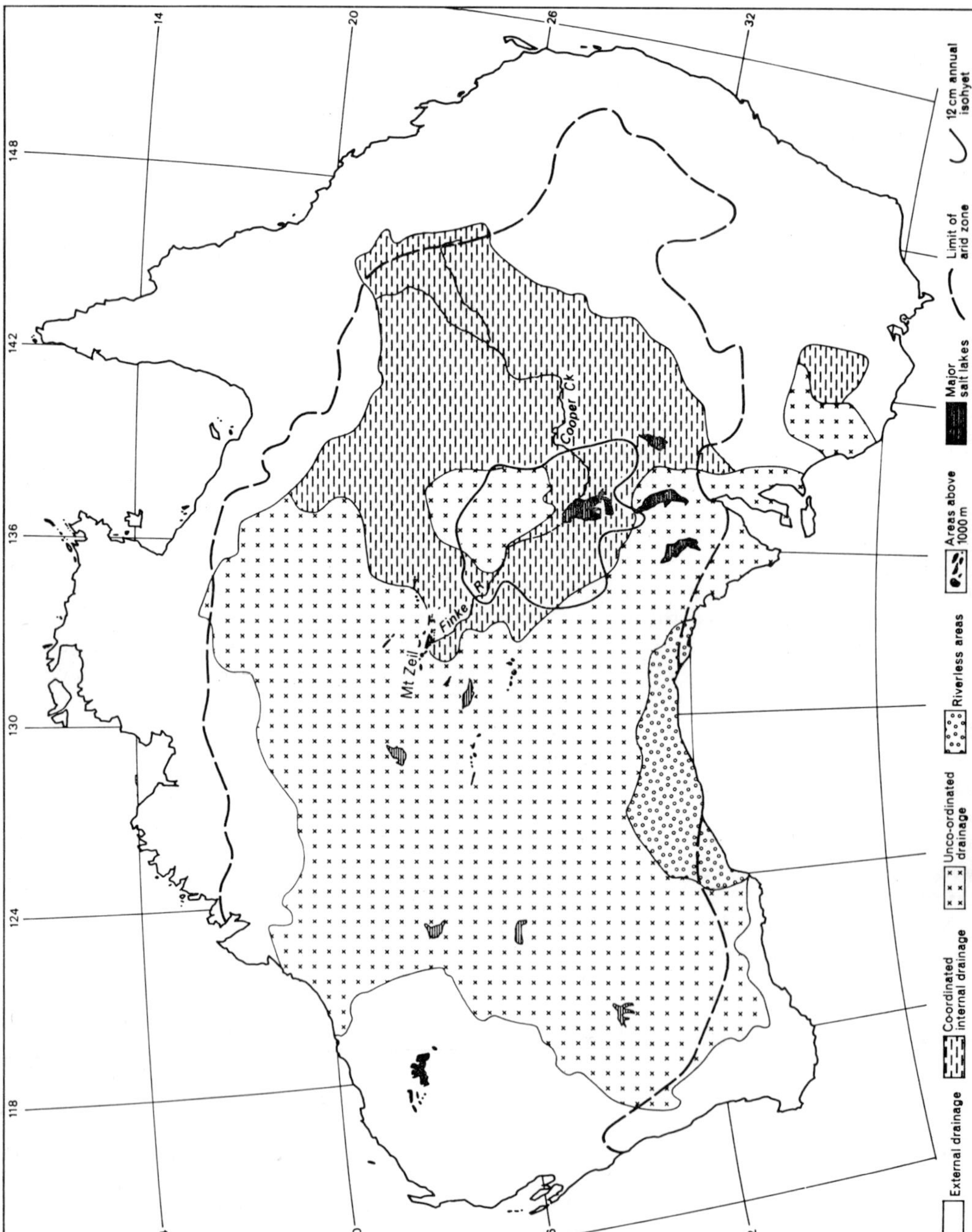

cm. As shown in Table 6:1 and in published maps of rainfall intensity (Jennings 1967), there is increasing probability of run-off northwards across the Australian arid zone as the summer rainfall increment increases, and the northeastern sector is especially favoured, particularly if the likelihood of very heavy falls is taken into account.

Such figures help destroy the myth that deserts are areas of infrequent but heavy rainfall. In arid Australia, as in other deserts, rainfall intensity decreases with the amount of rainfall received, with the result that average annual run-off is generally less than 1 cm (Dury 1968:1), representing a yield to streamflow of less than 5 per cent of rainfall. Despite such low overall yields imposed by the broad climatic setting, local factors of slope and rockiness determine local surpluses.

Much of the Australian arid zone consists of uplands set island-like in broad plains between 150 and 600 m above sea level. Characteristically, each such upland area of steep rocky slopes gives rise to a drainage aureole, but the channels so generated commonly fail to link into large systems before dying out on the adjacent plains where gentle slopes and permeable superficial deposits yield little discharge (Plate I). Downstream decrease in stream discharge resulting from influent seepage and evaporation is generally not compensated by tributary recharge on the lowlands, and the magnitude and frequency of flows normally decrease with distance from the upland sources. This tendency will of course be counteracted to the extent that integration of drainage systems continues across the lowlands, giving an increase in effective catchment area down-channel; but this is by no means the common case.

Three levels of organisation of surface drainage can be recognised in arid Australia (Fig. 6:1):

co-ordinated drainage
unco-ordinated drainage
no surface drainage.

These represent increasing stages of disintegration of Tertiary drainage systems initiated under effectively moister conditions, and it is clear from the map that the level of disintegration locally is dependent on the structural and geomorphic setting rather than on the degree of aridity.

The main areas of co-ordinated drainage are the Great Artesian Basin and its marginal intercratonic basins, in which mainly fine-grained sedimentary rocks have been gently warped, giving rise to a favourable disposition of bounding uplands and long centripetal slopes fashioned on uniformly soft and relatively impervious rocks. Here we find large catchments, such as that of Cooper Creek (250,000 km²), on a scale unique to the Australian arid zone. Only in the central part of the Basin have the river systems been disjointed through earth movements and subsequently obliterated by sand dunes.

Unco-ordinated drainage is typical of the western Australian plateau, on the uniform geologic base and stable tectonic setting of the Shield. Remnants of a lateritic Tertiary land surface form extensive low sandy divides which provide little run-off. Accordingly, broad Tertiary valleys of low gradient have degenerated into aligned 'river lakes' which serve as foci for local drainage systems of only a few hundred square kilometres. These lakes show their fluvial origin only when partly linked during infrequent extreme flooding.

The main areas not generating surface drainage are lowlands characterised either by highly pervious surface deposits, notably dunefields and sandplain, or by limestone at the surface as on the Nullarbor Plain. Sand deserts are shown in Fig. 6:1 as areas of unco-ordinated drainage, since ephemeral channels enter them from adjoining areas.

In some of the world's arid areas deficiencies of water supply are remedied locally through the presence of upland areas which yield an orographic increment to rainfall and which may also store moisture in the form of snow and thus maintain perennial or seasonal streams. Unfortunately, the Australian arid zone lacks such uplands. There are only limited areas above 1,000 m (Fig. 6:1), the highest point (Mt Zeil) barely exceeds 1,500 m, and although the Macdonnell Ranges of central Australia do cast a rainshadow to the south, their effect in increasing local rainfall is insignificant in this context.

A valuable addition to desert water supplies also occurs where streams enter an arid zone from the more humid borderlands, as in the case of the Nile or Colorado Rivers. In Australia, however, the boundary between external and interior drainage follows the limit of the arid zone so closely (Fig. 6:1) that there can be no important increment of this kind. The well-integrated Cooper Creek and Diamantina systems, draining into the heart of the arid zone from its better-watered northeastern rim, come closest to this category. Extensive catchments combined with what is, for the arid zone, moderately high and intense rainfall, yield large rivers which are intermediate between ephemeral and seasonal in régime. Substantial flows occur in most summers, whilst at irregular intervals of about five years there is extensive flooding of longer duration, such as the flow of Cooper Creek at Birdsville in 1954/5, with discharge estimated at 12 km$^3$.

NATURAL WATER STORAGES

In the absence of water-retentive soil and vegetation, run-off responds immediately to intense rainfall but is characteristically short-lived, and the stream systems of arid Australia are ephemeral in régime as well as geographically disconnected. Further, desert rainfall is often localised, even within catchments of moderate extent, and this is another factor in the limitation of streamflow. Under such conditions of low yield, high loss, and ephemeral run-off, a satisfactory natural water supply must depend on the local concentration of run-off allied with suitable surface or underground storage. The first of these conditions is satisfied by the steep rocky catchments and channels of upland and piedmont areas; it is the second factor which determines the localisation of supply.

*Rock holes.* Rock holes of various forms intercept overland flow directly or accept discharge from hillslope rills. Pitting through differential weathering of rock surfaces is particularly characteristic of the desert, with the combination of case-hardening of rock surfaces through capillary return of salts and patchy weathering resulting from the localisation of available moisture. It is commonest on granite hills, for example the circular pans and pits known as *gnammas* in Western Australia (Twidale and Corbin 1963:1). The smooth steeply rounded surfaces of granite domes are most efficient watershedders into such weathering pits.

*Channel storage.* Rock pools of various types characterise upland channels and range from small potholes and plunge pools to scour pools of large dimensions at bends, channel constrictions, and rock bars. Like weathering pits, such storages lose little water through seepage, and evaporative loss may also be reduced in shaded gorges; for example, the transverse canyons of the Macdonnell Ranges maintain such favourable microclimates as to have preserved plants and animals representative of a generally more humid past.

In the lowland sectors, most large channels are cut in sandy alluvium and are of bed-load type (Schumm 1963:8) with relatively large width-depth ratios. The coarse bed materials are scoured out during floods and sizeable holes are formed, particularly at bends in braiding or meandering channels. These may be filled only at irregular intervals and are subject to losses through seepage and evaporation, but near-permanent waterholes of this type occur in the largest channels; for instance there are fifteen such waterholes along the meandering course of the Finke River south of the James Range.

Very large waterholes occur in straight reaches between anastomosing sectors of the deep suspended-load rivers of the Channel Country in the Great Artesian Basin (Plate II). These are subject to recharge in most years, and with reduced seepage through the fine-textured alluvium they are commonly perennial.

*Storage in pans and flood-outs.* A striking feature of the desert map is the abundance of lakes—mainly ephemeral pans of various sorts, reflective of the persisting derangement of surface drainage under aridity rather than of abundant surface water. They include pans formed behind alluvial or aeolian barriers as a result of irregular deposition, and pans scoured by wind erosion. Both types occur mainly in the lowlands: in floodplains, where drainage channels enter dune country, and on stony plains. They are commonly associated with terminal zones and flood-outs of partially disintegrated drainage systems, for instance the lunette chains of the lower Darling region. They are filled only at times of general abundance of surface water, are commonly shallow, and are subject to strong evaporation; hence they are characteristically short-lived and tend to become brackish in the later stages of drying out. Most of the smaller ones are claypans; salt crusts are more characteristic of larger and well-enclosed terminal basins with restricted groundwater drainage.

Of larger dimensions are the basin lakes which serve as evaporative terminals of hydrologic systems involving groundwater. These tend to occupy major lowland axes and may be partly of tectonic origin. They are generally too saline to be of value as surface water supplies.

*Shallow groundwater supplies.* The longer water remains at the surface in an arid area, the more it is subject to evaporative loss and deterioration in quality; for this reason, shallow groundwater supplies are of the greatest importance.

In an arid area, much recharge to shallow groundwater is from ephemeral streams, where channels cross intake beds; except in constricted upland valleys, the water table tends to lie closest to the surface here and to slope away from the channel. Shallow underflow may persist through a sandy channel fill for some time after a flood, long after the smaller waterholes have dried out, and man and animals have learned to dig and tap such supplies. In general, such groundwater decreases in supply with time following the last flood, whilst quality falls off down-channel.

Springs and soaks occur at geologic and topographic breaks, as at the piedmont junction of rock face and alluvial plain, steps in stream courses, or where a river crosses a rock bar. Springs and shallow groundwater supplies occur on a large scale in fluvial and lacustrine limestones in central and western Australia, and particularly in karst areas, as in the underground lakes of the Nullarbor Plain.

## Desert Habitats

In terms of natural water supplies, upland and piedmont zones and riverine tracts across the plains clearly constitute the most favourable environments, firm bases from which man might range more widely after general rains. One can best explain the disposition of such habitats in the framework of the main geomorphic types of Australian desert (Mabbutt 1965:105).

*Mountain and piedmont desert.* Mountain desert, whether the sandstone ranges of the Macdonnells or the granitic uplands of the Musgrave Ranges, together with the adjacent piedmont plains, constitute the most favourable combination of desert habitats, with advantages for water and food supply, both animal and vegetable. Most run-off and water storage occurs in these areas. The hills bear a varied cover of useful trees and shrubs, and where the piedmont plains include grasslands on fine-textured soils, as in the Mitchell grass downs north of the Macdonnells, they provide valuable pastures for the larger macropods, both the plains kangaroo and the hill-inhabiting euro.

*Riverine desert.* Rivers and floodplains are valuable elements of mountain and piedmont desert, but under this heading are included the major riverine plains and drainage systems of the arid zone. They are characterised by open grasslands on fine-textured alluvial soils and by gallery woodlands along the anastomosing channel tracts, which find their fullest development in the Cooper Creek system of southwest Queensland. The combination of perennial grasslands, as a source of game and seeds, with the enduring waterholes of the deep channels constitutes a particularly favourable desert habitat (Plate I).

*Stony desert.* Stony deserts are best represented in the outer part of the Great Artesian Basin. The main relief is formed by silcrete cappings on soft-weathered Mesozoic and Tertiary rocks. In the west these give rise to extensive stony tablelands, open and inhospitable; in the east the typical upland form is the duricrusted mesa, with sandier soils and generally bearing mulga (*Acacia aneura*) woodland. The plains below the escarpments comprise two main elements, each supporting valuable open grassland. These are the poorly watered 'Rolling Downs', with stony clay soils and many of the shallow depressions known as 'gilgai', and river plains as described above, but on a smaller scale.

*Desert clay plains.* These are mainly formed on the fine-grained rocks and derived fine-textured alluvium, and the typical vegetation is perennial grassland, as on the Barkly Tableland. They may have provided good hunting country, but away from the widely spaced river frontages they offer an exposed habitat with poor water supplies.

*Sand desert.* The sand deserts mainly occupy lowlands in which sandy alluvia have been abandoned to wind action through the retraction of drainage; hence they tend to lie beyond the main active drainage terminals, whilst the combination of low relief and porous sand inhibits local run-off. They comprise sandplain, particularly in areas of clay-rich sand derived from igneous rocks, and dunefields typical of longitudinal ridges. They are mainly spinifex (*Triodia* spp.) grasslands, readily hunted by firing but yielding only small game, with plant food mainly restricted to tubers.

The most favourable environments within the sand deserts are the margins of the dunefields. These are commonly zones of interaction between fluvial and aeolian processes, with pans included in the dune networks which here tend to replace the parallel ridges. Generally, the alternation of dune and swale leads to some concentration of surface run-off, and dunefields are better-watered than open sandplain, with larger areas of trees and shrubs.

*Shield desert.* The shield deserts as represented on the south interior plateau of Western Australia are broad plains with two main elements, an upper surface capped by Tertiary laterite and a lower level etched on relatively fresh granite. What little relief there is consists of hill belts of metamorphic rocks, scattered groups of granite hills on the lower surface, and long low breakaways which separate the two plateau elements and which contain shelters and minor rock holes.

Aeolian sand mantles much of the shield desert, as sandplain rather than as dunes, and gives rise to spinifex grassland. The lower plains bear extensive mulga woodland with shrubs, which become stunted on the hill-slopes. There are also areas of useful perennial grasses on deep coarse-textured soils, whilst saltbush (*Atriplex* and *Kochia* spp.) and eucalypt woodland characterise the river flats.

Drainage heads at the breakaways and leads to the river lakes by way of calcreted alluvial fills in the main valleys; these provide plentiful shallow groundwater which becomes brackish downstream.

On the high plains of the central Australian shield deserts, mulga woodland on red earth soils covers large areas, particularly in peripheral tracts traversed by active drainage.

Areas of these desert types are given in Table 6:2.

*Table 6:2 Extent of Australian desert habitats*

| | **Km²** | **Percentage of arid zone** |
|---|---|---|
| Mountain and piedmont deserts | 930,000 | 17·5 |
| Riverine desert | 210,000 | 4·0 |
| Stony desert | 640,000 | 12·0 |
| Desert clay plains | 690,000 | 13·0 |
| Sand desert | 1,680,000 | 31·0 |
| Shield desert | 1,200,000 | 22·5 |
| Total | 5,350,000 | |

## Desert Habitats and Aboriginal Man

The significance of the desert habitat and of contrasts within it for the distribution, movements, and way of life of the Australian Aborigines has been noted in many anthropological studies (see Lawrence 1968:41 ff.) and we may infer that the same environmental factors operated for prehistoric man. Aboriginal occupation of the Australian arid zone appears to have been notably sparse (1 person per 30 to 100 km²), even in a thinly peopled continent, probably between one-half and one-tenth the density of population in the better-watered northern and eastern margins (Meggitt 1962:32). The desert social group was generally very small, and larger assemblies were dependent on rain and on ensuing food and water supplies. In the north of the arid zone this led to a seasonal cycle with maximum dispersion into wide-ranging family groups by the end of the drier season (Meggitt 1962:52). Drought might have led to conflict for limited water supplies.

Distinction between upland and piedmont desert, more-watered and with richer and more varied food resources, and the poorer lowland desert appears to have been fundamental. Berndt and Berndt (1942:325) have contrasted hill people and spinifex people in the southwest of the arid zone, while in the northwest Meggitt (1962:32) has estimated that the hill territories of the Aranda and Ngalia peoples supported twice as dense a population as did the open sandplain.

Major riverine tracts formed distinct habitats, for instance the Cooper Creek area with its abundance of fish, game, and seeds (Lawrence 1968:53). The Aranda name for the Finke River (Larapinta) was given to the people who lived along and from it (Spencer and Gillen 1927:**1**, 8), and the Lander floodplains were similarly identified as Yalpari territory among the Walbiri people northwest of Alice Springs (Meggitt 1962:47). The river courses, with their soakages and waterholes, were important lines of Aboriginal movement inherited by early European travellers and graziers.

The spinifex sandplains appear generally to have been recognised as difficult habitats, little-watered and with restricted food resources, and were commonly abandoned in severe drought.

## Changes in Desert Habitat

To chronicle climate change is not a prime object of this contribution, but it is relevant to consider the significance to man of the related changes in desert environment during the 30,000 years or so of his habitation, a period which in other desert regions embraces fluctuations between greater and reduced aridity. In the Australian arid zone these will have been superimposed on a longer-term desiccation from the deep weathering, advanced pedogenesis, and outgoing drainage of the Tertiary Era.

### GREATER ARIDITY IN THE PAST

*Evidence of the dune systems.* The clearest evidence for past greater aridity is the dune systems now mainly stabilised by vegetation. The degree of fossilisation is greatest in the north, where longitudinal dune ridges transgress the present limits of the arid zone. In the Pindan of Western Australia and in the undulating sand country south of the Barkly Tableland there are completely vegetated, subdued and rounded dune forms comparable with those of the Goz of northern Nigeria (Grove 1958:531), which similarly indicate a former extension of desert conditions equatorwards. There is a transition southwards across the Australian arid zone, through live-crested ridges with vegetated flanks to the more mobile ridges of the driest southeast, and there is an increase in the dimensions of the dunes in the same direction.

The Australian dune ridges describe a giant ellipse, open to the west, about the centre of the desert belt (Fig. 6:2). Given the approximate coincidence of dune trend with resultant wind, the dunes record that the arid period or periods of their formation involved no major shift of the pressure and wind belts with which the arid zone is genetically associated. Further, the continuity of trend across discrete dunefields over a range of latitude, and the absence of important later realignment of the ridges, indicate that the dune system formed as one. This suggests that desert conditions were intensified during one or more periods in which the arid zone also expanded—particularly equatorwards—about its present core.

The age or ages of the central Australian dunes are not known, but there is circumstantial evidence for rejecting the hypothesis that they were entirely formed in a short mid-Recent arid period. The internal geometry of the dune systems reflects the attainment of equilibrium forms over an area of 500,000 km$^2$, a vast achievement which appears inconsistent with but three or so millennia of heightened aeolian activity. To this admittedly subjective assessment based on the magnitude of the dune systems must be added the stratigraphic evidence. In the Simpson Desert, for instance, each ridge typically has a hard core of clay-cemented sand which sits centrally beneath it and commonly rises to as much as 8 m above the level of the adjoining swales, although rarely outcropping. Such clay cores indicate long-

*Fig. 6:2 The Australian arid zone: desert habitats and dune ridges*

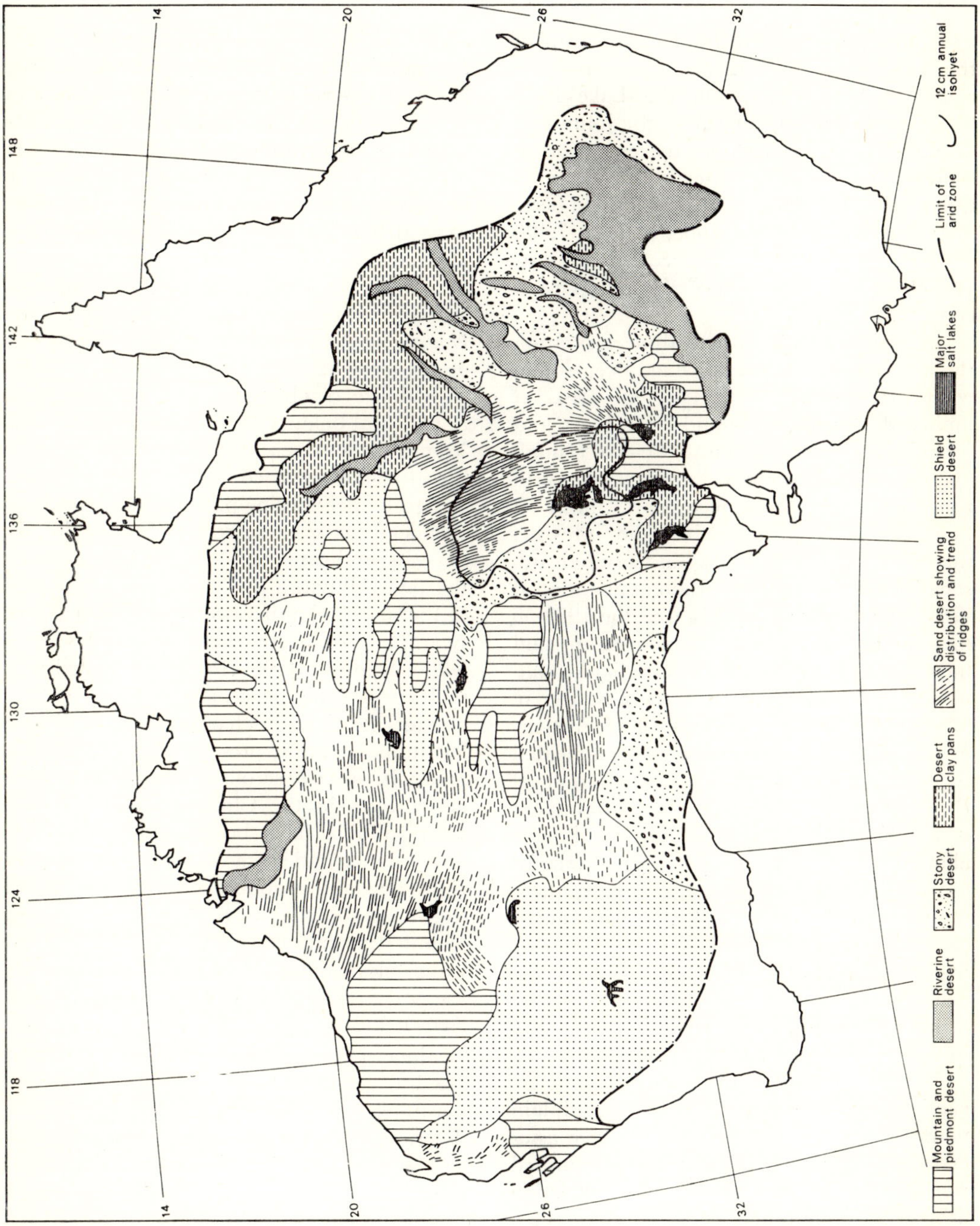

continued illuviation without lateral shift of the ridges. More specifically, there is some evidence that deflation in the Lake Eyre basin had begun 40,000 years ago (Johns and Ludbrook 1962:33), whilst the extension of dune ridges into the intertidal zone near Derby, W.A. indicates that they formed here, at the limit of the arid zone, at least 7,400 years ago and before the complete post-glacial recovery of sea level (Jennings pers. comm.).

At a time of widespread dune formation the arid zone must overall have been somewhat drier than are the present areas of limited dune growth around Lake Eyre, and probably received less than 10 cm of rainfall annually. For the south part of the arid zone this does not necessarily imply a fundamental contrast with the present; dunes are currently forming in favoured sites such as playa margins and along abandoned sandy channels, and in this area the 1959-66 drought brought a general increase in sand movement on vegetated dune flanks and much aeolian erosion and mobilisation of sand in apparently stable swales. Further, no great recovery of rainfall need be indicated by the subsequent partial stabilisation of the dunes in this area, since the attainment of equilibrium forms should in itself have allowed some colonisation by vegetation. In the north of the arid zone, however, there must have been a major decrease in annual rainfall of between 15 and 40 cm, tantamount to an equatorward shift of the isohyets by eight degrees of latitude and presumably bringing a significant increase in seasonality in the savannah zone to the north.

Transformation of the sandy deposits of earlier river systems into dunes and sandplain marks a retraction of surface drainage associated with less frequent and more limited run-off, and hence with geographic restriction as well as diminution of water supplies. Such a period of increased aridity must have accentuated the relative harshness of the lowland desert environment. Areas of sand desert would have been rendered uninhabitable and there would have been a shrinkage of the better riverine habitats. Even the favoured upland and piedmont desert habitats were affected, as shown by the encroachment of dunes on their southern and eastern margins (Plate I).

*Evidence of fluvial landforms.* In many riverine desert areas, as in the Channel Country, there is evidence of changes in river régime and in the local balance between flood and wind. Many prior river traces are braiding or broadly sinuous in contrast to the reticulate or tightly meandering patterns of the present. Further, coarse-textured prior accretionary deposits indicate bed-load régimes, whilst the finer alluvium and deeper cross-sections of the active channels are of suspended-load type. The contrast recalls that in the Riverina, where Schumm (1967:187) has claimed that such prior channels may indicate former greater seasonal discharges, but a climate less humid than at present in the catchment area. In many scrolls, braid-islands, and levee tracts in the Channel Country, these sandy alluvia have been wind-formed into dune islands and since vegetated. A bed-load régime would probably have constituted a less favourable riverine habitat in the Channel Country in that deep channels with perennial waterholes would have been fewer.

### PAST PERIODS OF GREATER HUMIDITY

Distinct periods wetter than at present are less readily identified in the arid zone inasmuch as any subsequent climatic deterioration would be in the sense of the general post-Tertiary trend already mentioned. Further, a period of climatic amelioration will not necessarily wipe out the effects of earlier aridity; for instance, the deterioration of habitat resulting from the obliteration of surface drainage by aeolian sand in the Australian arid zone has outlived the causal climatic stress in that run-off tends not to be re-established on lowland surfaces rendered pervious through the wind-sorting of prior alluvia.

The environmental evidence for more humid conditions in the past includes fluvial and lacustrine deposits beyond the present ambit of the formative agents, relict soils, and fossil remains of a less desertic fauna and flora. Naturally enough, this evidence is mainly forthcoming from the piedmont and riverine deserts.

*Alluvial evidence.* Increased rainfall leads to more vigorous and frequent run-off in piedmont tracts and to changes in the load-discharge balance of minor streams. A widespread succession of three piedmont terraces in the Macdonnell Ranges of central Australia has been interpreted as a result of alternating wetter and drier climate, more humid phases of hillslope weathering and channel incision being followed by relatively arid phases of stream planation, colluvial mobilisation, and the deposition of gravel fans (Mabbutt 1966:83). Litchfield (1969) has identified comparable alluvial cycles in piedmont plains near Alice Springs, each marked by the deposition of a layer of alluvium followed by soil formation.

The terraces and the alluvial layers were modified or invaded in a succeeding episode of intensified aeolian activity. The slewed courses of prior distributary streams north of Alice Springs indicate interference by wind transport in the lowermost sectors, culminating in the wind-sorting and piling of the alluvial trains (Mabbutt 1967:172). Such evidence points to there having been more than one fluctuation between wetter and drier before aeolian sand attained its maximum extent; to a progressive desiccation upon which such fluctuations were superimposed, as shown by contrasts between successive alluvial deposits and the soils formed in them; and to the contested and fluctuating boundary between the fluvial and aeolian provinces in the desert lowlands.

*Evidence of playa levels.* The firmest evidence of more abundant water supplies in the arid zone in the past is afforded by the relict high strandlines of desert playas. The palaeoclimatic record of such lakes is discussed elsewhere (Bowler, see ch. 5), but their significance for prehistoric man should be noted. River-fed pans are more important than saline terminal basins in this respect, and an extreme example of the former is Lake Yamma Yamma, a lateral sump lake west of Cooper Creek with an area of 450 km$^2$. Along its northwestern margin is a prominent shingle beach ridge capped by white sand dunes; it attains 3·5 m above the lake floor, on which faint sand rises mark present-day flood limits. The feeder entrance is partly blocked by sand dunes and the lake is now entered only by the largest floods, at intervals of a decade or two. It is noteworthy that none of the dune-rimmed lakes of central Australia has yet proved to have relict strandlines, suggesting that any such pluvial period preceded the main extension of dune sand.

*Fossil evidence.* A fossil fauna in Pleistocene alluvia in arid western Queensland has been cited as evidence for past pluvial conditions on a major scale (Whitehouse 1940:64). In fact, it would not be justified to conclude that it shows a zonal climate vastly different from that of today; the containing deposits are those of interior drainage to playa lakes and are characterised by evaporites such as lime and gypsum. Nevertheless, the remains of giant macropods, turtles, and freshwater crocodiles indicate that river channels and lakes were then more extensive and connected and that mound springs were more active. This implies an increase in the extent and richness of a favoured desert habitat, with improved and more assured water and food supplies for prehistoric man.

It appears unlikely that the climatic changes described above could fundamentally have altered the arid zone as a difficult human environment and as a barrier to movement. There is no sign of a breakdown or major latitudinal shift of the meteorological controls of aridity, but rather of periodic intensification and relaxation of desert climate and the associated expansion and contraction about a persistent arid core, involving changes of degree rather than of kind. Should the Australian dune systems date entirely from the late Pleistocene, their growth certainly constituted a deterioration of habitat, irreversible in large degree, but it is probable that the areas of dune formation were formerly poorly watered alluvial plains of low value to man.

It is unlikely that desert rainfall régimes have changed drastically; in the past, as now, the concentration and localised storage of run-off would have been essential for human existence, and as now this would have been controlled by relief and lithology and by the major drainage lineaments inherited from a more humid geologic past. The favoured environments of upland and piedmont and riverine tracts would have persisted throughout, although with fluctuating limits and changes in relative importance which would lead to associated changes in the range and frequency of human migration into adjoining desert lowlands.

## References


Berndt, R. M. and Berndt, Catherine H. 1942. A preliminary report of fieldwork in the Ooldea region, western South Australia. *Oceania* **12**: 305-30.

Dubief, J. 1953. *Essai sur l'hydrologie superficielle au Sahara.* Clairbois-Birmandreis, Algiers: Gouvernement général de l'Algérie, Direction du service de la colonisation et de l'hydraulique, Service des études scientifiques.

Dury, G. H. 1968. An introduction to the geomorphology of Australia. In *Studies in Australian geography* (eds. G. H. Dury and M. I. Logan), pp. 1-36. Melbourne: Heinemann.

Grove, A. T. 1958. The ancient erg of Hausaland, and similar formations on the south side of the Sahara. *Geogrl J.* **124**: 528-33.

Jennings, J. N. 1967. Two maps of rainfall intensity in Australia. *Aust.Geogr.* **10**: 256-62.

Johns, R. K. and Ludbrook, N. H. 1962. Investigation of Lake Eyre. *Rep.geol.Surv. S.Aust.* No. 24.

Lawrence, R. 1968. *Aboriginal habitat and economy.* Canberra: Australian National University, School of General Studies, Department of Geography Occasional Papers No. 6.

Litchfield, W. H. 1969. *Soil surfaces and sedimentary history near the Macdonnell Ranges, N.T.* Soil Publ. CSIRO Aust. No. 25.

Mabbutt, J. A. 1965. Landscapes of arid and semi-arid Australia. *Aust.nat.Hist.* **15**: 105-10.

——1966. Landforms of the western Macdonnell Ranges. In *Essays in geomorphology* (ed. G. H. Dury), pp. 83-119. London: Heinemann.

——1967. Denudation chronology in central Australia: structure, climate, and landform inheritance in the Alice Springs area. In *Landform studies from Australia and New Guinea* (eds. J. N. Jennings and J. A. Mabbutt), pp. 144-81. Canberra: Australian National University Press.

Meggitt, M. J. 1962. *Desert people. A study of the Walbiri Aborigines of central Australia.* Sydney: Angus and Robertson.

Meigs, P. 1953. World distribution of arid and semi-arid homoclimates. *Reviews of research on arid zone hydrology*, pp. 203-10. Paris: UNESCO.

Schumm, S. A. 1963. *A tentative classification of alluvial river channels.* Circ.U.S.geol.Surv. No. 477.

——1967. Paleohydrology: the application of modern hydrologic data to problems of the ancient past. In *Proceedings International Hydrology Symposium.* International Association for Scientific Hydrology, Fort Collins, Colorado, 1967. Vol. 1, pp. 185-93. Colorado.

Slatyer, R. O. 1962. Climate of the Alice Springs area. In *General report on lands of the Alice Springs area, Northern Territory, 1956-57.* Land Res.Ser. CSIRO Aust. No. 6:109-28.

Spencer, B. and Gillen, F. J. 1927. *The Arunta.* 2 vols. London: Macmillan.

Tricart, J. and Cailleux, A. 1960. *Le Modelé des régions sèches*: Fasc. I. Le milieu morphoclimatique. Les mécanismes morphogénétiques des régions sèches. Paris: Centre de Documentation Universitaire.

Twidale, C. R. and Corbin, Elizabeth M. 1963. Gnammas. *Revue Géomorph.dyn.* **14**: 1-20.

Whitehouse, F. W. 1940. Studies in the late geological history of Queensland. *Pap.Dep. Geol.Univ.Qd* **2** (1).

# 7 Man, Fauna, and Climate in Aboriginal Australia

J. H. Calaby

The first Aborigines arrived in the Australian-New Guinea continent by sea, perhaps 30,000 years ago, when the last glaciation was in progress and sea levels were much lower than at present. Australia and New Guinea formed a continuous land mass and the Timor Strait was less than 160 km wide. We can safely assume that the earliest people had a well-developed technology, and carried no other animals with them than their own parasites. At the time of man's arrival the tropical sea would have been a few degrees cooler and the climate also somewhat cooler. The fauna of the sea and land would have been at least as abundant in individuals as now. The land fauna of today was in existence, with a few exceptions that have arrived since or were arriving at the same time. These would include a few kinds of birds, and perhaps an odd species of bat and a few sorts of reptiles, mostly colubrid snakes. Faunal interchange between Australia and New Guinea was active across the broad land corridor bridging the two land masses. The essential differences distinguishing the Pleistocene vertebrate fauna from that of today were that it was more diverse and included giant herbivorous and carnivorous marsupials and cursorial birds, and a greater variety of chelonians, crocodilians, and lungfishes.

## The Land Vertebrate Fauna

Aspects of the composition and historical zoogeography of the modern vertebrate fauna have been discussed recently by Darlington (1957), Simpson (1961), Storr (1964), Main (1968), and Calaby (1969), from which the following account has largely been compiled. Australia and New Guinea have been separated by deep sea from other land masses since the Cretaceous. This long isolation is reflected in the composition of the fauna. It is a typical island one; in general there are few basic types (Orders) and some of these, such as the marsupials among the mammals and the family Leptodactylidae among the frogs, have radiated widely to fill available niches.

The faunal impoverishment at ordinal level is shown chiefly in the freshwater fishes, Amphibia, and mammals. The reptiles and birds are more varied taxonomically and represent a reasonably diverse segment of the fauna of the Oriental Region and Old World tropics from which they have largely been derived. The only discordant group among the reptiles contains the swamp and river tortoises (Chelidae) which otherwise occur only in South America. The birds and reptiles have a much higher capacity for dispersal over sea barriers than mammals (except bats) and frogs. There are a few old endemic elements, including the Queensland lungfish (*Neoceratodus forsteri*) and the barramundi (*Scleropages leichhardti*), the monotremes and marsupials, and the ratite birds. None of the large groups of primary freshwater fishes is present in Australia and all except the two species mentioned above are derived from marine families. Of the three orders and many families of living amphibians, only four families of one order, the Anura (frogs), are present in Australia and New Guinea, and two of these are poorly represented and geographically restricted in Australia. Only four orders of mammals are found in Australia. The bats (Chiroptera) are closely related to those of southeast Asia. The rodents belong to a single family, the Muridae, which reaches its greatest development in the islands and mainland of southeast Asia. They have had a long history in Australia and have radiated widely to produce a fauna of diverse morphology and habits. The other orders of mammals are the marsupials, which are discussed below, and the monotremes. The latter are unique to the Australian area and there is no evidence that they have ever occurred outside it.

Broadly speaking the vertebrate fauna of New Guinea is closely related to that of Australia. Except for the lungfish (Dipnoi) the same orders are present and mostly the same families. The barramundi occurs there. The frogs differ considerably in the relatively high development of Microhylidae and Ranidae, two families that just reach northern Australia, and the small intrusion of Leptodactylidae, the dominant family in Australia. A few families of birds are present in each country that are not in the other, and there is a higher proportion of birds more closely related to those of southeast Asia than Australia. New Guinea was an active centre for speciation; among the important groups that proliferated there were the water-rats (Hydromyinae), tree-kangaroos (*Dendrolagus*), and the birds-of-paradise (Paradisaeidae).

The ancestors of the majority of Australian vertebrates 'island-hopped' across the Indo-Malayan island chain throughout the Tertiary and Quaternary and their descendants have accumulated here and evolved into numerous endemic families, subfamilies, genera, and species. The process is still going on and occasional species of birds have established themselves as breeding species, without aid from man, within the past thirty years. There has been very little reverse traffic. A few birds of Australian origin have transgressed Wallace's Line. Of the marsupials, one cuscus (*Phalanger*) reaches as far west as Timor and two others are found on Celebes, one of which also occurs in Talaut.

Because of their dominant position in the fauna and the large size of many of them, the most important animals to the Aborigines were the marsupials. Their ancestors presumably came by sea around the beginning of the Tertiary. The radiation that produced marsupial adaptive equivalents of many kinds of eutherians of other regions is well known. Although the early Tertiary history of

Australian marsupials is unknown it is probable that evolution was initially rapid and the modern families, together with others that did not survive into modern times, came into existence fairly early in the Tertiary. As in other mammal groups elsewhere, species evolved and replaced one another throughout the Tertiary and in some of them there was a progressive increase in size culminating in the giant forms of the Pleistocene. This gigantism was a feature of most families of marsupials. Thus there were thylacines (*Thylacinus*) and devils (*Sarcophilus*) larger than existing species, and wombats about 1·5 m long (*Phascolonus*). There were many species of large Macropodidae. Some 'wallabies' (*Protemnodon*) were bulkier than existing kangaroos and a 'rat-kangaroo' (*Propleopus oscillans*) was as large as the modern grey kangaroo. Its nearest existing relative, *Hypsiprymnodon moschatus*, is much smaller than a rabbit. The largest kangaroos were members of the subfamily Sthenurinae; they were short-faced, heavily-built browsing forms, an adaptive type absent from the modern fauna. The greatest of them, *Procoptodon goliah*, must have been about 3 m high. The family Diprotodontidae, containing the largest marsupials that have ever lived, became extinct after the arrival of man. The family consisted largely of heavy quadrupedal browsers and the greatest and best-known of them, *Diprotodon optatum*, was the size of a large rhinoceros and had a head almost a metre long. There were also some smaller, more lightly built grazing forms. Unfortunately we have no clear idea of the external appearance and know nothing of the behaviour of these creatures. Another extinct family contained the 'marsupial lions' (*Thylacoleo*), large flesh-eating phalangeroids. We do not know whether they were active predators or subsisted on carrion (Woods 1956). Examination of skeletons could decide this point; such material has been discovered but has not been described. It seems to be a peculiarity of the Australian larger mammal fauna that the ratio of carnivorous to herbivorous species was and is low. If the mouse-sized, predominantly insectivorous predators and the bats are excluded, the ratio in the modern fauna is about 1:20, roughly one-third of the figure for North America.

Other giant animals present in the Pleistocene were echidnas of the genus *Zaglossus*, which still occur in the highland forests of New Guinea, birds of the family Dromornithidae, varanid lizards (*Megalania*), which grew to a length of probably 5 m, and the relict horned land turtle *Meiolania*. We do not know how much of the large Pleistocene fauna still existed when man arrived. The data, which have been reviewed by Merrilees (1968) and Jones (1968), are few and equivocal in some cases but there seems to be no doubt that at least some of the giant marsupials were still extant.

### Land and Freshwater Invertebrates

Little need be said of the land and freshwater invertebrates, as most of them have little interest to hunting and gathering people except in times of food shortage. The important ones that are actively sought in Australia are the occasional large palatable forms that occur in reasonable numbers, such as freshwater crayfish and some molluscs, and the few insects that provide ready sources of simple carbohydrates and fats. The sugar producers are the small bees of the genus *Trigona* (which also provide wax), the desert-dwelling honey ant (*Melophorus inflatus*) that stores nectar and honey-dew in the greatly distended bodies of special large workers, and the lerp-scales of the homopterous family Psyllidae. The important fat-producing insects are the large larvae, pupae, and adults of several species of hepialid and cossid moths which seem to be readily obtainable (Tindale 1966). The bogong moth (*Agrotis infusa*) of the southeastern highlands was no doubt popular because the adults were highly gregarious and in summer occurred in immense numbers, and a highly nutritious meal was easily obtained. These insects have a dry-weight fat content of 50 to 60 per cent (Common 1954). Of the invertebrates only mollusc shells are of importance archaeologically.

**Relative Abundance of Vertebrates**

The number of species of mammals and other vertebrates in Australia is considerably less than that of other areas of the world of comparable size. This is presumably a consequence of the low relief and lack of geographical barriers that could isolate populations and promote speciation, and lack of habitat diversity. Australia has roughly the same land area as the United States excluding Alaska, but contains only about two-thirds the number of species of mammals. New Guinea, which has only one-tenth of the land area of Australia, has approximately the same number of birds and more than two-thirds the number of species of mammals.

In the modern mammal fauna the ground density of animals is in general low when compared with other parts of the world. The density of small ground mammals as measured by trapping success is appreciably lower than that found in North America and Europe. Two mammalogists with considerable experience in both North America and Australia have informed me that in the southwestern deserts of the United States they usually obtained catches in 20 per cent, and often higher, of their traps, and in the eastern deciduous forests their average figure was roughly 10 per cent. In Australia their catches averaged 2 to 5 per cent, except in agricultural situations such as tropical rice and canefields, and at times of high population density in *Rattus villosissimus* (J. Mary Taylor and B. Elizabeth Horner, pers. comm.). Small islands represent another exceptional situation and, on these, populations of mammals and some other vertebrates may be quite high. There are some records of large groups of kangaroos at the beginning of European settlement in the better-watered country but none in the drier inland areas. The high densities of red kangaroos (*Megaleia rufa*) in western New South Wales and central Australia and euros (*Macropus robustus*) in northwestern Australia in recent years are a consequence of European settlement. In large parts of western New South Wales the original plant cover was basically an *Acacia* woodland with a saltbush shrub ground stratum. Pastoral settlement and over-grazing by sheep have thinned the tree density and replaced the ground vegetation with a short grass cover. This is apparently optimum habitat for the red kangaroo and locally its numbers have increased enormously (Frith 1964). The addition of watering facilities for stock has also probably helped the kangaroo. Cattle in central Australia and sheep in northwestern Australia have also changed the plant cover in favour of kangaroos (Newsome 1965; Ealey 1967).

In the arid and semi-arid parts of Australia several species of small mammals, particularly rodents, undergo large fluctuations in numbers depending on seasonal conditions. In drought years it is difficult to find even an occasional example, but following a run of good rains they may increase greatly, and the numbers of one or two species, such as *Rattus villosissimus*, may be described as 'plagues'.

We can have no idea of animal abundance in prehistoric times and deposits of fossils or human occupational debris yield no information on this point. Such deposits may not even give information on the relative abundance of different species. Human predators may be very selective or unable to catch some species, and the species found in natural deposits may not be a random selection. Among the larger mammals Macropodidae are the most abundant in species in both Pleistocence and Recent deposits (Merrilees 1968; Tedford 1967; Wakefield 1964a) as one might expect considering their dominance in the fauna.

### Late Quaternary Climates

It is generally agreed that during the late Pleistocene the climate was cooler and more humid than at present. At or towards the end of the Pleistocene conditions became warmer and drier. There is no agreement, however, on the detailed nature of these climates or the changes that have taken place since the end of the Pleistocene. The controversy has been reviewed in some detail by Merrilees (1968) and Jones (1968). Most authors believe that the Pleistocene rainfall was much higher than that of today, but Galloway (1965) has recently argued that the rainfall was lower. As a result of lower temperatures evaporation was reduced, which would lead to high lake levels and other effects that are usually attributed to higher rainfalls. Similarly there are diametrically opposed views on the nature of climatic fluctuations, if any, since the end of the Pleistocene, particularly whether there was a 'Great Arid Period' during the mid-Recent.

In general, vertebrate remains are of limited use in palaeoclimatic deductions and can provide only broad generalisations. The reasons for this are that broadly speaking homoiothermic animals and at least the larger poikilothermic ones such as lizards have wide climatic tolerances; and more importantly we know far too little about the geographical distribution and ecological requirements of existing species. Ride (1968) has pointed out our lack of knowledge of modern species and cautions the use of such inadequate data in palaeoclimatic speculations.

One thing appears obvious. There is such a large and well-adapted desert fauna and flora in Australia that one must assume that it has been in existence for a long time and predates the Pleistocene. One species, the marsupial mole (*Notoryctes typhlops*), is so unlike any other marsupial that its relationships are obscure and it must have evolved fairly early in the Tertiary. It is so highly specialised and adapted to its habitat of deep dune sand that it must be concluded that such conditions have prevailed somewhere in Australia since early Tertiary times.

### Distribution of Animals and Hazards of Interpretation Based on Remains

The great bulk of land animal remains in fossil deposits and human occupational sites are of mammalian origin. Except for a couple of amphibious species and a few others with special requirements, such as steep rocky slopes, the distribution of mammals can be correlated in general with vegetation communities in a structural but not necessarily in a floristic sense. For example, the scrub wallaby *Macropus dorsalis* needs dense scrub for daytime shelter. It may utilise rainforest edges in the wetter parts of its range or brigalow thickets (*Acacia harpophylla*) in the drier parts. One of the more widespread mammals, the grey kangaroo (*Macropus giganteus*), is found in woodlands with a grassy floor. It is equally at home in the tropical woodland around Cooktown, at 1,525 m in the Snowy Mountains, in north-eastern Tasmania, and in the narrow belts of woodland along watercourses north of Broken Hill. The most widespread and adaptable of all Australian mammals is the brush-tailed possum (*Trichosurus vulpecula*). It lives in a wide variety of habitats and climatic régimes from eastern coastal forests to desert. It survives in some treeless country by sheltering in holes in the ground or rocks. In some wet sclerophyll and rainforest communities in eastern Australia *T. vulpecula* is replaced by another species, *T. caninus*. These two species are very difficult to tell apart on skeletal material and would probably be impossible to distinguish on the fragments found in archaeological deposits. Although the above are extreme examples, many species with more restricted habitats show wide climatic tolerances.

The swamp wallaby (*Wallabia bicolor*) is normally found in damp gullies in wetter forested country from north Queensland to Victoria. In the Mitchell-Morven area of southern Queensland it has a surprising 'inland' distribution where it is common in shrubby patches, which may not be particularly dense, along watercourses, and on associated sandhills. Here it overlaps the red kangaroo in distribution, the latter animal being found on Mitchell grass flats. The annual rainfall in this area is about 50 cm. The koala (*Phascolarctos cinereus*), usually considered an animal of relatively high rainfall eucalypt forests, is also present here. Kikkawa and Walter's (1968) recent survey indicates that the koala occurred in areas of even lower rainfall within living memory. There are probably many apparently anomalous situations of this type in mammal distribution and knowledge of such distributions could be critical in palaeoclimatic interpretations.

Our knowledge of the geographical distribution of animals in historical times is most inadequate. The magpie goose (*Anseranas semipalmata*), a favourite food of Aborigines, is a breeding species only in the well-watered tropical north, but during the last century it bred in large numbers in coastal and inland New South Wales and southern Victoria (Frith and Davies 1961). *Rattus tunneyi*, a species known today only in the better-watered coastal parts of northwestern Australia, northern Australia, and Queensland, was collected as a living animal in inland New South Wales, southern South Australia, and southwestern Australia about 1840 or earlier. This knowledge has come to light only recently, from studies of old collections (J. Mary Taylor, pers. comm.). The Blandowski-Krefft expedition to the Murray-Darling junction in 1857 collected a number of species of mammals in Victoria which have not otherwise been recorded there (Krefft 1866; Wakefield 1966). Today these species are found only much further inland or in Western Australia.

Apparently many species of animals were sensitive to changes produced by European interference and perhaps long-term environmental changes, and had it not been for what amount to historical accidents we would never have known that their modern distributions were so broad (see also Ride 1968). There are other species represented by abundant bones in cave deposits that have been rarely collected as living animals in historical times. Some of these, for example *Gymnobelideus leadbeateri* and *Pseudomys novaehollandiae*, were for long periods considered to be extinct. It is apparent from recent work that the optimum habitats of these two species are unstable early successional stages of forest regenerating after destructive fires. At the present time we know of populous colonies of these animals in such communities (Warneke 1968; Keith and Calaby 1968). Possibly uncontrolled hunting fires of Aborigines had a lot to do with keeping them in existence in prehistoric times.

The fact that species characteristic of very

different habitats can be found living in close proximity is not generally appreciated and also could make difficult an ecological interpretation of the fossil or archaeological record. I have studied a living mammal fauna in northeastern New South Wales in which more than thirty native species of terrestrial, arboreal, and amphibious mammals of great taxonomic diversity, including ten species of Macropodidae, together with several species of bats of both suborders, and a number of feral exotics, including the dingo, could be found within a kilometre of each other. There were species characteristic of rainforest, wet and drier eucalypt woodland, stream banks, etc. This rich fauna is a consequence of the diversity of soil types, topography, and aspect, producing an interdigitation of vegetation communities in a small area with sharp boundaries between them (Calaby 1966). It is not difficult to imagine the bones of most of these animals coming together in a human occupational site or in a single fossil deposit, especially in a cave frequented by both carnivorous mammals and owls, as many caves seem to have been. Gill (1968) has listed a cave fauna from Tasmania containing a considerable number of mammal species and claims that as a variety of habitats is represented the animals were gathered from a range of localities by a predator, probably man. This may be so; however, all of these animals could have occurred close together around the cave, given a wet sclerophyll forest with an open area kept open by repeated burning by Aborigines, a not uncommon occurrence in Tasmania (see Jackson 1965; Jones 1967).

A practical problem for archaeologists and palaeoclimatologists that should be mentioned is that misdeterminations of faunal remains may be provided by less competent authorities. Wakefield (1964a, 1964b) has re-examined collections and corrected mistakes of previous investigators that have appeared in print. Another disturbing problem may be mentioned here—the publication without adequate documentation of determinations of species, based on fragmentary specimens, that represent very large extensions of range. Authors should describe such specimens fully as has been done by Tedford (1967) for *Lagorchestes conspicillatus*, and preferably figure them also (Calaby and White 1967). Wakefield (1963, 1964a) has recorded *Onychogalea unguifer* from southern Victoria, based on a few juvenile mandibles and a maxilla. Only a superficial comparison with authentic material is provided, and no comparison with a congeneric species, *O. frenata*, recorded in the same deposit. He may be right, but *O. unguifer* has never been found previously outside the tropics, and the Victorian record defies zoogeographical interpretation.

It might be instructive to see what ecological or climatic information can be deduced from a few representative recently published lists of animals from late Pleistocene and Recent deposits. Merrilees (1968) has given a list from the Mammoth Cave and other cave deposits in that area of Western Australia. Two carbon dates are available from Mammoth Cave, both greater than 30,000 years BP. The fauna consists of three components, a few giant Pleistocene forms which can tell us nothing, a number of species still living in the region in historical times, and a few species identical with or closely related to species still extant in southeastern Australia but extinct in Western Australia since prehistoric times. The latter group of species includes a koala, a wombat (*Vombatus*), and a *Wallabia* resembling *bicolor*. At least the koala and *W. bicolor* are found in eastern Australia in areas considerably drier in terms of rainfall than the Mammoth Cave area at the present time and if it was climatic deterioration that brought about their extinction, the rainfall would need to have fallen to less than half its present figure. On the other hand *Potorous gilberti* (a subspecies of *P. tridactylus* of New South Wales), which survived into historical times in southwestern Australia, is not known to occur anywhere except in moist situations with a rainfall no lower than 76 cm—about two-thirds of the present rainfall in the Mammoth Cave area. Merrilees (1967) has analysed the stratigraphical occurrence of two species of bandicoots in the Mammoth Cave deposit and thinks it possible that a slow increase in rainfall since the Pleistocene brought about the changes in abundance found.

Stirton *et al.* (1961) have listed an Upper Pleistocene fauna from the Lake Eyre area. There were lungfishes (Ceratodontidae) which argue for fresh water but there were also flamingoes of a species (*Phoenicopterus ruber*) still living in other parts of the world (Miller 1963). These are highly specialised, very gregarious birds, which need large expanses of shallow saline or brackish permanent water for feeding and nesting. No details have been published of the chelonians or crocodilians (including very large ones). The lake must have been salt (we know this anyhow, as it was a closed basin) and the streams presumably produced a much greater volume of water than now, when they scarcely run at all; however, it was not necessary to have permanently running streams to support lungfishes. The area must have been considerably moister than now to provide the vegetation to feed the kangaroos and other large herbivores. Perhaps they were associated with the riparian vegetation. There is no need to postulate 'lush' vegetation. Merrilees (1968) has mentioned that the great biomasses of large grazing herbivores are not found in wet tropical vegetation, but in the much drier communities such as the African savannahs. The Lake Eyre area had a considerably moister climate in the Upper Pleistocene than it has today, but whether this was due to a much higher rainfall is an open question.

Tedford (1967) records an Upper Pleistocene mammal fauna from Lake Menindee which shows some evidence of human association. Two carbon dates are available but they agree only in order of magnitude—say roughly 20,000 years ago. The fauna consists of a number of giant marsupials, a few species known in the area in historical times, and others identical with or closely related to modern species not known in the region in historical times. Of these latter a few (*Dasycercus*, *Caloprymnus*, and *Lagorchestes conspicillatus*) are known from farther inland or northwards but not necessarily from drier areas. We know so little of their distribution in early historical times that they may well have occurred in the Menindee area. Two species (*Phascolarctos* and *Wallabia*) are presently characteristic of considerably moister areas. Tedford's *Wallabia* material was fragmentary and he was unable to decide whether it was *bicolor* or *agilis*. On zoogeographical grounds one would expect it to be *bicolor*. It has been mentioned above that the koala and *W. bicolor* occur in a moderately dry situation in southern Queensland. To obtain a similar situation at Menindee one would need the rainfall to double and the appropriate woodland vegetation to 'migrate' westwards perhaps 500 km. Perhaps all that was needed was a somewhat moister climate and a broader riverine vegetation belt. Contrary to popular belief the koala is not particularly restricted in its choice of food trees; it will eat river redgum (*Eucalyptus camaldulensis*), which still occurs in the area, and we need not postulate the food trees it normally lives on in the forest and woodland farther east.

Wakefield (1964a) has listed the Recent faunas found in various caves and human occupational sites in southwestern Victoria. These contain only thylacines and devils representing the animals that disappeared from mainland Australia in prehistoric times. Much of the fauna still occurred in the area in historical times but a few species apparently did not. However, the Western District of Victoria was settled and stocked very rapidly; no zoologist visited the area during that period and we know nothing of the fauna that may have existed there. Wakefield makes some suggestions of climatic change in the area including the possibility of an 'arid period' because of specimens of *Bettongia lesueur* which he says is a 'desert species'. However, it occurred as a common animal in southwestern Australia in this century in areas with rainfall as high as 63·5 cm a year and Dahl (1926) found it abundant in 1895 at Hill Station, north of Broome, Western Australia, which also has a rainfall of about 63·5 cm. This figure is little different from those of the two sites in Victoria where its remains were found. *Aepyprymnus rufescens* is recorded from some deposits and the comment is made that it is an animal of the 'wetter more heavily vegetated areas of S.E. Australia'. But it was collected as a living animal at Echuca and the Gunbower district of Victoria by Blandowski and Krefft, as Wakefield (1966) himself has recorded. The rainfall of these localities is about 35 to 40 cm.

A more satisfactory climatic or ecological explanation of some of these faunal suites might be suggested if the data were better. Perhaps there were minor fluctuations in climate which caused a change in the vegetation and allowed some different animals to occupy the localities for varying periods. We need undisturbed sites covering a long time span, and many more dates. Only one such site, Kenniff Cave, appears to have been discovered, but unfortunately it contains no faunal remains (Mulvaney and Joyce 1965). It would also be useful to have some data on the rate of movement of vegetation communities and associated fauna following climatic changes.

## Impact of Aborigines on Fauna

When the first immigrants arrived in the northern part of the Australia-New Guinea continent they would have found plenty of familiar elements in the fauna. The rats and bats and some of the birds and reptiles would not differ greatly from or, in a few cases, were the same species as those of their homeland. It is unlikely that they found some of the marsupials and perhaps even the echidna particularly strange. The dasyurid marsupials are not very different in general appearance from such animals as shrews and viverrids. Porcupines and a pangolin occur in the Indonesian islands and in some features echidnas have a superficial resemblance to those creatures. No Sundaland birds resemble the giant flightless cursorial birds—the emus and cassowaries and the even larger Pleistocene dromornithids. The large diprotodontids may not have seemed particularly odd. The newcomers would have been familiar with large ungulates in their homeland such as elephants, cattle, tapirs, and rhinoceroses. The only really strange animals would have been the Macropodidae because of their structure and gait. It is perhaps appropriate to remember that although the marsupials were known from 1500 and a wide variety was available from 1788, the savants of Europe did not recognise the unity of the order until 1816 (Blainville 1816), and in their classifications distributed the marsupials known to them among various orders of mammals depending on external resemblances. Some conservative mammalogists continued this practice until as late as the 1830s.

We do not know how readily the immigrants took to hunting the animals of the new country. It is said that desert Aborigines were reluctant to eat rabbits for some years after these appeared in central Australia. Peoples in all cultures are conservative in their food habits but hunger can cause a radical revision of these ideas. We cannot know anything of the importance of large game compared with small land animals (reptiles, birds, invertebrates), marine and freshwater shellfish, fishes, etc., in the diet of the earliest Australians. The food habits of modern Aborigines could be useful as a guide but there are extremely few data although long lists of individual food items have been published (see review by Lawrence 1968). Many animals were not eaten or otherwise utilised by modern Aborigines, including numbers of birds, almost all frogs, venomous snakes, most invertebrates, etc. (see Meggitt 1962).

McCarthy and McArthur (1960) have studied the amount and types of food collected for a short period by two small groups of people using traditional methods and equipment, although their spears had iron blades which no doubt increased their efficiency as killers. This study remains unique; further and more extended studies of this nature are an urgent necessity. A more superficial account by Campbell (in Cleland 1966), taken down secondhand from one of the participants, of the food-gathering activities of a small group of desert people for a short period is of little use for comparative purposes. The party caught quite a number of kangaroos and goannas but their success was largely due to a pack of dogs, no doubt of modern European origin. Also there were no children and the women rode on donkeys, further factors increasing the hunting efficiency of the group.

Meggitt (1964) concluded that for the northern groups he studied, vegetable material was overwhelmingly important. On the other hand Hiatt's (1967) conclusion from her analysis of the literature of the Tasmanians was that vegetable material was probably of much less importance, owing to the paucity of nutritious or palatable plants in Tasmania.

Did the Pleistocene men hunt the large marsupials? A *Diprotodon* would certainly be a rich prize for a hunting group. Modern Aborigines have hunted buffalo and wild cattle, but they have had the advantage of possessing spears with large sharp iron blades propelled by spearthrowers. No doubt the diprotodontids had thick hides which were more than a match for the hunting weapons of the day. The structure of the feet of diprotodontids suggests that they were slow lumbering creatures but they may have been able to outrun men. Hunting peoples in other times and places have exploited large mammals by various means: snares and pit-traps, running gregarious species over cliffs, using powered projectiles (arrows), metal points, etc. Our early men may have used pits as modern Aborigines have for trapping wallabies (McCarthy 1963).

A gregarious species would be easier to hunt than a non-gregarious one. We can have no idea whether or not the diprotodontids were gregarious. The large number of skeletons of *Diprotodon* at Lake Callabonna suggests gregariousness, but these may be the accumulation of many years. Some large ungulates are gregarious while others are not. The modern large kangaroos are similar in this regard. The red and the grey are social but others such as the euros and some of the large wallabies are not, even though they may occur in high densities and appear to be gregarious. If they are disturbed, however, the gregarious species will coalesce into a group and generally behave as a unit while the non-social forms will disperse in all directions. The archaeological evidence that might give us some details of the direct predation of man on the giant Pleistocene animals has not so far been discovered.

Birdsell (1958) has estimated that the entire continent could have been peopled by the descendants of the first handful of immigrants in 2,000 years or less. This is a virtually instantaneous exposure of the whole fauna to a new predator. As the population density of Aborigines was very low and they generally did not collect more food than they could use at any particular time (Meggitt 1964), there seems to be no reason to doubt that they had little or no deleterious effect on the survival of most faunal species as a direct result of their predatory activities.

There is no archaeological or ethnographical evidence that predation by Aborigines has caused the extinction of any species. Large sedentary groups prey heavily on edible animals around their camps and perhaps cause local extinctions. In pre-European times, however, such groups would come together only in coastal localities where food could be said to be superabundant. Probably larger-than-normal groups would be tied to permanent waters in deserts in times of severe drought. In either place such abnormal concentrations would be widely spaced. Endemic species on small islands are especially vulnerable, and there are a few peculiarly vulnerable species in continental faunas. One special case has been reported by Jones (1966, 1967), who found that prehistoric Tasmanians probably caused the extinction of the southern elephant seal (*Mirounga leonina*) as a breeding species on the Tasmanian mainland. However, this is an exceptional situation involving a highly gregarious animal that is very conservative in its attachment to traditional breeding grounds.

A later colonisation of Aborigines brought the dingo, which would have spread over the whole continent within a short period of its introduction. Another canid, the European fox (*Vulpes vulpes*), reached the Wyndham area of Western Australia about seventy years after its release in southern Victoria. The dingo presumably came in Recent times as the only dated material from stratigraphically-controlled excavations is Recent (Macintosh 1964; Campbell *et al.* 1966). It is an opportunist predator with a catholic diet including virtually anything it can catch, vertebrate or invertebrate, together with carrion and even some vegetable material. A predator with such unspecialised food habits is not in general an especial hazard to other animals. Whether or not the dingo was responsible for the comparatively recent extinction of the thylacine and devil on the Australian mainland must remain an open question.

Animal populations are regulated by many factors and in general are well adapted to their environments, which include predators. They can usually cope with new predators providing that these are not unduly formidable and abundant. Decreases in populations caused by predators lead to such results as more available food, lessening of territorial intolerances, more security for the survivors, and greater survival of young (see Errington 1967). Probably the effect of both Aborigines and the dingo on the food species has been more in the nature of cropping, after initial adjustments had been made.

There remains the question of the responsibility of the Aborigines for the extinction of the Pleistocene giant fauna. Merrilees (1968) and Jones (1968) have independently reviewed the subject in detail, and the extinction of the Pleistocene fauna on a world-wide basis has recently received comprehensive symposium treatment (Martin and Wright 1967).

The customary view has been that climatic deterioration since the end of the Pleistocene was the primary cause of the extinction of the Pleistocene large fauna, although some authors admitted that Aborigines may have been a contributing factor. Tindale (1959) was the first to state clearly that the Aborigines probably had a profound effect on the vegetation, chiefly by their destructive and uncontrolled use of fire for hunting, and that this may have led to animal extinctions. The extensive burning of the countryside by Aborigines in all parts of Australia is attested by numerous observations in historical times. Merrilees's and Jones's analyses lead to the conclusion that continued burning caused habitat modification on a vast scale which led to the extermination of the giant fauna, although Jones believes that predation by Aborigines played a part in bringing about major ecological changes.

It would be premature to consider that the problem is now nearing solution. Observations on the association of man and the Pleistocene giant fauna are few and unsatisfactory, and much more information is needed on the chronology of extinction of the various species, especially in earlier Pleistocene times. One would like to see some measure of agreement, particularly among geomorphologists, on the nature of climatic fluctuations in the Pleistocene and Recent. Natural selection is a continuing process and in course of time populations of organisms become increasingly better adapted to their environments. Notwithstanding attractive alternative hypotheses on the fate of the giant fauna, more evidence is needed before rejecting the view that the fauna may have been well adapted to a relatively stable environment and was unable to respond to changes during and since the Pleistocene period.

## References


Birdsell, J. B. 1958. Some population problems involving Pleistocene man. *Cold Spring Harb.Symp.quant.Biol.* **22**:47-69.

Blainville, H. D. de. 1816. Prodrome d'une nouvelle distribution systématique du règne animal. *Bull.Soc.philom.Paris* 1816:105-24.

Calaby, J. H. 1966. *Mammals of the upper Richmond and Clarence Rivers, New South Wales.* CSIRO Wildl.Res.Tech.Pap. No. 10.

——1969. The vertebrates. In *The last of lands* (eds. L. J. Webb, D. Whitelock, and J. Le Gay Brereton), pp. 91-101. Brisbane: Jacaranda Press.

——and White, Carmel. 1967. The Tasmanian devil (*Sarcophilus harrisi*) in northern Australia in Recent times. *Aust.J. Sci.* **29**:473-5.

Campbell, T. D., Edwards, R., and Hossfeld, P. S. 1966. Archaeological excavations in the south east of South Australia. Canberra: Australian Institute of Aboriginal Studies (unpublished).

Cleland, J. B. 1966. The ecology of the Aboriginal in South and central Australia. In *Aboriginal man in South and central Australia.* Part 1 (ed. B. C. Cotton), pp. 111-58. Adelaide: Government Printer.

Common, I. F. B. 1954. A study of the ecology of the adult Bogong moth, *Agrotis infusa* (Boisd.) (Lepidoptera: Noctuidae), with special reference to its behaviour during migration and aestivation. *Aust.J. Zool.* **2**:223-63.

Dahl, K. 1926. *In savage Australia.* London: Philip Alan and Co.

Darlington, P. J. 1957. *Zoogeography: the geographical distribution of animals.* New York: John Wiley and Sons.

Ealey, E. H. M. 1967. Ecology of the euro, *Macropus robustus* (Gould), in north-western Australia I. The environment and changes in euro and sheep populations. *CSIRO Wildl.Res.* **12**:9-25.

Errington, P. L. 1967. *Of predation and life.* Ames: Iowa State University Press.

Frith, H. J. 1964. Mobility of the red kangaroo, *Megaleia rufa. CSIRO Wildl.Res.* **9**:1-19.

——and Davies, S. J. J. F. 1961. Ecology of the magpie goose, *Anseranas semipalmata* Latham (Anatidae). *CSIRO Wildl.Res.* **6**:91-141.

Galloway, R. W. 1965. Late Quaternary climates in Australia. *J.Geol.* **73**:603-18.

Gill, E. D. 1968. Aboriginal bone implement from fossil bone bed, Tasmania. *Rec.Queen Vict.Mus.* No. 31.

Hiatt, Betty. 1967. The food quest and the economy of the Tasmanian Aborigines. *Oceania* **38**:99-133.

Jackson, W. D. 1965. Vegetation. In *Atlas of Tasmania* (ed. J. L. Davies), pp. 30-5. Hobart: Lands and Surveys Dept.

Jones, R. 1966. A speculative archaeological sequence for north-west Tasmania. *Rec. Queen Vict.Mus.* No. 25.

——1967. Middens and man in Tasmania. *Aust.nat.Hist.* **15**:359-64.

——1968. The geographical background to the arrival of man in Australia and Tasmania. *Archaeol.&phys.Anthrop.Oceania* **3**:186-215.

Keith, K. and Calaby, J. H. 1968. The New Holland mouse, *Pseudomys novaehollandiae* (Waterhouse), in the Port Stephens district, New South Wales. *CSIRO Wildl.Res.* **13**:45-58.

Kikkawa, J. and Walter, Margaret. 1968. Distribution of koalas in Queensland. *Wildlife in Aust.* **5**:100-3.

Krefft, G. 1866. On the vertebrated animals of the lower Murray and Darling, their habits, economy, and geographical distribution. *Trans.phil.Soc.N.S.W.* 1862-1865: 1-33.

Lawrence, R. 1968. *Aboriginal habitat and economy.* Canberra: Australian National University, School of General Studies, Department of Geography Occasional Papers No. 6.

McCarthy, F. D. 1963. Kangaroos and wallabies in Aboriginal life and art. *Aust. nat.Hist.* **14**:174-8.

——and McArthur, Margaret. 1960. The food quest and the time factor in Aboriginal economic life. In *Rec.Am.-Aust.scient.Exped. Arnhem Ld* **2** (Anthropology and Nutrition) (ed. C. P. Mountford), pp. 145-94. Melbourne: Melbourne University Press.

Macintosh, N. W. G. 1964. A 3,000 years old dingo from shelter 6. App. 2 to D. J. Mulvaney, G. H. Lawton, and C. R. Twidale. *Proc.R.Soc.Vict.* **77**:498-507.

Main, A. R. 1968. Ecology, systematics and evolution of Australian frogs. *Adv.ecol.Res.* **5**:37-86.

Martin, P. S. and Wright, H. E. (eds.) 1967. *Pleistocene extinctions. The search for a cause.* New Haven: Yale University Press.

Meggitt, M. J. 1962. *Desert people. A study of the Walbiri Aborigines of central Australia.* Sydney: Angus and Robertson.

——1964. Aboriginal food-gatherers of tropical Australia. In *The ecology of man in the tropical environment.* International Union for the Conservation of Nature and Natural Resources, Ninth Technical Meeting, Proceedings and Papers, pp. 30-7. Morges: IUCN (Publ.Ser.4).

Merrilees, D. 1967. Fossil bandicoots (Marsupialia, Peramelidae) from Mammoth Cave, Western Australia, and their climatic implications. *J.Proc.R.Soc.West.Aust.* **50**: 121-8.

——1968. Man the destroyer: Late Quaternary changes in the Australian marsupial fauna. *J.Proc.R.Soc.West.Aust.* **51**:1-24.

Miller, A. H. 1963. The fossil flamingos of Australia. *Condor* **65**:289-99.

Mulvaney, D. J. and Joyce, E. B. 1965. Archaeological and geomorphological investigations on Mt. Moffatt Station, Queensland, Australia. *Proc.prehist.Soc.* **31**:147-212.

Newsome, A. E. 1965. The abundance of red kangaroos, *Megaleia rufa* (Desmarest), in central Australia. *Aust.J.Zool.* **13**:269-87.

Ride, W. D. L. 1968. On the past, present, and future of Australian mammals. *Aust.J. Sci.* **31**:1-11.

Simpson, G. G. 1961. Historical zoogeography of Australian mammals. *Evolution, Lancaster, Pa.* **15**:431-46.

Stirton, R. A., Tedford, R. H., and Miller, A. H. 1961. Cenozoic stratigraphy and vertebrate paleontology of the Tirari Desert, South Australia. *Rec.S.Aust.Mus.* **14**:19-61.

Storr, G. M. 1964. Some aspects of the geography of Australian reptiles. *Senckenberg.biol.* **45**:577-89.

Tedford, R. H. 1967. The fossil Macropodidae from Lake Menindee, New South Wales. *Univ.Calif.Publs geol.Sci.* **64**:1-156.

Tindale, N. B. 1959. Ecology of primitive Aboriginal man in Australia. In *Biogeography and ecology in Australia* (eds. J. A. Keast, R. L. Crocker, and C. S. Christian), pp. 36-51. Monographiae biol. **8**. The Hague: W. Junk.

——1966. Insects as food for the Australian Aborigines. *Aust.nat.Hist.* **15**:179-83.

Wakefield, N. A. 1963. Sub-fossils from Mount Hamilton, Victoria. *Victorian Nat.* **79**:323-30.

——1964a. Recent mammalian sub-fossils of the basalt plains of Victoria. *Proc.R. Soc.Vict.* **77**:419-25.

——1964b. Mammal remains. App. 1 to D. J. Mulvaney, G. H. Lawton, and C. R. Twidale. *Proc.R.Soc.Vict.* **77**:494-8.

——1966. Mammals of the Blandowski expedition to north-western Victoria, 1856-57. *Proc.R.Soc.Vict.* **79**:371-91.

Warneke, R. M. 1968. Leadbeater's possum. In *Wildlife in southeastern Australia* (ed. K. Dempster), pp. 23-8. Sydney: Australian Broadcasting Commission.

Woods, J. T. 1956. The skull of *Thylacoleo carnifex*. *Mem.Qd Mus.* **13**:125-40.

# 8 Cave Sediments as Palaeoenvironmental Indicators, and the Sedimentary Sequence in Koonalda Cave

R. Frank

Any discussion of the palaeoenvironmental significance of cave sediments in archaeological sites must in the present context be brief. The analytical approaches and methods mentioned here are far from exhaustive. For a good introduction to the subject the reader is referred to Kukla and Ložek (1958), Schmid (1958, 1963), Warwick (1961), Butzer (1964), and Brain (1967) from which most of the more important literature can be reached.

## Cave Sediments and the Palaeoenvironment

Cave sediments can be as varied, texturally, structurally, mineralogically, and chemically, as almost any other superficial terrestrial deposits and interpreting palaeoenvironments from them is more a study by comparisons with analogous surface deposits than by rigorous deduction from intrinsic characteristics. Fortunately for the archaeologist, humans have avoided most of the depositional environments that occur in caves. The entrance facies type of sediment defined by Kukla and Ložek (1958) is by far the most common in archaeological cave sites. It is composed mostly of talus (breakdown) and local soil-derived material that has been deposited by slope wash or by gravity through creep and other soil movement processes. Wind blown material may also contribute significantly to the deposits.

I gratefully acknowledge the assistance of J. N. Jennings, P. W. Williams, and M. A. J. Williams, Department of Biogeography and Geomorphology, Australian National University, for valuable discussions and for critical reading of the manuscript. R. V. S. Wright, Department of Anthropology, University of Sydney, also gave helpful suggestions while in the field and during subsequent informal discussions. Chloride analyses were done by K. Fitchett, Department of Biogeography and Geomorphology, Australian National University.

The depositional environment, sediment type, sediment producing processes, and degree of preservation, will vary within the entrance facies type of sediment and will be controlled by the morphology of the cave, bedrock type, topographic position of the entrance, topography of the cave floor, and many other factors. As Schmid (1958, 1963) pointed out, the position of the sediment relative to the cave entrance is one of the most important controlling factors. It should be a guide to the analytical methods used and to the level of interpretation of the past environment.

Generally there will be a decrease in the intensity of weathering processes from the entrance towards the interior of the cave. Also, there will be less alteration of the deposits due to human activity since occupation is usually concentrated near the entrance. For these reasons the sediments in the interior will be better preserved in relation to their original chemical and mineralogical constituents.

On the other hand, selective decrease of the allochthonous (externally derived) material towards the interior of the cave makes the interior sediments less representative of the surface source material and consequently of the surface environment. Autochthonous (internally derived) material inherited from the bedrock may also be mixed with the allochthonous material as well as with selected small sized portions from any horizon in the surface soil brought in through small cracks and fissures not related to the main entrance. Hence, palaeoenvironmental interpretations of the interior sediments may use chemical and mineralogical criteria with a fair degree of confidence provided that the different portions can be distinguished.

The sediments deposited nearer the entrance will be more complete in terms of the source material but weathering processes and human activity may have altered them considerably. These sediments can, however, prove useful when textural and other sedimentologic techniques are used to interpret the palaeoenvironment.

### SOME SPECIFIC METHODS

Besides the obvious palaeoenvironmental inferences that can be drawn from a study of the faunal and floral remains preserved in the cave sediments, an analysis of the detrital and precipitated constituents can also be of value. An examination of the gross character of the sediment in terms of the type of soil from which it was derived and a comparison with the present day surface soil may be a clue to the past environment. More detailed analysis of the type and amount of clay minerals and the type and degree of weathering of the larger detrital grains can give information on the soil processes that took place during the formation of the source material.

Schmid (1958, 1963), Sabels (1960), and Martin, Sabels, and Shutler (1961) have attempted trace element analyses of cave sediments, and although the interpretations based on this method depend heavily on the amount and degree of post-depositional alteration of the sediments, the results are often useful when coupled with other evidence and may even indicate a relative age (Sokoloff and Carter 1952).

Angular breakdown with the appearance of fresh fracturing has been interpreted by many European workers (e.g. Lais 1941; Schmid 1958, 1963; Bonifay 1956) as due to shattering by alternate freezing and thawing, and the relative amounts of this frost-produced breakdown and the more rounded breakdown due to chemical solution have been used as indicators of the past environment. However, Renault (1968) has indicated that significant amounts of angular breakdown may result from mechanical adjustment of the bedrock to the cave void. Crystal wedging (e.g. *Salzsprengung*) and hydration may also contribute angular breakdown (Wilhelmy 1958) as may even simple impact of falling rock. Angular breakdown can also be rounded by post-depositional processes.

The presence of flowstone in the sedimentary sequence has also been used as a climatic indicator, but there is as yet no agreement as to whether it signifies a cold or warm period. Most workers agree that it means at least increased wetness, whether local, seasonal, or regional, but even this is not necessarily true. The rate of sedimentation could be fast enough to preclude chemical precipitation regardless of how much water is present, especially, as is often the case, when the sediment is a fairly impermeable clay.

Palaeotemperature information may be gained through a study of the type of mineral comprising the secondary precipitate. The proportionate occurrence of calcite and aragonite was used by Moore (1956) as a palaeothermometer but, as Curl (1962) pointed out, the relationship between the two polymorphs is not due to temperature alone but depends on many other factors, such as amount and kind of additional ions and organic molecules. Following laboratory experiments on the precipitation of calcium carbonate, Siegel and Reams (1966:246) have even stated that, 'temperature apparently does not affect the polymorph formed in cave environments'. Recent pioneer work by Hendy and Wilson (1968) on $^{18}0/^{16}0$ indicates that this method may, in fact, be more definitive in establishing past absolute temperatures.

Other methods that have not been applied widely enough to cave deposits to assess their value, such as palaeomagnetism (Reams 1968), porosity measurements (Brain 1958), $^{13}C/^{12}C$ measurements (Galimov and Grinenko 1965), may eventually prove useful in archaeological cave sites in the future, depending on the particular conditions.

PALAEOENVIRONMENTAL STUDIES OF ARCHAEOLOGICAL CAVE SEDIMENTS IN AUSTRALIA

Work on palaeoenvironmental interpretations of cave sediments in archaeological sites in Australia has been limited, but the investigations by Walker (1964), Twidale (1964), Mulvaney, Lawton, and Twidale (1964), and Joyce (in Mulvaney and Joyce 1965) have produced interesting and worthwhile results.

Walker's analysis of the sediments in sandstone shelters and the associated surface soils in eastern New South Wales shows that even the sediments in shallow shelter caves can differ texturally from the associated surface soils forming at the same time. His work also suggests the influence of human occupation on soil development processes and he was able to make tentative correlations of the cave sediments with K-cycles (Butler 1959) evidenced in the surrounding surface soils.

Twidale's and Mulvaney, Lawton, and Twidale's work on the calcarenite shelters along the Murray River in South Australia correlated stages of human occupation with increased sedimentation rates which resulted in stepped erosion of the floor of the shelters. Joyce used methods similar to those of Schmid (1958, 1963) to analyse the sediments in sandstone shelters in eastern Queensland and arrived at the interpretation of an arid period 5,000 to 10,000 years ago.

**Koonalda Cave Sediments**

The investigations in Koonalda Cave have provided the first opportunity for the application of some of the principles and methods previously outlined to the context of an archaeological site not in a shallow shelter, but well into the environment of a large and deep cave (Fig. 9:2). For a more complete account of this work, including tables of data and illustrations, the reader is referred to Frank (n.d.).

Jennings (1961, 1962, 1963, 1967a, 1967b) has ably discussed the geomorphology of the Nullarbor Plain and its caves. The following brief notes on the geology and geomorphology and on the origin of the deep caves are taken almost entirely from his work.

The 195,000 km$^2$ that make up the geomorphic unit of the Nullarbor Plain is largely covered by the Lower Miocene Nullarbor Limestone, a hard, crystalline, well-jointed biosparite which is 15 to 30 m thick and is essentially flat lying. Beneath the Nullarbor Limestone is the Upper Eocene Wilson Bluff Limestone, a friable, highly permeable, thick-bedded biomicrite which has a thickness of over 200 m in the centre of the Plain. It is also horizontally bedded.

From the 50 m high sea cliffs along the southeast coast, the Nullarbor Plain rises gradually at about 1 in 5,000 to reach an altitude of about 200 m at the northwest edge where the limestone meets older rocks. Local relief is slight and consists primarily of straight, parallel, shallow troughs with intervening low rises which have a wavelength of several kilometres and an amplitude of a few metres. 'Claypans', circular depressions a kilometre in diameter and up to a few metres deep, are also numerous throughout the Plain. This monotonous landscape is occasionally punctured by blowholes (vertical shafts 1 to 2 m in diameter and 7 to 10 m deep which generally open up into small, one-room caves), and large collapse dolines such as the entrance to Koonalda Cave, which occasionally lead to deep caves.

The climate in the vicinity of Koonalda Cave is arid to semi-arid with a mean annual temperature of about 17°C and with a seasonal range of about 20°C. Occasional frosts occur and a few days reach a temperature above 38°C. The rainfall is about 200 mm and is concentrated in the winter months. Evaporation is approximately 2,000 to 2,500 mm annually and exceeds rainfall in every month of the year. Relative humidity is never high and varies between about 25 and 50 per cent.

#### ORIGIN OF KOONALDA CAVE

Koonalda Cave is one of seventeen known deep caves in the Nullarbor Plain which are conspicuous by their large size and are developed mainly in the Wilson Bluff Limestone (Plates III, IV). Jennings (1967a) concluded that these deep caves were probably formed as 'master caves' according to the theory of Glennie (1954). That is, they were formed by solution in the zone of a fluctuating water table after initial deep phreatic cavities had been partly drained.

#### DEPOSITS IN KOONALDA CAVE

*Clastics.* Autochthonous material in the form of breakdown is by far the most abundant of the deposits in the cave, though soil-derived material does occur in considerable quantity in a number of different situations throughout the cave. The floor of the entrance doline is composed of breakdown ranging in size from a few metres in diameter to silt-sized calcite. This is partially covered with recent soil. There is also an indurated, calcite-cemented, soil-derived sediment that fills cavities in the Nullarbor Limestone and which is exposed in the walls of the doline. This sediment is texturally and mineralogically similar to the present surface soil.

Within the cave, breakdown covers almost all of the floor, although an area near the beginning of the north passage is floored with recent soil-derived material washed in during flood times. Older soil-derived sediment also fills bedding plane cavities truncated by the cave walls, within the Wilson Bluff Limestone, and this material is partly cemented with calcite. One of these cavities occurs directly above the archaeological excavations. Sparsely distributed phreatic tubes within the Wilson Bluff Limestone are filled with clay.

*Precipitates.* Speleothems are rare in the deep caves developed within the Wilson Bluff Limestone. The majority of those that are present are composed of sulphates and chlorides. In Koonalda there are no decorative carbonate speleothems and the only carbonate precipitate present is small amounts of calcite cement. Gypsum occurs throughout the cave as gypsum flowers and as crusts on the walls, and cements some of the silt-and sand-sized breakdown on the floor. Glauber's Salts (hydrous sodium sulphate) have also been reported by King (1950) as occurring on the floor in Koonalda.

### SURFACE SOILS

The soils of the Nullarbor Plain have been only briefly mentioned by pedologists and few soil analyses are available. Crocker (1946:45) mapped them as desert loams but concluded that the Nullarbor Plain and its associated soils have very close affinities with the solonised brown soils, and are probably better considered as such. The analysis of a surficial sample of the soil from near Koonalda Cave by Prescott and Skewes (1938) showed a high percentage of total soluble salts. They also report that their samples were highly alkaline. Teakle (1938:174) described the Nullarbor Plain soils as, 'pinkish brown calcareous material of a loamy to clayey texture . . . '. His analyses of four surface samples taken near Forrest, 150 km northwest of Koonalda, showed high calcium carbonate and sodium chloride with pH ranging from 7·94 to 8·88. Northcote *et al.* (1968) mapped the Nullarbor Plain soils in the vicinity of Koonalda as shallow calcareous loams.

### DESCRIPTION OF THE SEDIMENTS EXPOSED BY THE EXCAVATIONS

The white unit (0 to 1·80 m). Chiefly white breakdown composed of Wilson Bluff Limestone with small amounts of clay and chert (Plate V). The top 2·5 to 10 cm is partly cemented by gypsum. The size ranges from silt to blocks up to 30 cm in diameter. The larger blocks become more abundant towards the centre of the passage. Thin, pink lenses of breakdown with small amounts of well disseminated admixed clay dip at various angles up to 20° and in varying directions, becoming more abundant towards the centre of the passage. The direction and angle of dip is determined by breakdown landing on larger tilted blocks. Sparsely scattered small lumps of red, silty clay up to 2·5 cm in diameter are distributed randomly throughout the breakdown. Three radiocarbon dates on charcoal from hearths in this unit were 15,850 ± 320 BP (ANU-70), 19,300 ± 350 BP (ANU-71) (Polach *et al.* 1968), and 22,000 ± 700 BP (ANU-245) (Wright, see ch. 9).

The intermediate zone (1·80 to 1·95 m). This unit is composed of lenses of white to pink breakdown interbedded with lenses of red silty clay ranging in thickness from a few millimetres to a few centimetres. The red, silty clay lenses are more abundant towards the centre of the passage.

The red unit (1·95 to 5·90 m). This unit is composed of red, sandy gravel to silty clay, soil-derived material with small amounts of white, Wilson Bluff Limestone breakdown (Plates V, VI). Seventy-five to 100 individual graded beds up to 30 cm thick occur as lenses with length-width ratio of 50-100:1. The silty clay between the graded beds is laminated (dark and light red) and occasionally contains desiccation cracks. The Wilson Bluff Limestone breakdown is sparsely distributed throughout as thin lenses of silt a few millimetres thick and as larger individual sand, pebble, cobble, and boulder sized pieces. The bedding of the red material is distorted around larger blocks of limestone breakdown. Slight cementation by calcite occurs about midway through the red unit. There are sparsely distributed small animal bone fragments throughout. The contact between the red unit and the underlying Wilson Bluff Limestone breakdown is irregular with a total relief of 1 m.

Radiocarbon dates on this unit establish that it was deposited about 20,000 years ago.

The basal breakdown (5·90 to 6·80 m). This unit is composed of Wilson Bluff Limestone breakdown with red, soil-derived material filling interstices between the large limestone blocks. The top of the unit dips steeply towards the cave wall. A similar sequence of sediment is exposed in the 4·5 m deep Trench I, a few metres west of the described section.

Surface soils. The soil is a pink to orange-red, silty clay with abundant calcium carbonate nodules throughout but slightly concentrated at 60 to 90 cm. The top 2·5 to 5 cm is light brown with roots and other organic matter.

## Methods and Results of Analyses

METHODS

*pH determinations.* These were made with an N.L. Jones model B pH electrometer, with calomel half cell and Type A glass electrode model 100. A soil-water mixture consisting of 2·5 grams of sample and 12·5 ml of distilled water was shaken for 30 minutes in a Baird and Tatlock flask shaker, poured into small plastic containers suitable for use in the pH electrometer and let stand for 30 minutes. The mixture was then introduced into the electrometer, which had been calibrated with pH4 and pH7 buffer solutions, and the pH was read directly after 1 to 3 minutes of equilibration. The electrometer was checked for drift and recalibrated with pH7 buffer solution after every third or fourth sample. The mean drift was found to be less than 0·025 pH units per sample, so no correction factor was employed since the electrometer can only be read to 0·05 pH units.

The pH of the five surface soil samples ranged from 9·25 in the top 5 cm to 8·15 at — 105 cm. There was a steady decrease in the pH with depth. The present day soil-derived material in Koonalda Cave had a pH of 8·20. The pH of fourteen samples from the red unit ranged from 8·80 at the top to 9·40 at the bottom. There was slight fluctuation through the sequence but a general increase with depth.

*Chloride and total soluble salts.* These were determined by the Mohr titration method and the conductivity method respectively, as outlined by Metson (1956:142-7). In both analyses a 1:5 soil:water extract was used. Chloride and total soluble salts were high in both the surface soil and the cave sediments. In the three surface soil samples tested the chloride was 335, 5,150, and 6,200 ppm at the top, −45 cm and −105 cm respectively. The corresponding total soluble salts were 1,200, 13,500, and 17,300 ppm. The present day soil-derived material in the cave showed a chloride value of 4,935 ppm and total soluble salts were 19,000 ppm. Three samples of the red unit at the top, −195 cm, and −390 cm, gave chloride values of 4,675, 1,740, and 1,670 ppm. Total soluble salts for the same samples were 10,700, 4,800, and 4,800 respectively.

*X-ray analysis of the clays.* The less than 2 micron fraction was prepared for X-ray analysis using the method of carbonate extraction described by Jonas (1959) and Horn (1967). The samples were analysed as oriented slides on a Philips model PW1051 X-ray diffractometer, using Ni filtered CuK$\alpha$ radiation through slits of $\frac{1}{4}° — 0\cdot2° — \frac{1}{4}°$.

X-ray diffraction analysis of the clays showed the surface soils and cave sediments to be identical in that they all contained b-axis disordered kaolinite and illite (1M and 2M polymorphs, at least). The Nullarbor Limestone contained kaolinite (form not determined), montmorillonite, and illite (polymorph not determined). The Wilson Bluff Limestone contained montmorillonite and 2M illite. Visual inspection of the diffraction peaks showed no change in the amounts of each clay type through the surface soil and red unit sequences.

*Thin sections.* Thin sections of selected samples of the surface soil, the present day soil-derived material in the cave, and the material in the red unit show that they are all similar and are made up of individual soil particles composed of clay and quartz silt cemented with calcite. The soil particles in the cave sediments have been slightly rounded during their transportation into the cave, but apparently no other physical changes have occurred.

A point count of 100 detrital grains in each thin section and of grains extracted from the Nullarbor and Wilson Bluff Limestones showed quartz ranging from 88 to 100 per cent. Most of the grains were single crystals with straight extinction. About one-fifth showed undulose extinction. Few grains had inclusion other than vacuoles and these were generally not abundant. The grains were generally round to subround and embedded grain argillans were observed on 75 to 80 per cent of the grains in the surface soil and cave sediment. Median size of the quartz grains ranged from 0·05 to 0·07 mm, sorting ($\sigma_G$) was from 0·67 to 0·87 ø (moderately to moderately well sorted). There was no significant difference between the quartz grains in the surface soils and in the cave sediments in respect of size, sorting, morphology, and grain argillans.

#### MODE OF DEPOSITION OF THE CAVE SEDIMENTS

The sedimentary sequence in Koonalda Cave represents simultaneous accumulation of two different sediment types whose deposition rates were drastically different. The steady, continual accumulation of the breakdown contrasts with the sporadic dumping of soil-derived material brought into the cave by a periodically shifting, ephemeral stream.

The breakdown. The mechanism of emplacement of the breakdown in the Nullarbor Plain deep caves has been discussed by Jennings (1961), Lowry (1964), and Hunt (1966). Hunt considered the breakdown in Mullamullang Cave to be due to collapse of the ceilings of underlying cavities. Lowry (referring to Cocklebiddy Cave) and Jennings suggested that the breakdown is due to structural adjustment of the weak Wilson Bluff Limestone to produce an arched cross-sectional form that is mechanically stable. Lowry also suggested that the breakdown followed removal of the hydrostatic support resulting from a lowering of the water table. Davies (1960) and more recently Renault (1968) have indicated that mechanical adjustments of the limestone surrounding a cavity will result in breakdown formation.

It therefore seems probable that the formation of most of the breakdown is due to adjustment of the galleries towards a mechanically stable geometric form. Hunt (1966) also pointed to solutional enlargement of the walls of Mullamullang Cave at the level of the lake surface. The implication is that this will produce additional structural instability due to the change in cross-sectional form and result in more breakdown.

Recent evidence of crystal wedging of limestone in the Nullarbor Plain caves suggests that at least some of the breakdown (especially the more recent material) has been assisted by this process. Lowry (1964, 1967) and Wigley and Hill (1966) showed crystal wedging to be operating in some of the Nullarbor Plain caves. Crystal wedging of chert by gypsum first noted by Bridges (pers. comm.), indicates that this process is taking place in Koonalda Cave. Gypsum crusts on the walls of the cave and on the back side of partly displaced sections of the wall with horizontal slickensides on their tops also suggest that crystal wedging is affecting the limestone walls and ceilings.

The white unit exposed by the archaeological excavation has incorporated in it lumps of red, silty clay as well as smaller red clay particles disseminated through the breakdown. This clay was probably derived from the bedding plane cavity fill which occurs in the ceiling directly above the excavation and was re-deposited contemporaneously with the breakdown.

The top few centimeters of the white unit have been partly cemented by gypsum. This cement could have resulted from water entering the cave by infiltration through the limestone. The absence of additional soil-derived material on top of the white unit precludes precipitation of the cement from water entering as a stream via the entrance.

The soil-derived sediment. The red unit has a thickness of 3·5 to 4 m at Trench III (Plates V, VI) but a few metres west, at Trench I, the thickness is only 30 to 60 cm. The top of the red unit in both trenches is at the same level. It thus appears that the stream bringing in the sediment was not through-flowing, but probably came to an end and was ponded at the west side by the breakdown in the northwest passage.

The size of this pond is indicated by the near pinch-out of the red unit exposed in Trench I and the cut and fill structures in the red unit exposed in Trench III. The velocity of the water bringing in the red material was sufficient in the vicinity of Trench III to erode the previously deposited sediments, and so the southeast edge of the pond must have been within a metre or two of the present location of Trench III. The total size of the pond then was probably about 15 m in diameter. The desiccation cracks in the laminated silty clays between graded beds show that the pond intermittently dried up—probably more as a consequence of water escaping through the extremely permeable breakdown substrata than of evaporation.

It is not necessary to invoke any climatic or other external change to account for the arrival and departure of the stream supplying the pond in the vicinity of the archaeological excavations. The stream must have flowed across or through about 100 m of breakdown before reaching the excavation area. Its course could have been easily altered by slight shifts or new falls of breakdown, of which there is adequate evidence. The stream itself may have aided in this course change by sedimenting part of its path through the breakdown on occasions when it did not have sufficient water to flow all the way to the excavation area.

The intermediate zone between the red and white units, where the two sediment types are about equally represented, suggests that at this time the stream was only occasionally reaching the excavation area and that most of its water had already been directed along another route—probably into the northwest passage.

PALAEOENVIRONMENT

The material of the white unit cannot be used for palaeoenvironmental interpretations with any degree of dependability or level of precision since it has never been exposed to the surface environment. The soil-derived material of the red unit exposed by the excavations was deposited during a comparatively short time, so palaeoenvironmental inferences from it are restricted to that period of time.

There is a marked similarity in the mineralogy and grain morphology of the present-day surface soils and the red unit material, indicating that the surface soil from which the red unit was derived 20,000 years ago was much the same as the present day surface soil. Whether or not the present surface soil is a relic is an academic point, for it is an arid type of soil which is concordant with the present climatic conditions in the area. According to Stace *et al.* (1968), desert loams occur where the rainfall is generally less than 250 mm and solonised brown soils where the rainfall is generally from below 250 mm to about 430 mm. The present surface soils at Koonalda are definitely closer to desert loams than solonised brown soils despite Crocker's (1946) generalisation.

There is little chance that the red unit sediments could have been altered after deposition in the cave to attain their present morphologic and mineralogic characteristics. Kukla and Ložek (1958) and numerous other authors have implied that material deposited in caves is generally altered very little after its deposition. According to the radiocarbon dates, the red unit was buried shortly after its deposition and the limits of cementation at the top of the overlying white unit indicate that subsequent percolating water capable of precipitation did not reach the red unit. Except for the slight calcite cementation near the middle of the section, there are no textural additions to the red material such as overgrowths or other illuviation phenomena.

Thus, on the basis of the mineralogy and texture of the red unit there is no evidence for a substantial difference between the environment 20,000 years ago and the present environment.

**References**

Bonifay, E. 1956. Les sédiments détritiques grossiers dans les remplissages des grottes. *L'Anthropologie, Paris* **60**:447-61.

Brain, C. K. 1958. The Transvaal ape-man-bearing cave deposits. *Transv.Mus.Mem.* No. 11.

——1967. Procedures and some results in the study of Quaternary cave fillings. In *Background to evolution in Africa* (eds. W. W. Bishop and J. D. Clark), pp. 285-301. Chicago, London: University of Chicago Press.

Butler, B. E. 1959. *Periodic phenomena in landscapes as a basis for soil studies.* Soil Publ. CSIRO Aust. No. 14.

Butzer, K. W. 1964. *Environment and archaeology.* Chicago: Aldine Publishing Co.

Crocker, R. L. 1946. Post-Miocene climatic and geologic history and its significance in relation to the genesis of the major soil types of South Australia. *Bull.Coun.scient. ind.Res., Melb.* No. 193.

Curl, R. L. 1962. The aragonite-calcite problem. *Bull.natn.speleol.Soc.* **24**:57-73.

Davies, W. E. 1960. Origin of caves in folded limestone. *Bull.natn.speleol.Soc.* **22**:5-18.

Frank, R. n.d. The sediments in Koonalda and Weekes Caves, Nullarbor Plain, South Australia. In *Archaeology of the Gallus site* (ed. R. V. S. Wright), in press. Australian Aboriginal Studies No. 26 (Prehistory and Material Culture Ser. 5). Canberra, Australian Institute of Aboriginal Studies.

Galimov, E. M. and Grinenko, V. A. 1965. Age effect in the isotopic composition of carbon of stalactite annulations of the Crimea Highlands. *Geokhimiya* 1965 (6): 661-7.

Glennie, E. A. 1954. The origin and development of cave systems in limestone. *Trans. Cave Res.Grp Gt Br.* **3**:75-83.

Hendy, C. H. and Wilson, A. T. 1968. Palaeoclimatic data from speleothems. *Nature, Lond.* **219**:48-51.

Horn, D. R. 1967. Recent marine sediments and submarine topography, Sverdrup Islands, Canadian Arctic Archipelago. Ph.D. thesis, University of Texas, Austin, Texas (unpublished).

Hunt, G. 1966. The geomorphology of Mullamullang Cave and its significance for the deep caves of the Nullarbor. Paper presented at the Sixth Conference of the Australian Speleological Federation, Mirboo North, Victoria (unpublished).

Jennings, J. N. 1961. *A preliminary report on the karst morphology of the Nullarbor Plains.* Cave Exploration Group (South Australia) Occasional Papers No. 2.

——1962. The limestone geomorphology of the Nullarbor Plains (Australia). In *Actes du deuxième Congrès International de Spéléologie.* Vol. 1, pp. 371-86. Castellana-Grotte: Instituto Italiano di Speleologia.

——1963. Some geomorphological problems of the Nullarbor Plain. *Trans.R.Soc.S.Aust.* **87**:41-62.

——1967a. Some karst areas of Australia. In *Landform studies from Australia and New Guinea* (eds. J. N. Jennings and J. A. Mabbutt), pp. 256-92. Canberra: Australian National University Press.

——1967b. The surface and underground geomorphology. In *Caves of the Nullarbor* (eds. J. R. Dunkley and T. M. L. Wigley), pp. 13-31. Sydney: The Speleological Research Council Ltd.

Jonas, E. C. 1959. Clay-size minerals in Ellenburger rocks. In *Stratigraphy of the Pre-Simpson Paleozoic subsurface rocks of Texas and southeast New Mexico* (eds. V. E. Barnes *et al.*), Vol. 1, pp. 131-43. Univ. Tex.Publs No. 5924 (Bureau of Economic Geology, University of Texas).

King, D. 1950. Geological notes on the Nullarbor cavernous limestone. *Trans. R.Soc.S.Aust.* **73**:52-8.

Kukla, J. and Ložek, V. 1958. K problematice výzkumu jeskynních výplni. *Ceský Kras* **11**:19-83.

Lais, R. 1941. Über Höhlensedimente. *Quartär* **3**:56-108.

Lowry, D. C. 1964. The development of Cocklebiddy Cave, Eucla Basin, Western Australia. *Helictite* **3**:15-19.

——1967. The origin of blow-holes and the development of domes by exsudation in caves of the Nullarbor Plain. Rec.geol. Surv.West.Aust. No. 1967/12 (mimeographed).

Martin, P. S., Sabels, B. E., and Shutler, R. 1961. Rampart Cave coprolite and ecology of the Shasta ground sloth. *Am.J.Sci.* **259**:102-27.

Metson, A. J. 1956. Methods of chemical analysis for soil survey samples. *Soil Bur. Bull.N.Z.* No. 12.

Moore, G. W. 1956. Aragonite speleothems as indicators of paleotemperature. *Am.J.Sci.* **254**:746-53.

Mulvaney, D. J. and Joyce, E. B. 1965. Archaeological and geomorphological investigations on Mt. Moffatt Station, Queensland, Australia. *Proc.prehist.Soc.* **31**: 147-212.

——, Lawton, G. H. and Twidale, C. R. 1964. Archaeological excavation of rock shelter no. 6, Fromm's Landing, South Australia. *Proc.R.Soc.Vict.* **77**:479-516.

Northcote, K. H. *et al.* (eds.) 1968. *Atlas of Australian soils.* Sheet 10. Central Australia, with explanatory data. Melbourne: Melbourne University Press.

Polach, H. A., Golson, J., Lovering, J. F., and Stipp, J. J. 1968. ANU radiocarbon date list II. *Radiocarbon* **10**:179-99.

Prescott, J. A. and Skewes, H. R. 1938. An examination of some soils from the more arid regions of Australia. *Trans.R.Soc. S.Aust.* **62**:320-41.

Reams, M. W. 1968. Cave sediments and the geomorphic history of the Ozarks. Ph.D. thesis, Washington University, St Louis, Missouri (unpublished).

Renault, Ph. 1968. Contribution à l'étude des actions mécaniques et sédimentologiques dans la spéléogenèse. Deuxième partie. Les effets mécaniques à l'échelle de la cavité. *Annls Spéléol.* **23**:259-307.

Sabels, B. E. 1960. Trace element studies on cave strata. *Nevada St.Mus.Anthrop.Pap.* No. 3:17-23.

Schmid, Elizabeth. 1958. *Höhlenforschung und Sedimentanalyse. Ein Beitrag zur Datierung des alpinen Paläolithikums.* Basel: Schriften des Institutes für Ur- und Frühgeschichte der Schweiz, No. 13.

——1963. Cave sediments and prehistory. In *Science in archaeology* (eds. D. Brothwell and E. Higgs), pp. 123-38. London: Thames and Hudson.

Siegel, F. R. and Reams, M. W. 1966. Temperature effect on precipitation of calcium carbonate from calcium bicarbonate solutions and its application to cavern environments. *Sedimentology* **7**:241-8.

Sokoloff, V. P. and Carter, G. F. 1952. Time and trace metals in archaeological sites. *Science, N.Y.* **116**:1-5.

Stace, H. C. T. *et al.* 1968. *A handbook of Australian soils.* Glenside, South Australia: Rellim Technical Publications.

Teakle, L. J. H. 1938. A regional classification of the soils of Western Australia. *J.Proc.R.Soc.West.Aust.* **24**:123-95.

Twidale, C. R. 1964. Effect of variations in rate of sediment accumulation on a bedrock slope at Fromm's Landing, South Australia. *Z. Geomorph.Suppbd* **5**:177-91.

Walker, P. H. 1964. Soil and landscape history in the vicinity of archaeological sites at Glen Davis, New South Wales. *Rec.Aust. Mus.* **26**:247-64.

Warwick, G. T. 1961. Cave deposits and palaeoclimatology. In *Atti del Symposium Internazionale di Speleologia* Varenna, 1960 (ed. S. Dell'Oca), Vol. 1, pp. 127-51, 182-6. Como.

Wigley, T. M. L. and Hill, A. L. 1966. Cave decoration. In *Mullamullang Cave expeditions 1966* (ed. A. L. Hill), pp. 37-8. Cave Exploration Group (South Australia) Occasional Papers No. 4.

Wilhelmy, H. 1958. *Klimamorphologie der Massengesteine.* Braunschweig: Georg Westermann Verlag.

# 9 The Archaeology of Koonalda Cave

R. V. S. Wright

Generalised maps of the Australian natural environment show the desert conditions of the interior reaching south to touch the southern coastline of the continent in only one place, along the cliffs of the Great Australian Bight, north of which is spread the Nullarbor Plain. Koonalda Cave lies on this limestone plain 23 km from the sea.

What these generalised maps do not show is the ameliorating effect that the sea has on a belt reaching some 16 km from the coast. Compared with the inland, this coastal belt receives more rainfall and, in the summer, reaches lower temperatures. Trees and shrubs flourish and the bush still maintains numerous marsupials.

This is, however, a luxuriant environment only when it is compared with the contiguous Nullarbor Plain. The rainfall is unreliable, averaging 25 cm a year and with a winter peak. This rain rapidly escapes from the surface into the limestone karst. Although some surface water is retained for a few days on scattered claypans and for longer in the infrequent rock holes, water cannot be found in holes in the beds of dried up rivers for there are no rivers or creeks on the Nullarbor; neither are there any springs.

The mean maximum temperature for January is about 25·5°C. Travellers' tales emphasise the scorching day temperatures of summer, which result from northerly winds. However, since the prevailing winds in summer are southerly, such days are less frequent than the cool ones when the wind blows off the Bight. In June the mean maximum temperature is about 7°C and frosts can occur. Annual evaporation, a measure of humidity, is around 152 cm.

Apart from its lower rainfall, this coastal belt has a climate not unlike that of Adelaide. Some 83 per cent of South Australia has an average annual rainfall of under 25 cm a year, while 50 per cent of the state has a rate of annual evaporation greater than 254 cm. We can concede, therefore, that the area under discussion is one of the damper parts of South Australia. Even so, I must emphasise two factors important in the human environment—the dearth of surface water and the rapidly increasing severity of the climate northwards.

The Nullarbor Plain is virtually treeless, as the translation of its Latin name implies. But the coastal belt is by no means so. Between the coast and the cave there is a scrub, dense in parts, with various species of eucalyptus and acacia. Near the coast these are dwarfed and trained by the prevailing southerly winds. Towards the cave there is a progressive reduction in the number and density of species until at the cave itself the most conspicuous plants are bluebush and saltbush. North of the cave the vegetation decreases until the treeless portion of the plain is reached in some 10-16 km. Though some fruits, nuts, and leaves would have been available to food collectors the area is comparatively unfavourable in its yield of plant foods and the existence of edible grass grains and roots has still to be demonstrated. (D. Symon pers. comm.)

It appears that the richer vegetation of the coastal belt is due to its proximity to the sea and not to a difference in pedology. Since during the late Pleistocene the sea would have been far to the south, and assuming that other factors remained constant, we must assume that the vegetational environment and its subsidiary effects on animal life would have been far less favourable for man.

**The Ethnography of the Present**

No single and satisfactory ethnography of the area exists, but a somewhat pinched picture can be reconstructed from scattered published and unpublished accounts. W. Williams stated (Curr 1886-7:401) that when the country was first occupied by Europeans in 1872 the population figures were much the same as in 1880, when he wrote his account. If his figures are accepted we can estimate a figure of about one person to 490 km². This figure would relate to the area from the coast to about 65 km inland, which appears to have been the greatest distance that people ventured inland in their day-to-day activities. The Nullarbor Plain itself was uninhabited. A consensus of opinion amongst commentators is that the Aborigines were almost as much penned in by the Plain to the north as they were by the sea to the south.

Apart from Williams there was no other serious observer of these Aborigines, whose comments have survived, except Daisy Bates, who died in 1951. Her writings are dispersed through newspapers and manuscripts and appear to have been largely unread. While she has a reputation for unreliability, from my close reading of her work I judge this to be undeserved, and being wise after the event can only regret that, for whatever reason, her knowledge was left untapped.

Knowledge of the Mirning tribe (centred on Eucla but with territory extending eastwards past Koonalda) died with Daisy Bates. Writing of the period 1912-14, when she lived at Eucla, she says, 'About thirty natives were camped in the vicinity but only one of the Eucla tribe . . . was still living. There were a few whose connections had been Eucla people . . . '. In 1918 she described the inhabitants of the area around the Bight as almost extinct. An undated MS. note appended to a Eucla vocabulary list says, 'the last Eucla Mirning [man] died at Albany S.W. Australia about 1918'.

An ethnographic jig-saw of the area can be assembled, though many pieces are missing. A fuller and documented account will be presented in the forthcoming monograph on Koonalda Cave to be published by the Australian Institute of Aboriginal Studies. Here I will provide merely a summary of the chief points of environmental significance.

The 1:250,000 (Coompana) map of 1967 shows nine rock holes within a radius of 48 km of Koonalda. Though drinkable water could be obtained by digging out and draining the roots of mallee trees, rock holes were the only semi-permanent supply of free surface water. As far as I am aware no study of the yield and permanence of these sources has been made. Since they are randomly scattered solution pockets in the limestone, with no surrounding pointers to their presence, they were mostly missed by the early European explorers, who consequently exaggerated the waterless nature of the area. They are in fact insignificant sources for the demands of white settlement; nevertheless for small bands of hunters and gatherers they were critical. These dispersed and unreliable sources of water must have closely controlled the movements of the Aborigines. Daisy Bates mentions the squeezing effect that periods of drought had on their flexibility of movement, forcing them to visit the larger waterholes as the smaller and shallower ones ran out.

The low population for the area suggests that food was scarce. It is likely, however, that the peculiarities of the surface water resources meant that periods of drought had a more severe pruning effect on the population than they would have had in other arid zones of a more conventional geomorphological nature. In other words, this could be an area where the normal pressures annually exerted on population by the rigours of the summer season bore little relation to the effects of the worst drought of the decade and even less so that of the century.

Williams (in Curr 1886-7:401-2) describes the principal food as wallaby hunted with spear. He also mentions snakes, lizards, and fish speared in the sea. Howitt (1904:761) adds wild dog and native cat, and also mentions berries, quandong nuts, and the baked bark of mallee trees pounded up with white ants. Daisy Bates fills out the picture. Unlike earlier writers, she stresses the seasonal usefulness of various plants and their influence on Aboriginal movement. 'Each season has its own products, so that a person knowing these, and the time of the year when they are at their best, will know where to find the natives.' She repeatedly refers to seafoods. It is worth noting that the low sea level of the late Pleistocene would have meant that in this region neither seafoods nor the belt of coastal vegetation, valuable environmental assets of the present, would have been then available.

#### TECHNOLOGY

The basic cutting edge in the technology of the area was produced from a superb form of silicate rock having the appearance and flaking properties of European flint. Its source was in the sea cliffs at Wilson Bluff, 13 km northeast of Eucla, where nodules lie in seams in the chalky limestone. Alluding to its value in trade for the Mirning, Bates describes this quarry as 'one of their great hereditary commercial assets'. The same seams outcrop in the deep caves, such as Koonalda, but there is no evidence that the Mirning exploited these.

This abundance of excellent flint was not associated with an elaborate stone industry. Only two types have been described for the Mirning—a hand-held knife and a mounted adze flake. Judging by the surface scatters of flints we examined, the latter took the form of a tula (cf. Fig. 9:1). Gum for hafting was obtained from various plants. In confirmation of these statements on the stone industry,

*Fig. 9:1 Selected specimens of tula adze flakes, from Ingaladdi, N.T. a, newly trimmed specimen; b, a flake worn through use and re-trimming of the working edge.*

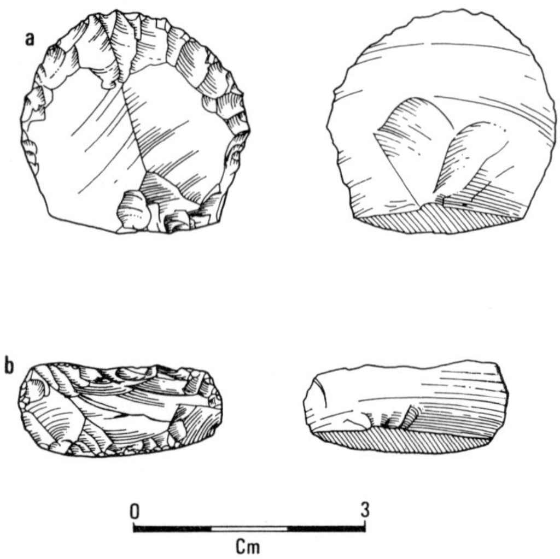


except for a few thumb-nail scrapers and a backed blade (cf. Fig. 10:3), I have not seen any other type of stone implement among surface collections of the area.

We can conclude that the technology and material culture of the Mirning had objects normally found with Aborigines of the central desert areas. Other artifacts included spears, spearthrowers, boomerangs, clubs, shields, wood and bark vessels, woven fur string, bunches of feathers, ochre, pearl shell, and torches for transporting fire. Some of these objects were all or partly traded in from surrounding tribes. Bates gives the meaning of a few simple artistic design elements but although she does not mention rock art, hand stencils are known from many rock shelters in the area.

EXTERNAL CONTACTS

Aborigines apparently moved freely in the coastal belt between Eucla and the Head of the Bight. Koonalda was thus within a well-traversed zone. Movement on a north-south axis was on the contrary very restricted. Such as took place was not for day-to-day economics but due to the more occasional needs of trade and ceremony. Writers on the Aborigines agree that the vast area of the plain was unused and uninhabited by the Aborigines. The general truth of these assertions is supported by accounts of white explorers of the last century.

According to Bates, 'the southern tribes did not know the water holes of the northern edge, nor did the northern tribes know of the rock holes in the southern portion . . . '. There was a route crossing the plain which Bates claimed connected Boundary Dam, to the north, with Eucla and the Head of the Bight. It passed near Koonalda Cave, while another track crossed to Ooldea. However, Bates says of these routes that the Aborigines' 'agents and messengers traversed these, never the entire local groups. The rest of the Plain was a "No Man's Land" of treeless waterless Plain, over which there is not even legendary record of any native traversing'.

These evidences of Aboriginal movements suggest that we should consider the prehistoric use of Koonalda Cave in the context of settlement of the southern coastal belt rather than of the Plain itself, or areas of the Great Victoria Desert further north. This sandy desert was undoubtedly occupied, but its people were only in sporadic and specialised contact with the southern coast.

When we consider the geomorphology of the late Pleistocene, it is evident that a walk to the coast from the cave was not a day's trip across 23 km of scrub, as it is now. It would have been across more than 160 km of flat land now submerged by the shallow waters of the Bight.

This brief summary of what is in any case a very unsatisfactory ethnography of the Mirning tribe is enough to show that the foundation of the contemporary ecology is

*Fig. 9: 2 Section through Koonalda Cave*

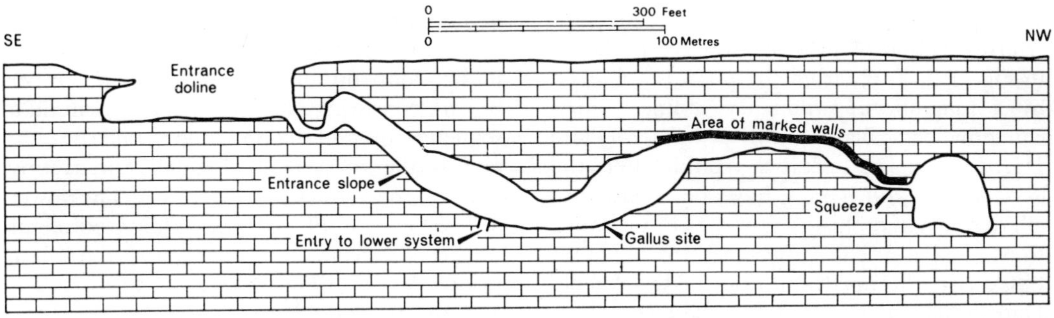


simply that the sea maintains the coastal belt, and the coastal belt maintains man. With the sea removed by eustatic lowering it seems probable that the present coastal belt would resemble the uninhabited and inhospitable plain to the north. Yet evidence from Koonalda Cave shows that man was there. It is tempting, therefore, to look to increased rainfall in the past to explain his presence.

Before accepting this easy explanation, however, two reservations are necessary. Human settlement patterns may have been different. Perhaps the general area was not regularly occupied and Koonalda Cave was a place of occasional resort for special purposes. There is also the implication of the now submerged coastal plain as additional territory in the late Pleistocene.

## Koonalda Cave

What evidence of purpose do we have in the use of Koonalda Cave? What was the form and history of the late Pleistocene coastal plain? The first scientific account of the cave appears to be that of Wells and Hunt (1919) who in 1904 examined and reported on the salinity of its underground lakes. Graffiti of dates and names indicate that the cave was subsequently the resort of casual explorers. There is no hint that it was ever considered to have an archaeological potential—indeed this is likely to have remained hidden to this day had it not been for the archaeological work of Dr Alexander Gallus, who first excavated there in 1956. It is fair to say that most archaeologists, accustomed as they are to the value of deposits in rock shelters, would not have given a second thought to the deposits in the floor of Koonalda Cave, which lie at the base of a precipitous 60 m descent underground, and in darkness (Fig. 9:2). Dr Gallus considered that he had found artifacts dating from the late Pleistocene, and in the face of considerable scepticism organised four further expeditions to the site.

A gradual vindication of his views led the Australian Institute of Aboriginal Studies to send an expedition to investigate the cave in 1967. This was led by myself and included Dr Gallus and a group of specialists in art, fauna, flora, and sediments. However, the vindication of Dr Gallus's broad claims for the site has not been based on data with which he would always agree, nor are the details of our conclusions the same. What I am giving here is my interpretation, and I outline one radical disagreement below, in dealing with the typology of the flint artifacts. The cave is now a 'prohibited area' under the South Australian legislation for the protection of relics—a fitting climax to Dr Gallus's persistence.

FORM OF THE CAVE

A cross-section of part of the cave is shown in Fig. 9:2. Entry from the plain is achieved by a magnificent crater-like doline, which can now be scaled only by a ladder (Plate III). Slight falls of rock could alter the accessibility and it may have been easier to climb down in prehistoric times. The entrance to the cave itself is by an inconspicuous passage in the northwest wall of the doline. Dim light is available down the steep slope to the 'floor' of the cave, immediately beyond which is the main area of excavation—or the 'Gallus site'. This area is in almost complete darkness due to the low ceiling immediately behind forming a final light-trap. Traces of prehistoric human activity extend beyond here to a total distance of 275 m into the cave, and include the extensive area of wall markings referred to by Edwards (see ch. 24) and illustrated in Plates XXIII and XXIV. This is the maximum distance possible in the upper systems of the cave, where the 'squeeze' comes out on a platform in the ceiling of a large chamber.

A passage leading to the lower cave system starts near the base of the main entrance slope. This system still receives rainwater and sediments from the surface; it includes saline lakes. It may have been used as a source of water in prehistoric times. As no excavations have been carried out there, I shall not refer to it again.

THE GALLUS SITE (Plate IV)

This is an area against the southwest wall of the cave. It is a surface relatively free of massive rocks, though it shows evidence of the fall of flint nodules from seams in the wall above. Dr Gallus had concentrated his excavating efforts here, and my purpose in excavating a column alongside his trench was to get a duplicate collection of artifacts and to enable excavations to be safely taken to a greater depth in the enlarged area.

Frank describes the nature of the deposits in his contribution (see ch. 8). The upper 2·1 m, consisting of chalky rubble, contain a sequence of flint flaking floors and hearths and is the most technologically interesting part of the sequence (Plate V). The lower 3·7 m consist of red water-lain sediments washed in from the surface of the plain (Plates V, VI). They contain a few artifacts and animal bones.

Miss Anne Bermingham, Institute of Applied Science of Victoria, and Mr Henry Polach, Australian National University, have given generously of their time and interest in radiocarbon dating charcoal from various positions within the excavations in order to clear up ambiguities within the sequence. To present the actual dates obtained, without a detailed discussion, would do their solicitude an injustice, and this discussion is included in the forthcoming monograph.

Suffice it here to say that on the basis of the radiocarbon dates the cave was used between some 15,000 and 22,000 years ago. There is no reason to think that the stratigraphically earliest deposits found at the Gallus site actually mark the first traces of human activity within the cave. We were prevented from going deeper by what seemed to be massive rock fall, rather than rock-bottom. Similarly there is no reason for thinking that the cave was not in use more recently than the time mentioned, although there are reasons for thinking that use of the cave was abandoned a few thousand years ago.

THE NATURE OF THE ARTIFACTS

Dr Gallus has published interpretations of the artifacts recovered from Koonalda Cave (e.g. Gallus 1964, 1966; Pretty and Gallus 1967). He has tended to use names long established in European terminology for the types of stone implements that he claims for Koonalda. These names include Mousterian side scrapers, prismatic cores, polyhedric burins, laurel leaf bifaces, Levalloisoid cores, and Gravettian-like knives.

These names summon up in the mind of an archaeologist, familiar with the European Palaeolithic, certain forms (or an ideal of such forms). From Koonalda Cave I have not seen such forms either in my own excavated material or in that of Dr Gallus which he has shown me.

This disagreement is not one based on a chauvinistic preference for Australian terminology, nor is it based on an alarm at the diffusionist implications inherent in a European terminology applied to Australian artifacts. In fact I would intellectually welcome the excitement of finding in Australia a site which contained the variety and number of types that he mentions: 'The oldest level investigated so far contains a stone-industry which can be estimated as final Mousterian in style and characterised by *racloirs* (scrapers), leaf shaped bifacially worked *racloirs*, knife-flakes, points, prismatic cores and hand-axes' (Pretty and Gallus 1967:48). A qualificatory note to this sentence adds 'Mousterian in the sense that it is technologically comparable to the Palaeolithic industry of this name which is known from excavations in Western Europe. The use of this term here arises from the necessity to give a name to the style of the industry and cannot be held to imply any necessary historical connection with the comparable European implements'. I appreciate the caution in this qualification and have no objection to the pragmatics of such a descriptive procedure. What I do find discordant is the lack of coincidence between the actual material from Koonalda and the Mousterian industries of Western Europe.

To make my disagreement less assertive I shall outline how I have examined flaked stone material from Koonalda. When Dr Gallus uses his terminology he is referring to artifacts produced by flaking. Basically, the form of any flaked stone artifact is recognised by observing on its surface the arrangement of conchoidal flake scars. Types are diagnosed by seeing on certain specimens a repeated pattern in the arrangement of these flake scars.

In most technologies it happens that the flaking which we take note of is itself done on a flake already struck from a core. Where this is the case we can distinguish between primary and secondary flaking. This analysis is applicable only to 'flake tools'. Where we are dealing with lumps of stone that have been shaped by flaking we have merely a single sequence of flaking, with no absolute distinction into primary and secondary; these objects may be cores or they may be implements.

We must expect (to select for the purpose of exposition merely two examples) that a 'prepared uniface Levalloisoid core' would be a piece of flint flaked invasively around most of the circumference and from which a final large flake has been removed, transecting the earlier preparatory work; a 'laurel leaf biface' would be a thin oval form pointed at both ends, almost certainly once a flake and shaped by shallow and invasive bifacial secondary flaking.

I have examined all the flint from the sequence in Area B, which is the column I excavated alongside Dr Gallus's main excavation and from which all pieces of flint were retained after sieving. Taking all 1,932 pieces of flint which are larger than 1·5 cm in at least one dimension, they can be divided into

| | | |
|---|---|---|
| Flakes | (accepting only those with positive bulbs of percussion; both complete and broken) | 206 (10·7%) |

Flaked pieces (flaked lumps of flint which were not originally flakes themselves) 55 (2·8%)

Residue (any bits of flint which are unflaked and are not themselves primary flakes) 1,671 (86·5%)

I conclude, therefore, that in my excavation nearly nine out of ten of the pieces of flint within the deposit showed no indubitable traces of conchoidal fracture in the form of a positive or negative bulb of percussion. Many of these pieces classed as residue look superficially like fragments of flakes (and some must be the broken off distal ends of genuine primary flakes). Yet caution is necessary, because some of the flint nodules *in situ* in the wall of the cave are already naturally fragmented into flat pieces that in all but their lack of conchoidal fractures have the general shapes of flakes.

Of the 206 pieces that are undoubtedly flakes or parts of flakes only nine have any secondary flaking at all, that is to say the vast majority are unmodified primary flakes. None of these nine pieces has sufficient secondary flaking to make me classify it as a special type of implement. The same interpretation applies to the flaked pieces, except for a couple of objects that I think might be picks for digging amongst the rubble of chalk and flints.

My conclusion about the artifacts that I have studied is that they represent the residue of quarrying within the cave. The absence of regularly worked cores and the mere 5·4 per cent of flakes less than 1·5 cm in maximum dimension suggest that the quarry workers were shaping lumps for transport to the surface and not making their implements on the spot.

Furthermore I feel that we can get no appreciation of the ultimate technology for which this raw material was intended from the residue of this late Pleistocene quarry. In this respect we have a picture comparable with the amorphous residue left in the Neolithic flint mines of Europe, which caused so much typological confusion in the early part of this century with those who wished to see in the residue affinities with the flaked stone material from Palaeolithic living sites.

I am confident that the more conventional archaeological debris of these people can be found in rock shelters round about. Meanwhile we have to be content with an intriguing glimpse into one aspect of their behaviour. To visit this magnificent cave is to feel a sort of disquiet about the motives that compelled men to seek their flint in the dark and hazardous passageways. Flints are such mundane archaeological materials; here they take on a significance that transcends the baldly technological.

### The Late Pleistocene Submerged Plain

The effects of eustatic change during the late Pleistocene, in both providing and removing land bridges, have for long been recognised as a significant factor in human dispersal and cultural history. Standing on the tall sea cliffs 23 km south of Koonalda, one has the impression of looking into a great depth of ocean water, but this is an illusion. The chart of ocean soundings prepared by the Hydrographic Office, R.A.N., Sydney, shows that the sea reaches a depth of about 46 m close inshore. But then there is a submerged plain stretching to the south with the sea getting deeper so gradually that as far away from the coast as 160 km it is still only about 82 m deep. At this point the sea bed plunges over the continental shelf.

Given the radiocarbon chronology for sea level heights in stable areas during the late Pleistocene (e.g. Shepard and Curray 1967:

283-91) it is inescapable that the prehistoric men of Koonalda Cave had available to them a vast coastal plain. If we take this curve for the rise in sea level literally, and plot this rise on the contours of the sea bed of the Bight, we can make conjectures about when this territory was being removed.

If the low point of the last glaciation was – 90 m, there would have been this 160 km of plain. At about 16,000 BP inundation by the rising sea would have begun, so that by 14,000 BP about 56 km would have been submerged. A critical period for man was between 13,000 BP and 12,000 BP, when the sea virtually reached the base of the present cliffs. The plain had then disappeared and the subsequent rise of some further 46 m to its present level was effectively vertical.

The plain is so nearly level that even a slight rise in the sea meant a substantial horizontal loss of land. From what we know of human territoriality (and assuming this had become stabilised in the Pleistocene) the effects on human society would have been disturbing. It appears that between 14,000 BP and 13,000 BP they were losing their land at the rate of some 1 m a week. Even if such precision in the chronology is rejected, I can see no reason to doubt that the men of Koonalda did have their coastal plain and that their descendants lost it rapidly. Thus the Mirning tribe around Koonalda were occupying a miserably restricted territory compared with that available to their late Pleistocene antecedents.

## References


Curr, E. M. 1886-7. *The Australian race.* 4 vols. Melbourne: John Ferres.

Gallus, S. A. 1964. Two Australian stratigraphic sequences. *Curr.Anthrop.* **5**:127.

——1966. Comment on Transition from Mousterian to Perigordian: skeletal and industrial. *Curr.Anthrop.* **7**:39-40.

Howitt, A. W. 1904. *The native tribes of south-east Australia.* London: Macmillan.

Pretty, G. L. and Gallus, S. A. 1967. The anthropology and archaeology of the Nullarbor Plain. In *Caves of the Nullarbor* (eds. J. R. Dunkley and T. M. L. Wigley), pp. 47-9. Sydney: Speleological Research Council Ltd.

Shepard, F. P. and Curray, J. R. 1967. Carbon-14 determination of sea level changes in stable areas. In *Progress in oceanography.* Vol. 4. The Quaternary History of the Ocean Basins, pp. 283-91. Oxford: Pergamon.

Wells, L. A. and Hunt, G. W. 1919. Nullarbor Plain. *Proc.S.Aust.Brch R.geogr.Soc.Aust.* **19**: 137-40.

# 10 Coastal Aborigines of Southeastern Australia

R. J. Lampert

*Fish is their chief support . . . The woods, exclusive of the animals which they occasionally find in their neighbourhood, afford them but little sustenance; a few berries, the yam and fern-root, the flowers of the different banksia, and at times some honey, make up the whole vegetable catalogue.*

*Collins 1798: 556-7*

At present the richest evidence available for Aboriginal life on the coast of southeastern Australia comes from the central-southern coast of New South Wales and it is with this area that I am chiefly concerned, though I refer below to relevant work by prehistorians and ethnographers in eastern Australia generally.

**The Environment**

In the southern part of this area, east of the Dividing Range, the coastal plain, where it exists, is extremely narrow. Often the foothills of the range fall directly into the sea, near Eden for example, but there are fairly extensive plains around the estuaries of such rivers as the Shoalhaven. The sea bed near the coast follows this pattern, the continental shelf falling away rapidly with deep water never far from shore. Thus the varieties of sea creatures most commonly preyed upon by man are contained within a narrow strip which can support only a small modern fishing industry. However, for the wide ranging diet of the coastal Aborigines (Lawrence 1968), available marine foods should have been plentiful and varied despite the narrowness of the zone. There are the river estuaries, and also many intermittently

I am particularly indebted to J. Golson, my constant adviser since I began work on the N.S.W. coast, and to J. V. S. Megaw, Department of Archaeology, University of Sydney, who made available so much of his unpublished data from excavations just south of Sydney.

*Fig. 10:1 Excavated sites in eastern New South Wales*

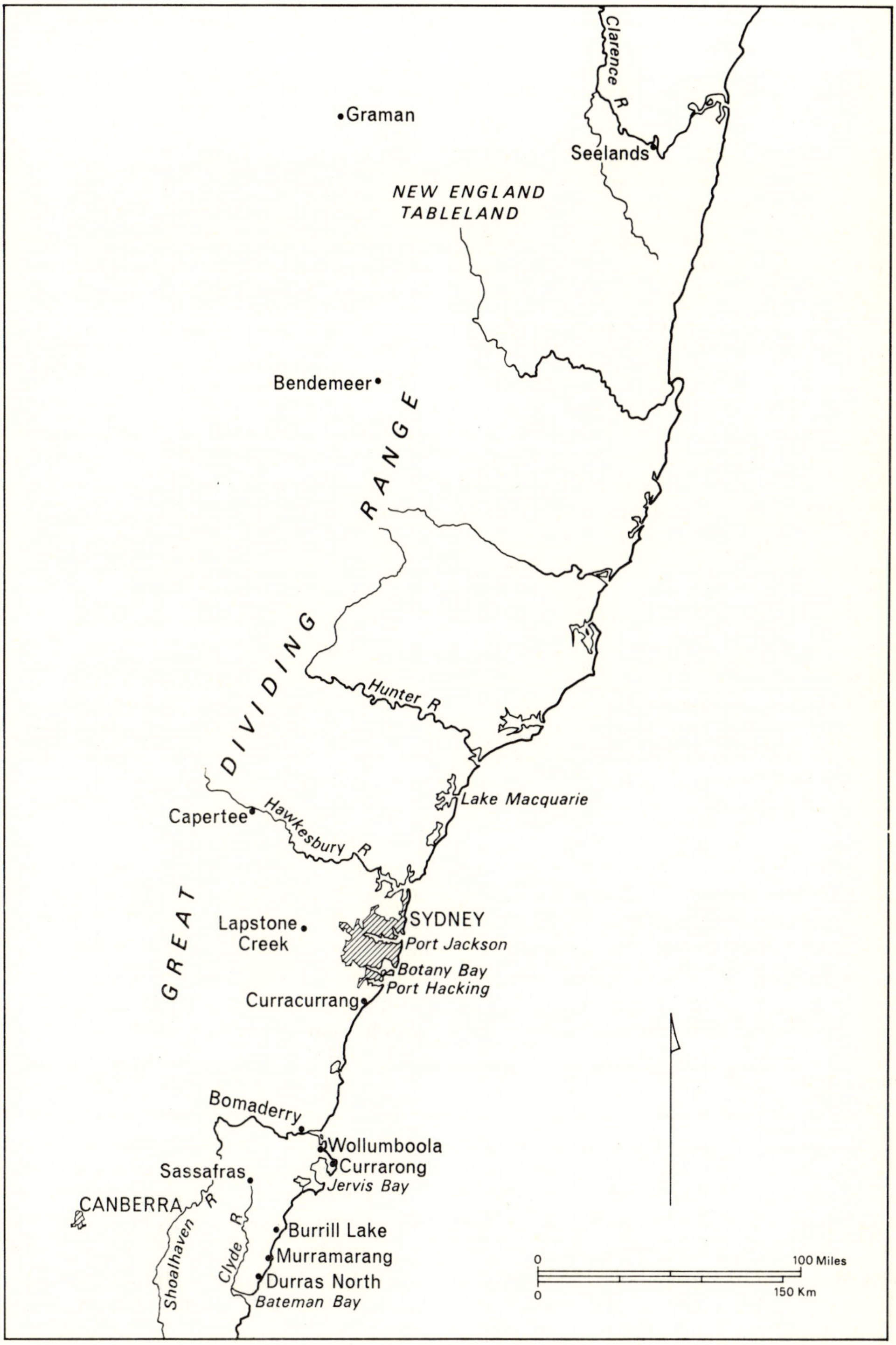

tidal coastal lakes with estuarine conditions, while the rocky coastline provides underwater rock platforms which, though limited in size, are rich in marine fauna. Intertidal and near shore environments are of three basic types: estuarine, rock platform, and sandy beach. Each is widely represented and is characterised by its own suite of marine life.

On shore the vegetation is mainly eucalypt forest of both wet and dry types. Minor landforms such as beach dunes and small areas of low rocky plateau are colonised by shrubs and heath plants. This varied environment supports a wide variety of land mammals. P. W. Thompson, who has been working on excavated faunal remains from the area at the Department of Prehistory, Australian National University, has compiled a list demonstrating this wide variety. He suggests that there are numerous small environmentally diverse areas each of which could support much of the total mammal range represented, exemplifying Calaby's conclusion that species characteristic of very different habitats can be found living in close proximity (see ch. 7).

The climate in the south is typically mid-latitude marine with no marked seasonality either in rainfall or temperature (Kendall *et al.* 1958:148). Moving northward, seasonal climatic conditions become more apparent as tropical north Queensland is approached.

## Prehistory: Historical and Ethnographic Sources

A source for the reconstruction of the most recent prehistory is the body of accounts containing observations of Aboriginal life made by explorers, settlers, and missionaries. The first are those made by members of Cook's expedition during their visit to Botany Bay in 1770 (Beaglehole 1955, 1962; H.R.N.S.W. 1893); these are reasonably detailed considering the party's short stay. The next, and perhaps collectively the most important, body of records are those made around Port Jackson by early observers, particularly members of the First Fleet. Their observations, contained in journals (especially Bradley 1786-92; Collins 1798-1802; Hunter 1793; Stockdale 1789; Tench 1789; White 1790), and letters and drawings (e.g. see Megaw 1967), were based on fairly long contact with the Aborigines, though not long enough for traditional native activities to have been markedly influenced by Europeans.

This phase was followed by more sporadic records kept by explorers, settlers, and missionaries, as people moved out from the first settlement. Such sources include the journals of Bass and Flinders who explored the coast in 1797-8 (Flinders 1946); an account of a trek along the coast from near the Victorian border to Port Hacking made by crew members of the wrecked ship *Sydney Cove* in 1797 (H.R.N.S.W. 1895); observations made by missionary John Harper at Jervis Bay and Bateman Bay (Harper 1826); detailed records kept by the Rev. Lancelot Threlkeld, based on a long stay at Lake Macquarie (Threlkeld 1855). Somewhat later there is G. A. Robinson's diary of his journey through southeast Australia, when he visited Twofold Bay and Bega on the far south coast of New South Wales and the Monaro district on the Tableland immediately inland (Mackaness 1941).

The first systematic anthropological work was carried out late in the nineteenth century, too late for many valid observations on the southeast Australian coast, though much information was collected from informants, particularly by A. W. Howitt (1904). At places more remote from those favoured by Europeans valid observations could be made until comparatively recently. W. E. Roth (1901, 1904) and Donald Thomson (1939) both recorded the technology and

*Fig. 10 : 2 Selected flake tools (scrapers) from deposits older than 5,000 years. a-b, Ingaladdi, N.T.; c, Green Gully, Vic.; d-e, Mt Burr, S.A.*

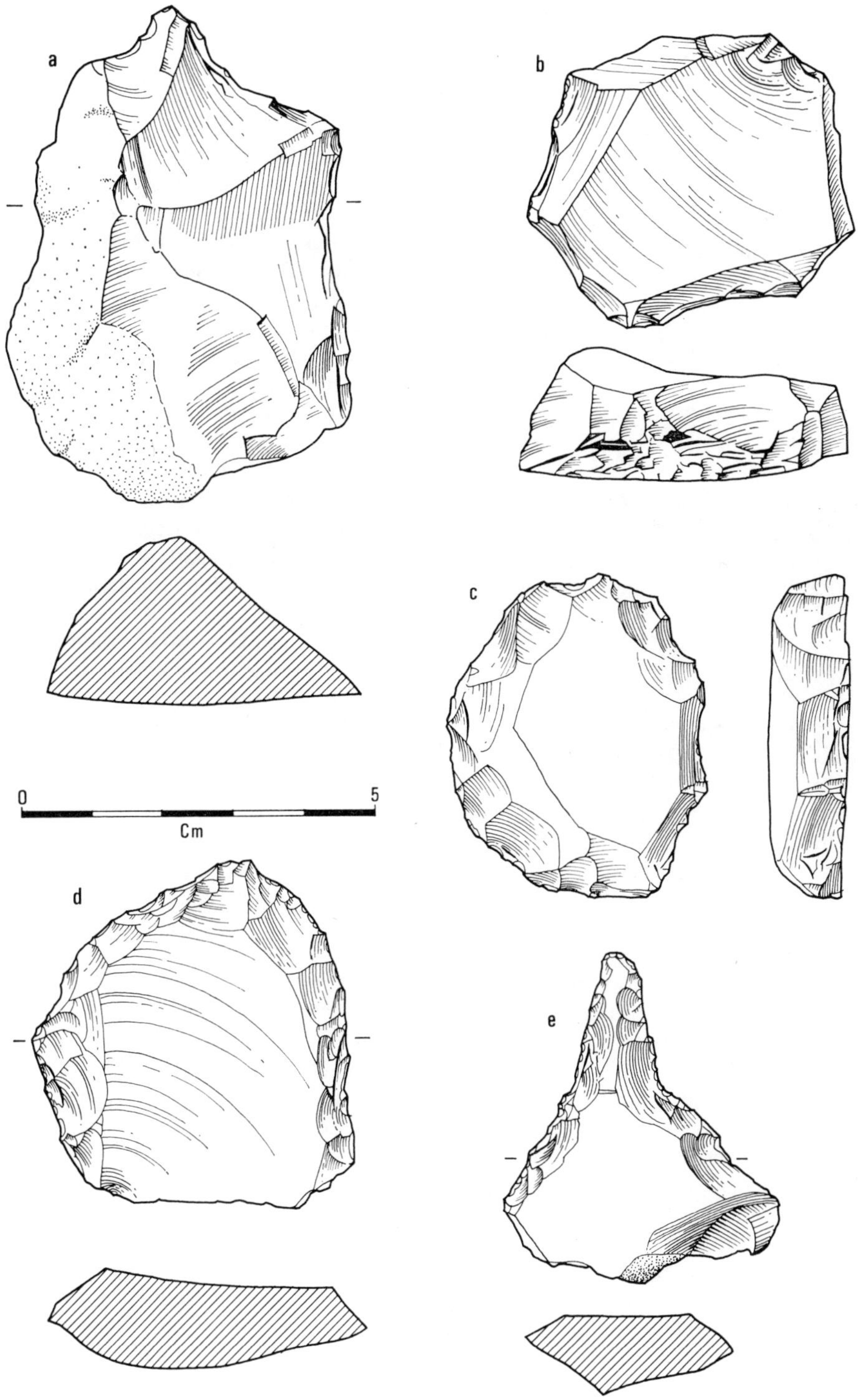

*Fig. 10: 3 Selected backed blades, excavated at Mt Burr, S.A. d-e, geometric microliths; remainder asymmetric points, including Bondaian type (f-g).*

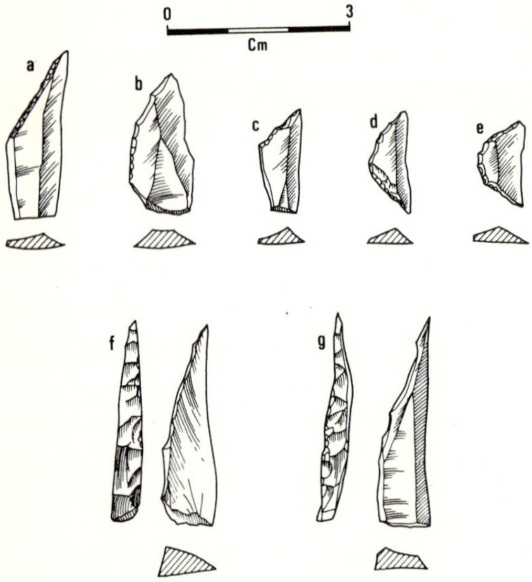


seasonal subsistence patterns of north Queensland groups.

A second ethnographic source of information lies in various items of material culture in current use at time of collection. Here the pattern of survival of the evidence is similar to that for verbal records: by the time systematic collection began the southeast coastal material had almost entirely disappeared and most of the museum items are from northern Queensland. An example is the multipronged fish spear. According to the descriptions, made mainly around Port Jackson, this spear was almost 4 m long; the shaft was the flower stalk of the grass tree (*Xanthorrhoea* sp.), at the end of which were fixed three or four prongs of hardwood, each tipped and sometimes barbed with pieces of bone. According to the frequency and widespread distribution of such accounts, these spears must have been in use in thousands along the southeast coast at the time of European contact. Yet not one complete example, positively from the area, has been located in any museum in Australia, though there are two sets of prongs only in the Australian Museum, Sydney. After searching abroad, Megaw (1967:287) located, in the Cambridge University Museum of Archaeology and Ethnology, two complete spears collected by Cook in 1770, almost certainly from Botany Bay. These might be the only two surviving southeast coast multipronged spears in the world, though there are many from north Queensland.

From these ethnographic sources a general picture of Aboriginal economy on the southeast Australian coast emerges. Relevant to the archaeological evidence are the following main points:

**1** The almost complete dependence by coastal groups on seafoods for the protein portion of their diet. Only a few items, mainly vegetable foods, were derived from the bush.

**2** The use of well-developed, specialised equipment for fishing, i.e. multipronged fish spears and crescentic fish hooks of shell.

**3** The strict sexual division of labour in using this equipment: women always fished with hook and line, men with spears.

## Prehistory: the Archaeological Sources

Before discussing interim results from recent excavations on the New South Wales south coast I will summarise relevant archaeological work previously undertaken elsewhere in eastern Australia.

First there are the rock shelter sites at Capertee and Lapstone Creek (McCarthy 1964, 1948) where from stratigraphic evidence McCarthy proposed a three-phase Eastern Regional Sequence which provided a temporal order for some of the stone implement types which he had previously recognised from surface collections. In

*Fig. 10:4 Selected eloueras. a, Mt Burr, S.A.; b, Merewether, N.S.W.; c, Burrill Lake, N.S.W.*

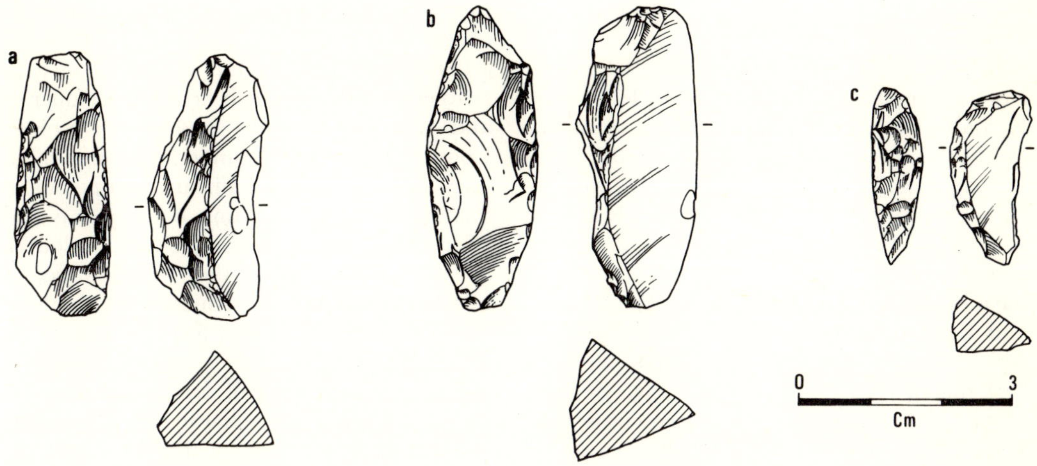


chronological order these phases are as follows:

**1** *Capertian*, an industry hard to characterise because most implements are simple scrapers made on flakes, common to all stone industries (cf. Fig. 10:2). The definition of such an industry is approached today by taking a large sample of the implements and assessing a wide range of characteristics by statistical means; so far this has not been done for the type site, Capertee. However, there are a few saw-edged flakes which promise to have a more obvious diagnostic value. The basal date for the Capertian at Capertee is 7,360 ± 125 BP (Bermingham 1966:513).

**2** The middle phase of the sequence is characterised by a backed blade industry known as the *Bondaian*, named after the asymmetric bondi point, though geometric microliths are sometimes present (cf. Fig. 10:3). In contrast to the Capertian this industry is instantly recognisable. It is widespread and thickly distributed in southeast Australia.

**3** The latest industry claimed by McCarthy is the *Eloueran*, named after the elouera which, though technically still a backed blade, is larger and probably functionally different from the bondi points and geometric microliths of the middle phase (cf. Fig. 10:4). Other tools often associated with eloueras on sites are fabricators and stone fish hook files.

From the Graman and Bendemeer rock shelters excavated in the New England District, McBryde (1966a, 1968) demonstrated that the Bondaian was a long period, at least from 5,450 ± 100 BP to *c.* 500 BP, though no single Bondaian site covers the whole of this time range. For sites in eastern Australia outside New England the Bondaian period falls somewhere within this range, the only exception being a not completely secure date of 6,550 ± 100 BP obtained by Coutts (1967a:28, 1967b:215) from Wilson Promontory in Victoria.

The Seelands rock shelter in New England has a basal date of 6,444 ± 74 BP (McBryde 1966a:285). The stone industry associated with this date contains no recognisable Bondaian traits but whether it is truly pre-Bondaian is questionable because of the small size of the sample (1966b:392).

Included in this basal industry consisting of scrapers there are also uniface pebble tools which persist through the later, demonstrably Bondaian, and more recent levels. At Kenniff Cave in Queensland a scraper industry beginning around 16,000 BP, but lacking a pebble tool component, was joined by backed blades and other small tools about 4,000 BP (Mulvaney and Joyce 1965).

On the coast, several sites south of Sydney have been excavated by Megaw. The largest and the longest occupied of these was also the most prolific in cultural material. At rock shelter ICU5/- at Curracurrang, Megaw (1965, 1968a) compares the three culturally and stratigraphically distinct levels excavated with McCarthy's Eastern Regional Sequence and sees a *general* similarity. The bottom level produced a stone industry consisting mainly of large flake and pebble tools, some of which were unifacially worked, and also horsehoof core-scrapers (cf. Fig. 12:4) and saw-edged flakes. The presence of saw-edged flakes suggests the possibility of links with the Capertian; Megaw points also to the similarity of basal carbon dates from Capertee (7,360 ± 125 BP) and Curracurrang (7,450 ± 180 BP).

The middle level at Curracurrang, with geometric microliths, bondi points, eloueras, fabricators, and burins is a typically Bondaian assemblage. Only in the top level are organic materials preserved. These are mostly food remains—shells and bones—but there are also artifacts, particularly double ended points of split mammal bone. From the high calcium and phosphate content of the soil in the underlying Bondaian level Megaw concludes that the absence of organic materials in early levels results from poor preservation; it does not mean that the occupants avoided eating in the shelter. However, the absence of free carbonate in those occupation areas not subsequently overlain by midden does suggest a low or absent shell content.

Stone artifacts in the top level include scrapers, fabricators, fish hook files, eloueras, and a few microliths and bondi points. Compared with the underlying Bondaian level there is a significant proportional increase in eloueras against bondi points and microliths, giving some support to the elouera as a type tool for the final phase of McCarthy's cultural sequence. The three carbon dates from this level are modern, i.e. $<$ 200 BP.

Other sites south of Sydney show that the final phase is characterised by the fabricator rather than by the elouera. Most convincing of these is Gymea Bay on Port Hacking (Megaw and Wright 1966), which has one cultural level with a date towards its base of 1,220 ± 55 BP. From this deposit came fifty-five fabricators but only one elouera. Megaw therefore has emphasised the need for revision of the word Eloueran as a blanket cultural term for the post-Bondaian in southeastern Australia (Megaw and Wright 1966:45).

At other sites the disappearance of backed blades in favour of fabricators or eloueras—or both—is less sudden. For example, Curracurrang shelter 2CU5/- has, in a lower level of black sand dated at 1,930 ± 80 BP (GaK-898), 27 backed blades, 16 eloueras and 37 fabricators. In the top level of shell midden (undated) there are 3 backed blades, 3 eloueras, and 21 fabricators. That the decline in popularity of backed blades was gradual receives better statistical support from a rock shelter (Sassafras I) on Sassafras Mountain, near Nerriga, where from large stratified samples of both backed blades and fabricators, a gradual change is shown from the former to the latter (Hume 1965).

Further afield, at Cape Otway on the Victorian coast, recent deposits in two adjacent rock shelters (370 ± 45 BP for shelter 2) produced well-finished bone points, hammer and anvil stones, and many stone flakes but not one convincing secondarily worked stone implement. Faunal remains indicate a heavy emphasis upon a seashore economy. Mulvaney (1962) suggests that though the sites might thus be specialised, the stone industry indicates a more general abandonment in recent times of the manufacture of finely made stone implements; this he supports with other evidence from Victoria.

### SITES ON THE N.S.W. SOUTH COAST

The foreshore site at Durras North, already published (Lampert 1966), needs only brief comment. The deposit, which was not culturally divisible, has a basal date of 480 ± 80 BP and evidences occupation into the period of European contact. In McCarthy's terms its wider cultural affiliation is Eloueran, since two eloueras were present, but no other secondarily worked stone; even unworked flakes were rare. The site produced a large number of bone points, mostly shaped only at one end, although some are pointed at both. They are smaller and lighter than those previously found at other Australian sites. The lateness of the site suggested that ethnographic analogy could safely be used to interpret the excavated material. From the historical works mentioned above, the most likely interpretation for such a large number of small bone points was that they were tips, or tips and barbs, of multipronged fish spears. This is supported both by the similarity of the points to those hafted on the few extant fishing spears and prongs and by the abundant fish bone in the deposit. Fishing apparatus of another type was also present: crescentic fish hooks made of shell, together with the stone files used in their manufacture, and partly made hooks.

Besides fish and shellfish, a major food source was the short-tailed shearwater (*Puffinus tenuirostris*), better known as the 'mutton bird', apparently caught during its summer migration along the coast. Nuts of the plant *Macrozamia communis* were also eaten. Land mammals, if eaten at all, do not appear to have been a significant food source. The archaeological evidence indicates that while occupying this site people virtually practised a marine based economy using well-developed, specialised fishing apparatus; the historical evidence suggests that this might have been their only means of subsistence, while there is no marked seasonality in the environment to induce periodic movement away from the coast.

The Burrill Lake rock shelter is under a sandstone ledge at the head of a small valley leading to the southern shore of the lake. Surrounded by eucalypt forest, it is in a sheltered position, near a small creek, in the bed of which rock pools persist in dry periods. The lake, which is intermittently tidal, supports estuarine forms of marine life.

The site had previously been excavated in 1931 under the direction of the late W. W. Thorpe, then Curator of Ethnology at the Australian Museum. Certain details in Thorpe's (1931-2) excavation report suggested that re-excavation might be profitable. He records that most of the floor area of the shelter—*c.* 43 m by 12 m—was excavated to a depth of approximately half a metre in fifteen days by a party of six. They used long-handled shovels and screened the deposit through a half-inch mesh. In view of the above method, the stone implements, described by Thorpe as being mainly 'Tasmanoid', including large scrapers and eloueras, seemed suspiciously large in size. The apparent absence of the smaller backed blades and the presence of eloueras later led

*Fig. 10:5 Vertical section of south face of trench, Burrill Lake shelter*

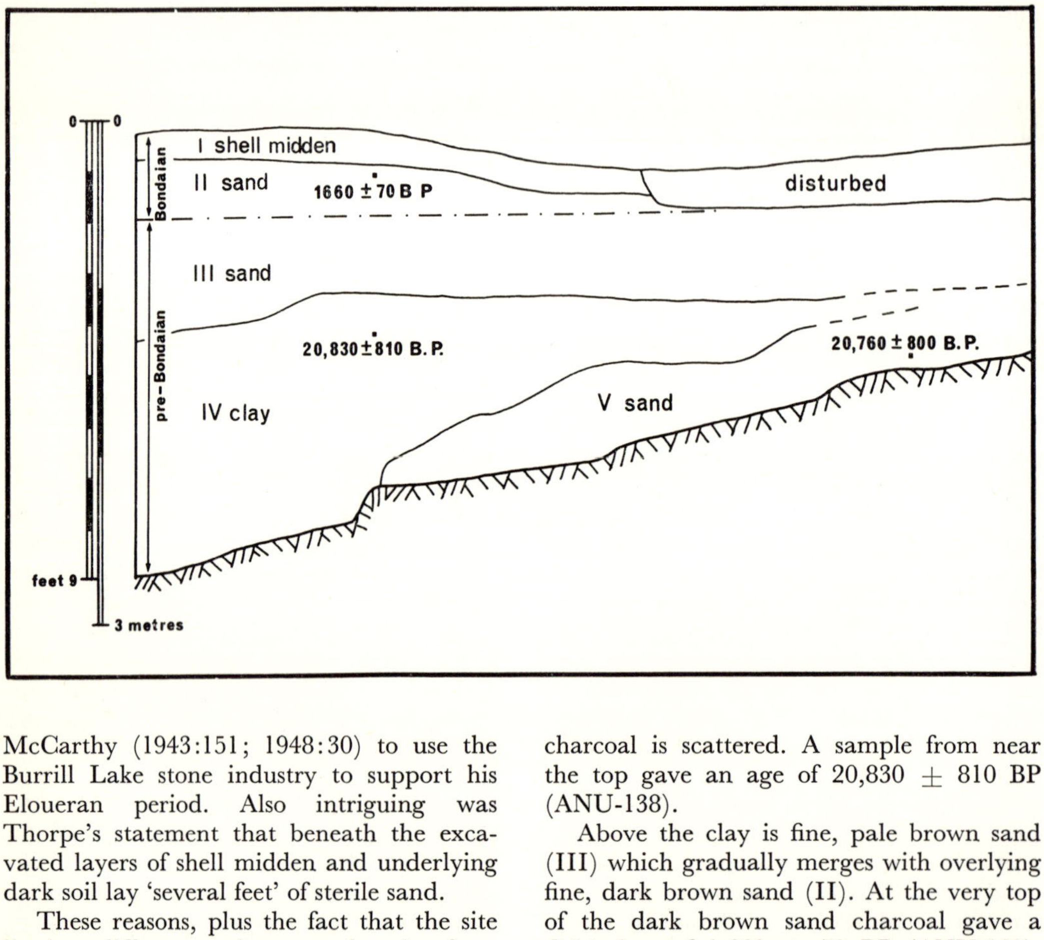


McCarthy (1943:151; 1948:30) to use the Burrill Lake stone industry to support his Eloueran period. Also intriguing was Thorpe's statement that beneath the excavated layers of shell midden and underlying dark soil lay 'several feet' of sterile sand.

These reasons, plus the fact that the site lies in a different environmental setting from Durras, prompted us to begin excavation. The results of three seasons of work concluded there are summarised on the section drawing (Fig. 10:5).

The deposit is stratigraphically divisible into five components, of which the earliest is a coarse yellow basal sand (V), scattered charcoal from which gave a C-14 age of 20,760 ± 800 BP (ANU-137). This is overlain by a wedge of reddish-brown sandy clay (IV) through which a large amount of charcoal is scattered. A sample from near the top gave an age of 20,830 ± 810 BP (ANU-138).

Above the clay is fine, pale brown sand (III) which gradually merges with overlying fine, dark brown sand (II). At the very top of the dark brown sand charcoal gave a C-14 date of 1,660 ± 70 BP (ANU-139). Immediately above this is a small remnant of undisturbed shell midden (I), Thorpe's excavation having disturbed both midden and some of the underlying dark sand over the rest of the area excavated by us. Flaked stone is present throughout the deposit but bone and shell occur only in the midden.

There is no necessary contradiction between the two basal dates (Levels IV and V) and their stratigraphic order. Even so, the deposition of the clay must have been a comparatively rapid event in the site's geological history.

The stone artifacts described from the bottom level upwards are as follows. Levels IV and V, which are provisionally combined here, together contain 2 pebble tools, 2 horse-hoof core-scrapers, 3 saw-edged flakes, and a large number of scrapers. Level III contains a similar assemblage, including 5 pebble tools, 2 saw-edged flakes, and a substantial number of scrapers. Level II is typically Bondaian, with bondi points, eloueras, fabricators, flakes which are glossy from use, and thumbnail scrapers. However, larger scrapers comparable in size to those in earlier levels are represented, as well as one horse-hoof core-scraper and a possible pebble tool. These implement types continue in Level I, but with a drop in the number of bondi points.

Among the Level I faunal remains are many shells of estuarine species found in the nearby lake, but a few are of rock platform species possibly from Burrill Headland, less than 2 km distant. Preliminary study of bones from Level I suggests that land mammals were a more important food source than fish, though neither is strongly represented. With the possible exception of a few bone points there is no evidence for the fishing gear so common at Durras North. The most recent occupants of the Burrill shelter practised a diverse economy, exploiting the resources of land, lakeside, and seashore.

Earlier than this there is only one piece of faunal evidence preserved: from the base of Level IV, towards the back of the shelter, came a cast, in the sandy clay, of the shell *Anadara* cf. *trapezia* whose habitat is estuarine flats (Dakin 1953:285). It lives nearby in the shallows of the present lake and is also common in the Level I midden. While its presence is unexpected in a horizon dated to 20,000 years BP, a time when the local sea level could have been 125 m lower and the shore 13 to 16 km farther away than at present (Bird 1968:44), there are other sites where shells have been found as far from their sources, and much further if the shell is a single one used as an artifact (Mulvaney 1969:95). Also, estuarine conditions could have extended some distance inland from the shore.

At Currarong, on the coast some 56 km north of Burrill Lake, there is a group of four rock shelters. They are near the upper tidal limit of a small creek which, before reaching its outlet on the sandy shore only a few hundred metres from the shelters, has a short lower stretch of estuarine mud flat. Just to the south of this outlet is a rocky headland. All three major types of intertidal zone are therefore represented: estuarine, sandy beach, and rock platform. The rock shelters are in a narrow, thickly wooded valley which continues upstream for some distance before leading up to an extensive low, open plateau, on which the vegetation is mainly low scrub. The environmental situation is thus a varied and sheltered one, near permanent fresh water.

Of the three shelters so far excavated, only one yielded enough stone implements to demonstrate cultural change. The results are summarised on the section (Fig. 10:6). Stratigraphically there are three main horizons. The basal layer consists of dirty white sand, devoid of any organic material including charcoal (8-10). It is overlain by a medium brown soil (7), apparently sand similar to the basal deposit plus humic material, gradually darkening upwards to merge with dark topsoil (6) just below surface. There are a few small unidentifiable bone fragments in this level, also some eroded shells near the surface. At the back of the shelter a pit has been cut from near the top of this horizon. The purpose of this pit is

*Fig. 10:6a Vertical section of west face of trench, Currarong shelter 1*
*Fig. 10:6b Distribution of implements in trenches 1 and 2, Currarong shelter 1*

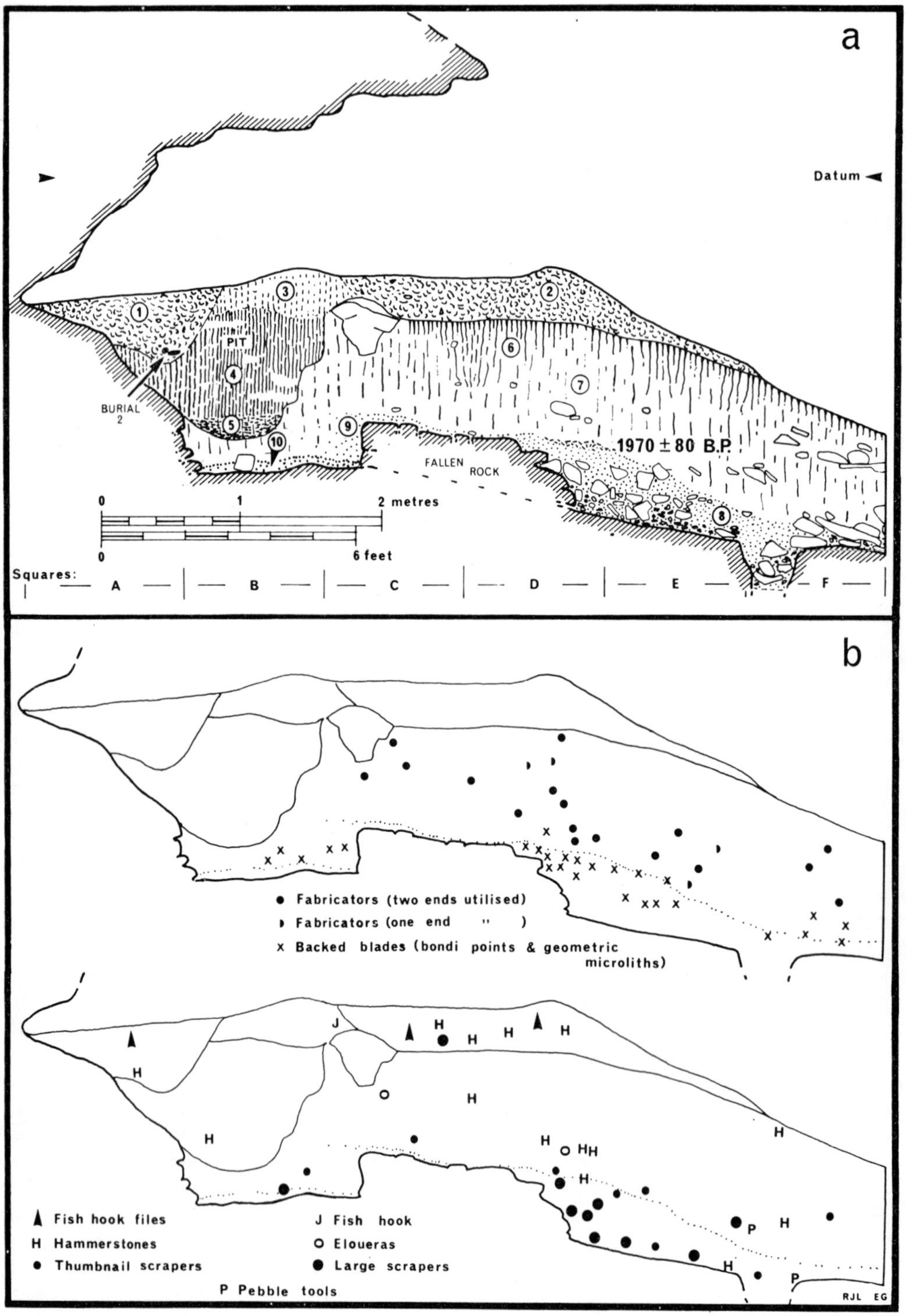

unknown; one of the three human skeletons found at the site was in the main infilling (4) but too near the top for the pit to have been primarily dug as a grave. For a pit towards the back of the Curracurrang ICU5/- rock shelter, Megaw suggests the clearing out of accumulated debris to increase the shelter's useful life (Megaw 1968a:326). Later than the infilling of the Currarong pit is a build-up of loose shell (1) partly infilling a smaller, later pit, probably dug as a grave (Burial 2). Some of this loose shell (2) has recently been thrown out from the back of the shelter, burying the topsoil just outside.

The section (Fig. 10:6b) shows the distribution of backed blades and fabricators. The point of change from backed blades to fabricators coincides with a stratigraphic division. From this point a carbon sample—from the lowest charcoal in the deposit—gave a C-14 age determination of 1,980 $\pm$ 100 BP (ANU-243). Very little can be said about the distribution of other stone implements represented on this schematic section, though we might note that the only two eloueras from the site are associated with the fabricators, while the only fish hook files are in the top midden.

Also from the midden came the only preserved faunal remains and artifacts of bone and shell. These include one crescentic fish hook of shell and seventeen bone points, of which all except two fall within the mode of the Durras North range for bone points, with an emphasis on short unipoints (Lampert 1966). However, the faunal content reflects a more varied economy than at Durras with a greater emphasis on land hunting. P. W. Thompson, who made the identifications, estimates for all mammals: 20 bandicoots, 11 wallabies, 8 potoroos (rat-kangaroos), 7 dogs (presumably dingo and not necessarily food remains), 6 fur seals, and 1 possum. Amongst fish remains, snapper predominate in number, followed by bream, then parrot fish and groper. Snapper, bottom dwelling reef fish easily hooked (Roughley 1953:77), were probably caught from the nearby headland using hook and line. Bream, which are difficult to hook and mainly estuarine in habit (Roughley 1953: 82), were probably taken in the tidal part of the creek, perhaps mainly by spearing. The presence of fish hooks and small bone points indicates that both line and spear fishing were practised.

The other two nearby shelters, excavated by Mrs E. Glover and J. V. S. Megaw, are much sparser in material useful to the archaeologist, but like the first shelter reflect wide exploitation of a diverse environment. Each contains evidence for both line and spear fishing, while Shelter 2 evidences the hunting of a suite of mammals remarkably similar in species proportion to Shelter 1.

I shall mention briefly a few sites, either worked on or examined, which show a narrower use of their local resources.

On the beach near the entrance to Lake Wollumboola, 8 km north of Currarong, is an open shell midden (Wooley 1966; Lampert 1966:111) containing material very similar to that from Durras North. There are no mammal bones, only those of fish and shearwater. While flaked stone is rare, there are crescentic fish hooks and a large number of small bone points. On the sand dunes at Murramarang Point, 16 km south of Burrill Lake, a small area of stratified shell midden again produced similar material: fish and shearwater bones but none of mammal; only a few stone flakes; several small bone points. However, this is not typical of the general range of surface material at Murramarang (McCarthy 1943), much of which appears to have eroded from earlier horizons.

A shelter on Bomaderry Creek a few hundred metres upstream from its junction with the Shoalhaven River at Nowra provides an interesting contrast to the seashore sites at Wollumboola, Murramarang, and Durras North. Though the Bomaderry Creek site is only a short distance from the tidal estuary of the Shoalhaven, test pits indicate a deposit rich in the remains of land mammals, but there are no fish and only a few estuarine shells; flaked stone seems abundant.

Just south of Sydney, several rock shelters recently excavated by Megaw have produced evidence for specialised exploitation of the seashore. Two of these are 1·6 km north of Curracurrang, in a small cove at Wattamolla (Megaw pers. comm.), immediately adjacent to the steep cliffed foreshore. The shell midden deposit in the larger shelter (WL/-) produced a large number of the bones of reef fish, minimum numbers being 91 snapper, 13 wrasse, 10 blue groper, 5 leatherjackets. Some bones of shearwater and seal were present but very few of land mammals. Cultural material included 7 bone points, 8 crescentic fish hooks, and 4 fish hook files. The stone industry is generally Bondaian, particularly in the lower midden level, while in the upper horizons stone is rare. In the smaller shelter (WB/-) the extremely shallow midden contained bones of reef fish (8 snapper, 1 groper, 1 leatherjacket) and a few bones of shearwater but none of mammal. Eight crescentic fish hooks were found.

A general cultural affinity of these sites in the Sydney area with those 160 km and more further south is indicated by the fish hooks, which are typologically similar if not identical. All are crescentic in shape (Massola 1956), unbarbed, notched in identical position on the outside tip of the shank leg, and made of the shell *Ninella torquata*. Recently a foreshore site at Inscription Point, Botany Bay (Megaw 1968b), produced, in addition to crescentic fish hooks, 48 bone points, most of which on preliminary examination withstand close comparison with those from Durras. For these reasons in the following discussion of their environmental and economic aspects, sites from the two areas are not separated geographically.

## Conclusions

### THE CULTURAL SEQUENCE

The first implication of the accumulated evidence is that man had arrived on the southeast Australian coast by 20,000 years ago. This counters an argument recently advanced by Jones (1968:189) for comparatively late settlement east of the Great Dividing Range. His argument was based on the currently known earliest carbon dates in Australia, the distribution of which shows a west-east descent in order of antiquity. This new evidence does not negate the hypothesis which Jones used this distribution to support, namely that man could have entered Australia somewhere west of the favoured Torres Strait route. It does mean that any ensuing west-east migration must have been more rapid.

In the earliest Burrill Lake levels, the tool kit, mainly rather amorphous stone scrapers plus a few pebble tools and saw-edged flakes, appears at first consideration to be similar to the stone industries that first appear at Capertee and Curracurrang some 12,000 years later, and continue at least until the beginning of the Bondaian. It is even possible that this stone working tradition continues into the Bondaian, and backed blades should be looked upon as a new technology adding to, rather than replacing, a pre-existing industry. At Capertee the saw-edged flakes found in the Capertian levels persist well into the Bondaian (McCarthy 1964:219), while at Seelands pebble tools occur possibly before, and certainly during and after the Bondaian phase (McBryde 1966a:285-6). Unfortunately most of the Bondaian level at Burrill Lake was removed in 1931. However, the implements, now in the Australian Museum, Sydney, were examined by McCarthy (1943: 151, 1948:30) who lists 17 eloueras, 18 fabricators, and 98 scrapers. This large number of scrapers suggests that, in terms of deposit volume, there is no great change in the density of their distribution compared with previous levels. Before it may be said that scrapers in the Bondaian belong to a tradition continuing from the earlier occupation of the site, however, detailed comparison between levels of all scrapers is necessary. Also closer comparison of the pre-Bondaian industries from Capertee, Curracurrang, and Burrill may show whether apparent similarities are real (Table 10:1).

The end of the Bondaian is shown at the Currarong site by the clear change in main stone tool type from backed blades to fabricators. Another site evidencing this same change is at Sassafras Mountain, not far inland on the scarp of the Great Dividing Range (Hume 1965), where the change appears less sudden than at Currarong. Perhaps this was because the Sassafras I shelter, in which the density of worked stone is much greater, was more frequently occupied. Fabricators also dominate throughout the Gymea Bay site on Port Hacking, where an early level has a carbon date some 800 years more recent than the point of change at Currarong. Recent evidence therefore strengthens Megaw's criticism of the term Eloueran for the latest of the three-phase Eastern Regional Sequence (Megaw and Wright 1966:45). Though at Lapstone Creek and Curracurrang the elouera was most popular in late Bondaian and subsequent times, except for these two sites it is too sparsely distributed to be a diagnostic type tool, and it is often numerically overshadowed by the fabricator. Even at Lapstone Creek the change from backed blades to eloueras is only marginally more significant than the change from backed blades to fabricators (McCarthy 1948:11-12), while at Curracurrang ICU5/- the change to fabricators is significant, though less so than the change to eloueras.

I am not suggesting, however, that the fabricator instead of the elouera should be taken as a broad cultural marker for the period following the disappearance of the backed blade in southeast Australia. Perhaps through cultural diversification, or because of adaptation to different environments, universal, easily recognisable elements do not persist in the post-Bondaian period. In the symposium discussion, Golson stressed that except for backed blades all Bondaian elements persisted into later times. Therefore the eventual disappearance of the backed blade was not a dramatic technological change, certainly not as dramatic as the introduction of these or, in other regions, other small 'hafted' tools (Mulvaney and Joyce 1965:172).

Looking at the whole range of excavated stone industries from southeast Australia, the only universal and distinctive industry is the Bondaian. It is well distributed and easily recognisable. It is possible, therefore, to argue from its absence. When we refer to the less distinctive earlier and later industries, as currently known, it might be preferable to use the non-associative terms 'pre-' or 'post-Bondaian', at least for this area.

### ENVIRONMENT AND ECONOMY

Much of our knowledge about the environment adjacent to a site and the economy pursued by its occupants comes from studying faunal remains in the deposit. Unfortunately these are preserved only for short periods in the open sites and sandstone shelters of the southeast coast. At Durras, fauna is preserved in all levels but occupation does not extend back beyond 500 years; at Burrill only organic material more recent than 1,680 BP is preserved; at Currarong the horizon dated around 2,000 BP is well below the earliest useful bones or shells preserved; while at Curracurrang ICU5/- levels dated 840 BP do not contain organic remains. Thus faunal evidence for environment and economy is likely to be available for about the past 1,000 years, i.e. the last 5 per cent of the time man is known to have inhabited the southeast coast. During the past 1,000 years there have been no widespread or dramatic environmental changes, while for the very time that the climate was different, there is no evidence preserved to indicate its nature.

This short survival period is shared by implements of shell and bone, which, being limited as archaeological evidence to recent times, can add little to the picture of cultural succession provided by stone implements. However, the manufacture of fish hooks of shell may be inferred from the presence of fish hook files, an example being at Curracurrang 2CU5/- where two files were recovered from a Bondaian level dated at 1,930 $\pm$ 80 BP.

For recent sites we can use the preserved faunal remains and a larger range of artifacts, both of stone and organic materials, to reconstruct recent economic activities in detail. Additionally there are numerous documentary ethnographic sources, which altogether make the latest period in this area extremely rich in source material. On the recent sites surveyed, if the stone industries were the only material available for study, at best little more than broad cultural inferences would be possible.

On an economic basis the recent coastal sites may be divided into three groups:

**1** Those such as Durras, Wollumboola, and Wattamolla situated on the foreshore, evidencing use of strictly coastal resources such as fish, shellfish, and marine birds; fishing equipment is specialised.

**2** The Bomaderry Creek site in the Shoalhaven Estuary indicates that the resources used were almost solely of inland type, while the nearby estuary was a food source of little consequence.

**3** Sites such as Currarong, Burrill, and Curracurrang, beside creeks or estuaries only a few hundred metres from the seashore. The whole range of the environmental setting has been utilised. While multipronged spears and crescentic fish hooks were used, they are not as numerous as on the specialised foreshore sites.

There are two possible explanations for this pattern of site economy. The first is that people were not tied to one particular system and periodically moved between coast and inland, taking advantage of favoured or the most easily procured foods in each. The second possibility is that there was an economic separation with specialised exploitation either by coastal or inland people, or perhaps by both.

The archaeological evidence obtained so far can be interpreted either way. That specialised sites such as Durras, Wollumboola, Wattamolla, and Bomaderry resulted from groups of people adapted entirely to specific environments seems the more attractive explanation, but it is possible that these sites represent specialised aspects of a wider economy. The mixed sites at Burrill, Curracurrang, and Currarong, on the other hand, can be neatly explained by people with diverse economic interests occupying a varied environment. They may be alternatively interpreted to indicate periodic visits by different groups of people, each with its specialised interest.

Several historical-ethnographic accounts tell of the strict division of Aborigines

*Table 10:1 Chronology and cultural sequence on excavated sites in eastern Australia*

| Period | Coastal or inland | Site | Level (if more than one period) | C-14 Age BP | Dominant implements | | | | if coastal | | | | Sources |
|---|---|---|---|---|---|---|---|---|---|---|---|---|---|
| | | | | | Backed blades | Elouera s | Fabricators | Saws | 'Small' bone points | 'Large' bone points | Fish hooks | F.H. files | |
| Post-Bondaian or comparable recent sites/levels | Inland | Lapstone Creek | D-F | 2,300 ± 100 (Lower/Mid-point) | | √ | √ | | | | | | McCarthy 1948; Polach *et al.* 1967 |
| | Coastal | Shelter 2, Cape Otway | | 370 ± 45 | Abundant stone flakes without retouch | | | | | √ | | | Mulvaney 1962 |
| | Coastal | Curracurrang 1CU5/- | 'Midden' | Modern | | √ | √ | | | √ | | √ | Megaw 1965, 1968a |
| | Coastal | Gymea Bay | | 1,220 ± 55 | | √ | √ | | | √ | | | Megaw and Wright 1966 |
| | Coastal | Inscription Point | | Undated | | | √ | | √ | | √ | √ | Megaw 1968b |
| | Coastal | Durras North | | 480 ± 80 | | √ | | | √ | | √ | √ | Lampert 1966 |
| | Coastal | Currarong: Shelter 1 | I-II | < 1,980 ± 100 | | √ | √ | | √ | | √ | √ | |
| | Coastal | Woolumboola | | Undated | | | | | √ | | √ | | Wooley 1966; Lampert 1966 |
| | Coastal | Murramarang | 'Midden' | Undated | | | | | √ | | | | |
| Bondaian or comparable sites/levels | Inland | Lapstone Creek | A-C | 3,650 ± 100 (Mid-point) | √ | √ | √ | | | | | | McCarthy 1948; Polach *et al.* 1967 |
| | Inland | Capertee Site 3 | A-F | 2,865 ± 60 (Upper; B) | √ | √ | √ | √ | | | | | McCarthy 1964; Bermingham 1966 |
| | Inland | Bendemeer | | Several Dates *c.* 500 (Upper) | √ | √ | √ | | | | | | McBryde 1966a |
| | Inland | Graman B1 | II | 5,450 ± 100 (Lower) | √ | √ | | | | | | | McBryde 1968 |
| | Inland | Kenniff | 11 | 3,830 ± 90 (Lower) | √ | √ | | | | | | | Mulvaney and Joyce 1965 |
| | Inland | Sassafras | | Undated | √ | | √ | | | | | | Hume 1965 |
| | Coastal | Curracurrang 1CU5/- | 'Bondaian' | 840 ± 90 (Upper) 2,360 ± 90 (Lower) | √ | √ | √ | | | | | | Megaw 1965, 1968a |
| | Coastal | " 2CU5/- | | 1,930 ± 80 | √ | √ | √ | | | | | √ | Megaw pers. comm. |
| | Coastal | Wattamolla WL/- | | 900 ± 150 (Upper) 1,800 ± 100 (Lower) | √ | √ | √ | | | | √ | √ | Megaw pers. comm. |
| | Coastal | Burrill Lake | I-II | 1,680 ± 100 (Mid-point) | √ | √ | √ | | | | | | |
| | Coastal | Currarong: Shelter 1 | III | > 1,980 ± 100 | √ | | | | | | | | |
| Pre-Bondaian or comparable early sites/levels | Inland | Capertee Site 3 | G-K | 3,625 ± 70 (Upper) 7,360 ± 125 (Lower) | | | √ | √ | | | | | McCarthy 1964; Bermingham 1966 |
| | Inland | Kenniff | 19<br>32 | 4,650 ± 100 (Upper) 16,130 ± 140 (Lower) | Scrapers only | | | | | | | | Mulvayney and Joyce 1965 |
| | Coastal | Curracurrang 1CU5/- | 'Bottom' | 2,500 ± 400 (Upper) 7,450 ± 180 (Lower) | | | | √ | | | | | Megaw 1965, 1968a |
| | Coastal | Burrill Lake | III-V | Two Dates *c.* 20,000 | | | | √ | | | | | |

inhabiting coastal areas of southeast Australia into hunters and fishers (Collins 1798-1802: 555-6; Townsend 1849:96; Howitt 1904:82; Lawrence 1968:113 for further evidence). Others explain the removal of part of the left little finger of women inhabiting the coastal fringe as a characteristic of the fishing branch of the tribe (see Howitt 1904:746-7 for collected references). Many writers describe the almost exclusive seafood diet of coastal people, or tell us that they spent most of their time fishing (H.R.N.S.W. 1892:132, 1893:194; Stockdale 1789:138; Hunter 1793: 62; Collins 1798:556; Becke and Jeffery 1896:120). Later writers contrast the comparatively sedentary habits of people practising a seashore subsistence with those who lived inland (Harper 1826:1563-4; Megaw and Wright 1966:44).

By contrast, accounts describing movement between coast and inland are rare. One suspiciously late writer (Kennedy 1932:106) speaks of annual migration of south coast people to Monaro during the bogong moth season. First Fleet writers tell of movement away from Port Jackson in 1788 during periods of poor fishing but do not say whether the destination of the emigrants was inland or another part of the coast, although they suggest the latter (Stockdale 1789:134-5; Hunter 1793:65). However, McBryde (1966a:289) has evidence for migration between coast and inland in northern New South Wales.

The general trend of this evidence is to support the existence of groups with specialised seashore subsistence. The case for inland groups tied to a specifically land hunting economy is not so well substantiated. Supporting evidence for this viewpoint is the well-developed nature of equipment used for marine exploitation. I find it hard to imagine that people with the skills necessary to manufacture and use such equipment should do so for part of the year only, especially in a mid-latitude marine coastal climate, where there is no marked seasonality to induce a cycle of changing subsistence. Evidence for multipronged spears of coastal type being employed for purposes other than fishing comes only from north Queensland, where seasonal change is marked. No such ethnographic evidence exists for southeastern Australia, nor has archaeological evidence been found on an inland site for either crescentic hooks or this type of spear. The materials used in their manufacture have their readiest source at the seashore, certainly for fish hooks, possibly for spears tipped with bird bone points; bird bones lie scattered along south coast beaches and are common in coastal middens. If these spears were periodically used inland where land mammal bones predominate in the faunal remains of archaeological sites, greater use of mammal bone would be likely.

This evidence suggests that some degree of specialisation did occur, but whether the division was as sharp as some early writers inferred is doubtful. If archaeology can provide an answer it would most likely be found in sites with a diverse economy—Currarong, Burrill, and Curracurrang—from which the excavated finds have not been fully evaluated. Meanwhile, because of their proximity to the shore, by their obvious associations with the more specialised foreshore middens, and by analogy with the omnivorous habits of Aborigines generally, I find it most attractive to interpret these as true coastal sites, exhibiting the full range of a coastal economy in which the fishing methods described are a specialised though perhaps major aspect.

## References

Beaglehole, J. C. (ed.) 1955. *The journals of Captain James Cook on his voyages of discovery.*

*I The voyage of the* Endeavour *1768-1771.* Cambridge: Hakluyt Society.

——(ed.) 1962. *The* Endeavour *journal of Joseph Banks 1768-1771.* 2 vols. Sydney: Angus and Robertson.

Becke, L. and Jeffery, W. (eds.) 1896. *A first fleet family.* London: T. Fisher Unwin.

Bermingham, Anne. 1966. Victoria natural radiocarbon measurements I. *Radiocarbon* **8**:507-21.

Bird, E. C. F. 1968. *Coasts.* Canberra: Australian National University Press.

Bradley, W. 1786-92. Journal. MS. A3631, Mitchell Library, Sydney.

Collins, D. 1798-1802. *An account of the English colony in New South Wales.* 2 vols. London: Cadell and Davies.

Coutts, P. J. F. 1967a. Coastal dunes and field archaeology in S.E. Australia. *Archaeol.& phys.Anthrop.Oceania* **2**:28-34.

——1967b. The archaeology of Wilson's Promontory. 2 vols. M.A. thesis, Australian National University, Canberra (unpublished).

Dakin, W. J. 1953. *Australian seashores.* Sydney: Angus and Robertson.

Fitzhardinge, L. F. (ed.) 1961. *Sydney's first four years.* Sydney: Angus and Robertson.

Flinders, M. 1946. *Narrative of his voyage in the Schooner Francis: 1798; preceded and followed by notes on Flinders, Bass . . . by G. Rawson.* London: Golden Cockerel Press.

Harper, J. 1826. Extracts from journal, Oct. 7-Nov. 10, 1826, giving an account of Aborigines met on a voyage from Jervis Bay to Batemans Bay. Bonwick Transcripts (5). MS. B. T. box 53, Mitchell Library, Sydney.

*Historical records of New South Wales: Phillip* (H.R.N.S.W.) **1** (2), 1892. Sydney: Government Printer.

——: *Cook.* **1** (1), 1893. Sydney: Government Printer.

——: *Hunter.* **3**, 1895. Sydney: Government Printer.

Howitt, A. W. 1904. *The native tribes of south-east Australia.* London: Macmillan.

Hume, S. H. R. 1965. The analysis of a stone assemblage to determine change. B.A. (Hons.) thesis, University of Sydney (unpublished).

Hunter, J. 1793. *An historical journal of the transactions at Port Jackson and Norfolk Island. . . .* London: J. Stockdale.

Jones, R. 1968. The geographical background to the arrival of man in Australia and Tasmania. *Archaeol.& phys.Anthrop. Oceania* **3**:186-215.

Kendall, H. M., Glendinning, R. M., and MacFadden, C. H. 1958. *Introduction to geography.* 2nd edn. New York: Harcourt, Brace and World Inc.

Kennedy, K. 1932. Collecting artifacts on the Monaro and the far South Coast. *Mankind* **1**:106-9.

Lampert, R. J. 1966. An excavation at Durras North, New South Wales. *Archaeol. & phys.Anthrop.Oceania* **1**:83-118.

Lawrence, R. 1968. *Aboriginal habitat and economy.* Canberra, Australian National University, School of General Studies, Department of Geography Occasional Papers No. 6.

McBryde, Isabel. 1966a. Radiocarbon dates for northern New South Wales. *Antiquity* **40**:285-92.

——1966b. An archaeological survey of the New England region, New South Wales. Ph.D. thesis, University of New England, Armidale (unpublished).

——1968. Archaeological investigations in the Graman district. *Archaeol.& phys. Anthrop.Oceania* **3**:77-93.

McCarthy, F. D. 1943. An analysis of the knapped implements from eight *Elouera* industry stations on the south coast of New South Wales. *Rec.Aust.Mus.* **21**:127-53.

——1948. The Lapstone Creek excavation: two culture periods revealed in eastern New South Wales. *Rec.Aust.Mus.* **22**:1-34.

——1964. The archaeology of the Capertee Valley, New South Wales. *Rec.Aust.Mus.* **26**:197-246.

Mackaness, G. 1941. George Augustus Robinson's journey into south-eastern Australia, 1844. *J.Proc.R.Aust.hist.Soc.* **27**:318-49.

Massola, A. 1956. Australian fish hooks and their distribution. *Mem.natn.Mus.Vict.* **22** (1):1-16.

Megaw, J. V. S. 1965. Excavations in the Royal National Park, New South Wales: a first series of radiocarbon dates from the Sydney district. *Oceania* **35**:202-7.

——1967. Archaeology, art, and Aborigines: a survey of historical sources and later Australian prehistory. *J.Proc.R.Aust.hist. Soc.* **53**:277-94.

——1968a. A dated cultural sequence for the south Sydney region of New South Wales. *Curr.Anthrop.* **9**:325-9.

——1968b. Trial excavations in Captain Cook's landing place reserve, Kurnell, N.S.W. *Aust.Inst.Aborig.Stud.Newsl.* **2** (9): 17-19.

——and Wright, R. V. S. 1966. The excavation of an Aboriginal rock-shelter on Gymea Bay, Port Hacking, N.S.W. *Archaeol.&phys.Anthrop.Oceania* **1**:23-50.

Mulvaney, D. J. 1962. Archaeological excavations on the Aire River, Otway Peninsula, Victoria. *Proc.R.Soc.Vict.* **75**:1-15.

——1969. *The prehistory of Australia.* London: Thames and Hudson.

——and Joyce, E. B. 1965. Archaeological and geomorphological investigations on Mt. Moffatt Station, Queensland, Australia. *Proc.prehist.Soc.* **31**:147-212.

Polach, H. A., Stipp, J. J., Golson, J., and Lovering, J. F. 1967. ANU radiocarbon date list I. *Radiocarbon* **9**:15-27.

Roth, W. E. 1901. Food, its search, capture, and preparation. *N.Qd Ethnogr.* **3**. Brisbane: Government Printer.

——1904. Domestic implements, arts, and manufactures. *N.Qd Ethnogr.* **7**. Brisbane: Government Printer.

Roughley, T. C. 1953. *Fish and fisheries of Australia.* Sydney: Angus and Robertson.

Stockdale, J. (ed.) 1789. *The voyage of Governor Phillip to Botany Bay. . . .* London: J. Stockdale.

Tench, W. 1789. *A narrative of the expedition to Botany Bay. . . .* London: Debrett.

Thomson, D. F. 1939. The seasonal factor in human culture. *Proc.prehist.Soc.* **5**:209-21.

Thorpe, W. W. 1931-2. A rock shelter at Lake Burrill, N.S. Wales. *Mankind* **1**:53-5, 78-9.

Threlkeld, L. 1855. From MS. Canberra: Australian Institute of Aboriginal Studies.

Townsend, J. P. 1849. *Rambles and observations in New South Wales.* London: Chapman and Hall.

White, J. 1790. *Journal of a voyage to New South Wales.* London. Reprinted 1962 (ed. A. H. Chisholm). Sydney: Angus and Robertson.

Wooley, P. H. 1966. Notes on an Aboriginal fishing site at Lake Wollumboola. *Mankind* **6**:318-19.

# 11 Prehistory in the Cape York Peninsula

R. V. S. Wright

The Cape York Peninsula lies entirely in the tropics. Most of it is open savannah woodland with monsoonal summer rains and dry winters. On the east coast, where winter rains occur, rainforest is common. The east coast is mountainous and the west coast low and flat. There is a great variety of rock types.

It would be presumptuous, on the evidence available, to try to present a synthesis of the relationship of prehistoric man to his environment, because there are a number of disparate environments. Rather than indulge in generalised ecological statements, I shall select two well separated and different types of site to exemplify the problems of reconstruction ahead.

## The Shell Mounds of Weipa

Weipa is on the west coast of the Cape York Peninsula, some 210 km south of the Cape itself. The shell mounds are on the banks of the shallow and muddy estuaries of the Embley and Hey rivers which flow into Albatross Bay.

### EARLY HISTORY OF THE COAST

The seventeenth-century Dutch explorers have left us with a brief account of the Aborigines of the west coast (Heeres 1899). Janz in the *Duyfken* sailed into Albatross Bay in 1606, but his journal does not survive. Carstenz in 1623 made observations up and down the coast, and from his account we can attempt a summary of Aboriginal life. We learn that the Aborigines lived in grass huts on the beach. Their weapons consisted of spears (some tipped with bone), shields, clubs, sticks, and a spearthrower with a hook on the end. Other artifacts included stone chips, bone scrapers, resin, woven containers, metal (thought by Carstenz to

I gratefully acknowledge the financial assistance of the Australian Institute of Aboriginal Studies in the field work here discussed.

be left over from the *Duyfken*'s visit), but no boats or canoes (a chance error of observation judging by later reports). Feathers were worn in the nose and the people painted their bodies. In some areas the Dutch saw quantities of human bones, which they attributed to cannibalism, but which perhaps can be explained as the relics of primary disposal of the dead. Roots were the only items of food mentioned. The Dutch saw smoke along the coast, sometimes so dense as to make the land invisible.

It is a tribute to Carstenz that there is nothing reported which does not square with the more detailed observations of later visitors, such as Roth (1901). Carstenz made frequent recordings of latitude. Since he was voyaging along a north-south axis it is possible to plot his position closely on the map. Along the 650 km of coast, Carstenz approached land, or actually landed, on seventeen occasions. On nine of these he met with Aborigines. These sightings are well spaced along the coast, indicating that in April and May, the start of the dry season, the whole length of it was occupied by Aborigines. This is an important point, because the distribution of middens, restricted as they are to the estuarine complex around Weipa, might be taken to suggest that much of the coast was not occupied.

### THEORIES OF ORIGIN

Roth (1901:7) visited Weipa at the turn of the century, and interpreted the heaps of shell as food refuse and mentioned huts on their tops. In his view they were spectacular examples of kitchen middens 'whose progress of formation has evidently been going on for several generations past'.

Stanner (1961), on the other hand, favoured a natural hypothesis for their origin. He based his conclusion on the absence of any signs of human activity within the mounds, their distribution around the estuary, and the disinclination of Aborigines to attempt such constructions. He therefore postulated some natural catastrophe as the cause, one which was probably associated with changing post-glacial sea levels.

In my opinion Stanner's arguments cannot be sustained. Such sampling excavation as I carried out in 1963 revealed artifacts throughout—the most common being ground bone points of the type used to point and barb spears in that area. Other evidence for human activity emerges from the fact that the mounds are made up almost exclusively of the single mollusc species *Anadara trapezia*, a selectivity from shoreline debris for which no natural explanation would seem adequate. I was unable to see the relevance of Stanner's arguments based on the distribution of the mounds. One that had potential point was the assumption that the mounds pre-dated the present drainage pattern; however, in an area where superficial observation did suggest this possibility, my excavation and core sampling satisfy me that the mound started to accumulate after the stream had become established.

I am wary of any argument about Aboriginal disinclination. In the U.S.A. it was argued that massive earth mounds could not have been produced by the American Indian tribes living in the area and must therefore have been constructed by a vanished civilisation. Subsequent archaeological excavation showed that they contained nothing that could not be squared with the prehistory of the recent Indians. The argument of disinclination seems unsound, because cultures change and the dispossessed remnants of social groups may not show some aspects of the cultural drive of their forebears. Of course Stanner has never suggested a lost civilisation—merely a natural origin. However, I am pessimistic enough to predict that someone will propose the theory of a past civilisation (lost or just in transit) to explain their origin. All I can say in the way of unimaginative sabotage is that the material that I recovered from the Weipa mounds can be matched with the ethnography of the area, and this ethnography is intelligible in terms of Australia as a whole.

Thus as far as origins are concerned, I can see nothing to negate Roth's theory that the shell mounds at Weipa are humanly constructed middens, albeit of most impressive size and form.

### AGE

By mapping and drilling I was able to demonstrate that the middens are entirely superficial in relation to the surrounding landform. There is, furthermore, evidence of the building out of land into the estuaries since the present post-Pleistocene sea level was attained, and the middens post-date this phase. This indicates that the middens must date well within the period of the last 8,000 years or so.

This theory of recent age is confirmed by their sharp state of preservation and the radiocarbon dates from one of the middens. Charcoal was taken from a cutting that exposed a midden from top to bottom. The section used for collecting was a freshly cut face of the trench originally exposed by a prospector's bulldozer. This midden was almost 5 m high and in the form of an elongated mound. The dates show that it started to accumulate some 800 years ago (I-1738: 810 $\pm$ 105 BP, nearly at the base). Sample I-1737: 235 $\pm$ 110 BP, from near the top, suggests that accumulation continued up to recent times. Although dates for only one midden are available, so similar are the other heaps in distribution and state of preservation that I doubt whether any are substantially older than the one dated.

### EXPLANATIONS FOR THEIR CONSTRUCTION

The mounds appear in many cases to be deliberate structures rather than randomly distributed kitchen waste. The most impressive are truncated cones some 10 m high with a slope of 1:1. But it must be stressed that these examples are structures that apparently took a few centuries to reach their present form. There are also many which are slight rises that have, so to speak, just got off the ground.

Many and various explanations of a functional nature might be postulated, but I have yet to meet one that convinces. Roth suggests that they were to escape sandflies and mosquitoes, but we know that this would be a cure long in accomplishment. There is in any case rising land close by. My personal experience shows in addition that escape is not achieved. Duration of accumulation defeats the explanation that they were built to see around the countryside; it would be quicker to climb a tree. Higher land behind the shoreline seems to make them pointless as a place on which to escape floods.

Finally, a suspicious point for any strictly functional explanation is their absence elsewhere on the coastline of Cape York. The Weipa mounds are conspicuous from the air. Discussions with pilots and those who have done aerial surveys have produced no reports of them elsewhere on this coast, although there are numerous other estuaries along it. As previously outlined, reports from the seventeenth century onwards show that the rest of the coast was occupied by Aborigines. Therefore it appears that Weipa does present us with an aspect of human behaviour unique to this estuary.

For the last few hundred years at Weipa there was an enduring but localised tradition whereby it was culturally desirable to dispose of shells in heaps, taking care to keep the area of disposal constricted. On the rest of the coast this was not a cultural tradition and the random disposal of shells may have led to their total disintegration in soils where, until a critical mass has been accumulated, individual shells do not survive at all.

**A Rock Shelter at Laura**

The rock shelter L-1 is some 80 km west-southwest of Cooktown and about 3 km from the present township of Laura. This is dissected sandstone country with numerous superbly painted shelters on the slopes of valleys and hillsides. The floors of these shelters tend to retain merely a few centimetres of loose dusty deposit with artifacts characteristic of the most recent levels at L-1. The reason for L-1 having a deep deposit is due to its being away from eroding slopes. The deposit in this shelter is not of the conventional talus type, but it is part of the accumulation of sand that has taken place on the surface round about. In other words the sheltering rock is protruding through a land surface that has become increasingly deep in its cover of sand.

The deposit must originate from the gently rising land behind the shelter and is being transferred down the slope 'grain by grain'. The surface is even, unmoved by wind and showing no stream action from unabsorbed rainfall. The present surface is covered with grass and open woodland. Erratic shifts of sand on such a surface would seem to have been unlikely. Rather I presume that with such superficial disturbances as rainfall, each grain of sand, whenever it has been disturbed, has tended to move downhill.

The rock is shaped like a mushroom, providing shelter all round. The maximum overhang is 5 m and the maximum clearance from floor to ceiling is 2·5 m. Paintings on the walls and ceiling are prolific, but do not survive below the groundsurface, where the rock below the surface of the deposit is covered with the accretions of termites.

Excavations were made on the western and eastern sides of this rock. The habitat of the western, down-slope side appears to be the more attractive for camping at present and the deposits there are more useful archaeologically in terms both of the depth of deposit (6 m) and the greater density of artifacts. Judged by the appearance of the deposits in section, and the analysis of the cultural sequence, the shallow eastern side of the shelter, with under 2 m of deposit, represents the same amount of time as the western side but in a 'compressed' version.

The deposit is predominantly coarse sand throughout. Its acidity of around pH 5·5 has destroyed all bone except for a few corroded fragments in the top few centimetres under the overhang. The shelter thus contains a sequence of stone artifacts without associated organic evidence. On the deeper western side the artifacts cease at a depth of under 4 m; below this is sterile sand reaching decayed rock floor at 5·5 m. Down this section the colour changes markedly from dark grey at the top, becoming browner, and lower still changing to orange. This colour change is due to a progressive diminution in the quantity of charcoal present. No charcoal at all occurs below 2 m. Since artifacts occur prolifically for another 1·5 m below this point, the early levels of occupation are undatable. Unfortunately it looks as though even extrapolated dates for the earliest occupation are ruled out because of a suspected disconformity in the sequence below the earliest datable horizon.

These broad changes in colour do not provide usable stratigraphy for the excavator. However, given the present flat surface of the deposit, it can be assumed that horizontal excavation gives a best estimate. A compari-

*Table 11:1 Changes in the size of implements over time*

| | Weight in grams | | | | | | | | | |
|---|---|---|---|---|---|---|---|---|---|---|
| | ·47 | 1·0 | 2·2 | 4·7 | 10 | 22 | 47 | 100 | 220 | 470 |
| Surface—1·2 m | 1 | 2 | 24 | 35 | 58 | 41 | 27 | 20 | 7 | 1 |
| 1·2 m—2·1 m | 0 | 0 | 1 | 3 | 9 | 22 | 44 | 27 | 15 | 1 |
| 2·1 m—3·4 m | 0 | 0 | 0 | 0 | 2 | 9 | 14 | 16 | 11 | 3 |

son of cultural stratigraphy in various columns of the deposit confirms that this is so, although the site cannot be made to yield assemblages or components. It can be used to show gross stabilities and changes in stone artifact traditions through time. The internal consistencies between changes in the various columns of the site, and the quantities of artifacts recovered, demonstrate that the sequence in the site is a very useful standard of reference for the area. The radiocarbon dates show that we have covered the last 10,000 years or so.

A basic observation on the industrial sequence is that there were two distinct and stable traditions. The changeover point was about 3,000 years ago. The upper industry consists of small artifacts. It contains implements made on flakes and to repeated patterns. The lower industry is larger and the flaked edges, irregular in form and disposition, are less frequently produced on flakes.

To illustrate these points I have presented data from the excavated areas E and F, some 10 per cent of the area actually excavated (Table 11:1). Comparisons with other columns show that the conclusions drawn from this limited volume are valid for the site as a whole.

The artifacts used in Table 11:1 include flakes with secondary retouch and any piece of stone which has itself been flaked. It excludes unmodified primary flakes. The two categories used include objects that can be interpreted as flake and 'core' implements and also as cores themselves. What this table represents, therefore, is the sizes of the pieces of stone that have been selected by the artisan for special flaking treatment—for whatever reason. Given the heterogeneity of the categories making up this group of *flaked* artifacts, their size in weight is expressed in a logarithmic progression. This gives equal emphasis to percentage variability in size along the whole scale.

There are three divisions of depth in the table. The one at 1·2 m is set by typology—it is at this point that certain new types of implements first occur, continuing then to the top of the site. The division at 2·1 m is typologically arbitrary but is included to show that the lower deposits are homogeneous in the characteristics here discussed.

Each of the three rows shows an approximately normal distribution, though the mode for flaked pieces from the top 1·2 m is considerably smaller than that from the lower part of the deposit. The modal size for the division, transferred back to arithmetical terms, is for flaked pieces about one-fifth the weight of those lower down. In addition, the whole distribution of the pieces near the top is shifted to the smaller end of the scale. It is in general a much smaller industry. As might be expected, there is a highly significant difference between the distributions of rows 1 and 2, but there is none between rows 2 and 3. The artifacts from the bottom 2·1 m thus seem homogeneous in size.

It is interesting to note that the other

side of the shelter shows the same results. A column there was divided into upper and lower on the same typological basis. In terms of size, the most recent industries on both sides cannot be distinguished from each other and neither can the bottom ones.

Another characteristic, correlated with this stratigraphic change, is the proportion of flakes making up the worked pieces. An inspection was made of all the specimens included in Table 11:1 to see which ones had been definitely made on flakes. In row 1 it is 67 per cent, whereas in rows 2 and 3 it is only 37 per cent and 35 per cent respectively.

What are the ages and typological characteristics of these two industries? On the western side the more recent one must have been the industry at the time of European contact, and, with a radiocarbon date of 1,830 $\pm$ 110 BP (I-1736) at 75 cm, can be estimated to have begun around 3,000 years ago. The most common type of implement is one made on a flake where a steeply worked straight edge runs parallel with the long axis of the flake—the whole implement being approximately 3 cm long. In some cases two parallel edges are treated in the same manner, the end result resembling a worn adze-flake slug (cf. Fig. 9:1). Rare traces of gum, so far located only on single-edge specimens, indicate hafting with the worked edge projecting. These objects were probably mounted like tula adze-flakes, the double-edged ones resulting from remounting at 180°. They differ formally from tulas inasmuch as tulas have a worked edge opposite the striking platform. In addition, distinct examples of backed eloueras occur, though rarely (cf. Fig. 10:4). There was a single ground axe and several ground flakes, presumably from axes. The larger flaked pieces present can be reasonably interpreted as cores for the production of flakes.

The elements I have mentioned are characteristic of the recent prehistory of the eastern coast of Australia, though in terms of the number of forms present this is a relatively impoverished industry. It is worth noting what is absent. Absence at Laura can be taken to be real and not due to sampling error, in view of the several hundred flaked pieces recovered. There are no backed blades (cf. Fig. 10:3) or pirri points or other trimmed points (cf. Fig. 12:5); on the basis of surface collections this is to be expected because the distribution of these in a northerly direction is already beginning to thin out when central Queensland is reached. Fabricators are absent; the surface distribution of these is unclear, however. Also unclear is the surface distribution of carefully worked semicircular edges, as seen on end-of-flake or thumb-nail scrapers; but the absence of these in the industry at Laura is notable. This stone industry of the last 3,000 years or so can perhaps be best summed up as a generalised version of the late stone industries of Australia, containing no unexpected elements.

What occurs below this industry can be readily distinguished. However, lack of stratigraphical acuity makes for a fuzzy interface and the site is therefore quite unsuitable for answering the questions of evolution or replacement. The lower industry's larger size and less frequent use of flakes has already been mentioned. Edges are characteristically heavily flaked, often having an under-hanging effect. All shapes and sizes are treated in an apparently random manner. Flakes which show grinding marks, and which can be interpreted as having come off axes, occur to the bottom of the deposit. One piece seems to have preserved on it the intersection between two ground surfaces. A few unifacially flaked pebbles occur; one of them is so carefully flaked that its worked side looks like an Acheulean hand-axe.

In the upper industry I interpreted some of the larger flaked pieces as cores. I am loath to do this with the lower industry because there are not the smaller implements made on flakes that could have originated from such cores. In addition, the amount of subsidiary crushing present on the edge of the platforms of these flaked pieces seems to be excessive if due solely to flaking mis-hits, and therefore they must have been used as tools.

Indications of the age of this industry come from I-1736, providing a terminal date of about 3,000 years. It is also dated to 6,870 ± 150 BP at a depth of 2 m, which is the lowest position in the sequence from which charcoal can be collected. The industry continues to a depth of almost 4 m. Because of a suspected disconformity, it is not possible to estimate by extrapolation when the shelter was first occupied.

Ground ochre of various shades is profusely distributed from top to bottom of the deposit and the age of the paintings present on the shelter walls cannot therefore be postulated from the occurrence of ochre. It is worth noting, however, that the shallow deposits of nearby painted shelters always yield the recent industry of L-1.

**Further Evidence**

It seems fair comment to say that the rock shelter at Laura provides only a reasonable sequence of stone implements. It does at least provide a long sequence, which is more than that provided by the middens at Weipa. I also have excavated two rock shelters in other areas of the Peninsula. While I am not going to outline their contents here, I have incorporated some of their characteristics into Table 11:2, in order to draw attention to some of the interesting problems still to be faced in writing even an elementary prehistory of the Cape York Peninsula.

*Table 11:2 Characteristics of Cape York archaeological sites*

| | Weipa | Laura | Bare Hill | Chillagoe |
|---|---|---|---|---|
| Rock shelters | 0 | 1 | 1 | 1 |
| Stratigraphically separable long sequence | 0 | 1 | — | — |
| Paintings | 0 | 1 | 1 | 1 |
| Stone implements | 0 | 1 | 1 | 1 |
| Shells | 1 | 0 | 1 | 1 |
| Bones | 1 | 0 | 1 | 1 |
| Charcoal | 1 | — | 1 | 0 |

0 = not present
— = inadequate
1 = adequate

The two additional sites are Bare Hill, a granite rock shelter on the borders of the rainforest near Mareeba on the Atherton Tableland, and Chillagoe, a limestone rock shelter at Chillagoe.

Table 11:2 is a rough schematic attempt to show the possibilities of making comparisons between sites. The least comparable are Weipa and Laura. It can be seen that Laura is the only site that is sequentially 'adequate' and yet it is just the one that has acid soils leaving no bone and shell residue. How is one to integrate Laura and Weipa analytically, when the only points of comparison are radiocarbon dates?

The three rock shelters present a better picture with their stone implements, which permits some useful comparisons. For example, flaked stone implements are rare at Bare Hill though there are quantities of unmodified quartz flakes. At Laura, quartz flakes are also plentiful and are rarely trimmed; they account for only 2·5 per cent of the flaked pieces. The people at Bare Hill, apparently short of a tractable stone material, imported marine shells for flaking. At Chillagoe, where adequate stone was available, marine shells were not so used. Unfortunately, however, the sequence at Chillagoe is probably undatable.

While the student of stone implement sequences can have a field-day with sites like the Laura rock shelters, the analyser of kitchen refuse can feel at home with the Weipa shell mounds. What we have not got at present in the Cape York Peninsula is a long datable sequence of artifacts and organic remains that would allow us to write a prehistory in terms of the relationship of man to his environment.

## References

Heeres, J. E. 1899. *The part borne by the Dutch in the discovery of Australia 1606-1765.* London: Luzac and Co.

Roth, W. E. 1901. Food, its search, capture and preparation. *N.Qd. Ethnogr.* **3.** Brisbane: Government Printer.

Stanner, W. E. H. 1961. The Weipa shell-mounds. *The Etruscan* **11** (2):8-12. Sydney: Bank of New South Wales.

# 12 Man and Environment in Northwest Arnhem Land

Carmel White

*The Manus delight in facts and argument. . . . They debate heatedly whether or not a report (received by one man from a geologist) that New Guinea was once joined by a land bridge to Australia is true. . . . They debate questions about entirely useless matters for the love of truth.*

*R. F. Fortune 1935:v*

Since 1964 I have completed a research project in the Northern Territory, some implications of which are discussed here. Much of the data is available in my doctoral thesis (White, C. 1967a) and elsewhere (White, C. 1967b, 1967c), but since these were published some fresh information has become available and I have reconsidered some of my earlier conclusions. I first outline briefly the ecology of the research area and the main results of the prehistoric investigations that have been made there. Then, using these data, I discuss the climate and environment of prehistoric times and their relationships to the cultural remains we find today, both in this area and in other parts of the continent.

## The Natural Setting

### GEOLOGY

The most important geological features of the area, which lies in the northern part of the Northern Territory and includes the country falling within a radius of some 30 to 50 km from Oenpelli Mission, are the plateau and the adjoining plain (Fig. 12:1).

The Upper Proterozoic Buldiva plateau is made up of near-horizontal strata of quartzite,

I should like to thank the following people for their helpful comments on this paper, though the opinions expressed in it are my own: R. W. Fairbridge, Columbia University; A. D. L. Hooper, Department of Primary Industry, N.T.; D. J. Mulvaney; J. Peter White; M. A. J. Williams, Department of Geography, A.N.U.

*Fig. 12:1 Archaeological sites in northwest Arnhem Land*

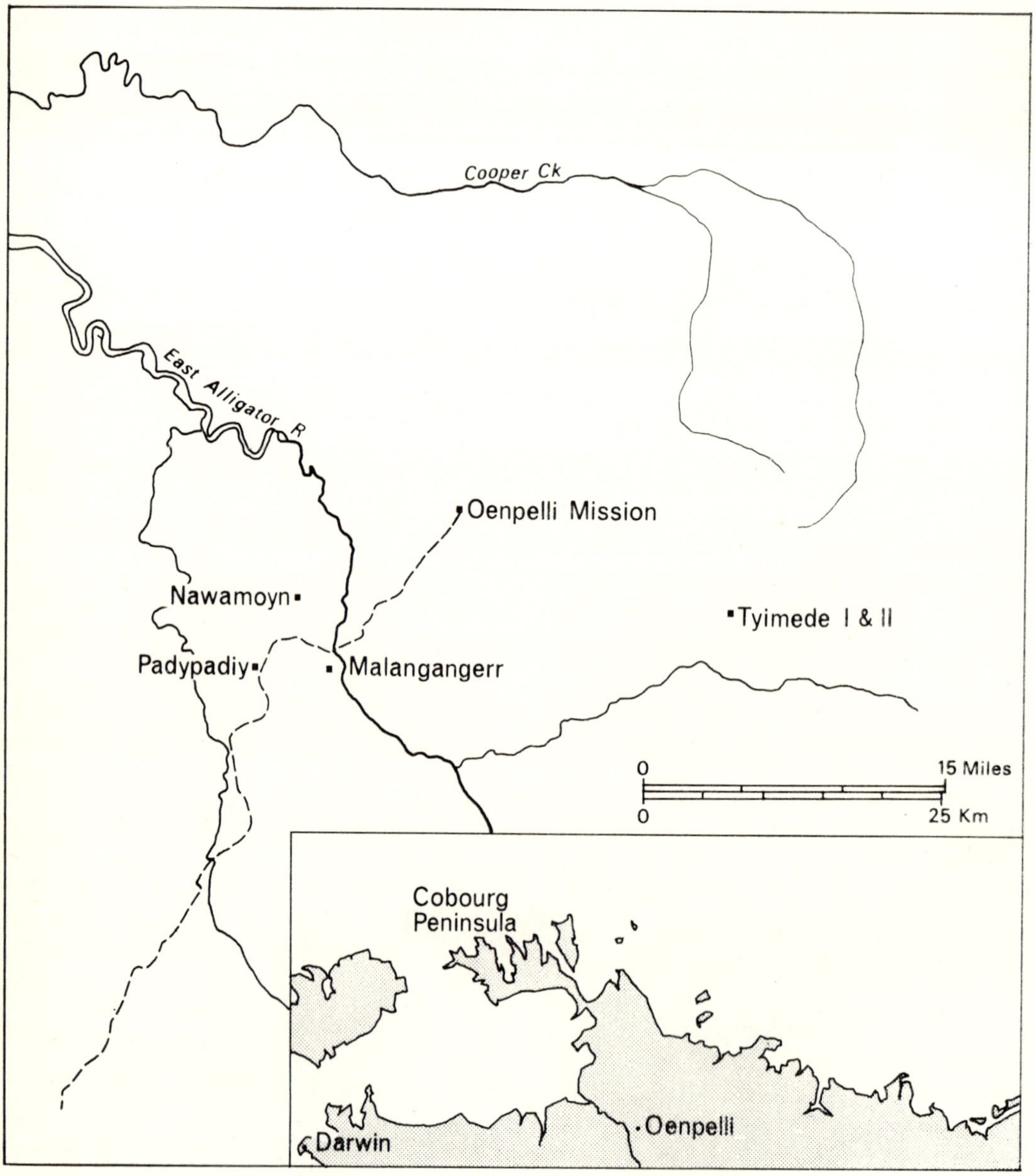

sandstone, and quartz conglomerate, which show no signs of any major post-depositional tectonic warping. The escarpment occurs as a series of steep cliffs that rise up out of the plain to a height of 245 m in places and which form the northwest border of the main Arnhem Land plateau. Further south, the plateau is cut by wide valleys watered by seasonal streams and pools, whilst in the north erosional outliers form a series of mesas and buttes.

The plain can be subdivided into an alluvial plain composed chiefly of Tertiary sands and silts derived from the adjoining plateau, and a coastal plain which is made up mainly of Quaternary muds and clays. This is a particularly ill-drained feature, lying only about 6 m above sea level, 50 km inland from the coast at Oenpelli. Specht (1958:343-4, 349) suggested that it represents the in-filled estuaries of north flowing rivers, drowned in the last major marine transgression. Relict dunes (R. Story pers. comm.), raised beaches (Phipps 1966:320), and submarine shorelines (van Andel *et al.* 1967) all occur along the Arnhem Land coast, but there is, to date, no better documented evidence of the extent, effect, and age of eustatic fluctuations there.

Large rivers, which originate in the plateau, flow towards the Arafura Sea, widening out on the alluvial flats and meandering in the lower reaches. Many large rivers are tidal for a considerable distance inland because of the strong 7·9 m tidal range in the northern sea, and estuarine conditions prevail on the upper reaches of the East Alligator River, where it emerges from the plateau some 50 km inland from the coast.

#### CLIMATE

The climate is monsoonal with marked seasonality. The rain comes from a number of sources (see Southern 1966) and most of it falls between October and March. The current rainfall distribution pattern in Arnhem Land shows a gradient towards the south, and has been described as follows:

Annual rainfall is a maximum in the vicinity of Darwin (over 60 inches [152 cm]), decreasing by approximately 15 inches [38·1 cm] for each 100 miles distance south-eastward to 25 inches [63·5 cm] per annum along the southeastern border. The isohyets have a general south-west to north-east orientation across the western portion of the region, but over central Arnhem Land turn sharply east then south-east, a secondary maximum of over 50 inches [127 cm] appearing on the west coast of the Gulf of Carpentaria. (Bureau of Meteorology 1961:2)

Specht (1958:335) cites a 35-year record of rainfall at Oenpelli, where 94 per cent of the average annual rainfall of 134·5 cm falls in the wet season. Despite the marked variations in humidity, the temperature stays fairly constant, ranging from 35·8°-31·1°C to 24·4°-15·4°C at a station about 190 km west of Oenpelli.

The seasonal changes have a profound effect on the behaviour of men and animals living in the region (see Thomson 1939, 1948-9, 1949; Specht 1958:333-9; Worsley 1961:161-2). It could be said that the seasonal abundance of water in Arnhem Land is as much an ecological determining factor as is the scarcity of water in central Australia. For example, in the wet season in Arnhem Land the plains are inundated and swamps overflow. Men and beasts retreat to higher ground where they congregate in the same way as fauna cluster around permanent waterholes in arid regions. During the wet, men cannot reach swamp plants or estuarine fauna, but as the rains slacken off and the dry season advances, movement becomes far easier, and men and animals are free to move over a wider area and to exploit a wider range of ecological zones for food.

FLORA AND FAUNA

The vegetation of Arnhem Land has been described as climax communities where only some mangrove forests, swamps, and marshes are not as yet developed to maturity (Specht 1958). Each of the chief physiographic zones has a complex and varied flora. The plateau is mainly under spinifex grass, with clumps of savannah woodland on the gentler slopes and patches of monsoon forest in deep gorges and valleys. The alluvial plains are covered with tall trees such as eucalypts and cypress pines, as well as palms, annual grasses, and herbs. The coastal plain supports wild rice and sedge on the estuarine flats, whilst bamboos and reeds grow along stream-beds and a wide range of water plants such as lotus lilies grow in swamps and lagoons. Many of the plateau plants grow on the borders of the scarp and on rocky outliers on the plain, but the aquatic plants are restricted to permanent stretches of water on the coastal flats.

The fauna of the area includes a wide range of indigenous terrestrial and aquatic forms. The land fauna is a typical, tropical Australian savannah-woodland one, which lacks the large representation of New Guinea elements found in Cape York. All animal phyla are represented, and a diversity of genera occurs, including bandicoots, possums, kangaroos, wallabies, flying foxes, and rats. These animals live on both the plateau and plain though many retreat to higher ground at the height of the wet season.

Many types of birds are present, and large flocks of water fowl congregate around permanent lagoons on the plain. The aquatic fauna is restricted mainly to the flats where both fresh and brackish water forms occur, including fish, tortoises, turtles, crocodiles, shellfish, and crabs. Marine phyla are present but are poorly represented owing to the distance from the open sea. Some freshwater fish and molluscs live in streams and pools in the plateau valleys but there are no estuarine forms there.

To summarise, therefore, we are dealing with a region of marked physiographic, climatic, and biotic contrasts which affect human behaviour today and also presumably influenced it in prehistoric times. Of course these contrasts are not absolute: for instance, plateau and plain are not exclusive features, for there are tongues of broad, flat valleys in the escarpment, and the rocky country sprawls in a broken series of outliers over the plain. Again, although there is marked seasonality, the intensity and duration of the cycles may vary considerably even from year to year. Finally, although many similar kinds of plants and animals live in both zones, the plain supports a richer and more varied biota, mainly because it is an ill-drained feature with many tracts of permanent water.

## The Archaeological Data

An initial survey of the region, followed by detailed excavations at five rock shelters, provided a broadly consistent and coherent picture of the prehistory there (White, C. 1967a). Generally speaking, two successive industrial traditions can be distinguished on stratigraphical and typological grounds. The earlier one is present both on the plain and in the nearby plateau valley and is dated from about 22,000 to 6,500 years BP. The later tradition occurs as two marked regional variants lasting from about 6,500 BP until the ethnographic present, some 100 years ago. Fig. 12:2 is a purely diagrammatic representation of the contents of the excavated sites to show the positions of dated charcoal

*Fig. 12:2 Diagrammatic presentation of contents of excavated sites*

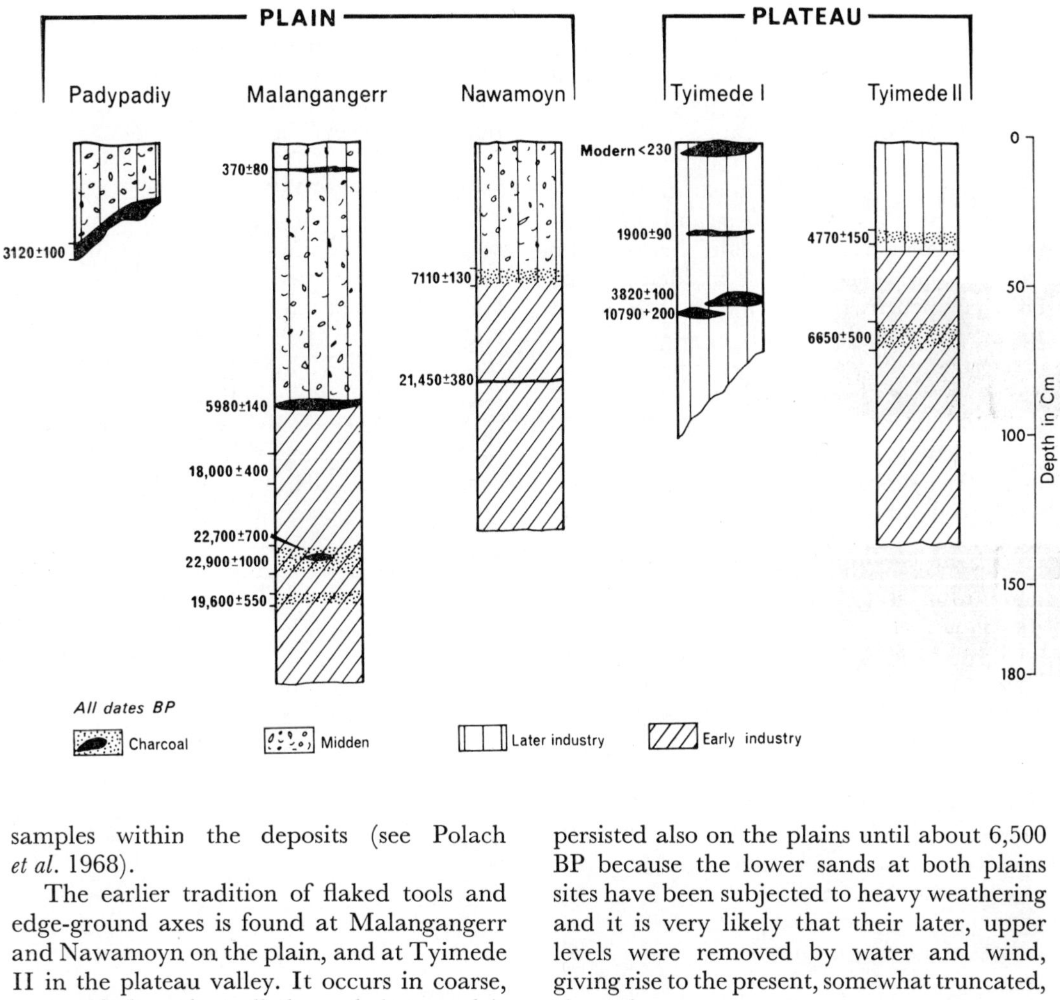


samples within the deposits (see Polach *et al.* 1968).

The earlier tradition of flaked tools and edge-ground axes is found at Malangangerr and Nawamoyn on the plain, and at Tyimede II in the plateau valley. It occurs in coarse, unstratified sand at all three shelters and is dated by a consistent series of radiocarbon dates at both plain sites to about 22,000 to 18,000 years BP, but only to around 6,500 BP at Tyimede II. Given these dates, it might be tempting to draw broad conclusions about the relative antiquity of occupation in these two adjacent ecological areas, but it should be borne in mind that the industrial tradition of the early layers could have persisted also on the plains until about 6,500 BP because the lower sands at both plains sites have been subjected to heavy weathering and it is very likely that their later, upper levels were removed by water and wind, giving rise to the present, somewhat truncated, chronology.

The industry is characterised by stone artifacts only, because extensive post-depositional weathering has destroyed any evidence of the economy, physical type, and supplementary artifacts of the people who occupied the shelters at that time.

The later industrial tradition occurs directly above the earlier flaked and ground axe series at the three sites already mentioned, and it is also found at another plain site, Padypadiy, and another shelter in the plateau valley, Tyimede I. It can be distinguished from the preceding series by the presence of small stone points and scrapers, which are totally lacking in the earlier industries. The assemblages lie in coarse, unstratified sands at both the Tyimede sites and in well-stratified middens in the plain shelters. The earliest appearance of these middens, whose chief component is estuarine shell, is dated to about 6-7,000 years BP at both Malangangerr and Nawamoyn, but there is some discrepancy in the earliest dates of the 'point-scraper' industries at Tyimede.

Tyimede I and II are only about 100 m apart. At Tyimede II, the point-scraper series lies directly above an earlier industry dated to 6,650 ± 500 BP (ANU-18), and the lowest levels of the later industry there are dated to 4,770 ± 150 BP (ANU-50). Only the later series has been recognised at Tyimede I, and the lowest levels there yielded two inconsistent radiocarbon dates of 10,790 ± 200 BP (GaK-632) and 3,820 ± 100 BP (ANU-52). According to the evidence from Tyimede II, men were *not* making and using stone points and small, rectangular scrapers in the vicinity some 10,000 years ago, but unfortunately those characteristic elements of the early industry (such as chunky, thick-sectioned scrapers, utilised flakes, and edge-ground axes) also are all present in association with points and small scrapers in the later series, making it virtually impossible to distinguish the earlier tradition on the basis of 'type fossils'.

It seems likely, therefore, that the people who camped at Tyimede I some 10,000 years ago neither made nor used stone points and small scrapers, but that these tools were trodden and scuffed down into the more ancient hearths by later occupants. In short, it appears advisable to treat the early date of points and small scrapers at Tyimede I both as anomalous and with some reservation until we have less equivocal evidence that such tools were current in the area earlier than 6-7,000 years ago.

**Climate and Environment**

As a general rule, attempts are made to reconstruct prehistoric environments from evidence in occupied rock shelters by analysing such elements as pollen grains, sediments, and food debris. Pollen is present throughout the deposit at Malangangerr (W. Litchfield pers. comm.) and probably occurs at all the other excavated sites too, but unfortunately at this stage there is no detailed key available to identify and interpret the remains to the extent that climatic inferences are possible. Most of the excavated deposits are composed of locally derived sand and, given the current knowledge of soils in the area, it is impossible to say whether sedimentary changes reflect the climatic conditions prevalent during deposition in the shelters or in the original parent rock.

We therefore have to rely most heavily upon the analysis of food debris to reconstruct prehistoric climates, and unfortunately these data are present only in association with the later industry on the plain sites. Of course, many faunal and floral remains in archaeological sites have passed through a cultural filter so that they often do not reflect the whole biotic pattern in the surroundings, but fortunately there are many fairly detailed ethnographic observations in Arnhem Land which allow us to assess the implications of dietary remains more fully and coherently.

Now that I have set the somewhat daunting limitations to this discussion, I should like to examine the extent of possible deductions, by drawing not so much on the immediate evidence as on general premises.

Men lived near Oenpelli more than 20,000 years ago. It was roughly at this time that the Wurm glacial advance in the Northern Hemisphere reached its peak. Movius (1960) places the peak at 26-15,000 BP; Fairbridge (1961) estimates 25-17,000 BP and Butzer (1964), 28-16,000 BP. Whilst it is difficult to estimate quite how much water was locked up in icesheets then, there was unquestionably a major drop in sea level at that time. Most authors (Fairbridge 1961; Cotton 1962; Curray 1961; Butzer 1964; van Andel *et al.* 1967; Jennings, see ch. 1) agree that it dropped some 100-150 m below its present level. Land bridges linked Tasmania and New Guinea to the Australian mainland at that time, and the point I wish to stress is the extent of the northern land bridge. Both the Arafura Sea and the Gulf of Carpentaria are relatively shallow bodies of water: Arafura is an average of 50-80 m deep and the Gulf is only 46-65 m deep (Tjia 1966:44; Phipps 1966:317). The main Wurm marine regression must have exposed a plain in the north about 1,300-1,600 km long and 640-800 km wide. Under those conditions, sites like Malangangerr and Nawamoyn probably stood some 90-120 m above sea level and possibly up to 320 km inland from the nearest coastline (see Fig. 1:1).

There are two points which follow. The first concerns the climate and environment in northwest Arnhem Land at the peak of the Wurm. We have already noted the change in land/sea relationships. Furthermore, palaeoecological studies in Australia (Galloway 1965; Dury 1967) and elsewhere in the Southern Hemisphere (van Zinderen Bakker 1967) suggest a drop of at least 5°C in land temperatures at that time. This was associated with a corresponding reduction in surface sea temperatures of the order of about 5°C for tropical waters (Emiliani 1958; Fairbridge 1961:133; van Zinderen Bakker 1967:134). Van Zinderen Bakker applies the principle of 'ecological actuality' to interpret climatic changes in sub-Saharan Africa over the past 70,000 years, because he assumes that the wind and sea current systems remained substantially similar to present conditions through that time (van Zinderen Bakker 1967:126). Using the same assumption, I shall speculate upon the changes in the north Australian environment during the last Ice Age.

The present distribution of rainfall in Arnhem Land is related both to the distance from the sea and from the equator. The land/sea relationships changed during the Wurm glaciation, and also presumably the latitudinal belts shrank at that time, thus altering the effective distance from the thermic equator. Brookes has suggested (Fairbridge 1953:9) that less rain fell on the present north Australian coast at that time, because the summer winds blew over rises in a broad expanse of land exposed by the marine regression and dropped some of their moisture before reaching the present coastal plain. However, this need not necessarily have been the case: the effective annual rainfall near Oenpelli may have been very similar to present conditions, because reduced land temperatures probably allowed less evaporation, compensating for any slight drop in mean annual rainfall. (The whole question of evaporation is complex: less evaporation may lead to a reduction in the cloud cover, but this in turn will allow more sunlight to reach the earth and thereby promote more evaporation.) Nor does it seem that the region under discussion was necessarily any better drained 20,000 years ago than it is today. Assuming that no major warping has taken place along the coast since this time, and making allowance for the accumulation of submarine sediments near

*Fig. 12: 3 Pleistocene edge-ground axes from Arnhem Land. a, Malangangerr; b and c, Nawamoyn.*

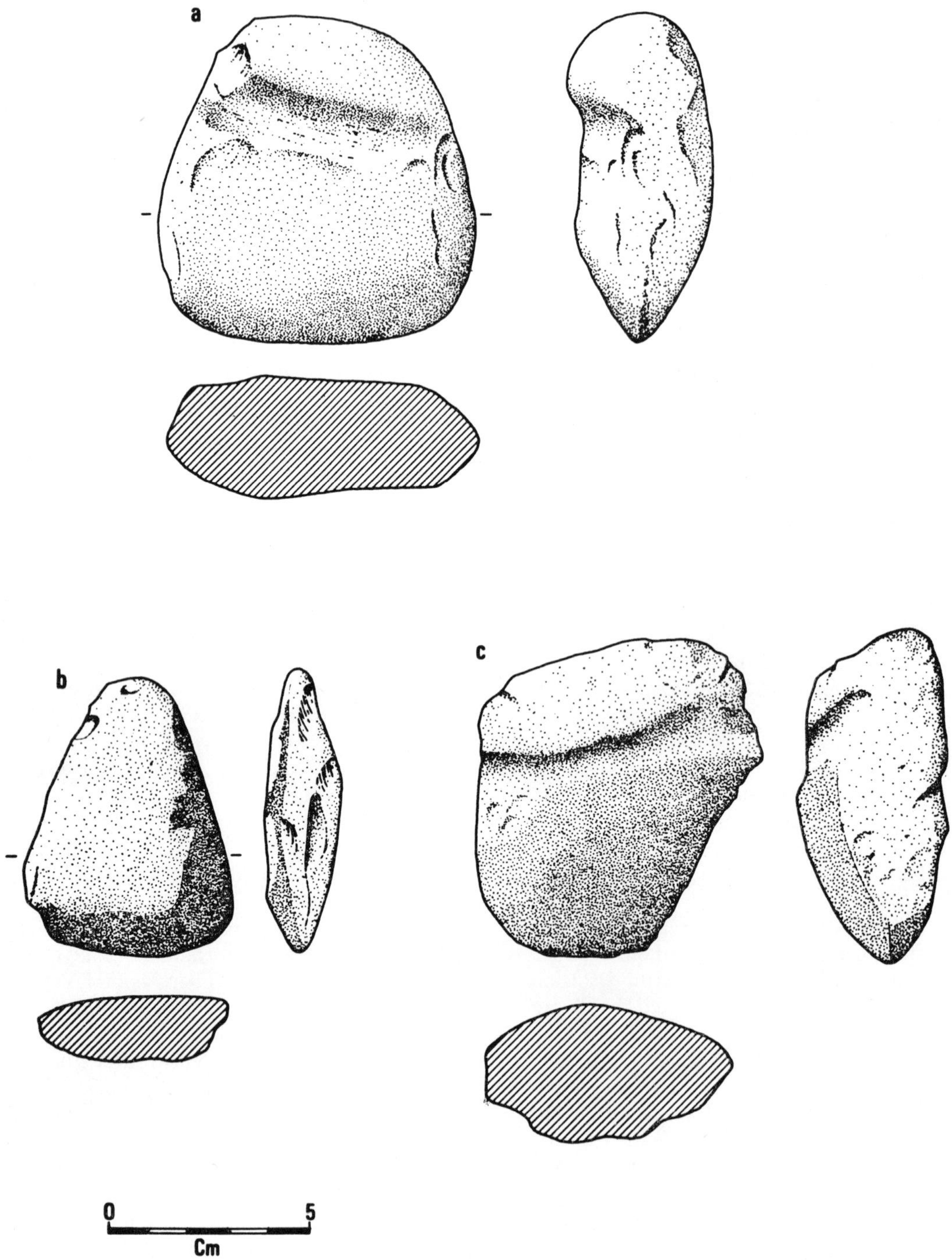

mangroves in post-glacial times, there is no evidence of a steep slope off the Arnhem Land coast, and the marine charts show a very gradual decline. The most striking difference in the environment near Oenpelli 20,000 years ago was probably the absence of estuarine conditions in the area due to the great distance from the sea.

I suggest therefore that during the peak of the last glaciation the climate and economy of the plain near Oenpelli were most like those found today in broad, well-drained plateau valleys. This theory would serve to explain why the archaeological deposits in plain shelters, dated to 18-20,000 BP, are so similar to those found in plateau valley shelters, dated as recently as the ethnographic present. All contain stone artifacts and occasional lumps of charcoal in coarse, weathered sand, with scarcely a trace of bone or shell. This does not mean that food debris was never present, but rather that it was destroyed more rapidly because the preservative agent, estuarine shell, was never there.

The second point is that the evidence of the earliest landfall on the Australian-New Guinea landmass was drowned by the rising post-glacial sea. The 20,000-years-old occupational debris in Arnhem Land was probably deposited by people who had already penetrated over 160 km into their new homeland. Therefore these remains already possibly reflect some degree of adaptation to the new environment. Quite how much adaptation was needed is a moot point, for it seems that once men had passed through Wallacea, the most striking difference they encountered in the new lands was neither the climate nor the vegetation (Burbidge 1960), but the peculiar marsupial fauna. One might argue (Calaby, see ch. 7) that no particular new adaptation was required for hunter-gatherers from Southeast Asia to hunt marsupials, since most extant forms probably could be treated in much the same way as their mammalian counterparts west of Wallacea. But it is nonetheless interesting to examine the archaelogical remains of the early settlers in north Australia to try to assess what degree of adaptation might have taken place.

In order to discuss this question at the primary level, we need the association of artifacts used to hunt and/or dress the prey, and the beasts themselves. I have indicated already that there are no faunal remains associated with the earliest industries in northwest Arnhem Land, and unfortunately the same may be said also of all similar early industries in other parts of Australia. We are forced to examine therefore the only available data, the nature and significance of the stone assemblages from the lower sands at Malangangerr, Nawamoyn, and Tyimede II, which are dated from about 22,000-6,500 BP. The artifactual assemblages from all three sites include a flaked element—steeply retouched, thick-sectioned scrapers, some of which resemble horsehoof core-scrapers, and utilised flakes—as well as a flaked and ground element of edge-ground axes made mostly of igneous/metamorphic rocks, several of which have apparent hafting devices such as grooves, stems, or nicks (Fig. 12:3). In addition to these forms, stone hammers, grinders, and fragments of ochre also occur.

Assemblages containing a similar flaked element have been found stratified (cf. Fig. 12:4) beneath assemblages with points, backed blades, and small scrapers at several other sites throughout Australia (cf. Fig. 12:5). These include Sleisbeck (Golson 1964) and Ingaladdi (Mulvaney and Joyce 1965: 207) in the Northern Territory; Laura (Wright 1963, and see ch. 11), Kenniff Cave, and the Tombs (Mulvaney and Joyce 1965) in Queensland; Seelands (McBryde 1966), Curracurrang (Megaw 1965), Noola (Tindale

*Fig. 12:4 Selected examples of one thick, steeply retouched flake (a) and 3 core-scrapers (b-d), including the horsehoof type (c). a and b, Mt Burr, S.A.; c, Kenniff Cave, Qld; d, Ingaladdi, N.T.*

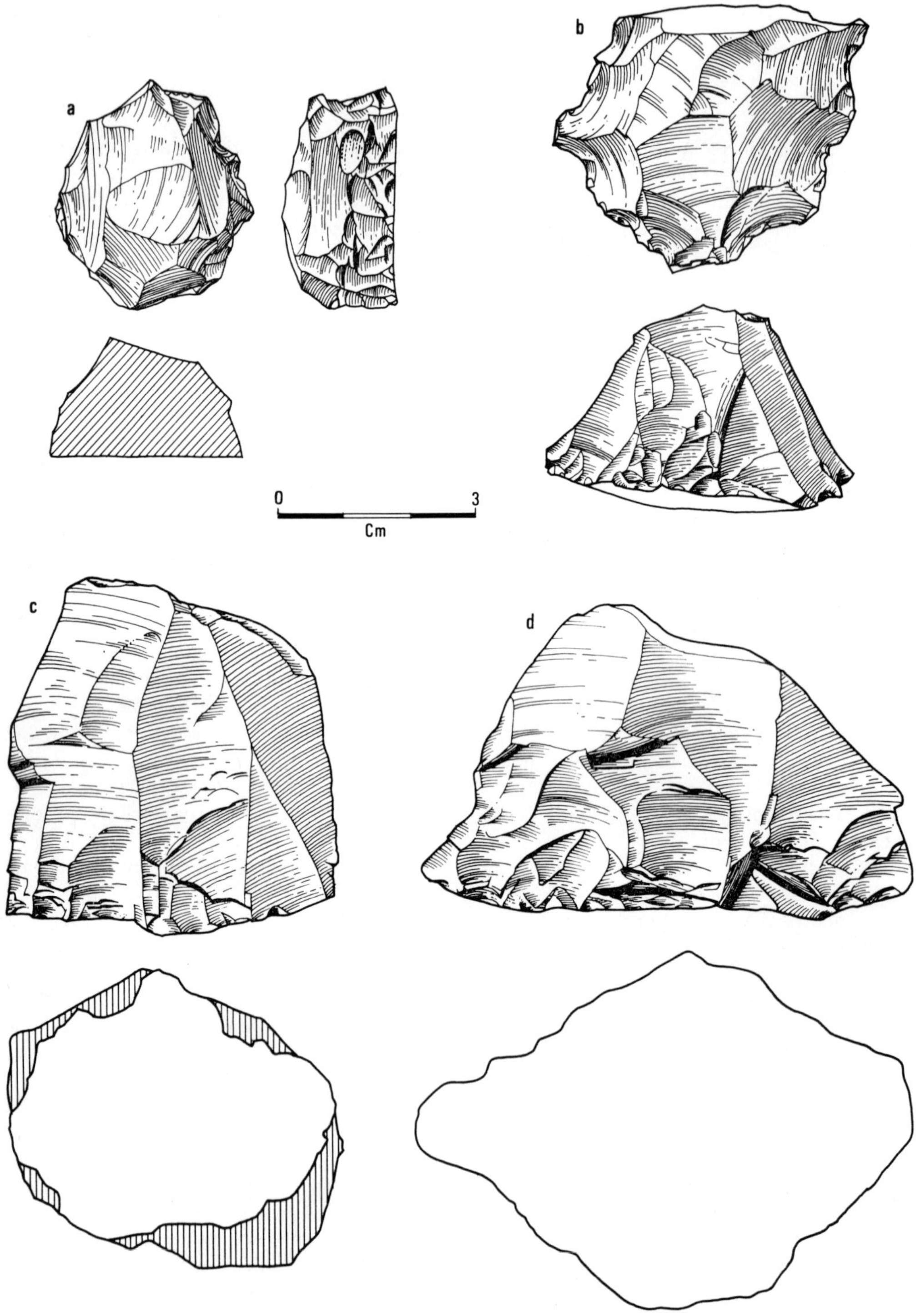

*Fig. 12: 5 Selected examples of bifacial (a-c) and unifacial (d-g) points. a, c-d, Ingaladdi, N.T.; b, Padypadiy, N.T.; e-f, Mulka, S.A.; g, Adelaide, S.A. e-g are pirri points.*

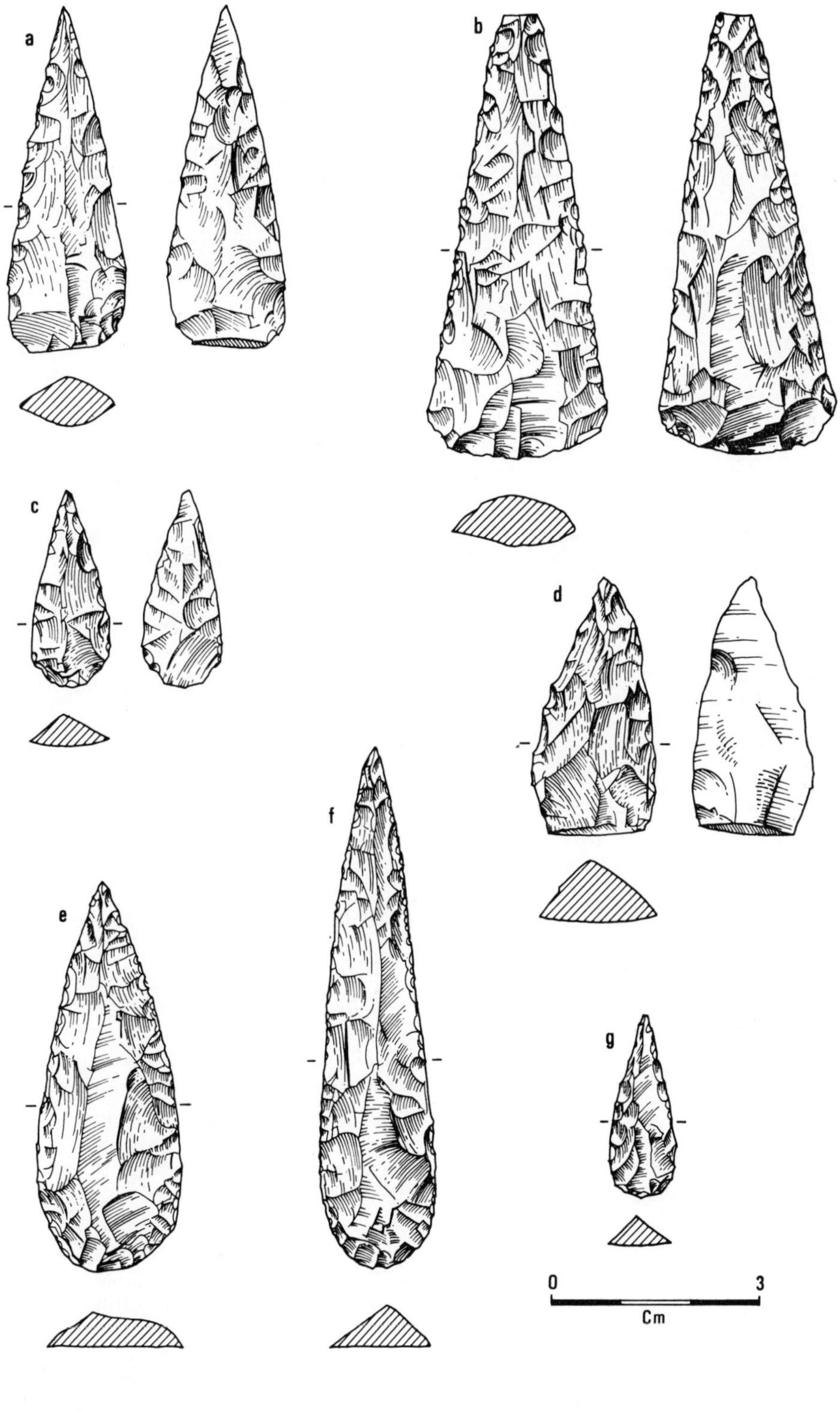

1961), Capertee (McCarthy 1964), and Lake Burrill shelter (Lampert, see ch. 10) in New South Wales; and Puntutjarpa shelter (Gould 1968) in Western Australia. Their rarity in Western Australia is probably due more to the relative lack of work there than to any other cause.

According to the published reports, none of these early assemblages contained edge-ground axes, so that the Oenpelli material appeared to be unique. Recently, however, some new analyses have been made which suggest that this is not the case.

**1** At Sleisbeck, the earliest assemblage of stone tools contains flaked implements that are very similar to those found in the earlier series at Oenpelli, including numerous step-flaked scrapers and horsehoof core-scrapers. These are closely associated with several edge-ground axes and with numerous flakes of igneous/metamorphic rock, many of which have ground surfaces. Charcoal associated with an axe from the base of the occupational level has been dated to about 6,500 BP (J. Golson pers. comm.), which agrees well with the date of a similar industry at Tyimede II, about 160 km north of Sleisbeck.

**2** At Laura, intact axes occur in the upper 1·2 m of the deposit together with small rectangular scrapers, parallel-sided adze flakes, and eloueras. The axes are the only tools made of igneous/metamorphic rock and they are associated with numerous fragments of such rock, some of which have ground surfaces. Below 1·2 m, the flaked stone tools include thick-sectioned scrapers and horsehoof core-scrapers. There are no intact axes in this collection, but about thirty flakes of igneous/metamorphic rock are present here, all of which show signs of grinding, especially one from the lowest occupational level which has two intersecting ground surfaces and possibly represents part of the blade of an edge-ground axe. Unfortunately the lower levels of Laura are undated and the only radiocarbon estimate from 4 m of occupation debris comes from a level 2 m below the surface and is about 7,000 years old (Wright, see ch. 11).

**3** At Kenniff Cave, two main cultural levels were distinguished (Mulvaney and Joyce 1965). The lower one contained stone tools such as scrapers, cores, and waste flakes, whilst the upper level, which extended to a depth of 1·2 m below the surface, contained forms similar to those found in the lower level, as well as new types of implements such as points, backed blades, tula adzes and possibly pieces of an edge-ground axe (1965:188-9). On the basis of this change in stone tool types, the material was analysed in terms of the introduction of the hitherto unknown technique of hafting, which was said to appear in this deposit 4-5,000 years ago (cf. Mulvaney 1966).

There are no intact axes in this collection, but the report notes that fragments of volcanic rock are confined to the top 90 cm of the deposit, and adds that two of these fragments 'are ground, and appear to be broken from the same tool, which is presumed to have been an edge-ground axe' (Mulvaney and Joyce 1965:189). However, the position is not as unequivocal as this, because the report also observes that basalt pebbles with ground surfaces, termed 'millstones', are present in the lower 'non-hafted' levels (1965:192). A re-examination of the collection showed that basalt flakes occur in small numbers throughout the deposit: there are thirty-three in the upper 1·2 m and twelve below this depth. Both of the two ground basalt flakes from the upper level (1965:189) and a broken millstone from level 30 are all ground in one plane only, and none has the intersecting ground planes characteristic of a broken axe blade, so that all three could equally well be broken parts of axes *or* millstones. In short, given the total absence of definite signs of axe making or use, there is as much inferential evidence for the presence of edge-ground axes in the lower levels as in the upper ones. The rarity, if not total lack of axes at Kenniff is perhaps unusual, but it seems wrong in view of the evidence from other sites in Australia to use the Kenniff data to deny the presence of edge-ground axes in Australia before 5,000 BP.

**4** Finally, at Curracurrang, Megaw has delineated three main cultural and stratigraphical levels. The earliest industry is contained in the 'Bottom' level in coarse sand, and is dated to *c.* 7,500 BP (Megaw 1965). Pebble tools and heavy scrapers predominate here, but bondi points and axes are also present in the upper part of the sand, which has now been termed the 'Middle-Bottom' level. It is very likely that the backed blades were derived from the rich Bondaian level above, but it is less certain that the axes came from the same source, since axes are very rare in the Bondaian assemblage at this particular site.

Of the four cases listed, only the first two duplicate the Oenpelli evidence. The Kenniff findings neither support nor negate it and the cultural sequence at Curracurrang is so far unclear. Given the present evidence, therefore, it seems that edge-ground axes were made and used by the earliest men living in tropical north Australia, but not by those living further south. Evidence from the Niah Cave in Borneo suggests that edge-ground tools were in use there about 15,000 years ago (Golson 1970), and waisted or ground axes were present in New Guinea during the earliest occupation there (White, J. P. 1967:329, 408-9). Possibly this fragmentary evidence indicates that axes were tools adapted for use in tropical rather than temperate regions, and that one of the adaptive mechanisms used by men penetrating southwards through central Australia was to discard this particular trait rather than to develop it.

The absence of axes in Tasmania supports this contention to some extent. Tasmania was cut off from the mainland by the rising post-glacial sea by 10,000 years ago (Jennings, see ch. 1), so that it is assumed that the inhabitants had reached the island before that date, and that they were ignorant of axe making before they reached Tasmania. In contrast to this, there is evidence that prehistoric Tasmanians about 3,500 years ago abandoned, for no discernible reason, the practice of fishing and we might argue, therefore, that they abandoned the use of axes in a similar fashion too (see Jones 1966, 1968).

This is all very speculative, and there are simpler alternative explanations, such as absence of research, rather than absence of evidence. At this stage, however, all that can be deduced from the stone kit of the earliest inhabitants of Arnhem Land is that they selected certain kinds of stone to make specific types of tools and that they probably pulverised vegetable foods (anvils, pounders, grinders), chopped wood or trees to fashion other tools or to catch arboreal prey (axes) and cut and scraped various organic substances (utilised flakes).

A more detailed picture is available of the occupation of the Oenpelli area in post-glacial times, mainly because more evidence of this age has been preserved. The chief preservative agent is estuarine shell, which appears quite suddenly in the deposits in two shelters on the coastal plain, Malangangerr and Nawamoyn, about 6-7,000 years ago. Clearly, by then estuarine conditions were prevalent in the vicinity, and the age of the earliest middens accords well with the date of a peak in the Flandrian transgression of Europe (Fairbridge 1961:156-62) and a high point in the post-glacial marine transgression in Australia (Fairbridge 1961; Gill 1961; Shepard 1961; Hails 1965; Butzer 1964:210; Jennings, see ch. 1). It is worth noting that estuarine fauna are present in these sites until modern times, suggesting that the local environment has not changed substantially for the past 7,000 years.

The analysis of flora and fauna in three plain sites of this period shows that the inhabitants exploited an economy similar to that found on the plains today. They ate freshwater and estuarine shellfish, fish, and tortoises as well as macropods and small marsupials like bandicoots and possums. They also ate birds such as palmated geese, and probably ate plants like lotus lily rhizomes and cycad nuts, though macroscopic plant remains are preserved only at Padypadiy (White, C. 1967c). The economy remained broadly the same throughout this period, though some change can be seen in the relative proportions of different shell species over time. Unfortunately too few marsupials are present to infer any changes in hunting practices or diet over time, but one specimen of the Tasmanian devil (*Sarcophilus harrisi*), now extinct in mainland Australia, was found at Padypadiy (Calaby and White 1967).

The Tyimede shelters were occupied whilst the middens accumulated on the plains, but these plateau valley shelters contain hardly any organic remains so that the economy of the inhabitants has to be deduced from less direct sources. Two lines of evidence are present. Firstly, estuarine fauna—notably shellfish—are entirely absent, suggesting that the inhabitants did not collect food on the plains at the same time as they camped at Tyimede, but probably hunted and gathered most of their food in the immediate vicinity. The second line of evidence is more oblique and stems from the contrast between the assemblages of stone and bone artifacts found in shelters on the plain and in the plateau valley.

To put it briefly, stone points, small scrapers, utilised flakes, and edge-ground axes occur in both areas, but there are considerably fewer stone tools in the plain shelters and those that are present do not appear to have been made *in situ*. Prolific stone industries are found at both Tyimede sites and the analysis of the ratio of waste flakes to secondarily retouched implements indicates that many tools were manufactured on the spot. In contrast to this, there are many bone points, shell tools, and even wooden ones in the middens on the plain, but very few bone tools and no shell or wooden ones in the Tyimede shelters. Remembering that Tyimede and the plain sites are only about 22·5 km in a direct line apart, and that the raw material for making stone and bone tools is, and presumably always was, present in both areas, we are faced with the problem of interpreting a cultural dichotomy that lasted from about 6,500 years ago until the ethnographic present.

It was originally argued (White, C. 1967c) that this dichotomy reflected the presence of two distinct sub-cultures in the area, but since then more detailed work has led to the suggestion that it probably reflects the seasonal movements and accompanying dietary changes that forced some prehistoric hunter-gatherers to leave the flooded plains during the wet season, and retreat to the hills until the rains eased and the plains dried out, so that they could return again (White and Peterson 1969). What emerges from this work is the extent to which modern ethnographic studies can be used to interpret Australian prehistoric situations.

There is a marked contrast between the

quality of the evidence present of the two occupational levels in the Oenpelli area. It seems very unlikely that faunal remains will be preserved in deposits in sandstone rock shelters unless a neutralising preservative agent such as shell is also present. One might therefore be tempted to search for early deposits in limestone caves in order to discover the associated faunal remains, though Mulvaney's 1963 excavations at the Kintore Cave, a limestone cave near Katherine, yielded disappointing results. He discovered two successive cultural levels, similar to those found near Oenpelli and at Sleisbeck, but he found no associated faunal material (D. J. Mulvaney pers. comm.). On a more optimistic note, recent work on the soils of the coastal plain southwest of Oenpelli has revealed a sequence of non-marine sands overlain by a few metres of clay that probably register the arrival of estuarine conditions in that particular area. It has also been suggested that borings be made in old land-locked swamps in order to study the botanical conditions that prevailed when men first arrived in north Australia (A. Hooper pers. comm.).

These kinds of data will undoubtedly amplify the picture that I have tried to draw in this paper, but until more research is done, the inferences I have made remain largely speculative.

**References**


Burbidge, Nancy T. 1960. The phytogeography of the Australian region. *Aust.J.Bot.* **8**:75-209.

Bureau of Meteorology, Australia. 1961. Climatological survey region I—Darwin-Katherine Northern Territory. Melbourne: Director of Meteorology.

Butzer, K. W. 1964. *Environment and archaeology.* Chicago: Aldine Publishing Co.

Calaby, J. H. and White, Carmel. 1967. The Tasmanian devil (*Sarcophilus harrisi*) in northern Australia in Recent times. *Aust. J.Sci.* **29**:473-5.

Cotton, C. A. 1962. Low sea levels in the Late Pleistocene. *Trans.R.Soc.N.Z.—Geology* **1**:249-52.

Curray, J. R. 1961. Late Quaternary sea level: a discussion. *Bull.geol.Soc.Am.* **72**: 1707-12.

Dury, G. H. 1967. Climatic change as a geographical backdrop. *Aust.Geogr.* **10**: 231-42.

Emiliani, C. 1958. Paleotemperature analysis of core 280 and Pleistocene correlations. *J.Geol.* **66**:264-75.

Fairbridge, R. W. 1953. The Sahul shelf, northern Australia; its structure and geological relationships. *J.Proc.R.Soc.West. Aust.* **37**:1-33.

——1961. Eustatic changes in sea level. *Physics Chem.Earth* **4**:99-185.

Fortune, R. F. 1935. Manus religion. *Mem.Am. phil.Soc.* **3**.

Galloway, R. W. 1965. Late Quaternary climates in Australia. *J.Geol.* **73**:603-18.

Gill, E. D. 1961. Changes in the level of the sea relative to the land in Australia during the Quaternary Era. *Z.Geomorph.* Suppbd **3**:73-9.

Golson, J. 1964. Australia. In *COWA surveys and bibliographies* **3**, Area 22. Cambridge (Mass.): Council for Old World Archaeology.

——1970. Both sides of the Wallace line: New Guinea, Australia, Island Melanesia and Asian prehistory. In *Early Chinese art and its possible influence in the Pacific Basin: a symposium* (ed. N. Barnard), in press. New York: Intercultural Arts Press.

Gould, R. A. 1968. Preliminary report on excavations at Puntutjarpa rockshelter, near the Warburton Ranges, Western Australia. *Archaeol.& phys.Anthrop.Oceania* **3**:161-85.

Hails, J. R. 1965. A critical review of sea-level changes in eastern Australia since the Last Glacial. *Aust.geogrl Stud.* **3**:63-78.

Jones, R. 1966. A speculative archaeological sequence for north-west Tasmania. *Rec. Queen Vict.Mus.* No. 25.

——1968. The geographical background to the arrival of man in Australia and Tasmania. *Archaeol.& phys.Anthrop.Oceania* **3**: 186-215.

McBryde, Isabel. 1966. An archaeological survey of the New England Region, New South Wales. Ph.D. thesis, University of New England, Armidale (unpublished).

McCarthy, F. D. 1964. The archaeology of the Capertee Valley, New South Wales. *Rec.Aust.Mus.* **26**:197-246.

Megaw, J. V. S. 1965. Excavations in the Royal National Park, New South Wales: a first series of radiocarbon dates from the Sydney district. *Oceania* **35**:202-7.

Movius, H. L. 1960. Radiocarbon dates and Upper Palaeolithic archaeology in central and western Europe. *Curr.Anthrop.* **1**: 355-91.

Mulvaney, D. J. 1966. The prehistory of the Australian Aborigine. *Scient.Am.* **214** (3):84-93.

——and Joyce, E. B. 1965. Archaeological and geomorphological investigations on Mt. Moffatt Station, Queensland, Australia. *Proc.prehist.Soc.* **31**:147-212.

Phipps, C. V. G. 1966. Gulf of Carpentaria (northern Australia). In *The encyclopedia of oceanography* (ed. R. W. Fairbridge), pp. 316-24. Encyclopedia of Earth Sciences Ser. 1. New York: Reinhold Publishing Corp.

Polach, H. A., Golson, J., Lovering, J. F., and Stipp, J. J. 1968. ANU radiocarbon date list II. *Radiocarbon* **10**:179-99.

Shepard, F. P. 1961. Sea level rise during the past 20,000 years. *Z.Geomorph.* Suppbd **3**:30-5.

Southern, R. L. 1966. A review of weather disturbances controlling the distribution of rainfall in the Darwin-Katherine region, N.T. Bureau of Meteorology Working Paper 65/3203 of May 1966. Melbourne: Director of Meteorology (mimeographed).

Specht, R. L. 1958. The climate, geology, soils and plant ecology of the northern portion of Arnhem Land. In *Rec.Am.-Aust.scient.Exped.Arnhem Ld* **3** (Botany and Plant Ecology) (eds. R. L. Specht and C. P. Mountford), pp. 333-413. Melbourne: Melbourne University Press.

Thomson, D. F. 1939. The seasonal factor in human culture. Illustrated from the life of a contemporary nomadic group. *Proc.prehist.Soc.* **5**:209-21.

——1948-9. Arnhem Land: explorations among an unknown people. *Geogrl J.* **112**: 146-64; **113**:1-8; **114**:53-67.

——1949. *Economic structure and the ceremonial exchange cycle in Arnhem Land.* Melbourne: Macmillan.

Tindale, N. B. 1961. Archaeological excavation of Noola rock shelter: a preliminary report. *Rec.S.Aust.Mus.* **14**:193-6.

Tjia, H. D. 1966. Arafura Sea. In *The encyclopedia of oceanography* (ed. R. W. Fairbridge), pp. 44-7. Encyclopedia of Earth Sciences Ser. 1. New York: Reinhold Publishing Corp.

van Andel, T. H., Heath, G. R., Moore, T. C., and McGeary, D. F. R. 1967. Late Quaternary history, climate, and oceanography of the Timor Sea, northwestern Australia. *Am.J.Sci.* **265**:737-58.

van Zinderen Bakker, E. M. 1967. Upper Pleistocene and Holocene stratigraphy and ecology on the basis of vegetation changes in sub-Saharan Africa. In *Background to evolution in Africa* (eds. W. W. Bishop and J. D. Clark), pp. 125-47. Chicago, London: University of Chicago Press.

White, Carmel. 1967a. Plateau and plain:

prehistoric investigations in Arnhem Land, Northern Territory. 2 vols. Ph.D. thesis, Australian National University, Canberra (unpublished).

——1967b. Early stone axes in Arnhem Land. *Antiquity* **41**:149-52.

——1967c. The prehistory of the Kakadu people. *Mankind* **6**:426-31.

——and Peterson, N. 1969. Ethnographic interpretations of the prehistory of western Arnhem Land. *SWest.J.Anthrop.* **25**:45-67.

White, J. P. 1967. Taim bilong bipo: investigations towards a prehistory of the Papua-New Guinea Highlands. 2 vols. Ph.D. thesis, Australian National University, Canberra (unpublished).

Worsley, P. M. 1961. The utilization of food resources by an Australian Aboriginal tribe. *Acta Ethnogr.Hung.* **10**, Fasc. 1-2: 153-90.

Wright, R. V. S. 1963. Report on archaeological reconnaissance work in Cape York. Canberra: Australian Institute of Aboriginal Studies, Doc. 63/83 (mimeographed).

# 13 Prehistoric Research in Timor

I. C. Glover

In island Southeast Asia, and to some extent in Melanesia too, even the broad outlines of prehistory are not clearly understood. Archaeological research has often been concentrated on detailed studies of 'interesting' or unusual artifacts and pottery styles, rather than on establishing firmly dated regional sequences.

In 1966 and 1967 I spent ten months locating and excavating archaeological deposits in the eastern part of Timor, the largest of the Lesser Sunda Islands, 475 km long by 100 km wide, and situated approximately 9°S by 125°E, and 400 km north of the Kimberley region of Western Australia.

The principal aim of the excavations was to produce sequences of stone and pottery artifacts and bone food remains from the main environmental zones, which would enable the principal technical and economic changes in the eastern end of the island to be dated, and to provide a chronological framework to which surface finds and undated excavated material could be related.

Knowledge of the prehistory of the region was not such that problems could be narrowly defined and sites sought to answer them, but it was expected that some light might be thrown on the following questions:

**1** The timing of the first movement of the Aborigines into Australia and the anatomical status of these first migrants. Today Timor is

I wish to thank H. A. Polach for radiocarbon dates, J. A. Mahoney, J. H. Calaby, D. Witter, C. F. W. Higham, and P. Thompson for their advice with the faunal identifications, and numerous colleagues for their helpful comments and discussions, including F. J. Allen, Emily Glover, J. Golson, C. A. Key, R. J. Lampert, H. H. E. Loofs, D. J. Mulvaney, J. N. Peterson, R. P. Soejono, Sutjipto Wirjosuparto, Carmel White, J. P. White, and R. V. S. Wright. I acknowledge the assistance of the Portuguese Government and the people of Timor for their help in facilitating the field work on which this paper is based.

basically Indonesian in culture and race, but the island has often been cited as a possible migration route for Pleistocene man into Australia and Melanesia (Macintosh 1963: 248; Mulvaney 1961:62; Birdsell 1958:60, 1967:145-50) and a number of anthropometrists have claimed to find Melanesoid and Australoid characteristics among the peoples of the interior of the island (Bijlmer 1929: 19-20; Keers 1948; Nyessen 1945:96-7; van Bork-Feltkamp 1951).

**2** The dating and immediate origin of a series of technical innovations which appear in Australian stone industries about or before 7,000 BP and for which there is some evidence to indicate an origin outside Australia (Mulvaney 1969: 140-53; White, J. P., see ch. 14).

**3** The spread of pottery, horticulture, and domesticated animals such as pigs into eastern Indonesia and Melanesia from, presumably, western Indonesia and mainland Southeast Asia.

**4** The dating and more exact definition of the prehistoric culture found in western Timor by Bühler, Willems, and Verhoeven (Sarasin 1936; *Oudheidkundig verslag* 1939: 12-13; Verhoeven 1959: 970-2).

**5** The composition of the endemic land fauna of Wallacea, and the dating of the various introduced wild species: *Macacus irus*, *Cervus timorensis*, *Phalanger orientalis*, *Paradoxurus hermaphroditus*, etc. (Laurie and Hill 1954).

In summary, it can be said that as no deposits older than 13,500 years have been excavated, and practically no human skeletal remains have been found, no light has been thrown on the first problem.

Evidence relating to the second problem is entirely negative, as the stone industry found in Timor shows no relationship to any well-defined Australian stone industries.

Abundant faunal remains were found in the limestone caves of Timor which, together with the appearance of pottery, provide the basis for identifying and dating the transition from a hunting and gathering economy exploiting the endemic fauna and flora to one increasingly relying on introduced animals and crops.

The sequences of cultural remains found in Portuguese Timor are generally similar to those revealed by Bühler and Verhoeven in Indonesian Timor, and these can be dated with some confidence.

The faunal remains, which include at least four extinct large murids, support the view that all the existing larger land mammals of Timor have been introduced into the island by man within the past few thousand years and that the endemic land mammals in terminal and post Pleistocene times were mainly, if not exclusively, murids.

## The History of Archaeological Research in Timor

Archaeological work in Timor has been carried out intermittently since 1935 when Alfred Bühler visited the island and excavated a number of sites in then Dutch Timor and a cave near Baguia in Portuguese Timor (Sarasin 1936:1-19). One cave in particular, Abri I at Nikiniki in central west Timor, produced a rich assemblage of stone, pottery, shell, and bone artifacts including 22 tanged flint projectile points and 6 bifacially-worked flint adzes made on flakes struck from prepared cores in the Levallois manner. Except for a possible tanged point found at Kalumpang in west central Celebes (van Heekeren 1949: Fig. 32), such implements have still no clear parallels in Indonesia and Sarasin compared them with finds from Japan, Vietnam, and Melanesia (Sarasin 1936:23, 38-41). Although Bühler's excavations were done with care, the material was published by Sarasin while Bühler was still in the field and the stratigraphical details noted by the excavator have not been published in full (Bühler pers. comm.).

Sarasin divided the material from Nikiniki I into three levels, 0-25, 25-75, and 75-135 cm. However, nearly all the cultural materials relate to the lowest level, which thus contains pottery and flint tools and both wild and domesticated fauna including various small rats, a giant rat, *Coryphomys bühlerei* Schaub, species of *Pteropus*, a bovid probably buffalo, and horse, pig, and dog (Sarasin 1936:31; Schaub 1937:1).

In Nikiniki II further remains of *Coryphomys* were found at a depth of 70-80 cm (Schaub 1937:1) and in Nikiniki III bones identified as either sheep or goat were found at 70-80 cm below the surface.

Both Bühler and Sarasin ascribed these finds to a 'neolithic' culture and Sarasin interpreted the presence of horse and buffalo bones at some depth and the fact that some flint tools occurred towards the top of the cave deposits to mean that this neolithic culture continued in Timor into the recent past (Sarasin 1936:32-3). In the absence of any radiometric means of dating the deposits and of firm links with dated prehistoric cultures outside Timor, this was a reasonable supposition, although the geologist Wanner had already, and more correctly in my opinion, noted that the use of stone tools had long since been abandoned by the Timorese (Wanner 1913:148-9).

In 1938 and 1939 Willems made a number of excavations in west Timor which have only been reported briefly (*Oudheidkundig verslag* 1939: 12-13). A stone assemblage with tanged points was found at Ulnam cave (11 on Fig. 13:1) in the Mutis Mountains and these seem to have been absent in caves towards the north coast.

In 1954 Father Th. Verhoeven excavated two caves in the eastern part of Indonesian Timor: Liang Leluat II (7 on Fig. 13:1) on the Maubesi River in the south, and Liang Djenilu (8 on Fig. 13:1) in the north (Verhoeven 1959: 970-2). At Liang Leluat II a stone industry similar to Nikiniki I was found, with notched scrapers and 10 tanged points which occur only in the top 30 cm of the deposit. At Liang Djenilu no tanged points were found but, although not specifically mentioned, it seems that the notched scrapers were present there.

In 1953 the Portuguese anthropologists, Antonio de Almeida, Mendes Correa, and Ruy Cinatti found flaked stone tools at a number of surface sites on the north and south coasts of Portuguese Timor some of which they assert to be of late Pleistocene age and classified these according to the terminology of European Lower and Middle Palaeolithic stone industries (Correa, Almeida, and Cinatti 1953:52-3; Correa, Almeida, and Camarate 1956:295-8).

At one site (9 on Fig. 13:1) near the salt lake Gassi Issi, 6 km east of Laga on the north coast, they claim some stratigraphic separation for three typologically distinct assemblages.

In 1963 Almeida excavated a small trench in Lene Hara cave (10 on Fig. 13:1) at the far eastern tip of the island (Almeida and Zbyszewski 1967: 55-65) and found a stone assemblage unmixed with pottery which is described as 'typically pre-Neolithic' and related by Almeida to the lower or Proto-Toalian of the Celebes (Almeida and Zbyszewski 1967:64). However, the analysis of neither assemblage is sufficiently detailed in my opinion to permit such comparisons.

In 1964 Father Verhoeven again visited Timor and, in an eroding landsurface near Atambua in the northeast of Indonesian Timor, found remains of two species of stegodonts, placental mammals of Middle to Upper Pleistocene times (Hooijer 1969). He also reported the discovery of retouched flakes in the same area, although the association with the bones may be fortuitous (Verhoeven 1964: 634; see also Glover and Glover 1970).

## The Physical and Cultural Background

### CLIMATE

Timor's climate is ruled by its proximity to Australia. It is strongly seasonal and is generally similar to that of the coastal zone of northern Australia from Arnhem Land to the Kimberleys. However, the high mountain spine of Timor, lying at right angles to the main wind directions, produces strong local modifications.

In most areas there is a dry season (a rainfall of less than 100 mm per month) of 4 to 6 months during the southeast monsoon (April to October) although the south facing slopes get some rain then (Soares 1957: Table XVII). Mean annual rainfall ranges from less than 600 mm on the north coast to over 3,000 mm in the mountainous region near the south coast of Indonesian Timor (Ferreira 1965: map 39; Ormeling 1957: Fig. 5).

Mean daily temperature ranges from 31-23°C at sea level to 24-15°C at 1,000 m (Ferreira 1965: map 38) with some seasonal variation. It is usually a little colder and much windier during the dry season when wind speeds of up to 45 km per hour are not uncommon.

Soares has produced a simple but useful classification of the climate which does not distort the data too much in which he divides Portuguese Timor into three zones, a north, a mountain, and a south zone, each marked by different aspects, elevations, rainfall, temperatures, and vegetation patterns (Soares 1957:111).

### LANDFORMS

Fig. 13:1 shows a very general geological sketch map of Timor. As with climate, three principal zones can be found in eastern Timor which coincide to some extent with the climatic zones.

**1** A number of narrow depositional coastal plains and recently elevated coral plateaux in the north.

**2** A mountain zone reaching up to 2,980 m along the centre of the island.

**3** An area of foothills and wider alluvial coastal plains along the south coast. The alluvial plains on the north and south coasts are Recent and the coastal plateaux are mostly Pleistocene in origin and are found as high as 1,300 m on the Larigutu pass near Venilale (Veevers 1967:3). The central range is a complex of thrust sheets of Permian to Miocene, sedimentary, metamorphic, and occasionally eruptive rocks. The southern foothills are lower thrust sheets and Tertiary block clays.

The two formations which seemed most likely to contain caves with archaeological deposits were the uplifted coral reef limestones of which there are two main areas shown on Fig. 13:1 at Los Palos and Baucau, and the Miocene 'Fatu' limestones in the central mountain zone.

### VEGETATION

There is little primary monsoon forest left in Timor except in the extreme southeast and on some of the steeper and more inaccessible mountains in the central ranges, for most of the useful land has been cleared at one time or another for cultivation. In these areas only a few of the useful forest trees survive, such as *Pterocarpus indicus*, *Sterculia* sp., *Aleurites moluccana* and *Tamarindus indicus*.

Below 1,000 m repeated clearing and burning at the end of the dry season have generally produced a savannah with little of the original flora. Above that height, where there is a shorter dry season, and on

*Fig. 13:1 Timor, showing the main geological formations and archaeological sites. Details from Gageonnet and Lemoine (1958).*

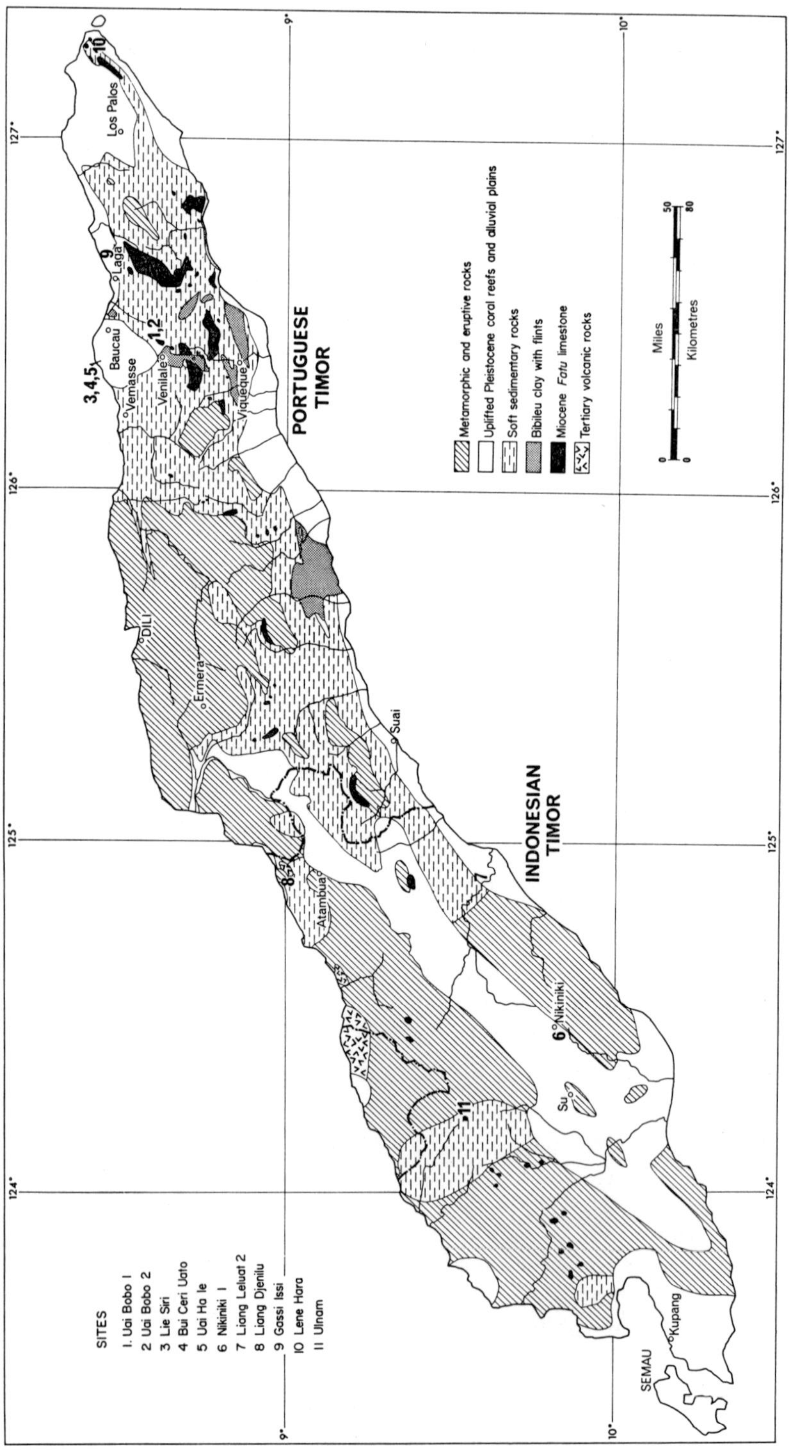

non-calcareous soils, a secondary forest develops where *Eucalyptus decaisneana* is a prominent tree species (Ormeling 1957:59; Gomes 1950:51).

FAUNA

Laurie and Hill (1954) list the living wild mammals of Timor as far as they are recorded. However, their list is almost certainly incomplete for rodents and bats.

Banteng cattle (*Bos sundaicus*) are rare in eastern Timor, but buffaloes, horses, pigs, goats, and dogs are numerous (goats and pigs alone outnumber the human population —Agencia geral do Ultramar 1965:71). Of the larger wild mammals, the Timor or rusa deer (*Cervus timorensis*) is the most common game animal, and monkeys, civet cats (*Paradoxurus hermaphroditus*), and the cuscus (*Phalanger orientalis*) are all plentiful in the better wooded areas, which generally means the mountain zone.

CULTURE

Timor is basically Indonesian in culture and not Melanesian, although it was not directly influenced by the great cultural traditions of western Indonesia, Buddhist, Hindu, and Islamic, which have developed there during the past 2,000 years.

Now a poor and isolated island, Timor was famous in the past as the main source of the finest quality white sandalwood (*Santalum album*) which appears to have been traded to western Indonesia and perhaps to India when the first relevant historical records appear in the second or third centuries A.D. (Wolters 1967:65; and see Burkill 1935:1954-6). Tomé Pires mentions Gujerati merchants who sailed south of Sumatra and Java into eastern Indonesia for wax, spices, and sandalwood before the Portuguese reached that area in 1512 (Cortesãa 1944:46).

Starting in the sixteenth century the Portuguese and Dutch established posts in Timor (Boxer 1948:174-98) and although trade declined in the eighteenth century they extended their control and the island was divided between these colonial powers in the nineteenth century.

Timorese culture can be distinguished from that of Melanesia by a number of traits such as a developed class system with hereditary chiefs, cattle keeping, the predominance of cereals, including wet rice, over root crops, metal working (iron for swords and gold and silver for ornaments), weaving of cotton, and a partial market economy with the production of goods for sale and export.

Most Timorese languages are Austronesian although some, mainly in the far east and in the mountain regions, are not (Capell 1944:311-37).

The Makassai people, 100,000 strong, in whose territory my excavations were made, live mostly in the mountain zone and depend largely on wet rice agriculture, maize and root crops being of secondary importance. But on the north coast maize, and on the south coast maize and dry rice are the main crops. In this area linguistic and other cultural boundaries coincide to some extent with those of the ecological zones and neighbouring groups are linked together in a network of market and marriage alliance exchange relationships (*feto-sa-umane*), which involve as well as livestock and food, pottery and other domestic utensils, many of which have distinct styles in the different cultural regions. Some of these, especially pottery, one might expect to find in archaeological deposits and this could prove valuable in correlating the sequences between sites in different areas.

## Excavations 1966-7

The information available on potential

archaeological sites in Timor was meagre and it was decided to confine excavations to caves—sites most likely to provide long sequences of artifacts and food residue. Other remains of great archaeological interest exist in Timor: rock painting sites, stone burial platforms, terraced field systems, abandoned fortified villages, and remains of the colonial period. However, these were unlikely to yield long sequences of artifacts and food remains, while the burial platforms are in many cases still in use.

Trial excavations were made in twelve caves of which five were selected for more extensive excavation. The locations of these are indicated on Fig. 13:1. There are three sites in the north coastal zone, Lie Siri (TL), Uai Ha Ie (TU), and Bui Ceri Uato (TB), all within 3 km of each other and less than 2 km from the sea, on the western edge of the Baucau Plateau between 175 and 260 m above sea level. These sites are discussed in detail in Glover 1970 (**1**, ch. 5, 6) and radiocarbon dates relating to them in Glover (1969) also.

In the mountain zone two caves were excavated near the hamlet of Uai Bobo, TO1 and TO2. They are a little over 600 m above sea level, in the older, Miocene limestone, approximately 126°27′E, 8°37′S and face east over the upper Seiçal Valley—the largest river system in the eastern part of the island. No useful sites were found in the south coastal zone in the short time spent there for reconnaissance.

At Uai Bobo 1 (TO1) an area of some 13 m² was excavated, most of the available floor space. The maximum depth reached was 1·40 m and the total excavated volume only 10·9 m³.

Five main stratigraphic layers could be seen (Fig. 13:2), though not always sharply defined. From the surface to about 10 cm was a layer of loose goat dung, sticks, leaves, corn cobs, and loose greyish-brown earth. Below this to about 25 cm was a greyish-brown deposit with occasional lenses of ash and charcoal. From 25 to 45 cm there was a layer of light brownish-grey earth, then a grey to dark grey layer to between 70 cm and 1·0 m where there was a fairly abrupt change to a hard-packed, reddish-brown to light yellowish-brown deposit, with little charcoal and an increased number of rocks and limestone rubble. Where the cave floor was exposed to heavy rain storms, all visible stratification had been lost, and a hard, crumbly, dark greyish-brown soil had developed.

The deposit was excavated in m² units and 10 cm spits, except where the stratigraphic boundaries could be recognised—though, as always, these were much clearer after digging than at the time. The maximum depth reached was 1·40 m, in thirteen spits, and rock was reached over about one half of the trench. These spits have been combined horizontally, following the trend of the stratigraphy, and the deposit has been divided into eight main horizontal units, or horizons, which form the basis for analysing the distribution of the cultural remains over time. Where there is a sufficient density of cultural materials, these horizons have been subdivided. They are numbered from bottom to top and correlate with the main stratigraphic layers, described in terms of the Munsell Soil Color Code, as follows:

| | | |
|---|---|---|
| Horizons I and IIa | are within the basal yellowish-brown layer | 10 YR 6/4—7·5 YR 6/8 |
| Horizon IIb | marks the transition from a dark grey layer | |
| Horizon III | is within the dark grey layer | 10 YR 4/1 |
| Horizons IV and V | are within a grey layer | 10 YR 5/1 |

*Fig. 13:2 Uai Bobo 1 (TO1). a, Plan of cave showing the excavated squares; b, section through cave at B-B¹; c, section at A-A¹ showing division of deposit into horizons for analysis; d, section at A-A¹ with details of stratigraphy; colour code is from Munsell Soil Color Charts.*

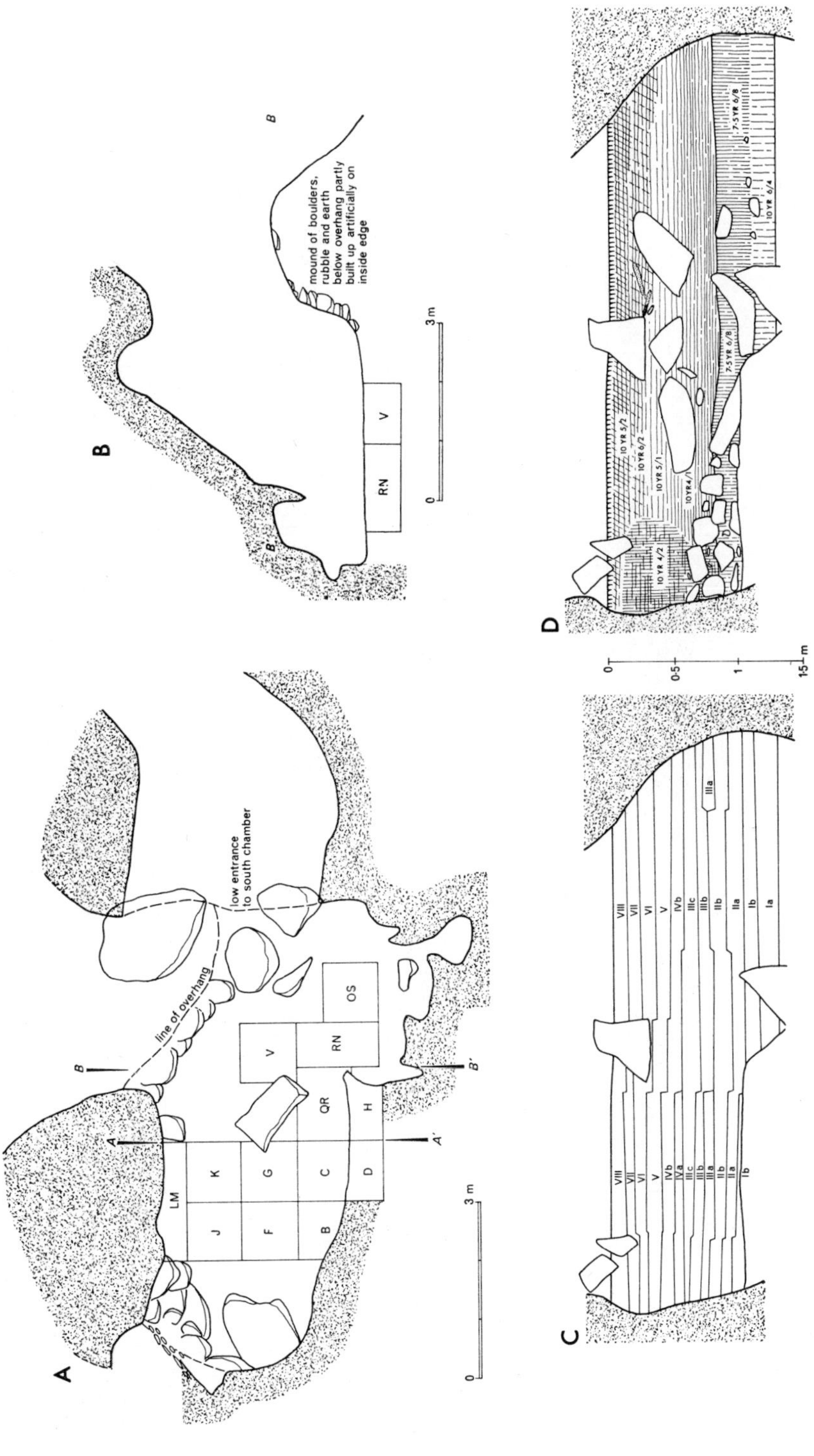

*Fig. 13:3 Uai Bobo 2 (TO2). a, Plan of cave showing the excavated squares at the surface; b, section through cave along north face of squares A, B, and C, with division of deposit into horizons for analysis; c, same section with details of stratigraphy. Deposit at back of upper chamber was loose and dry and collapsed before it could be recorded; colour code is from Munsell Soil Color Charts.*

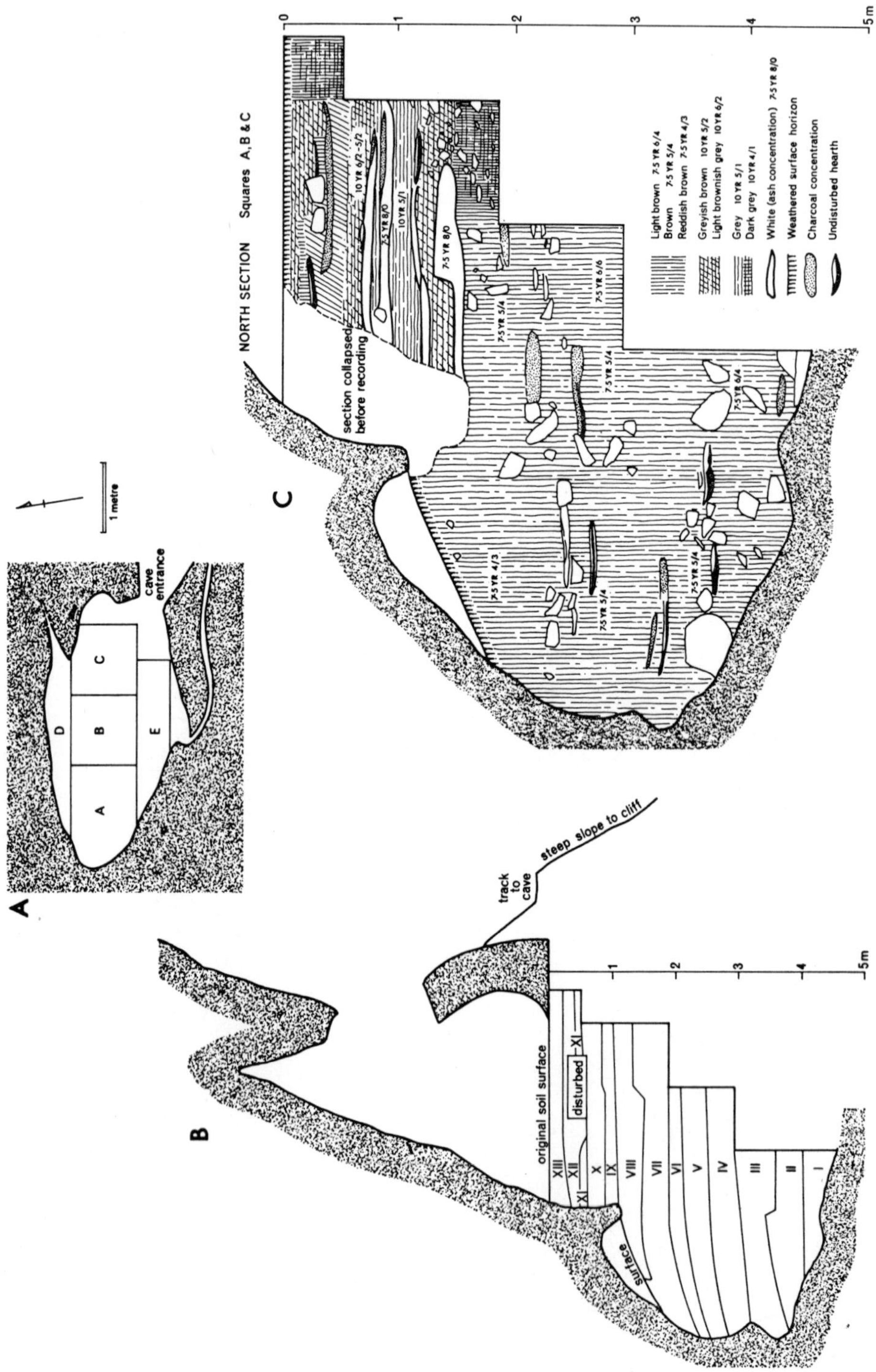

*Fig. 13:4 Provisional depth-age calculations for TO2, based on three C-14 estimations, correlated with horizons in TO1*

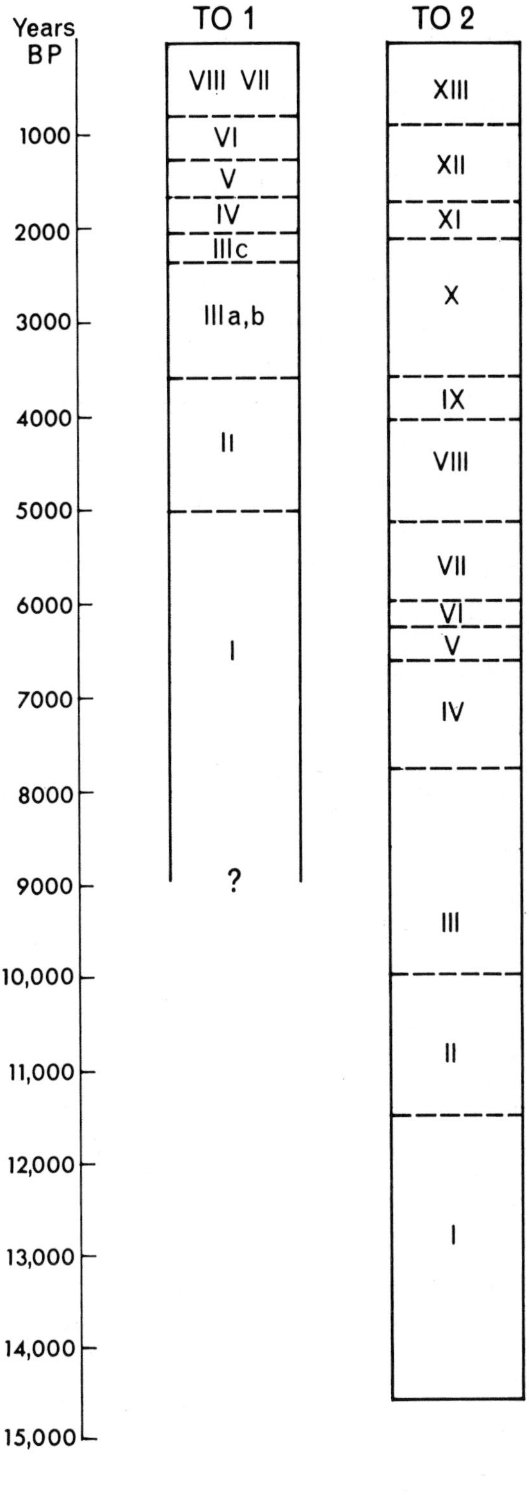


Horizon VI mostly comprises the light brownish-grey layer 10 YR 6/2

Horizon VII forms the transition from a greyish-brown layer

Horizon VIII comprises the top greyish-brown of soil with dung 10 YR 5/2

Cave Uai Bobo 2 (TO2) is only some 50 m northwest of Uai Bobo 1 and is much smaller. It is more of a solution fissure than a true cave, and the dimensions of the modern surface were only 3 m front to back by 2 m wide (Fig. 13:3). The entrance was blocked by a large fallen boulder and protected by tree roots which made the interior dark but dry and well protected.

Despite the small area of the cave the deposit was surprisingly productive and deep. The maximum depth reached was 4·9 m below the surface and the cave broadened out and extended backwards as the deposit was dug out.

The upper 2 m of deposit were more clearly and finely stratified than at Uai Bobo 1, with beds of white ash and interleaving hearths often no more than 5 cm thick. Below 2 m the deposit was a fairly uniform brown colour interrupted only by occasional hearth lenses. The site was dug in the same way as TO1 and the forty-three vertical excavation units have been grouped into thirteen horizons.

RADIOCARBON DATES

Up to late 1968, when this paper was prepared, one C-14 date had been obtained for Uai Bobo 1 and three for Uai Bobo 2.

**1** Uai Bobo 1: ANU-237, 2,190 ± 80 BP, for a concentration of charcoal at the top of Horizon IIIc, 50 cm below the surface and closely associated with a small copper ornament (Glover 1969: Plate 1·1), the only piece of prehistoric metal found in a datable

*Fig. 13:5 Numbers of waste flakes per m³. The thickness of each bar is proportional to the duration of the respective horizon calculated from the C-14 dates for TO2 only.*

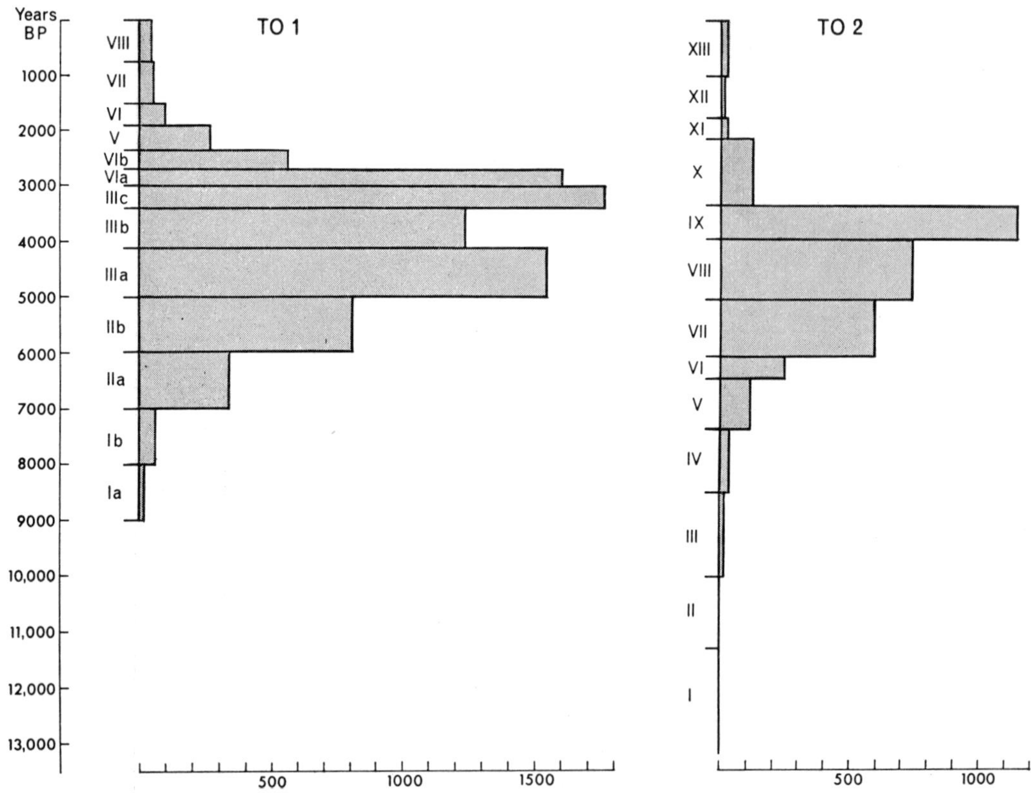


context in the excavations. Dates received subsequently confirm the reliability of ANU-237 (Glover 1970: **1**, 211-13).

**2** Uai Bobo 2: ANU-238, 13,400 ± 520 BP, for a sample comprising a mixture of charcoal and the insoluble fraction of bone from a hearth 20 cm above bedrock, together with 12 g of seedcases scattered throughout Horizon I.

ANU-187, 5,520 ± 60 BP, for a charcoal sample from a hearth in Horizon VII, 1·6 m below the surface.

ANU-239, 3,740 ± 90 BP, for a hearth in Horizon IX 1 m below the surface.

Subsequent to the 1968 seminar, an age of approximately 7,000 years was obtained for a sample from Horizon IV (Glover 1970: **1**, 268-70).

Fig. 13:4 illustrates the results of depth-age calculations for TO2 based on the four dates obtained (which give a figure of approximately 320 years for every 10 cm of deposit) and shows the correlation of horizons between this site and TO1.

### THE FLAKED STONE INDUSTRY

Fig. 13:5 shows the distribution of waste flakes in the two sites, as the number per m³ in each horizon. The volumes used were those of the deposit *in situ*. The probable increasing compaction of the deposit towards the bottom has not been allowed for, nor

*Table 13:1 Flaked stone: numbers in each horizon of TO1 and TO2*

| Site TO1 | Volume excavated, $m^3$ | Waste flakes | Utilised flakes | Flakes with edge gloss | Flakes with secondary working | Cores and trimming flakes |
|---|---|---|---|---|---|---|
| VIII | 1·1 | 45 | 0 | 0 | 8 | 1 |
| VII | 0·8 | 41 | 5 | 0 | 6 | 1 |
| VI | 1·0 | 101 | 6 | 0 | 1 | 2 |
| V | 1·1 | 288 | 22 | 8 | 16 | 4 |
| IVb | 1·1 | 764 | 67 | 13 | 30 | 8 |
| IVa | 0·3 | 399 | 13 | 8 | 20 | 10 |
| IIIc | 1·2 | 2,073 | 139 | 30 | 70 | 20 |
| IIIb | 1·2 | 1,533 | 87 | 28 | 54 | 19 |
| IIIa | 0·4 | 667 | 38 | 12 | 23 | 12 |
| IIb | 0·9 | 698 | 48 | 9 | 31 | 5 |
| IIa | 0·9 | 299 | 10 | 4 | 2 | 5 |
| Ib | 0·6 | 39 | 2 | 0 | 0 | 1 |
| Ia | 0·3 | 4 | 4 | 0 | 3 | 0 |
| Total | 10·9 | 6,951 | 441 | 112 | 264 | 88 |
| **Site TO2** | | | | | | |
| XIII | 1·0 | 25 | 6 | 0 | 0 | 1 |
| XII | 0·8 | 8 | 1 | 0 | 1 | 0 |
| XI | 0·5 | 12 | 2 | 0 | 0 | 0 |
| X | 1·3 | 168 | 14 | 2 | 9 | 3 |
| IX | 0·9 | 1,058 | 36 | 27 | 40 | 3 |
| VIII | 1·6 | 1,205 | 61 | 23 | 45 | 15 |
| VII | 3·7 | 2,190 | 79 | 44 | 95 | 27 |
| VI | 1·3 | 324 | 17 | 9 | 18 | 0 |
| V | 2·4 | 273 | 14 | 3 | 19 | 5 |
| IV | 2·2 | 71 | 2 | 1 | 5 | 0 |
| III | 1·5 | 6 | 0 | 0 | 0 | 1 |
| II | 1·7 | 0 | 0 | 0 | 0 | 0 |
| I | 0·5 | 0 | 1 | 0 | 0 | 0 |
| Total | 19·4 | 5,340 | 233 | 109 | 232 | 55 |

has the greater proportion of organic material in the upper layers. However, large rocks where they occurred were measured and are not included in the horizon volumes.

In TO1, the shallower site, the sequence appears to be compressed. Although Horizon I still contains 43 waste flakes and 9 worked or utilised pieces, the density at the base of the trench is lower than the graph indicates, for only one flake was found in each of the three deepest spits.

In Table 13:1 the flaked stone has been divided into five classes, which will form the basis for further analysis: waste flakes, simple utilised flakes, utilised flakes with gloss on the working edges, flakes with secondary modification, and cores. The distribution of flaked stone tools through time follows the same lenticular pattern as that of the waste flakes, with maximum densities of both occurring in Horizon III in TO1 and Horizon IX in TO2 (Fig. 13:5).

### SITE OCCUPATION

The distribution of cultural materials suggests that the sites were very little used in the earliest period, although there is clear evidence for human occupation even at the bottom of TO2, in the form of hearth lenses and the charred and broken remains of both

*Fig. 13:6 Numbers of sherds per m³. There is no pottery below horizon III in TO1 or below VIII in TO2.*

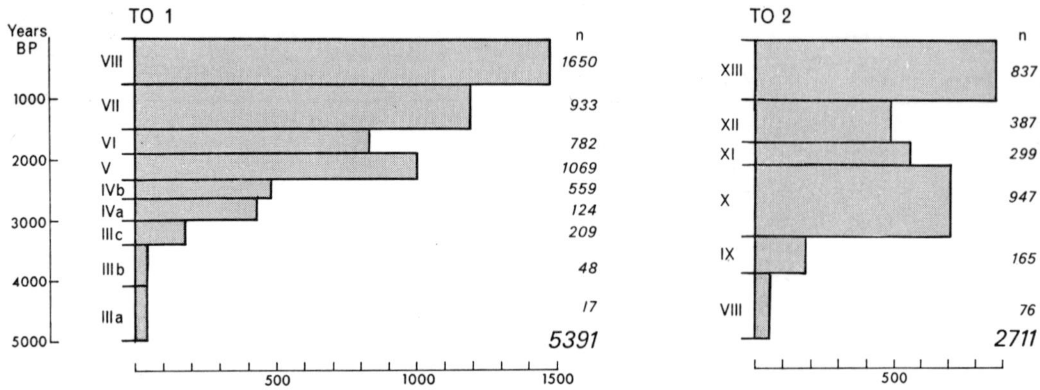


small and large rats, fruit bats, birds, and reptiles. There is also one clearly utilised flake almost on bedrock in Horizon I and one seed of the wild cereal *Coix lachryma-jobi*, which appears to have been pierced for use as an ornament. The use of Job's tears for food and ornament is documented for many places between Africa and Melanesia (Massal and Barrau 1956:43; Burkill 1935:629-31).

Whether the lack of use of the cave at this time is due to very low population density or merely to a preference for other locations is difficult to say. Starting perhaps 7,000 years ago, the sites came into more frequent use and the flaked stone reaches a maximum density in the levels which can be dated to about 3,500 years ago. Before this, in Horizons IIIa at TO1 and VII at TO2, the first pig bones appear, and also the first pottery (Horizon IIIa at TO1 and VIII at TO2), which must date to at least 4,500 years ago. It is tempting to see the increasing density of flaked stone as the result of a rapidly increasing population following the introduction of some form of agriculture. Whether this agriculture included rice or depended only on root crops and tree fruits, as did recent Melanesian culture and the earliest agriculture of Southeast Asia (Gorman 1969), it is impossible to say. Rice was certainly cultivated by the Lungshan peoples of north China at a date conventionally put in the middle of the third millennium B.C. and so it is conceivable, if unlikely, that it was introduced to Timor by the earliest agricultural immigrants. The apparent decline in the density of artifacts in more recent times does not necessarily mean that the caves were less frequented because, as I shall attempt to show later, stone may have been replaced by metal by 2,000 years ago.

Fig. 13:6 shows the density of pottery in the two sites. The maximum density occurs in the topmost levels, showing clearly that these sites have been occupied up to quite recent times. Thirty years ago, Bühler noted that caves were commonly used for overnight camps by hunting parties and by people preparing and harvesting gardens far distant from their houses (Sarasin 1936:4), and my own observations in east Timor suggest that caves are still used for these purposes today.

It is difficult to estimate the probable population of Timor in pre-agricultural times, because existing hunter-gatherer societies for which some—not always reliable—figures are available on population densities live in very different environments. But there appears to be general agreement that a density in excess of say three per km$^2$ is unlikely for a hunter-gatherer society except in very special circumstances such as in the fishing societies of the northwest coast of North America (Service 1966: 3; Lee 1963: 253).

The present population of Portuguese Timor is about 600,000 with a density of about 32 per km$^2$ (Agencia geral do Ultramar 1965:23) and estimates—perhaps not very reliable—suggest that it was stable at about 11 or 12 per km$^2$ from the seventeenth century to the early twentieth century, when it started to increase fairly rapidly (Hicks 1968; Castro 1862:470). By analogy with these other hunter-gatherer societies, the population of Portuguese Timor in pre-agricultural times would have been of the order of 15,000 to 30,000, and the probable increase following the adoption of agriculture may have been substantial and rapid. Such an increase in population could lead to a greater frequency in the use of caves, even though the caves were occupied only as temporary camps by hunting parties and people working nearby gardens.

### THE INTRODUCTION OF METAL

Fig. 13:5 and Table 13:1 show clearly that the density of flaked stone falls markedly in the top four horizons. Flint is used only occasionally for strike-a-lights today, and this evidence suggests that it was abandoned as an important tool-making material some time ago.

Although this is not shown in the tables, the regularly retouched 'type' tools are not found in the top two or three horizons; only flakes with a small amount of irregular working occur there, or those with a form of use wear which closely resembles that found on strike-a-lights. The flakes with edge gloss, which appear to have been cutting tools, perhaps for bamboo, also disappear well before the surface in both sites.

The caves excavated by Bühler at Nikiniki show an even more marked decline in the density of artifacts in the upper levels than do the sites at Uai Bobo. In his discussion (Sarasin 1936:13) Bühler recognised the possibility that metal tools, traded from centres of Asian civilisation, could have replaced stone long before the Portuguese arrived in the East. But accepting current ideas that Southeast Asia was a backward area, marginal to the Asian cultural centres of India and China (Linton 1955:173-4), Bühler invoked the Atoni and Belu myths of origin, which placed the coming of metal to Timor in the recent past (Sarasin 1936:10-13; Vroklage 1952:**I**, 147-52). The great depth of deposit that had to be excavated before stone age material was met was explained as due to continual deposition of earth rapidly eroded from nearby hillslopes in the monsoonal climate (Sarasin 1936:13).

If the decline in the use of flaked stone tools marks the introduction of metal into Timor, then the evidence for it remains largely negative. Apart from one copper ornament from Horizon IIIc in TO1 and a few nails on the surface, there was no metal in the sites. But knives and swords, brought to the island in trade, would be too valuable to leave lying in a cave and their absence is not altogether surprising. In addition to this negative evidence, there are two bones from Horizons XI and XII in TO2 which appear to have been cut by metal tools.

A date for the first use of metal in Indonesia is not well established; it mostly relies on comparisons and analogies with mainland Southeast Asia. Leaving aside any re-evaluation of the evidence there, which will be necessary following Solheim and Bayard's finds at Non Nok Tha in northeast

Thailand where a well-developed bronze industry has been dated to about 4,500 BP (Solheim 1968:59), a date of about 2,000 BP has generally been accepted for the introduction of bronze and iron into Indonesia (van Heekeren 1958:1-3). The evidence from the sites in Timor indicates that it may have been much earlier than this; that Timor, though far from the centres of Indonesian culture, may have been receiving metal tools by about 3,000 BP, when the popularity of flaked stone started to decline, and that by 2,000 BP metal was sufficiently common to replace stone for most purposes. There is, of course, the possibility that flaked stone was replaced not by metal, but by a perishable material which has not survived in the archaeological deposits. Dr Carmel White (pers. comm.) has suggested bamboo, commonly used for arrow and spear tips in New Guinea (Cranstone 1961:62) and Indonesia (Burkill 1935:295) in recent times. It is not unlikely that some of the large bamboos, common in Timor today, were introduced as useful plants at this time.

However, we do know that metal was in use in Timor when Western historical records for the area first appear in the sixteenth century A.D., and that, together with cotton cloth, it was commonly exchanged for sandalwood and beeswax by the early traders (Stanley 1874:153). In the light of the recent evidence from Thailand for a fairly sophisticated Southeast Asian metal industry by 2,500 B.C. (Solheim 1968), I think that it is reasonable to regard the decline in the density of flaked stone artifacts in these sites as the result of their gradual replacement by metal.

A preliminary examination of the distribution of stone in the coastal sites makes it clear that this pattern is not restricted to the caves at Uai Bobo and it is not, therefore, merely the reflection of a change in the use of these two caves.

*Table 13:2 Numbers of incised sherds*

| TO1 | | TO2 | |
|---|---|---|---|
| VIII | 0 | XIII | 0 |
| VII | 0 | XII | 0 |
| VI | 0 | XI | 2 |
| V | 8 | X | 13 |
| IVb | 6 | IX | 0 |
| IVa | 0 | VIII | 0 |
| IIIc | 0 | | |
| IIIb | 0 | | |
| IIIa | 0 | | |
| Total | 14 | Total | 15 |

DECORATED POTTERY

Small numbers of sherds decorated with bands of incised geometric motifs are found in all the excavated sites and well after the first appearance of pottery. The distribution of these sherds in TO1 and TO2 is given in Table 13:2.

These sherds can be related, in a very general way, to pottery in Bali (Soejono 1962: Fig. 1), the Celebes (van Heekeren 1957: Plates 37-8; van Stein Callenfels 1951: Plates XIV-XVIII) and the Philippines (Solheim 1964a, 1964b: 196-209), which, though as yet undated, seems to be associated with the early metal age in those areas. In another paper (Glover 1969) I have illustrated some of this pottery and discussed its dating and relationships in rather more detail.

CONTINUITY IN THE STONE INDUSTRY

In Fig. 13:7 the proportions of the four main classes of flaked stone tools are given for the middle levels of the two sites where reasonable numbers occur. Despite the differences in the density of stone tools over the 4,000 years or so represented by these horizons, and in the size of the samples, and despite the evidence of economic change during this period, the proportions of these four classes remain fairly constant within each site. In TO1 an average of about 30 per cent of all worked and utilised flaked stone has some

*Fig. 13:7 Percentages of the major classes of flaked stone tools. The height of the blocks is proportional to the horizon volumes and the width to the class percentages within each horizon.*

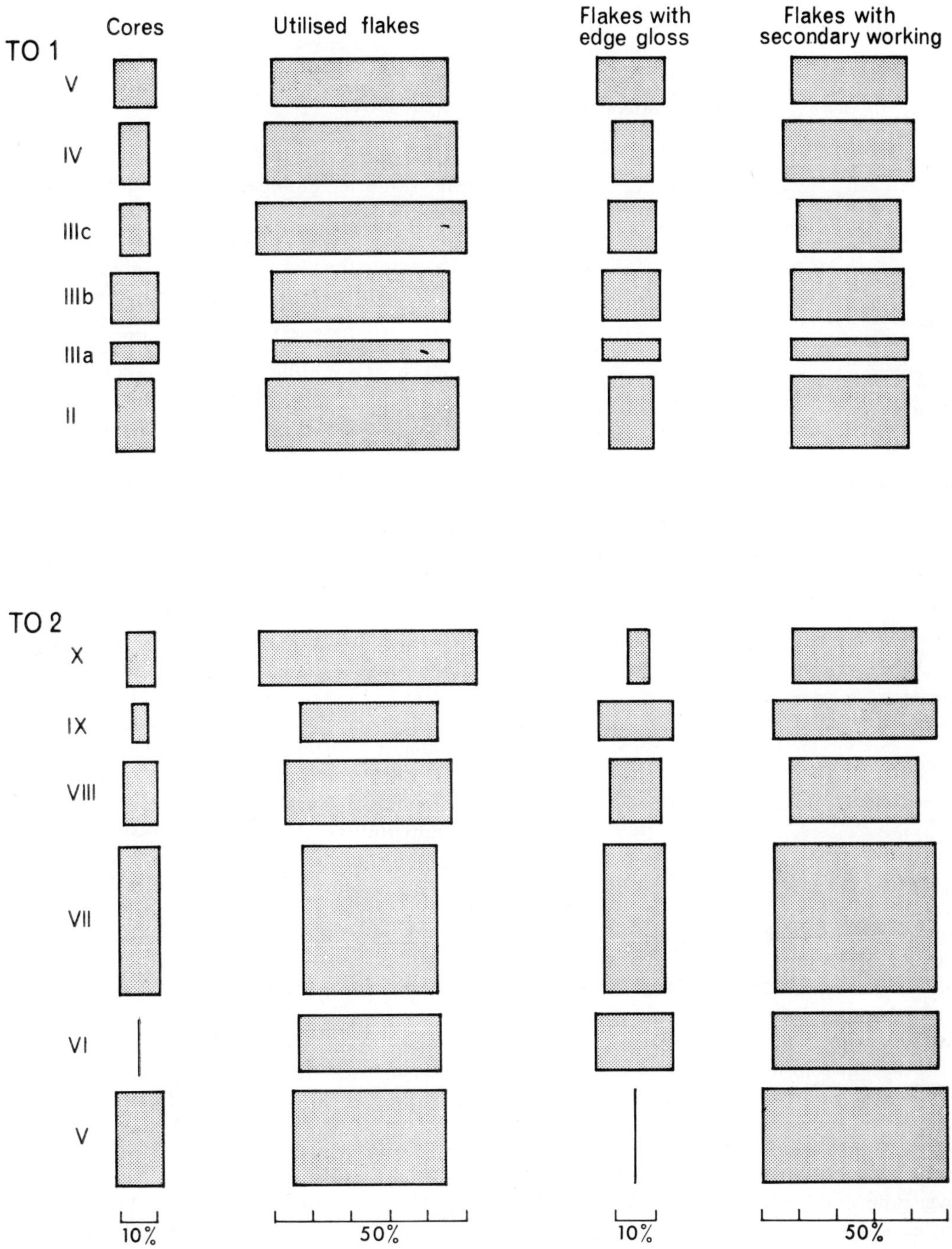

*Table 13:3 Numbers of shell ornaments*

| TO1 | Pierced discs | Olive shells | *Trochus* armlets | Pierced cockles |
|---|---|---|---|---|
| VIII | 0 | 0 | 0 | 0 |
| VII | 0 | 0 | 0 | 0 |
| VI | 0 | 0 | 0 | 0 |
| V | 1 | 0 | 0 | 0 |
| IVb | 0 | 1 | 0 | 0 |
| IVa | 0 | 0 | 0 | 0 |
| IIIc | 11 | 1 | 0 | 1 |
| IIIb | 8 | 1 | 2 | 0 |
| IIIa | 0 | 0 | 0 | 0 |
| IIb | 0 | 1 | 0 | 0 |
| IIa | 0 | 0 | 0 | 0 |
| Ib | 0 | 0 | 0 | 0 |
| Ia | 0 | 0 | 0 | 0 |
| Total | 20 | 4 | 2 | 1 |
| **TO2** | | | | |
| XIII | 0 | 0 | 0 | 0 |
| XII | 0 | 0 | 0 | 0 |
| XI | 0 | 0 | 0 | 0 |
| X | 1 | 0 | 0 | 0 |
| IX | 11 | 0 | 0 | 0 |
| VIII | 21 | 2 | 1 | 0 |
| VII | 4 | 3 | 0 | 3 |
| VI | 1 | 3 | 0 | 0 |
| V | 0 | 0 | 0 | 1 |
| IV | 0 | 0 | 0 | 0 |
| III | 0 | 0 | 0 | 0 |
| II | 0 | 0 | 0 | 0 |
| I | 0 | 0 | 0 | 0 |
| Total | 38 | 8 | 1 | 4 |

sort of secondary working, 12 per cent have edge gloss, 48 per cent are simply utilised flakes, and 10 per cent are cores. In TO2, 37 per cent have some form of secondary working, 17 per cent edge gloss, 37 per cent are utilised flakes, and 9 per cent are cores. Also, in both sites, the proportion of worked and utilised stone to waste varies only between 10 per cent and 17 per cent, except where the sample is very small. The fairly constant proportions of waste flakes, cores, and tools suggest that these are functionally related units in a tool kit which changed very little over time, and furthermore that the presumed introduction of agriculture did not radically affect the pattern of activities taking place in the caves.

It is not surprising that a change in the basic mode of subsistence should not affect the stone technology, for after all the same sorts of cutting tools would be needed for the same sorts of purposes (cf. Peterson, see ch. 16), while the cave sites would reflect hunting more than agricultural activities. We might expect to get a greater range of tools, and there are indeed some tanged points found only in the middle horizons in TO1, though they are very few. There is only one flake of what could be a ground axe, together with a stone ochre palette in Horizon VIII in TO2. Perhaps a slightly greater range of activities in the middle levels is also indicated by the numbers of shell ornaments found there and listed in Table 13:3.

In the New Guinea Highlands it seems also that the presumed introduction of agriculture was not marked by a change in the nature of the flaked stone industry (White 1967: **1**, 435, 450; see also ch. 14).

Between the sites there is a small but systematic difference in the smaller proportion of cores and larger proportion of secondarily-worked flakes at TO2. This might indicate that the preparation of tools was more frequently carried out at TO1; as a larger, more open, and better lit cave, it would be better suited for a more permanent camp than the cramped niche at TO2.

### THE ANALYSIS OF FAUNAL REMAINS

Table 13:4 gives the results of a preliminary analysis of the principal faunal material from the two sites. Identifications were made by Professor C. F. W. Higham, Department of Anthropology, University of Otago, Dunedin, N.Z., J. A. Mahoney, Department of Geology and Geophysics, University of Sydney, J. L. McKean, CSIRO Division of Wildlife Research, Canberra, Dan Witter, Department

*Table 13:4 Preliminary analysis of faunal remains; minimum numbers of individuals are given for each horizon, not corrected for unequal volumes*

| **Site TO1** | **Large murids** | **Small murids** | **Mega-chirop-tera** | **Micro-chirop-tera** | *Canis* | *Sus* | *Capra/Ovis* | *Phalanger* | *Paradox-urus* | *Macaca* | *Cervus* | *Bos/Bubalus* | *Equus* |
|---|---|---|---|---|---|---|---|---|---|---|---|---|---|
| VIII | 0 | 2 | 0 | 0 | 0 | 2 | 3 | 0 | 1? | 1 | 0 | 2 | 0 |
| VII | 0 | 2 | 0 | 0 | 1 | 1 | 1 | 0 | 1 | 0 | 0 | 1 | 1? |
| VI | 0 | 3 | 0 | 0 | 1 | 1 | 3 | 0 | 1 | 0 | 0 | 0 | 0 |
| V | 1 | 6 | 0 | 2 | 1 | 1 | 2 | 0 | 1 | 1 | 1? | 1 | 0 |
| IV | 8 | 34 | 3 | 2 | 0 | 1 | 0 | 1 | 0 | 0 | 0 | 0 | 0 |
| III | 22 | 46 | 11 | 2 | 0 | 2 | 0 | 2 | 0 | 0 | 1? | 0 | 0 |
| II | 12 | 239 | 8 | 5 | 0 | 0 | 0 | 0 | 0 | 0 | 0 | 0 | 0 |
| I | 2 | 585 | 3 | 2 | 0 | 0 | 0 | 0 | 0 | 0 | 0 | 0 | 0 |
| Total | 45 | 917 | 25 | 13 | 3 | 8 | 9 | 3 | 4 | 2 | 2? | 4 | 1? |
| **Site TO2** | | | | | | | | | | | | | |
| XIII | 0 | 6 | 0 | 0 | 0 | 2 | 0 | 0 | 0 | 0 | 1? | — | — |
| XII | 0 | 19 | 0 | 0 | 0 | 1 | 1 | 0 | 0 | 1 | 0 | — | — |
| XI | 0 | 15 | 0 | 0 | 0 | 2 | 0 | 1 | 1 | 1 | 0 | — | — |
| X | 1 | 151 | 1 | 1 | 1? | 1 | 1 | 1 | 1 | 1 | 0 | — | — |
| IX | 5 | 19 | 8 | 0 | 0 | 2 | 1? | 1 | 1? | 0 | 0 | — | — |
| VIII | 8 | 12 | 5 | 0 | 0 | 2 | 0 | 0 | 1 | 1 | 0 | — | — |
| VII | 16 | 92 | 5 | 0 | 0 | 1? | 1? | 1 | 1 | 0 | 0 | — | — |
| VI | 6 | 33 | 2 | 2 | 0 | 0 | 0 | 0 | 0 | 0 | 0 | — | — |
| V | 11 | 62 | 3 | 1 | 0 | 0 | 0 | 0 | 0 | 0 | 0 | — | — |
| IV | 3 | 318 | 2 | 1 | 0 | 0 | 0 | 0 | 0 | 0 | 0 | — | — |
| III | 5 | 545 | 0 | 1 | 0 | 0 | 0 | 0 | 0 | 0 | 0 | — | — |
| II | 18 | 701 | 9 | 3 | 0 | 0 | 0 | 0 | 0 | 0 | 0 | — | — |
| I | 18 | 337 | 15 | 2 | 0 | 0 | 0 | 0 | 0 | 0 | 0 | — | — |
| Total | 91 | 2,310 | 50 | 11 | 1? | 11 | 4 | 4 | 5 | 4 | 1? | — | — |

of Zoology, University of Texas, Peter Thompson, late Department of Prehistory, Australian National University, and the author. The numbers of small rats and bats are based on a count of mandibles only, except at the top, where the samples are small and other identifiable bone is taken into account. Numbers for other animals are based on all identified bone.

The main points to be made about this table are:

**1** That the number of small rats decreases as the intensity of human occupation increases. I would interpret most, if not all, of the small rats as the remains left by non-human predators.

**2** That the Microchiroptera, many of which are known to be cave dwelling species in Timor, may also have been included in the deposit partly as the result of natural agencies, together with small birds, frogs, snakes, and lizards not listed in the table. Some Megachiroptera have been recorded as cave-dwellers in Timor according to McKean, who has identified the excavated bat remains, and two of these, *Dobsonia peroni* and *Rousettus amplexicaudatus*, are also found in the deposits. The absence of bats, however, from the upper levels of both caves suggests that their presence lower down is due to man. Bats are still eaten in Timor but only rarely in this region, according to my observations and to the statements of my workmen. Neither cave contains bat colonies today.

**3** That all the other species represent human food remains, for the tables show a definite trend from exclusively wild fauna at the base—large rats and bats—to mostly domesticated species at the top—pig, *Capra/Ovis* (probably goat), dog, and a bovid, which is probably buffalo, although the bones appear to be rather small. The basic change takes place between Horizons III to V in TO1 and Horizons VII to IX in TO2, where pig, goat, and dog appear and the large rats and bats cut out.

This change, together with the appearance

of pottery, which I think we can reasonably take to mark the introduction of agriculture, must have started some 4,500-5,000 years ago.

About this time bones of four wild mammals first occur in the deposits—cuscus, civet cat, deer, and monkey; all probably were introduced to Timor through the agency of man.

Again, because of the small size of the sample, it is difficult to be certain about the order of appearance of the domesticates. Pig and goat appear first in TO2/VII, and pig alone in TO1/III. Dog and buffalo come in later in Horizon V at TO1, and dog alone in X at TO2. The appearance of a pig's tooth associated with a date of 5,520 ± 60 BP (ANU-187) in Horizon VII at site TO2 is not surprising, for the pig has already been recorded at sites in the New Guinea Highlands at a comparable date; in Kiowa Layer 3 between 6,100 ± 60 BP (Y-1370) and 4,840 ± 140 BP (Y-1371) (Bulmer 1966:504), and at Kafiavana Horizon IV before 4,690 ± 170 BP (ANU-42) (Brookfield and White 1968:47).

No dog bones appear below Horizons V at TO1 and X at TO2: that is to say, not before 2,000-2,500 BP and long after the presumed introduction of agriculture. A parallel situation exists at present in the New Guinea Highlands, where White found dog only in the most recent deposits, and at one site only out of five excavated (White 1967:**1**, 449). In both cases this may be due to the small samples obtained and to the fact that dog was never a common food animal. Nevertheless, it is a curious absence, as White points out, because dogs apparently accompanied the first settlers of eastern Polynesia nearly 2,000 years ago. Not only have different species of dogs been found closely associated with man over 9,000 years ago in western Europe, western Asia, Japan, and North America (Lawrence 1967:44) but man had introduced the dingo into Australia by 7,000 years ago (Campbell *et al.* 1966:10). This Australian evidence reinforces the long-held view that dog was the first animal tamed and domesticated by man.

The surprisingly early appearance of goat in TO2/VII (5,520 ± 60 BP) and IX (3,760 ± 90 BP) requires some comment. Both identifications are tentative and depend on a few, not absolutely diagnostic bones—a deciduous premolar in VII and an immature calcaneum and ulna in IX. Goat bones appear in quantity only in TO2/X and in TO1/V, which are dated to between 3,500 and 2,000 BP (Fig. 13:4). It may be that the two apparently earlier specimens in TO2 have been introduced to those levels through some unrecognised disturbances. On the other hand, a preliminary examination of the faunal remains from Lie Siri, one of the coastal sites, supports the early dating. There, samples ANU-173, 2,660 ± 110 BP, and ANU-172, 3,545 ± 120 BP, were associated with goat remains in the top 10 cm of the deposit, and a date of 3,530 ± 90 BP (ANU-235) was obtained for a hearth at 15 cm from the surface, which appears to be about the level at which the earliest goat bones are found in the site.

In western Indonesia and mainland Southeast Asia I have not been able to find any evidence for the presence of goats in prehistoric times. They are certainly absent from Pleistocene faunal sequences (van Heekeren 1957:10, 23, 39; Hooijer 1958: 71-6), although these may end well before Recent times (Golson 1970). Neither do they occur in Hoabinhian sites, which range from perhaps 20,000 to 7,000 BP (van Heekeren and Knuth 1967:106-7; Gorman 1969; Golson 1970), nor in the best recorded Javanese cave at Guwa Lawa, which probably belongs to this period (van Heekeren 1957:75-80; von Koenigswald 1955:16-17).

However, goats were present in Java by the mid-ninth century A.D., when the first historical sculptural records appear (e.g. on the Chandi Mendut, Krom 1923: 306). They are probably derived from the *Capra hircus* of western Asia (Zeuner 1963:130; Reed 1959: 1634) which was among the first of all domesticated animals, appearing in sites between Turkestan and the Mediterranean from about 9,000 BP (Reed 1959:1630-2). Goats appear in archaeological sites in West Pakistan before 5,000 BP—at Kili Gul Mohammad I (5,300 ± 200 BP, Fairservis 1956:356), and in the Lower Neolithic of Peninsular India before 4,000 BP—Utnūr (4,120 ± 150 BP, Allchin 1960:119) and Koedekal (4,410 ± 105 (TF-748), Sankalia pers. comm.). Consequently it is conceivable that they were introduced into Indonesia from India about this time, but only further work in the relevant areas can clarify the problem.

The faunal succession in the Timorese cave sites suggests that there might have been significant changes in the environment during the last 13,000 years. It is hoped that the detailed study of the numerous murid species at present being undertaken by J. A. Mahoney will throw more light on this issue. Mahoney's preliminary examination of the large rodents has revealed the presence of four species, which belong to four different genera. Only one of these, *Coryphomys bühlerei* Schaub, is already known. None of them occurs in the upper levels of any of the excavated sites and there is no record of them existing on the island today. The largest specimen was about the size of a cuscus or civet cat, both of which are common in Timor today, and it seems probable that they are all extinct.

As the largest native land mammals apart from the Pleistocene *Stegodon*, the large rats were probably always intensively hunted. This alone might be insufficient to cause their extinction; however, if we envisage a situation where, in addition, destruction of their habitat took place, following the adoption of agriculture through regular clearing and burning by an expanding population, then their disappearance is understandable. Some of the smaller rats, which were not regular game animals, have survived. A civet cat, probably *Paradoxurus hermaphroditus* (Pallas), may have been brought to Timor by the first agricultural settlers and, in the absence of any native mammalian carnivore, could also have played an important part in the extinction of the large rats.

## An Outline of Timorese Prehistory

The evidence suggests that Timor was occupied before 13,500 years ago by a small population of hunter-gatherers, exploiting at least the inland mountain zone of the island, the current earliest date for coastal occupation being some 7,500 BP (Glover 1969). Large rats and possibly Megachiroptera were the principal prey. Perhaps 4,500-5,000 years ago an agricultural people moved into the island, bringing pigs, civet cats, and pottery; goats, dogs, cuscus, and monkeys followed soon after. Population increased, leading to greater frequency in the use of caves and the extinction of the native giant rats through hunting pressure and wholesale clearance of the forests for agriculture.

About 3,000 years ago Timor may have come into more regular contact with South-east Asia, receiving metal tools which gradually replaced the traditional flint cutting tools. Small quantities of decorated pottery at this time show definite if unspecific links with the islands of Indonesia and the

Philippines to the north and west, as well as with mainland Southeast Asia. One may speculate that the export of sandalwood to Indonesia and China started at this time, although there is no certain documentary evidence for the sandalwood trade until after the seventh century A.D., when Timor was known as the source of the best quality white sandalwood (Wolters 1967:65-6, 205; Burkill 1935:1954-6).

A decorated pottery tradition never took firm root in Timor, as it did in Melanesia, where it was probably introduced about the same time. The flaked stone tool tradition, unsophisticated compared with those of western Asia, Europe, or America, continued almost unchanged by the introduction of agriculture.

So few ground stone tools are known from Timor that it is difficult to say much about them. No axes have been found in a pre-agricultural context in Timor, as they have in Australia (White, C., see ch. 12), New Guinea (White 1967:**2**, Appendix 10·1, 10·2; see also ch. 14), and perhaps in Borneo (Golson 1970). The early introduction of metal might explain their scarcity in ethnographic collections. However, only one fragment of a ground basalt tool, not necessarily an axe, was found in Horizon VIII at TO2 and one ground *Tridacna* adze was excavated in the coastal site of Bui Ceri Uato, but in a level postdating the appearance of pottery and domestic animals. As none appears to have been found in the sites excavated by Bühler, Willems, Verhoeven, and Almeida, it may be that ground stone cutting tools were never in common use in Timor.

There is no evidence for any significant contact with Australasia during the period under review, except for the introduction of the sole marsupial, *Phalanger orientalis*, or cuscus, which may have been the result of a single trade contact with either New Guinea, the Moluccas, or even Australia, where it occurs in Cape York.

The presence of shell artifacts which have parallels in Melanesia (Lewis 1929) as well as in Flores and Java (van Heekeren 1957: Plate 26B; van Heekeren and Knuth 1967: 157) and in Indochina (Mansuy 1923: Plates III and VII; Mansuy and Colani 1925: Plate XIV), shows the same sort of unspecific links with neighbouring areas indicated by the decorated pottery.

The colonial period, which started in the sixteenth century in Timor, is marked in the cave deposits only by a few fragments of broken glass, wire, nails, and Chinese ceramics. Important changes in the subsistence basis following the importation of maize, sweet potatoes, and manioc, have left no traces in the archaeological material.

Rice, after maize the most important crop today, may have been introduced during the first agricultural revolution, or later perhaps, when buffaloes arrived, but no direct evidence exists.

Given the paucity of any human skeletal remains, it is impossible to say whether the introduction of agriculture was accompanied by a migration of people, or merely by the diffusion of techniques. But considering the well documented reluctance of hunter-gatherers to take up agriculture, except under considerable pressure and continuous example (cf. White, J. P., see ch. 14), it seems probable that there was a considerable immigration of agricultural peoples into Timor some 5,000 years ago. Timor's island situation rules out the possibility of diffusion by continuous example on a broad front, which can effect such a transition in continental situations.

The absence of any remains earlier than 13,500 BP leaves the question open as to whether Timor was a corridor for the first migration of people into Australia (Mulvaney 1969: 160-5). The finds of *Stegodon*, made by Verhoeven near Atambua in Indonesian Timor (Verhoeven 1964; Hooijer 1969) and in Flores (Hooijer 1957:119-29), and the possible association of this Middle to Upper Pleistocene placental mammal with flaked stone tools in Timor, suggest that this is still a possibility (Glover and Glover 1970).

### References


Agencia geral do Ultramar. 1965. *Timor.* Lisbon.

Allchin, F. R. 1960. *Piklihal excavations.* Hyderabad: Andhra Pradesh Government Archaeological Ser. No. 1.

Almeida, A. de and Zbyszewski, G. A. 1967. A contribution to the study of the prehistory of Portuguese Timor—Lithic industries. In *Archaeology at the eleventh Pacific Science Congress* (ed. W. G. Solheim), pp. 55-67. 11th Pacific Science Congress, Tokyo, 1966. Honolulu: Social Science Research Institute, University of Hawaii (Asian and Pacific Archaeology Ser. No. 1).

Bijlmer, H. J. T. 1929. *Outlines of the anthropology of the Timor-Archipelago.* Weltevreden: G. Kolff and Co.

Birdsell, J. B. 1958. Some population problems involving Pleistocene man. *Cold Spring Harb.Symp.quant.Biol.* **22**: 47-69.

——1967. Preliminary data on the trihybrid origin of the Australian Aborigines. *Archaeol. & phys.Anthrop.Oceania* **2**:100-55.

Boxer, C. R. 1948. *Fidalgos in the Far East, 1550-1770.* The Hague: Martinus Nijhoff.

Brookfield, H. C. and White, J. P. 1968. Revolution or evolution in the prehistory of the New Guinea Highlands: a seminar report. *Ethnology* **7**: 43-52.

Bulmer, Susan. 1966. Pig bone from two archaeological sites in the New Guinea Highlands. *J.Polynes.Soc.* **75**: 504-5.

Burkill, I. H. 1935. *A dictionary of the economic products of the Malay Peninsula.* 2 vols. London: Crown Agents for the Colonies.

Campbell, T. D., Edwards, R., and Hossfeld, P. S. 1966. Archaeological excavations in the southeast of South Australia. Canberra: Australian Institute of Aboriginal Studies (unpublished).

Capell, A. 1944. Peoples and languages of Timor. *Oceania* **14**:191-219, 311-37.

Castro, A. de 1862. Résumé historique de l'établissement portugais à Timor . . . *Tijdschr. indische Taal-Land-en Volenk.* **11**: 465-506.

Correa, A. M., Almeida, A. de, and Cinatti, R. 1953. Preliminary notice of a Paleolithic station in the eastern Malaysian Archipelago (Portuguese Timor). In *Abstracts and messages fourth Far-Eastern Prehistory Congress and Anthropology Division of the eighth Pacific Science Congress.* Quezon City, pp. 52-3.

——, ——, and Camarate F. 1956. Nouvelles stations lithiques du Timor portugais et la préhistoire de l'Indonésie orientale. In *Actas de la IV sesion Congresos Internacionales de Ciências Préhistoricas y Proto Historicas.* Madrid, 1954 (ed. A. Beltran), pp. 295-8. Zaragoza: Libreria General.

Cortesãa, A. (ed.) 1944. *The Suma Oriental of Tomé Pires.* London: Hakluyt Society, Ser. 2, **89**.

Cranstone, B. A. L. 1961. *Melanesia: a short ethnography.* London: British Museum.

Fairservis, W. A. 1956. Excavations in the Quetta Valley, West Pakistan. *Anthrop.Pap. Am.Mus.nat.Hist.* **45** (2).

Ferreira, H. A. 1965. *O clima de Portugal; fascículo 12, Província de Timor.* Lisbon: Serviço Meteorológico Nacional.

Gageonnet, R. and Lemoine, M. 1958. *Contribution à la connaissance de la géologie de la province portugaise de Timor.* Lisbon: Estudos Ensaios Docum.Jta Invest.Ultramar No. 48.

Glover, I. C. 1969. Radiocarbon dates from Portuguese Timor. *Archaeol. & phys.Anthrop. Oceania* **4**:107-12.

——1970. Excavations in Timor: a study of economic change and cultural continuity in prehistory. 2 vols. Ph.D. thesis, Australian National University, Canberra (unpublished).

——and Glover, E. A. 1970. Pleistocene flaked stone tools from Flores and Timor. *Mankind* **7** (3): 188-90.

Golson, J. 1970. Both sides of the Wallace line: New Guinea, Australia, Island Melanesia and Asian prehistory. In *Early Chinese art and its possible influence in the Pacific basin: a symposium* (ed. N. Barnard), in press. New York: Intercultural Arts Press.

Gomes, R. C. V. M. 1950. *Explorações botanicas em Timor*. Lisbon: Estudos Ensaios Docum.Jta Invest.Ultramar No. 4.

Gorman, C. 1969. Hoabinhian: a pebble tool complex with early plant associations in southeast Asia. *Science, N.Y.* **163**:671-3.

Hicks, D. 1968. Eastern Timorese society Part 3, the political order. London: School of Oriental and African Studies (unpublished).

Hooijer, D. A. 1950. Man and other mammals from Toalian sites in south-western Celebes. *Verh.K.ned.Akad.Wet.Afd.natuurk.* 2 Sectie, **46** (2):7-158.

——1957. A stegodon from Flores. *Treubia* **24**:119-29.

——1958. The Pleistocene vertebrate fauna of Celebes. *Asian Perspect.* **2**:71-6.

——1965. Note on *Coryphomys bühleri* Schaub, a gigantic murine rodent from Timor. *Israel J.Zool.* **14**:128-33.

——1969. The Stegodon from Timor. *Verh.K. ned.Akad.Wet.* Ser. B, **72** (3):203-10.

Keers, W. 1948. *An anthropological survey of the eastern Little Sunda Islands.* Meded.K. Vereen.indisch.Inst.Amst. No. 74.

Krom, N. J. 1923. *Inleiding tot de Hindoe-Javaansche kunst.* 2nd edn, 1 dl. 's-Gravenhage: Martinus Nijhoff.

Laurie, Eleanor M. O. and Hill, J. E. 1954. *List of land mammals of New Guinea, Celebes and adjacent islands (1758-1952).* London: British Museum (Natural History).

Lawrence, B. 1967. Early domestic dogs. *Z.Säugetierk.* **32**:44-59.

Lee, R. B. 1963. The population ecology of man in the early Upper Pleistocene of southern Africa. *Proc.prehist.Soc.* **29**: 235-57.

Lewis, A. B. 1929. *Melanesian shell money in Field Museum collections.* Chicago: Field Mus.nat.Hist.Pub.Anthrop. Ser. **19** (1).

Linton, R. 1955. *The tree of culture.* New York: Alfred A. Knopf.

Macintosh, N. W. G. 1963. Origin and physical differentiation of the Australian Aborigines. *Aust.nat.Hist.* **14**:248-52.

Mansuy, H. 1923. *Contribution à l'étude de la préhistoire de l'Indochine III . . . Samrong Sen (Cambodge).* Hanoi: Mém.Serv.géol.Indoch. **10** fasc. 1.

——and Colani, M. 1925. *Contribution à l'étude de la préhistoire de l'Indochine VII Néolithique inférieur (Bacsonien) et Néolithique supérieur du Haut-Tonkin.* Hanoi: Mém. Serv.géol.Indoch. **12** fasc. 3.

Massal, E. and Barrau, J. 1956. *Food plants of the South Sea islands.* Noumea: Tech.Pap. S.Pacif.Commn No. 94.

Mulvaney, D. J. 1961. The stone age of Australia. *Proc.prehist.Soc.* **27**:56-107.

——1969. *The prehistory of Australia.* London: Thames and Hudson.

Nyessen, D. J. H. 1945. Anthropography of Portuguese Timor. *Bull.schweiz.Ges.Anthrop. Ethnol.* **21**:89-97.

Ormeling, F. J. 1957. *The Timor problem: a geographical interpretation of an underdeveloped island.* 2nd imp. 's-Gravenhage: Martinus Nijhoff.

*Oudheidkundig verslag*. 1939. Batavia: Oudheidkundige dienst in Nederlandsch-Indië.

Reed, C. A. 1959. Animal domestication in the prehistoric Near East. *Science, N.Y.* **130**:1629-39.

Sarasin, F. 1936. Beiträge zur Prähistorie der Inseln Timor und Roti. *Verh.naturf.Ges.Basel* **47**:1-59.

Schaub, S. 1937. Ein neuer Muride von Timor. *Verh.naturf.Ges.Basel* **48**:1-6.

Service, E. R. 1966. *The hunters*. Foundations of Modern Anthropology Ser. New Jersey: Prentice-Hall.

Soares, F. A. 1957. *O clima e o solo de Timor*. Lisbon: Estudos Ensaios Docum.Jta Invest. Ultramar No. 34.

Soejono, R. P. 1962. Indonesia. *Asian Perspect.* **6**:34-43.

Solheim, W. G. 1964a. *The archaeology of central Philippines*. Manila: National Institute of Science and Technology, Monograph 10.

——1964b. Further relationships of the Sa-Huỳnh-Kalanay pottery tradition. *Asian Perspect.* **8**:196-211.

——1968. Early bronze in northeastern Thailand. *Curr.Anthrop.* **9**:59-62.

Stanley of Alderley, H. E. J. 1874. *The first voyage round the world by Magellan, tr. from the accounts of Pigafetta, and other contemporary writers . . .* London: Hakluyt Society Ser. 1, **52**. Reprinted New York: Burt Franklin.

van Bork-Feltkamp, A. J. 1951. *A contribution to the anthropology of Timor and Roti . . .* Meded.K.Inst.Trop. No. 97 (Afd.Cult. phys.Anthrop. No. 41).

van Heekeren, H. R. 1949. Rapport over de ontgraving te Kamasi, Kalumpang (West Centraal-Celebes). *Oudheidkundig verslag* 1949: 26-46. Bandung: Oudheidkundige dienst in Nederlandsch-Indië.

——1957. *The stone age of Indonesia*. 's-Gravenhage: Martinus Nijhoff.

——1958. *The bronze-iron age of Indonesia*. 's-Gravenhage: Martinus Nijhoff.

——and Knuth, E. 1967. *Archaeological excavations in Thailand I, Sai-Yok*. Copenhagen: Munksgaard.

van Stein Callenfels, P. V. 1951. Prehistoric sites on the Karama River. *Jl E.Asiat.Stud.* **1**:82-97.

Veevers, J. J. 1967. Report on an official visit to Portuguese Timor, from 25th April to 9th May, 1967. Canberra: Bureau of Mineral Resources Records 1967/74 (unpublished).

Verhoeven, Th. 1959. Die klingenkultur der Insel Timor. *Anthropos* **54**:970-2.

——1964. Stegodon-Fossilien auf der Insel Timor. *Anthropos* **59**:634.

von Koenigswald, G. H. R. 1955. Remarks on a late Pleistocene or post-Pleistocene landbridge between Java and Asia. *Quaternaria* **2**:15-19.

Vroklage, B. A. G. 1952. *Ethnographie der Belu in zentral-Timor*. 3 vols. Leiden: E. J. Brill.

Wanner, J. 1913. Ethnologische notizen über die Inseln Timor und Misol. *Arch.Anthrop.* **40**:147-60.

White, J. P. 1967. Taim bilong bipo: investigations towards a prehistory of the Papua-New Guinea Highlands. 2 vols. Ph.D. thesis, Australian National University, Canberra (unpublished).

Wolters, O. W. 1967. *Early Indonesian commerce*. Ithaca, New York: Cornell University Press.

Zeuner, F. E. 1963. *A history of domesticated animals*. New York: Harper and Row.

# 14 New Guinea and Australian Prehistory: The 'Neolithic Problem'

J. Peter White

*It is strange that these people [Cape York Aborigines] have never learnt to cultivate the earth and build houses, but remain content to wander about, living precariously on wild fruits, grubs, a little chance fish, and such animals as they can spear, whilst their Papuan neighbours, in the near Torres Straits islands build good huts, supply themselves with constant vegetable food, and have fine canoes for fishing.*

*Captain John Moresby 1876: 18*

If Captain Moresby was less than accurate in his characterisation of a hunting and gathering economy, his comment nonetheless points to one of the larger problems in Australian prehistory and one of the greatest contrasts between the prehistory of Australia and that of Melanesia. This may be briefly put as: Why was there never a 'Neolithic Revolution' in Australia?

This is not a problem that has been frequently discussed in the literature. Elkin (1954:15) suggests that Aborigines did not know that plants grew from seeds, but this seems unlikely (Specht 1958; Worsley 1961). Worsley's own view seems to be the most favoured one (cf. Meggitt 1964:35). He observes (1961:178):

Their material cultural equipment is primitive, and they have been isolated for millennia in an inhospitable continent with hardly any animals suitable for domestication, and no grains . . . other than some wild rice in a few areas. These facts are explanation enough of the stationary nature of their society.

This seems to me to mis-state the problem.

For discussions and assistance I am indebted to F. Barnes, H. C. Brookfield, R. N. H. Bulmer, S. E. Bulmer, M. Dornstreich, I. C. Glover, J. Golson, L. Hiatt, R. M. Jones, M. F. Little, D. R. Moore, G. Pretty, G. Sampson, J. M. Wheeler, C. White. None of them is responsible for the result.

The Aborigines may have been unlikely to *invent* agriculture but why did they not adopt plants and animals from their external contacts and become agriculturalists in this way?

I wish to discuss two aspects of this problem:

**1** Some of the general explanations which have been given for the transition from hunting and gathering to agriculture in other parts of the world will be examined in the light of ethnographic and archaeological data from Australia and Melanesia, which suggest that the explanations are inadequate and more complex ones must be sought.

**2** On a more particular level, it will be suggested that some hitherto unrelated phenomena in Australian prehistory may, in fact, be linked to the economic transition in areas north of Australia.

It must be stressed here that this is primarily a speculative paper, designed to focus attention on some theoretical and practical problems in the prehistory of the area. With no more than a couple of dozen excavations spread over an area as large as that of the U.S.A., we cannot yet expect a detailed answer to many problems: it will be sufficient if they are recognised.

**The Adoption of Agriculture**

Apart from discussion of environment as a limiting factor, there seems to be little in the literature about the spread of agriculture. In general it has been regarded as 'natural' that prehistoric peoples should have become agriculturalists either because they saw it as a better way of life or because they were incorporated, more or less forcibly, into the new *oikumene* (cf. Turnbull 1965). In fact, northern Australia seems to be one of the very few areas of the world which could have become agricultural and did not. Perhaps, then, it may provide a testing ground for some of the theories about the adoption of agriculture.

Flannery (1965:1251) suggests that the development of a pattern of interchange of resources between groups exploiting contrasting environments was the single most important factor in establishing an agricultural economy. These exchanges were initially used to distribute rare items such as obsidian (in Mesopotamia), and gradually came to involve key species of edible grasses. As trade expanded and larger quantities were required it became much more efficient to concentrate grass plants in a smaller area than to collect them from their wild habitats. The areal expansion of trade would have led to the spread of this new economy to other non-agriculturalists.

Australian Aborigines of course maintained highly developed 'trading' systems (McCarthy 1939) but although, in the Torres Strait area at least, these included agricultural peoples, foodstuffs do not appear to have ever been significant items of exchange. One might suggest that in most areas there was insufficient environmental contrast to make the exchange of foods worthwhile and, in the coastal areas where such a contrast did occur, the problems of preserving seafoods militated against their use in exchanges.

Another factor often instanced is the development of sedentism (e.g. Braidwood and Willey 1962). Suttles (1951), discussing the adoption of potato cultivation among the Coast Salish, points out that the traditional root-gathering and fishing economy allowed these groups to live in permanent dwellings for half the year and stay within close range (2 days' walk) of them for the other half. Women were therefore able to care continuously for growing patches of the new root crop without disturbance (in the initial stages) to their traditional way of life.

Aborigines were not as sedentary as the Salish. However, in ecologically richer areas, such as the tropical coast, they were clearly semi-sedentary. Thomson, for instance, notes permanent wet season camps of four months or longer (1939:214), and their dry season nomadism tended to be confined to fairly small areas. Furthermore a considerable proportion of their vegetable diet consisted of roots of very similar kinds (often the same genera: cf. Golson, see ch. 15) to those being cultivated in New Guinea. Thus, although conditions were probably not quite as favourable as for the Salish, it might be argued that the general picture of permanently wandering Aborigines has been overstressed and will not completely serve to explain the economic pattern.

A third possible explanation derives from a recent statement by Frances Barnes (1970) that hunter-gatherer societies are characterised by the absence of a well-developed power hierarchy and a strong sense of territoriality and that both these are present in agricultural populations (see also Anon. 1907). Hiatt has suggested (pers. comm.) that the development of these social features may in some ways assist in the economic transition. While a complex interaction between social and economic factors is likely, both may also depend on demographic factors, such as a population increase beyond the limit which can be accommodated by current social and ecological arrangements. This would probably only result from the ability to iron out fluctuations inherent in the food supplies available to hunter-gatherer economies in highly seasonal environments. Aborigines in both Cape York and Arnhem Land did, at times, store food such as dried plums and yams (Thomson 1939:216, 1949: 23), but this seems to have been primarily for use on ceremonial occasions. Why this activity did not expand and allow ceremonies to become longer and more elaborate is not clear. While the absence of social competitiveness might be instanced, this of course simply brings the cultural argument back in a full circle.

The final explanation I wish to discuss derives from Caldwell's statement about the Adena and Hopewellian cultures of eastern North America. He suggests that 'the very efficiency of forest adaptation was a factor inhibiting the acceptance of food production as a major economic basis' (1962:305). Braidwood and Willey (1962:350) expand this to suggest that highly developed and culturally complex hunter-gatherer societies do not develop incipient food production. This, they show, seems to occur in some areas where, for various ecological reasons, this economy was not so successful (cf. Flannery 1969).

Was this in fact the case in northern Australia? Were Aborigines simply too well off there to bother about agriculture and husbandry, as suggested by R. M. and C. H. Berndt (1954:38)?

This explanation seems to me to have some merit. The data of Thomson (1939), Meggitt (1964), and Peterson (pers. comm.) all suggest that Sahlins's characterisation of many hunter-gatherers as belonging to the original affluent society might be applied in most coastal parts of northern Australia. In this context, Lee's (1965) study of the !Kung economy should also be recalled. Further, in relation to the failure of agriculture and husbandry to spread into sub-Saharan Africa until very recently, Clark (1964:180) points out that the same levels of subsistence could be maintained by hunting and gathering, and probably with less expenditure of labour.

Social as well as economic factors appear to favour this explanation for the Australian situation. As Meggitt points out (1964:35):

an important reason for the failure to devise a rational technique of improved plant production seems to have been the totemic religious philosophy which not only provided ritual non-technical substitutes for practical action but also morally discountenanced technological innovation in general.

Whether this statement covers all Australian societies remains, perhaps, for anthropologists to discuss. Nevertheless this, together with other factors, may have created a sufficiently inhibiting situation to ensure that Aborigines were still hunter-gatherers when Europeans arrived. We might remember, for instance, that the sea barrier of Torres Strait, which probably existed by 6,500-8,000 BP (cf. Jennings, see ch. 1), may have prevented accretionary advances of agriculturalists of the kind only too well documented in both ethnographic and archaeological records (Clark 1965; Butzer 1964:439-40)—not to mention in the later history of Australia.

Finally, it is worth pointing out that ecological arguments against the adoption of agriculture in northern Australia seem dubious. The geology and soils of the Cape York/Trans-Fly region are very similar. The seasonality and extent of the rainfall, the temperature régimes, and even the vegetation associations all show close parallels (Keast 1959; Brookfield and Hart 1966: map 9). On the New Guinea side Trans-Fly gardens provide food for much, if not always all of the year; agricultural life has become adapted to almost precisely the same environment as occurs in some parts of northern Australia. In Arnhem Land today Aborigines do plant some vegetables even when they are not residing on settlements, but still living primarily as hunter-gatherers.

Yet was the adoption of agriculture and husbandry in Indonesia and Melanesia totally without effect in Australia? In attempting to suggest that it was not, I will first discuss some of the ethnographic data, then look briefly at the archaeological evidence and finally propose a tentative and speculative framework within which to correlate the material currently available.

**Ethnographic Evidence**

At the time of European contact, Australian Aborigines were in communication with agricultural societies, or some members of them, in Cape York, the northern coasts of the Northern Territory, and northwestern Australia. In the Cape York/Torres Strait area the contact was between a series of functioning societies, some agricultural and some hunter-gatherer. By contrast, the Arnhem Land contact was with temporary, if frequent visitors, who came for highly specialised purposes. I shall therefore look first at the economic and technological aspects of societies in southwest New Guinea, Torres Strait, and Cape York.

By the nineteenth century A.D. nearly all New Guinean societies were economically dependent upon gardens and domestic animals, although the degree of dependence varied. In the Trans-Fly region the staple crop was yam. Taro, manioc, sweet potatoes, bananas, and sugar cane were also grown, while in swampy areas some groups depended on sago for part of the year (Williams 1936: 216-20). Most groups reared tame rather than domestic pigs, for Williams (1936:224) records that pigs were not bred in the villages, but wild piglets were captured and kept in the same way as cuscus and cassowary. Dogs were truly domestic (Williams 1936: 420). There were clearly economic variations within the Trans-Fly: some groups such as the Garamudi were heavily dependent on sago-making and hunting and were therefore semi-nomadic, while others, including the more sedentary Keraki, also occasionally moved seasonally in search of water (Williams

*Fig. 14:1 Australia and New Guinea at Torres Strait*

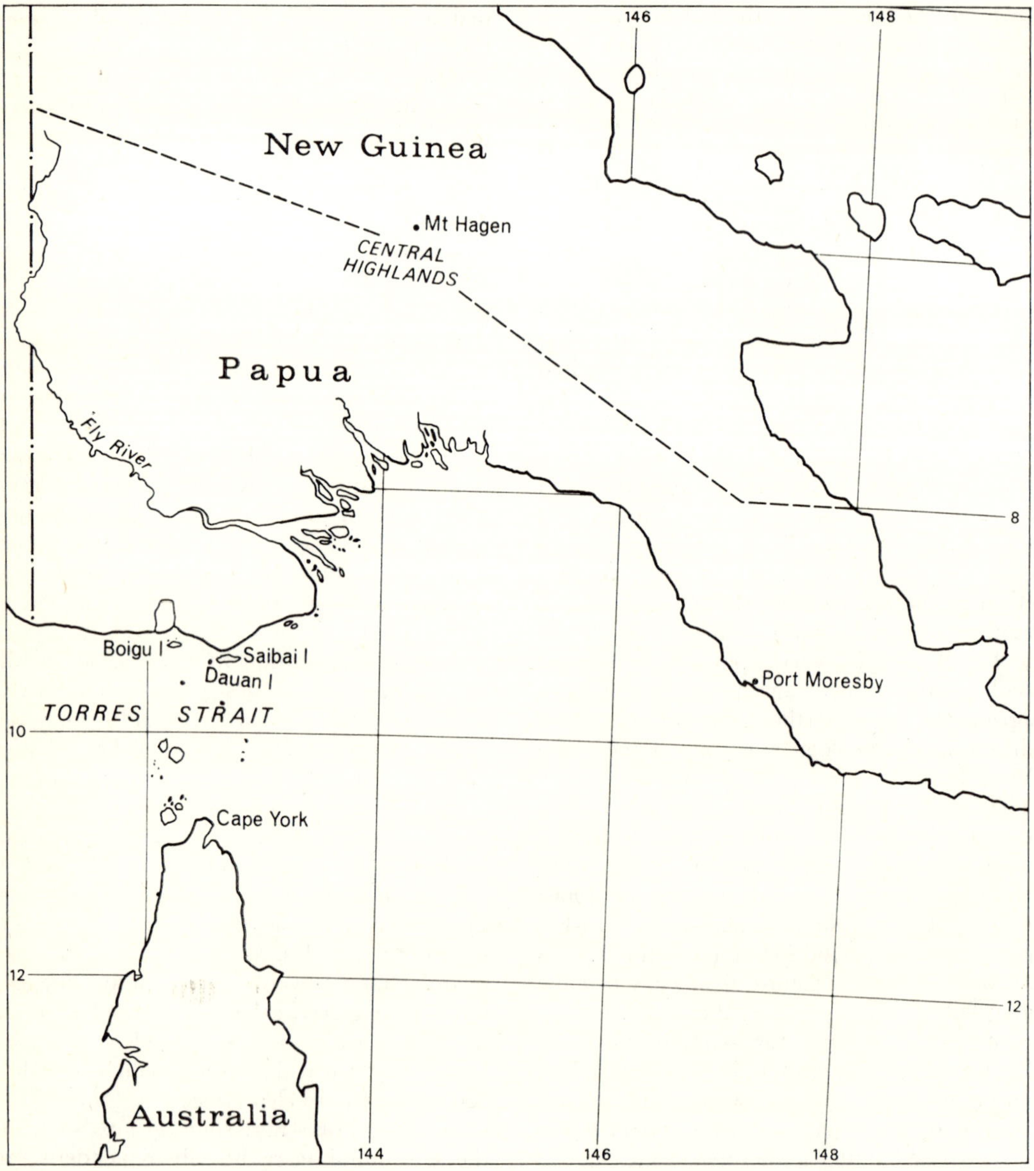

1936:12, 221). These differences, however, seem to be responses to different environments and do not appear to relate to an agricultural 'frontier' of any historical kind.

It is difficult to obtain a clear account of Torres Strait islanders' economy since there are few descriptions of it in the period before heavy European acculturation. Nonetheless, there are clear contrasts between the economic systems of the eastern and western islands. These are partly attributable to ecological variations, for the eastern group are volcanic in origin and have considerably better soil. The Miriam of the eastern islands were gardeners in the Papuan manner, while further west gardens were less common and food collection, especially of seafoods, was more prevalent. Within the central and western island groups, however, some gardens were cultivated, although close to Cape York the islanders seem to have been primarily hunter-gatherers (Haddon 1935: **1**, 410, 1912: **4**, 1-6). Crops in the Torres Strait islands were the same as those in the Trans-Fly, while the wild vegetable foods collected included cashew-nut, wild plum, arrowroot, mangrove, and matchbox or Queensland bean. There were only a few small land animals to hunt, but dugongs, turtles, birds, and fish were major sources of protein (Haddon 1912: **4**, 166-71). Pigs do not appear to have been kept or to have existed in a feral state on any of the islands, although some may have been traded from Papua to the closer islands of Saibai, Dauan, and Boigu; domestic dogs were present everywhere.

There is also some correlation of economy with the languages spoken by different groups, for only the Miriam of the east speak a Papuan language (Haddon 1907: **3**, 528). All other islanders, including those on Dauan and Saibai very close to the Papuan mainland, spoke Australian languages (Haddon 1907: **3**, 262; O'Grady *et al.* 1966). Physically, however, these islanders appear to be Papuans, 'and can be easily distinguished from Australians' (Haddon 1935: **1**, 410). More evidence on this matter would be useful.

Therefore the Torres Strait area might be characterised in many aspects as a transition zone, for there is clearly no sharp break between Papuan agriculturalists and Australian hunter-gatherers.

With Cape York, however, the contrast becomes complete: neither agriculture nor domestic animals other than the dog have ever been observed there. Yet the plants and animals which were being gathered show many similarities to those further north. Varieties of yam, arrowroot, mangrove, Queensland bean, and *Ipomoea* (but not the sweet potato, *I. batatas*) were all available in the area and were utilised—except for cultivation—in much the same way as further north (Thomson 1939; Lawrence 1968; Golson, see ch. 15). Taro has also been identified as a native food (Lawrence 1968: Table 12, quoting Roth and Cook). If the identification of *Colocasia esculenta* (*C. antiquorum*) is correct, the plant was probably introduced into Cape York as it presumably was into New Guinea, though it was apparently never cultivated in Australia.

There was considerable trade and contact between Cape York, the Torres Strait islands, and southwest Papua in historic times, and this was probably the end of a long tradition. This is demonstrated especially by the presence of many other elements of Melanesian culture, both material and non-material, in north Queensland, some of them being found several hundred kilometres from Torres Strait. These include complex hero cults and initiation ceremonies (Thomson 1933, 1934), drum types, outrigger canoes, and house forms, to name simply features which are most obviously anomalous in the Australian context (Moore 1965).

The contact was not all one way. The culture hero Kwoiam, who was of considerable importance in the western islands (e.g. Mabuiag, Haddon 1904: **5**, 67-83), was an Australian equipped with Australian weapons. Actual weapons seem to have been traded into several islands (Moore 1965) and routes certainly existed for the movement of objects from one mainland to the other (McCarthy 1939). There is also some blood group genetical evidence for movement of people (Booth and Oraka 1968:153).

In the other area where there was considerable contact—the northwest coast of Australia and Arnhem Land—the complex influences of 'Macassans' on art, myth, and ceremonial are well known, for the latter area at least (Berndt and Berndt 1954; Thomson 1949). This contact may have begun as late as the eighteenth century A.D. (Macknight and Thorne 1968:216), but it probably began in the sixteenth century (Berndt and Berndt 1954:38; Thomson 1949:82-3) or even earlier. The trade was a seasonal one, primarily to collect bêche-de-mer. There seems to be no record of the visitors making gardens, although they introduced the tamarind and possibly large bamboo species, both of which were then used by Aborigines (Berndt and Berndt 1954:43; Warner 1937:457). There is no certain record of animal introductions by Macassans.

The effect of this contact on Aboriginal culture was considerable. Thomson (1949) considers that Indonesian trade goods such as iron and glass stimulated and extended, if they did not create, the ceremonial exchange cycle in Arnhem Land. He also credits the introduction of dugout canoes, some house types and other items to this influence.

On the other hand, all recorded Indonesian visitors were males and, given their trade, they were probably not normally agriculturalists. Further, in Aboriginal societies women are responsible for collecting plant foods and it seems likely that they would have provided the Macassan visitors with some of these to supplement the basic diet of rice and seafoods. This may explain why Macassans did not cultivate crops in northern Australia even when they returned to the same locations every season. Therefore, although sexual and economic relationships occurred between potential agriculturalists and hunter-gatherer communities, these probably were not of a kind to change the economic status of the Aborigines. Economically speaking, Macassans visiting Australia became part-Aborigines rather than vice versa.

Also germane to this discussion is the direct contact which Aborigines had with agriculture. Both Warner (1937:458) and the Berndts (1954:50-68) record that some Aborigines went with the fleets to Indonesia and subsequently returned to Australia. Further, the bilingualism which developed as a result of the bêche-de-mer industry (Berndt and Berndt 1954:42-3) would have led to knowledge if not direct observation of agriculture. We may also note that one obvious source of experience would have been the occasional castaways who settled for some time in the area and, if modern examples are any guide, made gardens there. These in particular would be almost impossible to discover archaeologically after a few generations, apart from chance palaeobotanical discoveries. Why agriculture did not emanate from these contacts is not yet understood.

The contact situations which have been outlined above are known only from the ethnographic present. It seems very probable that similar forms of contact had been occurring over a long period, but this is difficult to document using synchronic evidence. Nevertheless the successful introduction of occasional plants in both areas of northern Australia indicates some long-term contact (cf. Carter 1963). Carter's case studies are, of course, all historical ones and deal with widely differing economies. Probably his more extreme statements about the difficulties of transferring crop plants need to be modified in relation to societies which have similar ecological bases, or to the transfer of plant genera which are cultivated in similar ways to those already present (Jones 1957; Orchiston 1968), though his thesis need not be abandoned entirely.

The relative lack of new plant introductions is paralleled in the faunal evidence. The history of highly successful feralisation in Australia during post-European times is a long one; the known animals include pigs, dogs, cats, rabbits, buffaloes, and foxes. However, except for the dingo, there is no definite evidence for the introduction of new animals into Australia in the prehistoric period. In some ways this presents more of a problem than the floral evidence, but the data are insufficient for proper discussion.

**Archaeological Evidence**

Any attempt to relate this ethnographic evidence to the archaeological record encounters a number of difficulties, the primary one being the paucity of archaeological data from New Guinea.

In relation to the general problem it would be appropriate to consider the archaeological data relating to coastal dwellers in highly seasonal environments, or even coastal dwellers generally. Unfortunately such data are not available. Apart from the 5,000-years-old Aitape skull (Hossfeld 1965), all material from coastal New Guinea is either from undated surface collections or is presumed to be recent on typological grounds (Golson 1968). In eastern Melanesia, pottery, ground stone tools, and pigs are dated to the second millennium B.C. (Golson 1968:12), but this material is only inferentially relevant.

The New Guinea Highlands, where better data are available, differ in many ways from the lowland areas. The physiography and soils, rainfall and temperature régimes, and vegetation associations are all markedly distinct (Howlett 1967: ch. 3) and settlers from coastal areas would have been forced to make wide-ranging adaptations to them. Further, there are good indications that the area's prehistory has been different. The presence of many Melanesian languages and the widespread use of pottery in coastal regions suggest that the Highlanders remained isolated from at least some changes in Melanesia during the last 10,000 years. We may also expect that external influences on New Guinea cultures will be visible in coastal societies first, and if they reach all societies, that they may have rather different impacts, depending on the ecological and cultural differences involved.

Since, however, the only available material is from the Highlands we must use it (Bulmer 1966a; White, J. P. 1967, 1968; Golson 1968). From the radiocarbon dates so far available, some Highlands areas of New Guinea were settled by at least 16,000-25,000 BP, although the evidence from this period is limited to flaked axe-adzes and waisted blades. At 11,000 BP rock shelter sites in the Central Highlands show people using a tool kit of amorphous flaked stones ('chunky scrapers'?), bone tools, and ground stone axes. There seem to be no introduced animals amongst the fauna. The wild fauna is almost entirely modern in type and, using a 'Pleistocene overkill' argument (Martin 1967), this suggests that these settlements are far from being the first. The economy of these groups is unknown. It is normally assumed to be based on hunting and gathering, but the evidence for this is negative—namely the absence of putative domestic fauna or specialised agricultural implements.

The evidence for the date of the introduction of agriculture is inferential, being based on faunal material. The first non-indigenous animals, pigs, appear in two shelter sites at about 5,000-6,000 BP (Bulmer 1966b; White, J. P. 1967), a date which correlates well with the introduction of pigs in Timor (Glover, see ch. 13). It has been argued by Bulmer (1966b) that *if* these pigs were domesticated this would almost certainly imply the presence of agriculture, since pigs need to be fed regularly if they are not to turn feral. However, no morphological tests of the domesticated status of New Guinea pigs are available, and pigs at this date in the Highlands might be feral relatives of lowland animals, simply being hunted and/or captured and reared in settlements to be eaten later. Neither is an uncommon practice in New Guinea today.

It should also be pointed out that pigs were not necessarily introduced into lowland New Guinea in their domesticated form for, as R. Bulmer emphasises (pers. comm.), it seems highly probable that tame (but not domestic) animals such as phalangers have been transferred between some Melanesian islands. Nonetheless, it is generally assumed that the presence of pigs in New Guinea does imply that agriculture was present somewhere in the island or nearby (cf. Reed 1969; White, J. P. 1970).

This is confirmed to some extent by dates of 3,000 BP for pig bones in eastern Melanesia (Golson 1968:12). The small islands there probably could not support both hunter-gatherer groups and a viable population of wild or feral pigs. It therefore seems safe to assume that agriculture began in New Guinea somewhere around 5,000 BP, if not earlier.

The first direct evidence of agriculture in New Guinea is dated to *c.* 2,300 BP (Golson *et al.* 1967), considerably later than the presumptive date given above. The developed drainage system of the prehistoric gardens discovered at the open Manton site near Mt Hagen was apparently well adapted to the Highlands environment and probably does not represent the first stages of agriculture in the area. It has been assumed that taro was the most likely major crop (Brookfield and White 1968:49). No animals have yet been found in dated open sites, but to judge from shelter sites it is only in very recent times that other introduced animals such as dogs and hens are to be found.

It should be stressed that Highlands sites show little change in stone and bone technology throughout the entire prehistoric period. Not only do the same traditions continue but no major new items are added to the assemblage. Wooden tools may have changed to some extent, but since these do not occur in shelter sites this is undocumented before *c.* 2,300 BP and there is no documented change subsequent to this. A parallel situation is found in Timor (Glover, see ch. 13).

There are several possible reasons for this uniform technological pattern. It may be that the Highlanders were agriculturalists in some form throughout the Holocene and that no major economic change occurred in this period. More probably, however, the shelter sites so far excavated were primarily for hunting and refuge and do not accurately reflect, in the later period, the total range of the technology. We know, for instance, that where Europeans have had the opportunity to observe them, flaked stone tools were used almost exclusively for wood-working (White, J. P. 1967:103) and there seems to be no good reason to expect a basic change in these techniques simply because an economic change occurred.

Against this background the Australian evidence may be briefly examined. In the last few years a continent-wide two-phase technological sequence has been recognised. Mulvaney (Mulvaney and Joyce 1965) pointed out that the change, as far as the flaked stone artifacts were concerned, was from an amorphous 'scraper' industry to one characterised by small, well-made tools including microliths. In the north, points and small scraper-adzes are common, while backed blades in various forms apparently occur only south of the 20° parallel (Glover 1967). Mulvaney interpreted the technological change as marking the introduction of the concept of hafting and pointed out that it appears around 5,000 BP (Mulvaney 1966). Evidence now available shows that the first interpretation, as originally proposed, is wrong and also suggests that Phase II may be present by about 7,000 BP (White, C. 1967 and see ch. 12).

Detailed explanations for the beginning of Phase II in Australia have not been sought. Mulvaney and Joyce (1965:209-10) refer to migration, diffusion, and local invention as possible causes but do not explore the possibilities (cf. Mulvaney, see ch. 25). One fascinating aspect of the change is that there are no obvious precursors: no similar industries are known from immediately adjacent areas. There is nothing in Melanesia, whether excavated or surface-collected, to parallel the later Australian assemblages, although it must be remembered that there are no sequences of stone technology from any coastal area. The same is true of Timor (Glover, see ch. 13). Closer parallels exist in some Indonesian islands, notably the southern Celebes and Java (van Heekeren 1957). None of these industries is radiometrically dated but the apparent absence both of domestic fauna (dogs, pigs) and of extinct wild forms suggests that they belong to terminal Pleistocene or early Recent periods, although some may be later. The situations in which these industries are found further support the idea that they were made by hunter-gatherers rather than agriculturalists.

Similar artifacts also occur around the Indian Ocean, in Ceylon, India, the Horn of Africa, and southern Africa. In India the upper levels of one microlithic (late Stone Age) site have been dated to *c.* 7,000 BP (Allchin and Allchin 1968:83). In southern Africa microlithic backed blades first appear in the Magosian and are dated to the late Pleistocene (13,000-20,000 BP). However, the full development of the microlithic tradition occurs only in the Late Stone Age, dated to perhaps 8,000 BP in Zambia and around 6,000 BP further south (Deacon 1966). It is clear that the transition was a gradual one (Clark 1959:183) and not relatively sudden as in Australia. It is nonetheless true that the tradition occurs in a fully developed form only after the development of a 'full' neolithic economy further north.

As well as the artifacts, there is one piece of faunal evidence in Australia which is important. The absence of the dingo from all records of terminal Pleistocene fauna (Merrilees 1968) suggests that its introduction was a good deal later than the advent of man (cf. Tindale 1959:45). The oldest published date for dingo in Australia is *c.* 3,000 BP (Macintosh 1964:507), but there are unpublished dates from Mt Burr in South Australia of around 7,000 BP (Edwards pers. comm.) which, if their stratigraphic relationship with the dingo is confirmed, conform closely to the earliest dates for the beginning of Phase II in Australian prehistory.

**Microliths and Agriculture**

Any attempt to draw the beginnings of a

prehistory out of this data must be both speculative and tentative, but may serve to provoke further discussion.

I have shown that the ethnographic data suggest a reasonably long-term and more than casual contact between parts of northern Australia and its northern neighbours. Although this did not lead to Aborigines adopting agriculture as a way of life, it did provide opportunities, both in Cape York and Arnhem Land, for the adoption of a wide range of objects and beliefs, some of which had major effects on Aboriginal society. In short, Aboriginal conservatism may be overstressed (cf. Mulvaney, see ch. 25).

It is also apparent that there is a striking temporal correlation between the probable beginnings of agriculture in Melanesia and Indonesia and the start of Phase II in Australia's technological history, which may be associated with the introduction of the dingo. The correlation is too marked to be simply coincidental.

I would therefore suggest that the introduction of the new elements of Phase II into Australia is related to the development and spread of agriculture into island Southeast Asia. Given the nature of the introductions, it seems more likely that they were the result of contact with hunter-gatherer fishermen than with agriculturalists, but this is conjectural. One must assume that sea travel to Australia and not in the reverse direction occurred in some form, and that selective adoption of new features would readily occur if modern ethnographic evidence is any guide (e.g. Thomson 1949). The precise form and context of this contact is, however, unclear, so that the Phase II technological distinctions between northern and southern Australia are still unexplained.

If this construct is provisionally accepted, it might be asked why there are no traces of the technological revolution in Melanesia. I suggest that it is because the change from hunting and gathering to agriculture occurred there at about the same time. If microlithic tools were primarily used by hunter-gatherers to improve hunting and gathering techniques, then they would be of little use to agricultural groups. And even if, on the other hand, these new tools had no obvious economic advantages they might have been adopted by Australian Aborigines who found them sociologically advantageous (cf. Macassan introductions discussed above), but rejected by Melanesians who regarded their owners as 'backward' (cf. Turnbull 1965: ch. 4). Whether there were other sociological or economic factors operating we cannot now determine, although the correlation between what may be the most recent dates of some giant herbivores (Jones 1968: 202) and the beginning of Phase II clearly requires further investigation.

**References**


Allchin, Bridget and Allchin, R. 1968. *The birth of Indian civilization.* Harmondsworth: Penguin.

Anon. 1907. Why did the Blacks not cultivate food? *Sci.Man* **9**:21-4.

Barnes, F. 1970. The biology of pre-neolithic man. In *The impact of civilisation on the biology of man* (ed. S. V. Boyden), pp. 1-26. Canberra: Australian National University Press.

Berndt, R. M. and Berndt, Catherine H. 1954. *Arnhem Land.* Melbourne: Cheshire.

Booth, P. B. and Oraka, R. E. 1968. Blood group frequencies along the south coast of Papua. *Archaeol.& phys.Anthrop.Oceania* **3**:146-55.

Braidwood, R. J. and Willey, G. R. (eds.) 1962. *Courses toward urban life.* Publs. Anthrop. Viking Fund No. 32. New York.

Brookfield, H. C. and Hart, Doreen. 1966. *Rainfall in the tropical southwest Pacific.*

Canberra: Australian National University, Research School of Pacific Studies, Department of Geography Publication G/3.

——and White, J. P. 1968. Revolution or evolution in the prehistory of the New Guinea Highlands: a seminar report. *Ethnology* **7**:43-52.

Bulmer, Susan. 1966a. The prehistory of the Australian New Guinea Highlands. M.A. thesis, University of Auckland, New Zealand (unpublished).

——1966b. Pig bone from two archaeological sites in the New Guinea Highlands. *J.Polynes.Soc.* **75**:504-5.

Butzer, K. W. 1964. *Environment and archaeology*. Chicago: Aldine Publishing Co.

Caldwell, J. R. 1962. Eastern North America. In *Courses toward urban life* (eds. R. J. Braidwood and G. R. Willey), pp. 288-308. Publs. Anthrop. Viking Fund No. 32. New York.

Carter, G. F. 1963. Movement of people and ideas across the Pacific. In *Plants and the migrations of Pacific peoples* (ed. J. Barrau), pp. 7-22. Honolulu: Bishop Museum Press.

Clark, J. D. 1959. *The prehistory of southern Africa*. Harmondsworth: Penguin.

——1964. The prehistoric origins of African culture. *J.Afr.Hist.* **5**:161-83.

Clark, J. G. D. 1965. Radiocarbon dating and the expansion of farming culture from the Near East over Europe. *Proc.prehist.Soc.* **31**:58-73.

Deacon, J. 1966. An annotated list of radiocarbon dates for sub-Saharan Africa. *Ann. Cape prov.Mus.* **5**:5-84.

Elkin, A. P. 1954. *The Australian Aborigines: how to understand them.* 3rd edn. Sydney: Angus and Robertson.

Flannery, K. V. 1965. The ecology of early food production in Mesopotamia. *Science, N.Y.* **147**:1247-56.

——1969. Origins and ecological effects of early domestication in Iran and the Near East. In *The domestication and exploitation of plants and animals* (eds. P. J. Ucko and G. W. Dimbleby), pp. 73-100. London: Duckworth.

Glover, I. C. 1967. Stone implements from Millstream Station, Western Australia: Newall's collection re-analysed. *Mankind* **6**:415-25.

Golson, J. 1968. Archaeological prospects for Melanesia. In *Prehistoric culture in Oceania: a symposium* (eds. I. Yawata and Y. H. Sinoto), pp. 3-14. Honolulu: Bishop Museum Press.

——, Lampert, R. J., Wheeler, J. M., and Ambrose, W. R. 1967. A note on carbon dates for horticulture in the New Guinea Highlands. *J.Polynes.Soc.* **76**:369-71.

Haddon, A. C. (ed.) 1904-35. *Reports of the Cambridge anthropological expedition to Torres Straits.* Vol. 1 General Ethnography (1935), Vol. 3 Linguistics (1907), Vol. 4 Arts and Crafts (1912), Vol. 5 Sociology, Magic and Religion of the Western Islanders (1904). Cambridge: Cambridge University Press.

Hossfeld, P. S. 1965. Radiocarbon dating and palaeoecology of the Aitape fossil human remains. *Proc.R.Soc.Vict.* **78**:161-5.

Howlett, Diana R. 1967. *A geography of Papua and New Guinea.* Melbourne: Nelson.

Jones, R. 1968. The geographical background to the arrival of man in Australia and Tasmania. *Archaeol. & phys.Anthrop.Oceania* **3**:186-215.

Jones, W. O. 1957. Manioc: an example of innovation in African economies. *Econ.Dev. Cult.Change* **5**:97-117.

Keast, A. 1959. The Australian environment. In *Biogeography and ecology in Australia* (eds. A. Keast, R. L. Crocker, and C. S. Christian), pp. 15-35. Monographiae biol. **8**. The Hague: W. Junk.

Lawrence, R. 1968. *Aboriginal habitat and economy.* Canberra: Australian National

University, School of General Studies, Department of Geography Occasional Papers No. 6.

Lee, R. B. 1965. Subsistence ecology of !Kung Bushmen. Ph.D. thesis, University of California, Berkeley (unpublished).

McCarthy, F. D. 1939. 'Trade' in Aboriginal Australia, and 'trade' relationships with Torres Strait, New Guinea and Malaya. *Oceania* **10**:81-104, 171-95.

Macintosh, N. W. G. 1964. A 3,000 years old dingo from shelter 6. App. 2 to D. J. Mulvaney, G. H. Lawton, and C. R. Twidale. *Proc.R.Soc.Vict.* **77**:498-507.

Macknight, C. C. and Thorne, A. G. 1968. Two Macassan burials in Arnhem Land. *Archaeol. & phys.Anthrop.Oceania* **3**:216-22.

Martin, P. S. 1967. Prehistoric overkill. In *Pleistocene extinctions. The search for a cause* (eds. P. S. Martin and H. E. Wright), pp. 75-120. New Haven: Yale University Press.

Meggitt, M. J. 1964. Aboriginal food-gatherers of tropical Australia. In *The ecology of man in the tropical environment.* International Union for the Conservation of Nature and Natural Resources, Ninth Technical Meeting, Proceedings and Papers. pp. 30-7. Morges: IUCN (Publ. Ser.4).

Merrilees, D. 1968. Man the destroyer: Late Quaternary changes in the Australian marsupial fauna. *J.Proc.R.Soc.West.Aust.* **51**:1-24.

Moore, D. R. 1965. The tribes of Cape York. Dip. Anthrop. thesis, University of Sydney (unpublished).

Moresby, J. 1876. *Discoveries & surveys in New Guinea and the D'Entrecasteaux Islands.* London: John Murray.

Mulvaney, D. J. 1966. The prehistory of the Australian Aborigine. *Scient.Am.* **214** (3): 84-93.

——and Joyce, E. B. 1965. Archaeological and geomorphological investigations on Mt. Moffatt Station, Queensland, Australia. *Proc.prehist.Soc.* **31**:147-212.

O'Grady, G. N., Wurm, S. A., and Hale, K. L. 1966. *Aboriginal languages of Australia (a preliminary classification).* Map. Department of Linguistics, University of Victoria, British Columbia.

Orchiston, D. W. 1968. On the human ecology of prehistoric South Island New Zealand. B.A. (Hons) thesis, University of Sydney (unpublished).

Reed, C. A. 1969. The pattern of animal domestication in the prehistoric Near East. In *The domestication and exploitation of plants and animals* (eds. P. J. Ucko and G. W. Dimbleby), pp. 361-80. London: Duckworth.

Specht, R. L. 1958. An introduction to the ethno-botany of Arnhem Land. In *Rec. Am.-Aust.scient.Exped.Arnhem Ld* **3** (Botany and Plant Ecology) (eds. R. L. Specht and C. P. Mountford), pp. 479-503. Melbourne: Melbourne University Press.

Suttles, W. 1951. The early diffusion of the potato among the coast Salish. *SWest.J. Anthrop.* **7**:272-88.

Thomson, D. F. 1933. The hero cult, initiation and totemism on Cape York. *Jl R. anthrop.Inst.* **63**:453-537.

——1934. Notes on a hero cult from the Gulf of Carpentaria, North Queensland. *Jl R.anthrop.Inst.* **64**:217-35.

——1939. The seasonal factor in human culture. Illustrated from the life of a contemporary nomadic group. *Proc.prehist.Soc.* **5**:209-21.

——1949. *Economic structure and the ceremonial exchange cycle in Arnhem Land.* Melbourne: Macmillan.

Tindale, N. B. 1959. Ecology of primitive Aboriginal man in Australia. In *Biogeography and ecology in Australia* (eds. A. Keast, R. L. Crocker, and C. S. Christian), pp. 36-51. Monographiae biol. **8**. The Hague: W. Junk.

Turnbull, C. M. 1965. *Wayward servants. The two worlds of the African Pygmies*. Garden City, New York: Natural History Press.

van Heekeren, H. R. 1957. *The stone age of Indonesia*. 's-Gravenhage: Martinus Nijhoff.

Warner, W. L. 1937. *A black civilization*. New York: Harper and Bros.

White, Carmel. 1967. Early stone axes in Arnhem Land. *Antiquity* **41**:149-52.

White, J. P. 1967. Taim bilong bipo. Investigations towards a prehistory of the Papua-New Guinea Highlands. 2 vols. Ph.D. thesis, Australian National University, Canberra (unpublished).

——1968. Studies of prehistory in the New Guinea Highlands. *Aust.nat.Hist.* **16**:83-6.

——1970. New Guinea: the first phase in Oceanic settlement. In *Studies in Oceanic Culture History* (eds. R. C. Green and M. Kelly), in press. Honolulu: Bishop Museum, Pacific Anthropological Records.

Williams, F. E. 1936. *Papuans of the Trans-Fly*. Oxford: Clarendon Press.

Worsley, P. M. 1961. The utilization of food resources by an Australian Aboriginal tribe. *Acta Ethnogr.Hung.* **10**, Fasc. 1-2: 153-90.

# 15 Australian Aboriginal Food Plants: Some Ecological and Culture-Historical Implications

Jack Golson

This paper sets out to explore the relationships in respect of plants used for food between different areas of Australia and between Australia and the Malaysian region whence the original settlement of Australia took place. Malaysia is defined in the same way as the plant geographical province served by *Flora Malesiana* (1948-67:**1**, XIII), comprising political Malaysia, Indonesia, the Philippines, and New Guinea. The incomplete inquiries of which it presents a partial record were prompted by the frequent appearance in Lawrence's recent tabulations (1968) of Australian Aboriginal food plants of genera familiar as food sources in Malaysia and Oceania.

In his Tables 2, 12, and 15 Lawrence (1968) lists food plants in central Australia and eastern and western Cape York. These lists are based on an exhaustive survey of the literature and record information derived from often specifically ethnobotanical inquiries, amongst traditionally oriented societies, by competent observers, at a time when the botanical exploration of the continent was well advanced. They may be regarded, therefore, as relatively complete. In addition to these I have used Specht's (1958d) list for Arnhem Land, compiled during the American-Australian Scientific Expedition of 1948. Though this is admittedly incomplete (Specht 1958d:479) and lacks the historical depth that the existence of

For help with problems of botanical nomenclature and synonymy, which I hope I have employed correctly, I thank Dr Nancy Burbidge, Curator of the Herbarium, Division of Plant Industry, Commonwealth Scientific and Industrial Research Organization, Canberra, and especially Mr L. Pedley, Research Botanist, Queensland Herbarium, Brisbane. Mrs Jocelyn Powell (née Wheeler), late of the Department of Biogeography and Geomorphology, Research School of Pacific Studies, ANU, helped me immeasurably with the botanical literature.

early observations gives to Lawrence's lists, it has the great merit of being the work of a professional botanist who provides an ecological context.

Between them these lists allow comparisons to be made between ecologically contrasted regions, between the tropical north and the arid and semi-arid centre and, in the north, between Cape York with its areas of luxuriant tropical vegetation and Arnhem Land with its less prolific flora (Specht 1958c:415). I use the regions as defined by these authors. Arnhem Land is the Arnhem Land Aboriginal Reserve, though Specht's ethnobotanical researches were largely confined to the coast and coastal plains. Cape York is the Cape York Peninsula. Central Australia, after Lawrence's study (1968:42), is a vast tract of arid and semi-arid country comprising the Northern Territory south of Newcastle Waters and the western half of South Australia.

Unfortunately no information of comparable value is available for the climatically temperate areas of the continent where traditional life was disrupted before the era of systematic ethnographic and botanical record. It ought nevertheless to be possible, by comparing the genera and species utilised for food in northern and central Australia with the botanical inventories of the temperate zone, to assess the magnitude of the adaptation required in the food-gathering activities of the original settlers of southern Australia.

My concern in this paper, however, is less for comparisons with southern Australia than with the Malaysian area to the north and for this purpose I have used Barrau's recent check-list (n.d.:47-73) of vascular plants of ethnobotanical importance in Malayo-Oceania. Since many of these are treated only at generic level, I have supplemented and extended Barrau by reference to Burkill's monumental two-volume *Dictionary of the Economic Products of the Malay Peninsula* (1935), a work of much wider area and subject coverage than the title suggests.

A full comparative assessment of plant exploitation in different areas requires the existence of floral check-lists for each region of concern, as well as information on plant distributions and ecology. In these respects I am in the present case limited to Specht's admirable coverage for Arnhem Land (1958a, b,c), in which the Australian and extra-Australian distributions of each Arnhem Land species of gymnosperm and angiosperm are set out and their Arnhem Land occurrences ecologically classified. Tindale (1958) supplies similar information for the ferns. For South Australia, and thus for the southern part of the central Australian region as defined by Lawrence, there is J. M. Black's up-to-date *Flora* (1943-57) revised by Eichler (1965), which gives some information about the further distribution of South Australian plants. There is nothing comparable for Queensland, so that for Cape York I have been virtually restricted to comparisons in terms of the recorded food plants themselves.

A major difficulty is posed by changes in plant names since the time when the utilised ethnobotanical records were made and, more seriously, by the validity of the original plant identifications themselves, particularly the earlier ones. Lawrence (1968: vii) appreciated the problem and was able to avoid it. Since, however, the present aim was to quantify the information derived from Lawrence and others, it was essential on the one hand to achieve accuracy on the question of nomenclature in order to avoid registering the same plant under different names, and, on the other hand, to establish confidence in the identifications, without which the whole exercise would be futile. For Arnhem Land Specht's work fulfils requirements on both counts. The labours of J. M. Black on the flora of South Australia and contiguous territories, which extended from 1902 to 1951, provided a sound systematic base for the thirty years of central Australian ethno-botanical research inaugurated by Cleland at the beginning of the 1930s. Of the 250 relevant observations recorded by Lawrence (1968: Table 2) of central Australian food plants, nearly 140 were made by Cleland and his Adelaide colleagues and rather more than 200 postdate 1920. Stirling's thirty-two records from the Horn Scientific Expedition of 1894 and Basedow's ten from the South Australian Government's North-west Prospecting Expedition of 1903 constitute the highly reputable earlier records.

For Cape York, the question really resolves itself into the accuracy of W. E. Roth's botanical information (1901). Of the 150 relevant citations in Lawrence's list (1968: Table 12) of food plants in eastern Cape York, no less than 133 are from Roth, and, of the remainder, only two refer to plants not in his compilation. As for the sixty relevant observations for western Cape York (Lawrence 1968: Table 15), thirty-eight are Roth's, while of the remainder only eight name plants not recorded by him. All eight new plant records derive from McConnell's and Thomson's field work amongst the Wik-Munkan in the 1930s. In making acknowledgment for botanical identifications to the Colonial and Assistant Colonial Botanists at Brisbane, F. M. and J. F. Bailey, Roth (1901: preface) says that they examined something like 400 plants collected by him: Mr L. Pedley, Research Botanist at the Queensland Herbarium, tells me there are still some Roth specimens in the Herbarium. As heir to the collections which the Baileys built up, Pedley has, by his answers to the specific queries I have addressed to him, in effect brought up to date for me the botanical identifications which they supplied to Roth.

There is no reason, therefore, to doubt the basic validity of the plant identifications recorded in the literature I am using. The alphabetically ordered plant list at the end of this paper (Appendix 15:1) is as accurate a consolidation of this information as my understanding of the botanical authorities, Specht, Black, Eichler, and Pedley, will allow. Dr Nancy Burbidge, Curator of the Herbarium, Division of Plant Industry, CSIRO, has helped me with unresolved problems in Eichler's revision of Black. The very few and minor disagreements that appeared as between Eichler, Specht, and Pedley are annotated in the consolidated list. I have followed Burbidge (1963) on the spelling of generic names. The appendixes as a whole contain in summary form all the data which the subsequent discussions have employed and reference is made to them where appropriate.

## The Indo-Malaysian Component of the Arnhem Land Flora

Specht (1958c: 416) records for Arnhem Land twenty-four species of ferns, two of gymnosperms, and 1,098 of angiosperms. Some 25-30 per cent of the 1,100 species in the two last groups are also recorded from New Guinea through island and continental Southeast Asia to the Indian peninsula (Specht 1958c: 417).

This strong Indo-Malaysian element is by no means evenly distributed through the Arnhem Land vegetation, though it is present in every major plant community (Specht 1958c: 420-1). It is, expectedly, most strongly represented in the strand vegetation and mangrove environment: some 70-80 per cent of the nineteen Arnhem Land species in the former category and some 60-75 per cent of the twenty-nine species in the latter are known from New Guinea to India. Of the eighty-one species making up the monsoon forests of Arnhem Land, 58 per cent are known in New Guinea, 49 per cent in island Southeast Asia, 37 per cent in continental Southeast Asia, and 35 per cent in India. Arnhem Land has seventy-four species in the coastal dune plant complex, whose representation in Indo-Malaysia ranges from 26-39 per cent, and 252 species in the freshwater swamp and stream complex, of which 29-37 per cent are found in Indo-Malaysia.

The actual land area occupied by these plant communities in Arnhem Land is very small (Specht 1958b: map 9, between pp. 384 and 385) and the vast bulk of the region is characterised by two major plant complexes where the Indo-Malaysian element is weaker. Of ninety species in the sandstone complex 23 per cent occur in New Guinea and 10-14 per cent in the rest of the Indo-Malaysian area, while of the 136 species of the *Eucalyptus* tall open forests 28 per cent are known in New Guinea and 15-17 per cent in the other regions.

It is important to note, however, that though the area occupied by the plant communities strong in Indo-Malaysian elements is small, some of them are widely distributed. By definition the strand, mangrove, and coastal dune communities are restricted to a coastal habitat, but streams, swamps, and marshes are widespread, though they may dry up in the dry season (Specht 1958c:424). Patches of monsoon forest are also to be found widely scattered in favourable habitats not only in Arnhem Land but throughout northern Australia, testifying to the previous existence of a widespread and relatively continuous monsoon forest flora (Specht 1958c: 421-2). In the Northern Territory patches of this vegetation are numerous in the Darwin-Katherine area and even perhaps as far south as Tennant Creek.

This pattern of plant distribution would have been modified, but not obliterated, whatever the climatic régime at the time of the late Pleistocene settlement of northern Australia by Malaysian man. The significance of that pattern for human colonisation is best illustrated by a consideration of Arnhem Land food plants.

*Table 15:1 Uses in Malaysia of shared species (A), and all shared genera (B) used for food in Arnhem Land*

| | 1 | 2 | 3 | 4 | Totals |
|---|---|---|---|---|---|
| A | 8 | 3 | 7 | 10(12) | 28(30) |
| B | 18 | 9 | 10 | 5 | 42 |

*Note*: 1 similar food use, 2 alternative food use, 3 medical, technological, etc. uses, 4 no recorded uses. In parentheses are the totals, counting as Malaysian two species of *Grewia* not certainly present in Malaysia.

If we exclude from Specht's list (1958d) the nectars and gelatinous foods (four species) and one fruit tree possibly introduced fairly recently by Macassan trepangers (*Tamarindus indica* L.), and if we ignore the purely varietal distinctions (in nos. 99a,b, 132a, 144a, 169a, 202a, 263a, and 272a of Appendix 15:1), we are left with sixty-five species of food plants in forty-eight genera, of which twenty-eight species are known widely throughout island and mainland Southeast Asia, while two more (154, 155) have been recorded in New Guinea and India but not as yet in between. A further twenty-seven species are confined to Australia (twenty species) or Australia and New Guinea (seven species) but belong to genera known on the Asiatic mainland and in the intervening islands. Only eight species of Arnhem Land food plant belong to (six) genera not occurring in the Malaysian region (Appendixes 15:1 and 15:2).

At generic or specific level, therefore, the vast majority of the plants used for food in Arnhem Land would have been familiar to immigrants from the Malaysian region. Indeed, as Table 15:1 shows, some of the same species and an appreciable number of the same genera have been recorded in recent

*Table 15:2 Distribution by Specht's ecological categories of 65 Arnhem Land food plants where A species, B genus only, C neither genus nor species is known in Malaysia*

| | **1** | **2** | **3** | **4/6** | **4** | **4/5** | **5** | **5/6** | **6** | **6/7** | **7** | **Totals** |
|---|---|---|---|---|---|---|---|---|---|---|---|---|
| A | 1 | 3 | 9(10) | 0 | 10(11) | 1 | 2 | 1 | 1 | 0 | 0 | 28(30) |
| B | 0 | 0 | 7(6) | 1 | 5(4) | 5 | 7 | 0 | 2 | 1 | 1 | 29(27) |
| C | 0 | 0 | 4 | 0 | 0 | 0 | 2 | 0 | 2 | 0 | 0 | 8 |
| Totals | 1 | 3 | 20 | 1 | 15 | 6 | 11 | 1 | 5 | 1 | 1 | 65 |

*Note*: 1 coastal dunes, 2 estuarine plains, 3 freshwater streams, marshes, and swamps, 4 monsoon forests, 5 savannah woodlands on coastal dunes and sandy fan deltas, 6 tall open forest, 7 quartzite and sandstone hills. In parentheses are the totals, counting as Malaysian two species of *Grewia* not certainly present in Malaysia.

*Table 15:3 Produce of 65 Arnhem Land food plants by season and ecological zone*

| | **Wet season** | | | | | | | | **Dry season** | | | | | | | | **No. of plants per use** | | | | **Totals of plants** |
|---|---|---|---|---|---|---|---|---|---|---|---|---|---|---|---|---|---|---|---|---|---|
| | **Oct.-Dec.** | | | | **Jan.-Mar.** | | | | **Apr.-Aug.** | | | | **Sept.-Oct.** | | | | | | | | |
| | **R** | **S** | **F** | **G** | **R** | **S** | **F** | **G** | **R** | **S** | **F** | **G** | **R** | **S** | **F** | **G** | **R** | **S** | **F** | **G** | |
| 1 | 1 | 0 | 0 | 0 | 1 | 0 | 0 | 0 | 1 | 0 | 0 | 0 | 1 | 0 | 0 | 0 | 1 | 0 | 0 | 0 | 1 |
| 2 | 0 | 0 | 0 | 0 | 0 | 0 | 0 | 0 | 1 | 1 | 0 | 0 | 0 | 0 | 1 | 0 | 1 | 1 | 1 | 0 | 3 |
| 3 | 4 | 0 | 0 | 0 | 4 | 0 | 0 | 0 | 12 | 2 | 2 | 1 | 13 | 3 | 1 | 0 | 16 | 3 | 2 | 1 | 20 |
| 4 | 1 | 0 | 6 | 1 | 1 | 0 | 6 | 1 | 0 | 0 | 7 | 1 | 0 | 2 | 6 | 1 | 1 | 2 | 20 | 1 | 22 |
| 5 | 0 | 2 | 10 | 0 | 0 | 1 | 7 | 0 | 2 | 1 | 1 | 0 | 1 | 3 | 3 | 0 | 2 | 3 | 14 | 0 | 18 |
| 6 | 1 | 2 | 2 | 1 | 2 | 0 | 3 | 1 | 1 | 0 | 2 | 1 | 1 | 2 | 2 | 1 | 3 | 2 | 4 | 1 | 8 |
| 7 | 0 | 0 | 1 | 0 | 0 | 0 | 1 | 0 | 0 | 0 | 2 | 0 | 0 | 0 | 1 | 0 | 0 | 0 | 2 | 0 | 2 |
| No. of plants per use | 6 | 4 | 15 | 2 | 7 | 1 | 14 | 1 | 16 | 4 | 12 | 3 | 15 | 9 | 10 | 2 | | | | | |
| No. of plants | 26 | | | | 22 | | | | 33 | | | | 35 | | | | | | | | |

*Note*: For ecological zones see Table 15:2. R rootstocks, S seeds (including nuts, kernels), F fruits, G greens (leaves, stems, shoots) but including stems, stem buds, and inner leaves of palms, etc.

*Table 15:4 Botanical affiliations of food plants in three Australian regions at generic level*

| | **Australasian** | **Other non-Malaysian** | **Malaysian** | **Totals** |
|---|---|---|---|---|
| Arnhem Land | 5 | 1 | 42 | 48 |
| Cape York | 12 | 0 | 94 | 106 |
| Central Australia | 10 | 14 | 32 | 56 |

times as food sources in Malaysia, despite the often highly specialised agricultural basis of economic life now characteristic of most of that region.

Specht (1958d:480) records the habitat of the food plants of Arnhem Land under seven heads and Table 15:2 uses these categories to set out their ecological distribution in terms of botanical relationships.

The Malaysian element in the utilised flora is seen to be well distributed through the different environments. We may also note the prominence as resource zones of the monsoon forest (4 in Table 15:2) and freshwater streams and swamps (3), whose small areal coverage but widely scattered distribution we have already noted, and whose richness in Malaysian species is evident. Together with the savannah woodlands (5), another formation of localised but repeated occurrence, they would provide a good subsistence base for the spread of settlement.

Table 15:3 underlines this point by showing the complementary nature of their produce in kind and seasonality. The totals of the plant numbers in the vertical and horizontal rows agree neither with each other nor with the sum of sixty-five species under analysis, because in the one case some plants are present in more than one ecological zone and in the other case produce at more than one season.

It would be difficult with the evidence available to translate into detailed terms of human subsistence the ethnobotanical information reviewed above. It is evident of some plants that they are so rare, their season so short, or their produce so unpalatable, that they can never have played more than a minor, specialised, or scarcity role in subsistence; of others that because of their number, productivity, prolonged availability, or food value, they became virtual staples in the diet. Specht touches on this aspect only in the most general of terms, remarking, for example, on the importance of *Cycas media*, whose kernels assumed the status of a staple during the latter half of the dry season and the beginning of the wet (1958d:482), and of *Boerhavia diffusa*, that in emergencies its fibrous tap roots would be utilised during the months of March to May (1958d:489).

No ecological information at all, of the type provided by Specht for Arnhem Land, is readily obtained for other areas, for which, therefore, a discussion of the type offered in this section is impossible. This being so, Arnhem Land must stand as the model of how Malaysian man could readily have established and expanded his settlement anywhere in the northern regions of Australia where it is reasonable to suppose his colonisation of the continent began.

## The Food Plants of Arnhem Land, Cape York, and Central Australia

It is nevertheless possible in a general way to examine and extend this proposition. Table 15:4 sets out the botanical affiliations of the food plants of the three Australian regions with which we are concerned. The aim in this and similar tabulations is to distinguish between genera that would have been familiar to Malaysian man, including a small number restricted to eastern Malaysia, and genera that he would have first met with in Australia, including a few examples present also in New Guinea. Within the genera in this second category there is a distinction made between those restricted to Australasia and those known also in other parts of the world. It is important to realise, however, that the botanical categories employed will mask the quite close resemblance one genus may bear to another. Thus the endemic Australian genus *Brachychiton* is so similar to the genus *Sterculia*, known in both Australia

*Table 15:5 Food plants shared at generic and specific level between three Australian regions*

| | **Arnhem Land** | | **Cape York** | | **Central Australia** | |
|---|---|---|---|---|---|---|
| | **Genus** | **Species** | **Genus** | **Species** | **Genus** | **Species** |
| Arnhem Land | *48* | *65* | — | — | — | — |
| Cape York | 35 | 28+[a] | *106* | *141* | — | — |
| Central Australia | 8 | 3+[b] | 12 | 4+[b] | *56* | *102*+[b] |

[a] If *Pandanus* sp. of Cape York was *P. spiralis* there would be an additional correspondence.
[b] The total of 102 species would be increased should an additional species of either *Eucalyptus* or *Ipomoea* be involved (nos. 122, 172 of Appendix 15:1). This in its turn might affect correspondences with Arnhem Land and/or Cape York, as might also the particular species of central Australian *Dioscorea* (no. 101).

*Note*: Figures in italics are the number of genera and species utilised in each region.

*Table 15:6 Numbers of species used in one area, present but unused in another*

| **used** \ **unused** | **AL/CA** | **AL** | **AL/CY** | **CY** | **CY/CA** | **CA** |
|---|---|---|---|---|---|---|
| AL/CA | | | | 2 | | |
| AL | | | | 25+[a] | 2 | 2 |
| AL/CY | | | | | | 1 |
| CY | 1 | 26+[a] | | | | X |
| CY/CA | | 1 | | | | |
| CA | | 4+[b] | 16+[b] | X | | |

[a] 25 would become 26 if *Pandanus* sp. of Cape York were not *Pandanus spiralis*, and 26 would become 27 if the Cape York species were one known in Arnhem Land.
[b] One or other of these totals would increase if *Aristida* sp., *Eriachne* sp., *Eucalyptus* sp., or *Ipomoea* sp. represented a species present in northern Australia.

*Note*: AL Arnhem Land, CA Central Australia, CY Cape York. X signifies that data are inadequate to make the computations.

*Table 15:7 Absence and presence in central Australia of Arnhem Land food genera by use*

| | **R** | **S** | **F** | **G** | **No. of plant genera** |
|---|---|---|---|---|---|
| Absent | 10 | 4 | 22 | 2 | 33 |
| Present unused | 3(1) | 2(1) | 2(2) | 0 | 7(4) |
| Present differently used | 1 | 1 | 0 | 0 } | 8(3+) |
| Present similarly used | 4(1+) | 1 | 2(2) | 1 } | |
| No. of plant genera | 17(2+) | 7(1) | 26(4) | 3 | 48(7+) |

*Note*: R rootstocks, S seeds, F fruits, G greens (further details Table 15:3). In parentheses the numbers of identical species: + means that *Dioscorea* sp. of central Australia may be *D. bulbifera* of Arnhem Land.

and Malaysia, that it was for a long time included within it. It is obvious, therefore, that from the point of view of human ecology the figures in the tables should not be interpreted too literally.

Though the table deals with genera, because relevant information on species is lacking, the marked degree of speciation characteristic especially of the monsoonal north (Specht 1958c:418, 419-20) would recommend the practice, even were the situation otherwise. Cape York, as invariably in this paper, is Lawrence's eastern and western Cape York areas combined. This is partly a matter of necessity, in that some shortly relevant information about the presence of Arnhem Land species in Cape York does not distinguish the region further. But it does no great violence to the ecological situation: in the matter of food plants, over half the forty-three species and three-quarters of the thirty-six genera recorded for western Cape York are present amongst the incomparably richer resources of the east (cf. Appendix 15:1 and suppressing the varietal distinction of no. 143a).

It is clear that, like Arnhem Land, Cape York derives its richer endowment of food plants preponderantly from Malaysian genera. In line with the general picture of floral distributions within Australia, where the Indo-Malaysian component drops off towards the arid centre and the temperate south (Specht 1958c:417), central Australia in Table 15:4 presents a somewhat different picture, characterised by the appearance of an appreciable number of plants of widespread distribution which are absent from Malaysia.

Table 15:5, which records the food plants shared between the three regions at generic and specific levels, brings out much more sharply the affinities between Arnhem Land and Cape York and their joint differences from central Australia.

There are numerous instances where a plant used for food in one region is not recorded as so used in another, though present there. Table 15:6 registers these occurrences.

Doubtless some of these occurrences are due to deficiencies in the ethnobotanical records. Absent from Specht's Arnhem Land list, though not from the Arnhem Land flora, are four mangrove species recorded as food sources by Roth for Cape York (nos. 39, 53, 54, 81 of Appendix 15:1; a possible exception is no. 54). Indeed Specht records no mangrove utilised for food in Arnhem Land at all and this may well be due in part to the lateness of his investigations.

On the other hand, of the twenty-eight shared species used for food in Cape York but not in Arnhem Land, no fewer than ten (nos. 21, 32, 38, 54, 65, 91, 117, 160, 195, 236 of Appendix 15:1) were not seen by Specht in the course of his field work, compared with only four unseen of the sixty-five species actually recorded by him as Arnhem Land food plants (nos. 151, 237, 266, 267). These ten species may therefore be so rare that if any of them were ever used for food, they were probably economically unimportant. The case of *Ficus* is instructive in another way. There are seven utilised species of fig in Arnhem Land and thirteen in Cape York. The use of only one species is shared, though five of the Arnhem Land figs are present in Cape York and two of the Cape York figs in Arnhem Land. This suggests that the possibility of choice may go some way to explain the features of Table 15:6. Confidence in the appropriateness of such ecological explanations is increased by the figure of sixteen species of central Australian food plant present but apparently unused in *both* Arnhem Land and Cape York.

Viewed another way, Table 15:6 confirms the low level of botanical correspondence between the two northern regions and

*Table 15:8 Utilised species in three Australian regions by food use*

| | **No. of species** | **No. of food uses** | **R** | **S** | **F** | **G** |
|---|---|---|---|---|---|---|
| Arnhem Land | 65 | 69 | 22 | 9 | 35 | 3 |
| Cape York | 141 | 149 | 46 | 19 | 73 | 11 |
| Central Australia | 102+[a] | 111 | 16 | 45 | 30 | 20 |

[a] See note b, Table 15:5.

*Note*: R rootstocks, S seeds, F fruits, G greens (further details Table 15:3).

*Table 15:9 Absence and presence in Arnhem Land of central Australian food genera by use*

| | **R** | **S** | **F** | **G** | **No. of plant genera** |
|---|---|---|---|---|---|
| Absent | 1 | 3 | 2 | 9 | 15 |
| Present unused | 6(6+) | 18(11+) | 11(6) | 4(2+) | 33(21+) |
| Present differently used | 1 | 1 | 0 | 1 | |
| Present similarly used | 4(1+) | 1 | 2(2) | 1 | 8(3+) |
| No. of plant genera | 12(7+) | 23(11+) | 15(8) | 15(2+) | 56(24+) |

*Note*: R rootstocks, S seeds, F fruits, G greens (further details Table 15:3). In parentheses the numbers of identical species: the + refers to the possibility of *Eucalyptus* sp. (R), *Dioscorea* sp. (R), *Aristida* sp. (S), *Eriachne* sp. (S) and *Ipomoea* sp. (G) corresponding to species known in Arnhem Land.

central Australia apparent in Table 15:5. Thus, while only three and at most five food species are shared between Arnhem Land and central Australia (Table 15:5), no more than five additional species of Arnhem Land food plants are actually present in central Australia, though they are not on record there as having been used (Table 15:6).

A more informative approach, given the high level of specific endemism already referred to, is to review the question of food plant relationships between northern and

*Table 15:10 Genera and species of grasses supplying seed foods in central Australia, present in Arnhem Land*

| **Genera** | **No. of species in Arnhem Land** |
|---|---|
| *Aristida* | 2 |
| *Brachiaria* | 4 |
| *Cymbopogon* | 3 |
| *Dactyloctenium* | 1 |
| *Enneapogon* | 1 |
| *Eragrostis* | 4 |
| *Eriachne* | 17 |
| *Ichnanthus* | 1 |
| *Panicum* | 7 |
| *Tragus* | Nil |

*Table 15:11 Absence and presence in Arnhem Land of central Australian food genera by botanical affiliation*

| | Australasian | Other non-Malaysian | Malaysian | Totals |
|---|---|---|---|---|
| Absent | 6 | 8 | 1 | 15 |
| Present unused | 3 | 5 | 25 | 33 |
| Present used | 1 | 1 | 6 | 8 |
| Totals | 10 | 14 | 32 | 56 |

central Australia at the generic level. Table 15:7 does this for Arnhem Land and central Australia: for Cape York and central Australia comparisons are less readily made.

Table 15:7 reveals the actual absence from central Australia of over two-thirds of the genera providing food plants for Arnhem Land Aborigines and is a measure of the degree of adaptation called for in this sphere of subsistence activities by the settlement of the Australian interior.

Table 15:8 reflects in part the nature of the adaptation made by setting out the food uses of all utilised species in the two northern regions and in the interior.

The pattern of food use for the two tropical regions is identical, with roots and fruits prominent. The contrasting importance of seed foods to interior Aborigines has often been noted (cf. Meggitt 1964) and is well reflected in Table 15:8, even though this records the number of contributing species and not their relative importance. The same is true in lesser degree of the green foods.

The important point to be made now, however, is that while two-thirds of the food genera of Arnhem Land are simply absent from central Australia (Table 15:7), over half the food genera of central Australia are present in Arnhem Land but do not figure in the food lists. Only in the case of greens is the actual absence of central Australian genera from Arnhem Land significant. The data are set out in Table 15:9.

On the whole, the resources of Arnhem Land evidently discouraged the exploitation of genera that were pressed into service in central Australia. We may illustrate the point by reference to seed foods. Of the forty-five species used for seed in central Australia, fourteen are grasses and nineteen are acacias. As set out in Table 15:10, Arnhem Land has forty species in nine of the ten genera of grasses that supplied seed foods in central Australia, at least three (nos. 94, 97, 210 of Appendix 15:1) being identical with central Australian food species. Not one is recorded as food. Arnhem Land has thirty-six species of *Acacia*, two (nos. 11, 15) identical with a central Australian species used for food. But no Arnhem Land species was so used. This situation may reflect both the importance of wild rice in the tropical north, to which its distribution is restricted, and the lesser relative importance of seed foods in subsistence.

The evidence of Tables 15:7-9 may be interpreted to suggest that the absence from central Australia of a whole series of food-providing tropical genera forced upon settlers of that area not only the exploitation of alternative sources of supply but a greater reliance on seed foods. In so far, however, as a large number of the genera providing food in

*Table 15:12 Uses in Malaysia of shared genera used for food in central Australia, by presence and absence in Arnhem Land*

| | 1 | 2 | 3 | 4 | Totals |
|---|---|---|---|---|---|
| Absent | 0 | 0 | 0 | 1 | 1 |
| Present unused | 8 | 5 | 4 | 8 | 25 |
| Present used | 4 | 1 | 1 | 0 | 6 |
| Totals | 12 | 6 | 5 | 9 | 32 |

*Note*: 1 similar food use; 2 alternative food use; 3 medical, technological, etc. uses; 4 no recorded uses.

central Australia are known in Arnhem Land, though not on record as being used there, it seems possible that man was in a position to obtain some of the ecological knowledge essential for successful settlement of central Australia during his prior sojourn in the more favourable conditions of northern Australia.

**Australian Food Plants in Malaysia**

Thirty-one of the thirty-three Arnhem Land food-providing genera absent from the central Australian flora are, understandably in the light of Table 15:4, of Malaysian affiliation. More surprising is the high proportion of Malaysian genera that figure in the total of central Australian food genera present but unused in Arnhem Land: twenty-five cases out of thirty-three (Table 15:11).

I want to look at this question by considering the uses in Malaysia of those Australian food genera which are common to both regions. I realise that the data on Malaysian food use which I am employing in these comparisons are not unexceptionable. With the aid of Burkill (1935) I have ranged widely through peninsular Malaya and island Southeast Asia for my examples. Information from a wide spectrum of economies has been utilised, from Malayan Aboriginal gatherers to Javanese rice farmers. A single recorded instance in Burkill has been accepted, though it is fair to say that such cases are not common. The nature of the decisions taken is noted in Appendix 15:2.

Bearing these points in mind we may now consider the data in Table 15:12, which registers the uses in Malaysia of those genera of central Australian food plants which are present in Malaysia. By setting out the genera according to their absence and presence, use and non-use in Arnhem Land, the table throws some light on the status of the twenty-five shared Malaysian genera which are used in central Australia and not in Arnhem Land.

*Table 15:13 Uses in Malaysia of shared genera used for food in two Australian regions*

| | 1 | 2 | 3 | 4 | Totals |
|---|---|---|---|---|---|
| Arnhem Land | 18 | 9 | 10 | 5 | 42 |
| Cape York | 46 | 15 | 19 | 14 | 94 |

*Note*: 1 similar food use; 2 alternative food use; 3 medical, technological, etc. uses; 4 no recorded uses.

Definitive interpretation of these figures is impossible because of the small numbers involved, but they could reflect a tendency for central Australian food plants generically shared with Malaysia to have lower identity of use with the Malaysian area when absent from or unused in Arnhem Land than when present there. It would be instructive to have the corresponding figures for central Australian food genera in Cape York, but the lack of up-to-date floral lists for that area has prevented their computation.

With respect to the twenty-five genera of central Australian food plants known but unused in Arnhem Land, the eight cases of identical use with Malaysia, representing 32 per cent of the total, need to be seen in the light of the corresponding figures for Arnhem Land (43 per cent) and Cape York (49 per cent) set out in Table 15:13.

It appears that similarity of plant exploitation with Malaysia is somewhat less for central Australia than it is for the tropical north, despite any tendency for the properties of a genus that render members of it suitable as food to recur wherever it is found.

The point I am trying to make is a

difficult one to prove. It is that the level of correspondence between food uses of the same genera in Malaysia on the one hand and Arnhem Land (43 per cent) and Cape York (49 per cent) on the other (Table 15:13) may well be due as much to ancient patterns of plant exploitation brought to Australia from the Malaysian region by its early inhabitants as to the properties of the flora itself. In both Arnhem Land and Cape York, Malaysian genera are predominant in the generic totals of plants used for food: 87·5 per cent in the former, 88·5 per cent in the latter (Table 15:4). Of Arnhem Land we may note that while species found in Malaysia account for about a quarter of the total Arnhem Land plant population (Specht 1958c:417), they provide nearly half (twenty-eight to thirty out of sixty-five) those used for food (Table 15:1).

The arid and semi-arid interior of the continent, colonised, it is reasonable to assume, after the tropical north, reveals not only a smaller component of Malaysian genera amongst its food plants (57 per cent, Table 15:4), but a lower level of correspondence of food uses with shared genera in Malaysia (37·5 per cent overall, Table 15:12). One of the thirty-two genera concerned is absent from Arnhem Land and twenty-five, though present there, are not on record as having been used. With these twenty-six genera the level of correspondence of food use with Malaysia falls to 31 per cent. As regards the six Malaysian genera which Arnhem Land and central Australia have in common as food sources, three have identical food uses in all three regions. It would appear that in central Australia a new pattern of plant exploitation, owing little directly to a Malaysian heritage, has emerged. Thirty-one Malaysian genera known as food providers in Arnhem Land, and 55 per cent of them in Malaysia also, are totally missing.

Since the argument I have developed for historical connections is based upon an evaluation of the closeness of Malaysian and Australian plant use, it is incomplete without some further consideration of plants common to the two regions which are used for food only in one or the other. To this end I have compiled a list, for which no claims for completeness are made, of eight genera and seventeen species providing food in Malaysia but not in Australia (Appendix 15:3). To this must be added two species, *Phaseolus mungo* L. and *Canavalia rosea* (Sw.) DC., whose Australian use is recorded (Maiden 1889:51, 12 under *C. obtusifolia* DC.) but not in the three regions with which we have been concerned. Conversely, of the Malaysian species used for food in Arnhem Land, at least seventeen have no recorded food uses in Malaysia at all (Table 15:1). The reasons for these disharmonies are no doubt as varied as those discussed in the identical case of the non-utilisation in Arnhem Land of species used in Cape York and the reverse (cf. Table 15:6). It is interesting to note that of the seventeen Arnhem Land species under discussion which have no food uses in Malaysia, ten are present but also are unused in Cape York (cf. Appendix 15:1).

Whereas food plant uses in Australia and Malaysia have been compared in the very general terms of roots, seeds, fruits, and green parts, a closer analysis might reveal detailed resemblances to provide more convincing support for the hypothesis of a historical relationship of food plant exploitation in the two regions. Thus in Malaysia, some Pacific islands, and eastern Cape York, the utilised portion of the coastal tree *Terminalia catappa* is the green rolled-up embryo of the fruit, contained in a tenacious husk (Burkill 1935:**2**, 2138). Some plants used in both Malaysia and Australia need special preparation, commonly involving both heating and

*Table 15:14 Uses in Malaysia of shared genera and species used for food in three Australian regions and to some degree domesticated in Malaysia*

| | 1 | | | | | | | | | | 2 | | 3 | | | | Totals of all food plants shared with Malaysia | | Totals of all food plants | |
|---|---|---|---|---|---|---|---|---|---|---|---|---|---|---|---|---|---|---|---|---|
| | R | | S | | F | | G | | Totals | | | | | | Totals | | | | | |
| | Gen. | Sp. | Gen. | Sp. | Gen. | Sp. | Gen. | Sp. | Gen. | Sp. | Gen. | Sp. | Gen. | Sp. | Gen. | Sp. | Gen. | Sp. | Gen. | Sp. |
| Arnhem Land | 3 | 2 | 3 | 1 | 3 | 0 | 0 | 0 | 9 | 3 | 1 | 0 | 2 | 1 | 12 | 4 | 42 | 28(30) | 48 | 65 |
| Cape York | 6 | 3 | 4 | 3 | 10 | 2 | 0 | 0 | 19 | 8 | 3 | 1 | 5 | 2 | 27 | 11 | 94 | X | 106 | 141 |
| Central Australia | 1 | 1? | 0 | 0 | 2 | 0 | 1 | 0 | 4 | 1? | 2 | 0 | 1 | 0 | 7 | 1? | 32 | X | 56 | 102+ |

*Note*: 1 similar food use; 2 alternative food use; 3 medical, technological etc. uses; R rootstocks, S seeds, F fruits, G greens (for further details see Table 15:3); ? means central Australia scores only if *Dioscorea* sp. is *D. bulbifera*; X signifies lack of data; for 28(30) see Table 15:1; for 102+ see Table 15:5, note b.

leaching, to rid their produce of toxic substances before it is fit for food. They include the seeds of various species of the mangrove *Avicennia* (Burkill 1935: **1**, 274; Lawrence 1968:208, 213), the seeds of *Entada phaseoloides* (Burkill 1935: **1**, 927; Lawrence 1968: 208, 213), the nuts of various species of *Cycas* (Burkill 1935: **1**, 720; Lawrence 1968: 208; Specht 1958d:482), and the tubers of some species of *Dioscorea* (Burkill 1935:**1**, 819-20, 823; Lawrence 1968:205, 212; Specht 1958d: 486).

A closely related practice is the use of the toxic parts of plants as fish poisons, on which Burkill (1935:**2**, 2398 under *tuba*) gives information for Malaysia (and elsewhere) and Lawrence (1968: Tables 10, 13) and Specht (1958d:491 under *Tephrosia arnhemica* C. T. White) for Australia. Plants from the same and related genera, and sometimes the same species, are used in the two areas: *Derris scandens* Benth. (nomenclature unchecked), *D. trifoliata* Lour. (*D. uliginosa* (Roxb. ex Willd.) Benth.) and *Pongamia pinnata* (L.) Merr. (*P. glabra* Vent.). The case of fish poisons, however, illustrates well the difficulties of making historical statements from contemporary cultural practice, particularly where the evidence is not wholly cultural but, as in the present case, depends on properties of the natural environment. Fish poisoning is a widespread practice and we may note that species of three of the genera used for the purpose in Malaysia and Australia were similarly employed elsewhere: *Tephrosia* in Africa and South America (Burkill 1935: **2**, 2133), *Derris* in South America (Burkill 1935:**1**, 783, 786) and *Diospyros* in the West Indies (Burkill 1935:**1**, 829).

The reliable evidence on historical relationships will be provided by archaeology and, in the particu r field that we have been exploring, palaeobotany. Though data from both for the connections I wish to establish are meagre, they lend strong support to the argument which I have developed. The discovery in northern Australia of edge-ground axes of late Pleistocene date (cf. White, C., see ch. 12) establishes relationships with the Hoabinhian-Bacsonian cultures of Southeast Asia, which on both circumstantial (Golson 1970) and more direct (Gorman 1969) evidence are probably to be credited with the development of horticulture, at a date perhaps as far back as the end of the Pleistocene. Recent archaeological discoveries thus tend to confirm the long held views of some ethnobotanists about the antiquity of plant domestication in Southeast Asia, based on considerations of the unparalleled floristic wealth of Indo-Malaysia and the nature of the plants involved in horticulture before the dominance of cereal agriculture (cf. Chang 1967).

Such an early development of horticulture argues a long and intimate prior knowledge of Indo-Malaysian plant life on the part of peoples related to the early settlers of the Australian continent, where in the most accessible northern regions Malaysian elements are prominent in the flora. In these circumstances the argument becomes more plausible that the high scores for Malaysian genera amongst the food plants of Arnhem Land and Cape York (Table 15:4) and the not insignificant degree of similarity in plant use between the two Australian regions and Malaysia (Table 15:13) have a historical explanation. The point is emphasised by Table 15:14, which presents information about the numbers and types of plants utilised in both Australia and Malaysia, which in Malaysia are accorded some degree of cultivation. All plants whose cultivation in Malaysia Burkill (1935) suspects, for whatever reason, may have been 'introduced', in 'historic' times, have been excluded.

Apart from the usual distinctiveness of the central Australian situation, Table 15:14 shows not only a notable component of the food genera of Arnhem Land and Cape York as containing species that are cultivated in Malaysia, but also in these genera a much higher degree of similar use between Australia and Malaysia than is exhibited by their shared food genera taken as a whole (Table 15:13). We may have isolated in this way in Table 15:14 some of the genera, and indeed some of the species, basic to early Southeast Asian horticulture (see Appendixes 15:1 and 15:2).

Should the plant exploitation of Australia's earliest settlers have been appreciably patterned by traditions established in Malaysia, expansion of early settlement may have been into areas typified by the greatest Malaysian element in their flora. These are less the interior regions, whose differences from the tropical north are so clearly registered for one large segment of those territories (Tables 15:4, 5, 7, 8, 9, 12, 13, 14), than the coast, where expansion may have been additionally recommended by the availability of seafoods. Furthermore, whereas at the northwest corner of the continent the coastal distribution of vegetation of the specified character ends within the tropics with a barren stretch of coast that must have been as inhospitable to man in the late Pleistocene as it is today (Jennings, see ch. 1), along the east coast it passes well beyond the tropics into the northern part of New South Wales (Specht 1958c: 419-20, 423).

I do not have the data to show in detail what this vegetation pattern may have meant in terms of food potential. However, a partial but perhaps significant insight is afforded by the final table (15:15), which records the distribution in other parts of Australia of Arnhem Land food plants, both as actually recorded by Specht and as increased by the sum of twenty-eight unused species utilised in Cape York.

*Table 15:15 Occurrence in other parts of Australia of Arnhem Land food plants*

| | **A** | **B** |
|---|---|---|
| Arnhem Land | 65 | 93 |
| Kimberley-Hamersley | 36 | 48 |
| Victoria River Downs | 31 | 40 |
| Darwin-Katherine | 40 | 56 |
| S Gulf of Carpentaria | 24 | 33 |
| Cape York | 58 | 86 |
| S coast Qld-N coast N.S.W. | 16 | 27 |
| New South Wales | 3 | 7 |
| Southern Australia | 2 | 3 |
| Interior | 8+[a] | 11[a] |

[a] Column A registers the possibility that the central Australian *Dioscorea* may be *D. bulbifera*; column B registers the fact that central Australia scores for *Dioscorea* whatever the species.

*Note*: A, food plants actually recorded for Arnhem Land, B, plus 28 species used in Cape York, which are present in, but not recorded as food plants for Arnhem Land.

# Appendix 15:1

## Species List of the Food Plants of Arnhem Land, Cape York, and Central Australia

The plants are arranged alphabetically and the family to which each belongs is specified in the right hand column.

An effort has been made to provide the correct names for the plants in question and these names form the sequentially numbered items in the list. A lower case letter attached to a number marks a variety, subspecies, etc. These are always merged with the species to which they belong in any calculations.

Only such synonyms, incorrect names, etc. are listed as appear in the ethnobotanical documents I have used or are necessary to their understanding. They are entered alphabetically in the list, prefaced by the number, in italics, of the correct plant name. In addition, under the correct name, all relevant synonyms, etc. are spelt out.

* before a plant name signifies a species domesticated to some degree in Malaysia.

The coded information on each plant is organised under five heads, not all of which may appear, though their relative order is maintained.

**1** *Introduced by the code AL (Arnhem Land)*

X signifies plant present but unused; when used (1-7) registers the habitat of the plant in terms of Specht's categories (1958d:480): 1 coastal dunes, 2 estuarine plains, 3 freshwater streams, marshes, and swamps, 4 monsoon forests, 5 savannah woodlands on coastal dunes and sandy fan deltas, 6 tall open forest, 7 quartzite and sandstone hills. R (rootstocks), S (seeds including nuts and kernels), F (fruits), G (greens comprising leaves, stems, shoots) and spelt out categories signify the type of produce.

(W) and (D) mark the season of production:

(W) wet, comprising W1, early wet (Oct.-Dec.) and W2, late wet (Jan.-Mar.)

(D) dry, comprising D1, early dry (Apr.-Aug.) and D2, late dry (Sept.-Oct.)

(W/D) means all year.

**2** *Introduced by the code CY (Cape York)*
X signifies plant present but unused; when used (E) means east (W) means west Cape York.
R etc. signifies the type of produce, as under Arnhem Land above.

**3** *Introduced by the code CA (central Australia)*
X signifies plant present but unused; when used R etc. signifies the type of produce, as under Arnhem Land above.

**4** *Introduced by the code M (Malaysia)*
X signifies present but unused; when used R etc. signifies the type of produce, as under Arnhem Land above, with the addition of M for medical and T for technological uses.
() containing one of these use symbols appears when a plant is domesticated but not (primarily) for the food use it has.

**5** *Australian distribution, as a set of numbers*
1 Hamersley Range and Kimberleys, 2 Victoria River Downs, 3 Darwin/Katherine, 4 Arnhem Land, 5 southern Gulf of Carpentaria, 6 Cape York, 7 south coastal Queensland, north coastal New South Wales, 8 New South Wales, 9 southern Australia, 10 interior Australia.

N.B.: Only for Arnhem Land and for Arnhem Land plants are presences and absences and Australian and extra-Australian distributions complete, thanks to Specht's coverage (1958a,c), which is not paralleled for any other region. The only exception is constituted by a few Cape York species, whose Malaysian presence is noted because they are domesticated there.

| Nos. | Names and synonyms | Common names | Codes | Families |
|---|---|---|---|---|
| 1 | *Acacia aneura* F. Muell. ex Benth. | Mulga | CA S | Leguminosae |
| 2 | *A. aulacocarpa* A. Cunn. ex Benth. | | 13467 | |
| 2a | *A. aulacocarpa* A. Cunn. ex Benth. var. *macrocarpa* Benth.<br>*A. crassicarpa* A. Cunn. ex Benth. | Lancewood | AL X/CY(E)R/456 | |
| 3 | *A. coriacea* DC. | Wattle | CA S | |
| *2a* | *A. crassicarpa* A. Cunn. ex. Benth. | | | — |
| 4 | *A. cuthbertsoni* Luehmann | | CA S | Leguminosae |
| 5 | *A. dictyophleba* F. Muell. | | CA S | |
| 6 | *A. estrophiolata* F. Muell. | Ironwood | CA S | |
| 7 | *A. farnesiana* (L.) Willd. | | CA S | |
| *12* | *A. frumentacea* Tate | | | — |
| *11* | *A. holosericea* A. Cunn. | | | — |
| 8 | *A. kempeana* F. Muell. | Witchetty bush | CA S | Leguminosae |
| 9 | *A. ligulata* A. Cunn. ex Benth. | Umbrella bush; broad leaf wattle | CA S | |

| Nos. | Names and synonyms | Common names | Codes | Families |
|---|---|---|---|---|
| 10 | *A. maitlandii* F. Muell.<br>*A. patens* F. Muell. ex Benth. | | CA S | |
| 11 | *A. mangium* Willd. var. *holosericea* (A. Cunn.) C. T. White<br>*A. holosericea* A. Cunn. | | AL X/CY(E)S/CA S/1234610 | |
| 12[1] | *A. murrayana* F. Muell. ex Benth.<br>*A. frumentacea* Tate | | CA S | |
| 13 | *A. notabilis* F. Muell. | Hickory; wattle | CA S | |
| *10* | *A. patens* F. Muell. ex Benth. | | | — |
| 14 | *A. pyrifolia* DC. | | CA S | Leguminosae |
| 15 | *A. salicina* Lindl. | Native willow | AL X/CY X/CA S | |
| *18* | *A. sentis* F. Muell. | | | — |
| 16 | *A. stipuligera* F. Muell. | | CA S | Leguminosae |
| 17 | *A. tetragonophylla* F. Muell. | Wattle | CA S | |
| 18 | *A. victoriae* Benth.<br>*A. sentis* F. Muell. | Prickly wattle; elegant acacia | CA S | |
| *19* | *Acrostichum aureum* auct. non L. | | | — |
| 19[2] | *Acrostichum speciosum* Willd.<br>*A. aureum* auct. non L. | | AL X/CY(E)R/ 1234567[3] | Pteridaceae ferns |
| 20 | **Aleurites moluccana* (L.) Willd. | Candlenut tree | CY(E)S/MS | Euphorbiaceae |
| 21 | *Allophylus ternatus* Lour.<br>*Schmidelia serrata* DC. | | AL X/CY(E)F/46 | Sapindaceae |
| 22 | **Alocasia macrorrhiza* (L.) Schott<br>*Colocasia macrorrhiza* Schott | Cunjevoi | CY(E)R/MG | Araceae |
| 23 | *Amaranthus grandiflorus* (Black) J. M. Black | | CA S | Amaranthaceae |
| 24 | *A. mitchellii* Benth. | | CA S | |
| 25 | *Amomum dallachyi* F. Muell. | | CY(E)F | Zingiberaceae |
| 26 | *Amorphophallus galbra* F. M. Bail. | | AL X/CY(E)R/F/G/346 | Araceae |
| 27 | *A. variabilis* Blume | | AL(3)R(W/D)/MR/34 | |

| Nos. | Names and synonyms | Common names | Codes | Families |
|---|---|---|---|---|
| 28 | *Ampelocissus acetosa* (F. Muell.) Planch.<br>*Vitis acetosa* (F. Muell.) Benth. | | AL(4/6)R/F(W)/CY(W) R(E)F/123456 | Vitaceae<br>native grapes |
| 29 | *Amyema linophylla* (Fenzl.) Tiegh.<br>*Loranthus linophyllus* Fenzl. | | CA F | Loranthaceae<br>mistletoes |
| 30 | *A. maidenii* (Blakeley) Barlow<br>*Loranthus maidenii* Blakeley | | CA F | |
| 31 | *A. pendula* (Sieb. ex Spreng.) Tiegh.<br>*Loranthus pendulus* Sieb. ex Spreng. | | AL X/CY X/CA F | |
| *94* | *Andropogon exaltatus* R. Br. | | | — |
| 32 | *Aneilema silicosum* R. Br. | | AL X/CY(W)R/46 | Commelinaceae |
| 33 | **Antidesma bunius* (L.) Spreng. | Bloomfield cherry | CY(E)/F/MF | Euphorbiaceae |
| 34 | *A. dallachyanum* Baill. | Herbert River cherry | CY(E)F | |
| *35* | *Aponogeton monostachyon* L. f. | | | — |
| 35 | *Aponogeton natans* (L.) Engl.<br>*A. monostachyon* L. f. | A freshwater herb | CY(E)R | Aponogetonaceae |
| 36 | *Archontophoenix alexandrae* (F. Muell.) H. Wendl. | Alexandra palm; bangalow palm | CY (E) base of growing shoot | Palmae<br>palms |
| 37 | *Aristida* sp. | Wiregrass | CA S | Graminae<br>grasses |
| *93* | *Aspidium unitum* Sw. var. *propinquum* (R. Br.) F. M. Bail. | | | — |
| 38 | *Atylosia reticulata* Benth. | | AL X/CY(E)R/46 | Leguminosae |
| 39[4] | *Avicennia marina* (Forsk.) Vierh. var. *resinifera* (Forsk.) Bakh.<br>*A. officinalis* auct. non L.<br>*A. tomentosa* auct. non Jacq. | White mangrove | AL X[5]/CY(E/W)S/ 123456789[6] | Verbenaceae |
| *39* | *A. officinalis* auct. non L. | | | — |
| *39* | *A. tomentosa* auct. non Jacq. | | | — |
| *217* | *Barringtonia careya* F. Muell. | | | — |
| *40* | *B. racemosa* auct. non Gaudich. | | | — |

| Nos. | Names and synonyms | Common names | Codes | Families |
|---|---|---|---|---|
| 40[7] | *Barringtonia racemosa* (L.) Spreng.<br>*B. racemosa* auct. non Gaudich. | | CY(E)F | Barringtoniaceae |
| 41 | *Beilschmiedia bancroftii* (F. M. Bail.) C. T. White<br>*Cryptocarya bancroftii* F. M. Bail. | Red or yellow walnut | CY(E)S | Lauraceae |
| 42 | *Blechnum indicum* Burm. f. | | AL(3)R(W/D)/CYX/ MX/2345678[8] | Blechnaceae<br>ferns |
| 43 | *B. orientale* L. | | AL X/CY(E)R/123456[8] | |
| 44 | *Blennodia eremigena* (F. Muell.) Benth. | | CA G | Cruciferae |
| 45 | *Boerhavia diffusa* L. | Hogweed | AL(5)R(D1)/CYX/ CAR/MM[9]/123467910 | Nyctaginaceae |
| 46 | *Bowenia spectabilis* Hook. ex Hook. f. | | CY(E)R | Zamiaceae |
| 47 | *Brachiaria miliiformis* (Presl) Chase<br>*Panicum distachyum* [non L.] Benth. | | CA S | Graminae<br>grasses |
| 48 | *B. piligera* D. K. Hughes | | CA S | |
| 49 | *Brachychiton diversifolium* R. Br.<br>*Sterculia caudata* Heward ex Benth. | Kurrajong | AL(6)S(W1/D2)/CY(E) F/12346 | Sterculiaceae |
| 50 | *B. gregorii* F. Muell. | Desert kurrajong; black kurrajong | CA R/S | |
| 51 | *B. paradoxum* Schott. | Red-flowering kurrajong | AL(5)S(W1/D2)/CYX/ CAX/12345610 | |
| 52 | *Bridelia monoica* (Lour.) Merr. | | AL(4)F(W)/CYX/MM/ 12346 | Euphorbiaceae |
| 53 | *Bruguiera gymnorrhiza* Lam. | | ALX/CY(E)S/134567 | Rhizophoraceae<br>mangroves |
| 54 | *B. rheedii* Blume | Red or orange mangrove | ?ALX/CY(E/W)S/ ?4567 | |
| 55[10] | *Buchanania arborescens* Blume<br><br>*B. muelleri* Engl. | Little gooseberry tree | AL(4)F(W)/CY(E)F/ MX/346 | Anacardiaceae |
| *55* | *B. muelleri* Engl. | | | — |
| 56 | *B. obovata* Engl. | Green plum | AL(5)F(W1)/CYX/346 | Anacardiaceae |

| Nos. | Names and synonyms | Common names | Codes | Families |
|---|---|---|---|---|
| 57 | *Calamus australis* Mart. | Large lawyer cane | CY(E)F | Palmae palms |
| 58 | *C. caryotoides* A. Cunn. | Ground palm; small lawyer cane | CY(E) young shoots | |
| 59 | *C. moti* F. M. Bail. | Large lawyer cane | CY(E) young shoots | |
| 60 | *Calandrinia balonensis* Lindl.<br>*Claytonia balonensis* F. Muell. | Broad-leafed parakeelya; munyeroo | CA S/G | Portulacaceae |
| 61[11] | *C. volubilis* Benth. | | CA G | |
| 62 | *Calostemma ?scott-sellickiana* F. M. Bail. | | AL(3)R(D2)/CYX/46 | Amaryllidaceae |
| *81*[12] | *Candelia* | | | — |
| 63[13] | *Canthium coprosmoides* F. Muell.<br>*Plectronia barbata* auct. non (Seem.) Benth. & Hook. f. ex Hemsl. | | CY(E)F | Rubiaceae |
| 64 | *C. latifolium* F. Muell. ex Benth.<br>*Plectronia latifolia* (F. Muell. ex Benth.) Black | Wild orange; wild lemon | CA F | |
| *65* | *C. lucidum* Hook. & Arn. | | | — |
| 65[14] | *C. odoratum* (Forst. f.) Seem.<br>*C. ludicum* Hook. & Arn.<br>*Plectronia odorata* F. Muell. | | AL X/CY(E/W)F/467 | Rubiaceae |
| 66[15] | *Capparis arborea* (F. Muell.) Maiden<br>*C. nobilis* auct. non F. Muell. | Native pomegranate; grey plum; caper tree | CY(E)F | Capparidaceae |
| 67 | *C. canescens* Banks ex DC. | Native pomegranate; native date | CY(E)F | |
| 68 | *C. humistrata* (F. Muell.) F. Muell. | | CY(E)F | |
| 69 | *C. lasiantha* R. Br. ex DC. | | CA F | |
| 70 | *C. mitchellii* Lindl. | Native orange; small native pomegranate | CY(E)F/CAF | |
| *66* | *C. nobilis* auct. non F. Muell. | | | — |
| 71 | *C. spinosa* L. var. *nummularia* (DC.) Bailey<br>*C. spinosa* L. | Caper bush | CA F | Capparidaceae |
| 72 | *Carallia brachiata* (Lour.) Merr.<br>*C. integerrima* DC. | | AL(4)F(D1)/CY(E)F/MF[16]/123456 | Rhizophoraceae mangroves |

| Nos. | Names and synonyms | Common names | Codes | Families |
|---|---|---|---|---|
| *72* | *C. integerrima* DC. | | | — |
| *217* | *Careya australis* F. Muell. | | | — |
| *73* | *Carissa brownii* F. Muell. | | | — |
| 73[17] | *Carissa lanceolata* R. Br.<br>*C. brownii* F. Muell. | Native scrub lime; konkleberry | AL(6)F(D)/CYX/CAF/ 12345610[17] | Apocynaceae |
| 74 | *Cartonema parviflorum* Hassk. | | AL(3)R(D)/CYX/12346 | Commelinaceae |
| 75 | *C. spicatum* R. Br. | | AL(3)R(D)/CYX/1346 | |
| 76 | *Castanospermum australe* A. Cunn. & Fraser ex Hook. | Bean tree; Moreton Bay chestnut | CY(E)S | Leguminosae |
| 77 | *Casuarina decaisneana* F. Muell. | Desert oak | CA F | Casuarinaceae |
| 78 | *Cayratia clematidea* (F. Muell.) Domin<br>*Vitis clematidea* F. Muell. | | CY(E)R | Vitaceae native grapes |
| 79 | *C. trifolia* (L.) Domin<br>*Vitis trifolia* L. | | AL(5/6)R(D)/CY(E/W) R/MM/12456 | |
| 80 | *Celtis philippensis* Blanco | | AL(4)F(D1)/CYX/ MM/12346 | Ulmaceae |
| 81[12] | *Ceriops tagal* (Perr.) C. B. Rob.<br>*C. candolleana* Arn. | Grey mangrove | ALX/CY(E)S/12346 | Rhizophoraceae mangroves |
| 82 | *Ceropegia cumingiana* Decne. | Yam | CY(W)R | Asclepiadaceae |
| 83 | *Chenopodium rhadinostrachyum* F. Muell. | | CA S | Chenopodiaceae |
| *192* | *Chilocarpus australis* F. Muell. | | | — |
| 84 | *Cissus opaca* F. Muell.<br>*Vitis opaca* F. Muell. | Pepper vine; round yam | CY(E)R | Vitaceae native grapes |
| *60* | *Claytonia balonensis* F. Muell. | | | — |
| 85 | *Clerodendrum floribundum* R. Br.<br>*C. ovalifolium* (A. Juss.) Bakheuzen | | ALX/CYX/CAR/F | Verbenaceae |
| 86 | *C. inerme* (L.) Gaertn. | | ALX/CY(W)F/34567 | |
| *85* | *C. ovalifolium* (A. Juss.) Bakheuzen | | | — |
| *87* | *Colocasia antiquorum* Schott. | | | — |

| Nos. | Names and synonyms | Common names | Codes | Families |
|---|---|---|---|---|
| 87 | **Colocasia esculenta* (L.) Schott.<br>*C. antiquorum* Schott. | Taro | CY(E)R/MR | Araceae |
| *22* | *C. macrorrhiza* Schott. | | | — |
| 88 | *Convolvulus erubescens* Sims | | CA G | Convolvulaceae |
| *41* | *Cryptocarya bancroftii* F. M. Bail. | | | — |
| *107* | *C. palmerstonii* F. M. Bail. | | | — |
| *89* | *Cucumis chate* [non L.] Hasselq. | | | — |
| 89[18] | *Cucumis melo* L. ssp. *agrestis* (Naud.) Greb.<br>?*C. chate* [non L.] Hasselq.<br>*C. melo* L. var. *agrestis* Naud. | Cucumber; Ulcardo melon | ALX/CAF | Cucurbitaceae |
| *89* | *C. melo* L. var. *agrestis* Naud. | | | — |
| 90 | *Curculigo ensifolia* R. Br. | | AL(3)R(D)/CY(E)R/MX/3467 | Amaryllidaceae |
| 91 | *Curcuma australasica* Hook. | Wild ginger | ALX/CY(E)R/46 | Zingiberaceae |
| 92 | *Cycas media* R. Br. | Cycad palm; nut palm; burrawang | AL(6)S(W1/D2)/CY(E)S/1346 | Cycadaceae |
| 93 | *Cyclosorus gongylodes* (Schkuhr) Link var. *hirsutus* (Mett.) Farwell<br>*Aspidium unitum* Sw. var. *propinquum* (R. Br.) F. M. Bail. | | CY(E)R | Thelypteridaceae |
| 94 | *Cymbopogon exaltatus* (R. Br.) Domin<br>*Andropogon exaltatus* R. Br. | Scented grass | ALX/CYX/CAS | Graminae grasses |
| 95 | *Cyperus bulbosus* Vahl. | Nalgoo | CA R | Cyperaceae sedges |
| 96 | *C. rotundus* L. | Nut grass | ALX/CYX/CAR | |
| 97 | *Dactyloctenium radulans* (R. Br.) Beauv. | Button grass | ALX/CYX/CAS | Graminae grasses |
| 98 | *Davidsonia pruriens* F. Muell. | Davidsonian plum | CY(E)F | Davidsoniaceae |
| 99[19] | **Dioscorea bulbifera* L.<br>*D. sativa* auct. non L. | | CY(E)R/MR | Dioscoreaceae yams |
| 99a[19] | *D. bulbifera* L. var. *elongata* F. M. Bail.<br>*D. sativa* auct. non L. var. *elongata* F. M. Bail. | Parsnip yam | AL(3)R(D)/CY(W)R/346[20] | |

| Nos. | Names and synonyms | Common names | Codes | Families |
|---|---|---|---|---|
| 99b[19] | *D. bulbifera* L. var. *rotunda* F. M. Bail.<br>*D. sativa* auct. non L. var. *rotunda* F. M. Bail. | Round yam | AL(3)R(D)/CY(E/W) R/346[20] | |
| *99* | *D. sativa* auct. non L. | | | — |
| *99a* | *D. sativa* auct. non L. var. *elongata* F. M. Bail. | | | — |
| *99b* | *D. sativa* auct. non L. var. *rotunda* F. M. Bail. | | | — |
| 100 | *D. transversa* R. Br. | Long yam | ALX/CY(W)R/34678[20] | Dioscoreaceae |
| 101[21] | *Dioscorea* sp. | | CA R[20] | |
| *102* | *Dolichos biflorus* L. | | | — |
| 102 | *Dolichos uniflorus* Lam.<br>*D. biflorus* L. | | CY(E)R | Leguminosae |
| 103 | *Elaeocarpus grandis* F. Muell. | Blue fig; Brisbane quandong | CY(E)F | Elaeocarpaceae |
| 104 | *Eleocharis dulcis* (Burm. f.) Trin. ex Hensch. | Spike rush; mat rush | AL(3)R(D2)/CYX/ MR/4567 | Cyperaceae sedges |
| 105 | *E. sphacelata* R. Br.<br>*Heleocharis sphacelata* R. Br. | Spike rush | CY(E/W)R | |
| *162* | *Elettaria scottiana* F. Muell. | | | — |
| 106 | *Enchylaena tomentosa* R. Br. | Ruby saltbush | CA F | Chenopodiaceae |
| 107 | *Endiandra palmerstonii* (F. M. Bail.) White & Francis<br>*Cryptocarya palmerstonii* F. M. Bail. | Black or Queensland walnut; walnut bean | CY(E)S | Lauraceae |
| 108 | *Enhalus acoroides* (L. f.) Rich. ex Steud.<br>*E. koenigii* Rich. | A marine plant | CY(E)F | Hydrocharitaceae |
| *108* | *E. koenigii* Rich. | | | — |
| 109 | *Enneapogon avenaceus* (Lindl.) C. E. Hubbard<br>*Pappaphorum avenaceum* Lindl. | | CA S | Graminae grasses |
| 110 | *Entada phaseoloides* (L.) Merr.<br>*E. scandens* Benth. | Matchbox bean; Queensland bean | CY(E/W)S | Leguminosae |
| *110* | *E. scandens* Benth. | | | — |
| 111 | *Eragrostis clelandii* S. T. Blake | | CA S | Graminae grasses |

| Nos. | Names and synonyms | Common names | Codes | Families |
|---|---|---|---|---|
| 112 | *E. dielsii* Pilger | Mulka grass | CA S | |
| 113 | *E. eriopoda* Benth. | | CA S | |
| 114 | *Eremophila freelingii* F. Muell. | | CA G | Myoporaceae |
| 115 | *E. latrobei* F. Muell. | | CA G | |
| 116 | *Eriachne* sp. | | CA S | Graminae grasses |
| 117 | *Eriosema chinense* Vog. | | ALX/CY(E/W)R/46 | Leguminosae |
| 118 | *Erodium cygnorum* Nees. | Wild geranium | CA R | Geraniaceae |
| 118a[22] | *Erythrina australis* | | | — |
| 119 | *Erythrina vespertilio* Benth. | Bean tree; (batswing) coral or cork tree | ALX/CYX/CA seedling shoots | Leguminosae |
| 119a[23] | *Eucalyptus camaldulensis* Dehnh. | | | Myrtaceae |
| 120 | *E. gamophylla* F. Muell. | | CA S | |
| 121 | *E. microtheca* F. Muell. | Coolibah; black, desert, dwarf, or swamp box | CA S | |
| 122[24] | *E.* sp. | | CA R | |
| *234* | *Eucarya acuminata* (R. Br.) Spr. & Summerh. | | | — |
| 123 | *Eugenia bleeseri* O. Schwarz | White love-apple | AL(5)F(W1)/34 | Myrtaceae |
| 124 | *E. carissoides* F. Muell. | | CY(W)F | |
| 125 | *E. cormiflora* F. Muell. | | CY(E)F | |
| 126 | *E. hislopii* F. M. Bail. | | CY(E)F | |
| 127 | *E. leptalea* Craib. *E. leptantha* Wight | | CY(E)F | |
| *127* | *E. leptantha* Wight | | | — |
| 128 | *E. suborbicularis* Benth. | Red love-apple | AL(5)F(W)/CY(E/W) F/346 | Myrtaceae |
| 129 | *Eupomatia laurina* R. Br. | Rose bush | CY(E)F | Eupomatiaceae |
| 130 | *Exocarpos cupressiformis* Labill. | Cypress cherry; native cherry | CY(E)F | Santalaceae |
| 131 | *Fenzlia obtusa* Endl. | | CY(E)F | Myrtaceae |
| 132 | *Ficus aculeata* A. Cunn. ex Miq. | | 145 | Moraceae |
| 132a | *F. aculeata* A. Cunn. ex Miq. var. *micracantha* Benth. | Fig | AL(5)F(W)/45 | |

| Nos. | Names and synonyms | Common names | Codes | Families |
|---|---|---|---|---|
| 133 | *F. albipila* (Miq.) King<br>*F. colossea* F. Muell. | Fig | CY(E)F | |
| 134 | *F. benjamina* L. | Fig | AL(4)F(D1)/CYX/<br>MX/46 | |
| *133* | *F. colossea* F. Muell. | | | — |
| 135 | *F. congesta* Roxb.<br>*F. fasciculata* F. Muell. | Fig | CY(E)F | Moraceae |
| *148* | *F. cunninghamii* Miq. | | | — |
| 136 | *F. drupacea* Thunb.<br>*F. pilosa* Reinw. ex Blume | Fig | CY(E)F | Moraceae |
| *147* | *F. ehretioides* F. Muell. | | | — |
| *149* | *F. esmeralda* F. M. Bail. | | | — |
| *142* | *F. eugenioides* F. Muell. | | | — |
| *135* | *F. fasciculata* F. Muell. | | | — |
| 137 | *F. fraseri* Miq.<br>*F. stenocarpa* F. Muell. | Sandpaper fig | CY(E)F | Moraceae |
| 138 | *F. henneana* Miq. | Fig | AL(4/5)F(D1)/CYX/46 | |
| 139 | *F. hispida* L. f. | Fig | CY(E)F/G | |
| *148* | *F. lacor* auct. non Buch.-<br>Ham. | | | — |
| 140 | *F. microcarpa* L. f. var.<br>*latifolia* (Miq.) Corner<br>*F. thynneana* F. M. Bail. | Fig | CY(E)F | Moraceae |
| 141 | *F. nesophila* (Miq.) F. Muell. | Fig | AL(4)F(D1)/CYX/1346 | |
| 142 | *F. obliqua* Forst. f.<br>*F. eugenioides* F. Muell. | Fig | CY(E)F | |
| 143 | *F. opposita* Miq. | Fig | CY(E)F | |
| 143a | *F. opposita* Miq. var.<br>*micrantha* (Miq.) Corner<br>*F. orbicularis* A. Cunn. | Fig | CY(W)F | |
| *143a* | *F. orbicularis* A. Cunn. | | | — |
| *136* | *F. pilosa* Reinw. ex Blume | | | — |
| 144[25] | *F. platypoda* A. Cunn. ex<br>Miq. | Fig | CY(E)F/CAF/<br>123456710[25] | Moraceae |
| 144a[25] | *F. platypoda* A. Cunn. ex<br>Miq. var. *cordata* Specht | Fig | AL(7)F(D1)/4 | |
| 145 | *F. racemosa* L. | Fig | AL(4)F(D2)/CYX/MX/<br>1246 | |

| Nos. | Names and synonyms | Common names | Codes | Families |
|---|---|---|---|---|
| 146 | *F. scobina* Benth. | Fig | AL(5)F(W)/CYX/1346 | |
| *137* | *F. stenocarpa* F. Muell. | | | — |
| *140* | *F. thynneana* F. M. Bail. | | | — |
| 147 | *F. variegata* Blume<br>*F. ehretioides* F. Muell. | Fig | CY(E)F | Moraceae |
| 148[26] | *F. virens* Ait.<br>*F. cunninghamii* Miq.<br>*F. lacor* auct. non Buch.-Ham. | White fig | ALX/CY(W)F/46 | |
| 149 | *F. virgata* Reinw. ex Blume<br>*F. esmeralda* F. M. Bail. | Fig | CY(E)F | |
| *238* | *Fluggea obovata* (Willd.) Wall. ex F.-Vill. | | | — |
| *150* | *Gahnia psittacorum* auct. non Labill. | | | — |
| 150[27] | *Gahnia sieberana* Kunth<br><br>*G. psittacorum* auct. non Labill. | | CY(E)G | Cyperaceae sedges |
| 151 | *Ganophyllum falcatum* Blume | | AL(4)F(W1)/CY(E)F/MT/46[28] | Sapindaceae |
| 152 | *Grevillea striata* R. Br. | Beefwood; silvery honeysuckle | ALX/CYX/CAS | Proteaceae |
| 153 | *Grewia breviflora* Benth. | | AL(4/5)F(W1)/14 | Tiliaceae |
| 154 | *G. orientalis* [non L.] Benth. | | AL(4)F(D1)/CYX/?MX[29]/46 | |
| *155* | *G. polygama* Roxb. | | | — |
| 155 | *G. retusifolia* Kurz.<br><br>*G. polygama* Roxb. | Plain currant | AL(3)F(D)/CY(W)F/CAX/?MX[29]/12345610 | Tiliaceae |
| *268* | *Hardenbergia retusa* Benth. | | | — |
| *105* | *Heleocharis sphacelata* R. Br. | | | — |
| 156 | *Hemicyclia australasica* Muell.-Arg. | | AL(4)F(W2)/CYX/467 | Euphorbiaceae |
| 157 | *Hibiscus brachychlaenus* F. Muell.<br>*H. microchlaenus* F. Muell. | | CY(W)R | Malvaceae |
| 158 | *H. brachysiphonius* F. Muell. | | CY(W)R | |
| *157* | *H. microchlaenus* F. Muell. | | | — |

| Nos. | Names and synonyms | Common names | Codes | Families |
|---|---|---|---|---|
| 159 | *H. radiatus* Cav. | | AL(3)S(D)/CYX/ MG[30]/2346 | Malvaceae |
| 160 | *H. rhodopelatus* F. Muell. ex Benth. | | ALX/CY(E)R/3467 | |
| 161 | *H. zonatus* F. Muell. | | AL(3)S(D)/145 | |
| 162 | *Hornstedtia scottiana* (F. Muell.) K. Schum. *Elettaria scottiana* F. Muell. | Wild ginger | CY(E)F | Zingiberaceae |
| 163 | *Ichnanthus australiensis* (Domin) Hughes *Panicum pauciflorum* R. Br. | | CA S | Graminae grasses |
| *194* | *Ipomoea angustifolia* N. J. Jacq. | | | — |
| 164 | *Ipomoea costata* F. Muell. | | CA R | Convolvulaceae |
| 165 | *I. eriocarpa* R. Br. | | CY(E)R | |
| 166 | *I. gracilis* R. Br. | | AL(3)R(D2)/CYX/MX/ 146 | |
| 167 | *I. graminea* R. Br. | | AL(3)R(D2)/CYX/MX/ 346 | |
| *253* | *I. grandiflora* (L. f.) Lam. | | | — |
| *170* | *I. heterophylla* R. Br. | | | — |
| 168 | *I. muelleri* Benth. | | ALX/CAS | Convolvulaceae |
| 169 | *I. pes-caprae* (L.) Sweet | | CY(E)R/MM | |
| 169a | *I. pes-caprae* (L.) Sweet ssp. *brasiliensis* (L.) Ooststr. | | AL(1)R(W/D)/134567 | |
| 170 | *I. polymorpha* Roem. & Schult. *I. heterophylla* R. Br. | | ALX/CYX/CAR/S | |
| 171 | *I. velutina* R. Br. | | AL(3)R(D1)/CYX/46 | |
| 172[24] | *I.* sp. | | CA G | |
| 173 | *Ixora timorensis* Decne. | | ALX/CY(E/W)F/346 | Rubiaceae |
| 174 | *Leichhardtia australis* R. Br. *Marsdenia australis* (R. Br.) Druce | Native pear; bush banana | CA G | Asclepiadaceae |
| 175[31] | *Lepidium fasciculatum* Thell. *L. ruderale* [non L.] F. Muell. | | CA G | Cruciferae |
| 176 | *L. muelleri-ferdinandi* Thell. | | CA G | |
| 177 | *L. oxytrichum* Sprague | | CA G | |

| Nos. | Names and synonyms | Common names | Codes | Families |
|---|---|---|---|---|
| 178 | *L. rotundum* (Desv.) DC. | | CA G | |
| *175* | *L. ruderale* [non L.] F. Muell. | | | — |
| 179 | **Lepironia articulata* (Retz.) Domin<br>*L. mucronata* Rich. | | ALX/CY(E)R/MT/ 345678 | Cyperaceae sedges |
| *179* | *L. mucronata* Rich. | | | — |
| 180 | *Licuala muelleri* H. Wendl. & Drude | Fan palm | CY(E) base of growing shoot | Palmae palms |
| *203* | *Limnanthemum geminatum* (R. Br.) Griseb. | | | — |
| 181 | *Livistona australis* (R. Br.) Mart. | Cabbage-tree palm | CY(E) growing stem | Palmae palms |
| 182 | *L. humilis* R. Br. | Fan palm | AL(6) central growing shoot (W/D)/CY(W) heart/23456[32] | |
| 183[33] | *L. mariae* F. Muell. | | CA basal parts inner leaves | |
| *184* | *Loranthus exocarpi* Behr | | | — |
| *29* | *L. linophyllus* Fenzl. | | | — |
| *30* | *L. maidenii* Blakeley | | | — |
| *185* | *L. mitchellianus* Blakeley | | | — |
| *186* | *L. murrayi* F. Muell. & Tate | | | — |
| *31* | *L. pendulus* Siev. ex Spreng. | | | — |
| *207* | *Lucuma galactoxyla* Benth. & Hook. | | | — |
| *226* | *Lucuma sericea* (Ait.) F. Muell. | | | — |
| 184 | *Lysiana exocarpi* (Behr) Tiegh.<br>*Loranthus exocarpi* Behr | | ALX/CYX/CAF | Loranthaceae mistletoes |
| 185 | *L. linearifolia* (Hook.) Tiegh.<br>*Loranthus mitchellianus* Blakeley | | CA F | |
| 186 | *L. murrayi* (F. Muell. & Tate) Tiegh.<br>*Loranthus murrayi* F. Muell. & Tate | | CA F | |
| 187 | *Maba humilis* R. Br. | Queensland ebony-wood | AL(4/5)F(D2)/CYX/ 124567 | Ebenaceae |

| **Nos.** | **Names and synonyms** | **Common names** | **Codes** | **Families** |
|---|---|---|---|---|
| 188 | *Malaisia scandens* (Lour.) Planch. | Crow ash | AL(4)F(W2)/CYX/3467 | Moraceae |
| 189[34] | **Manilkara kauki* (L.) Dubard<br>*Mimusops browniana* Benth.<br>*M. kauki* L. or R. Br. | | CY(E)F/MF | Sapotaceae |
| *174* | *Marsdenia australis* (R. Br.) Druce | | | — |
| 190 | *Marsilea mutica* Mett.<br>*M. quadrifolia* [non L.] Benth. | Clover fern; nardoo | ALX/CYX/CAS | Marsiliaceae ferns |
| *190* | *M. quadrifolia* [non L.] Benth. | | | — |
| 191 | *Melastoma malabrathicum* L. | | CY(E)F | Melastomataceae |
| 192 | *Melodinus australis* (F. Muell.) Pierre<br>*Chilocarpus australis* F. Muell. | Rubber vine | CY(E)F | Apocynaceae |
| 193 | *Melothria maderaspatana* (L.) Cogn. | | ALX/CYX/CAF | Cucurbitaceae |
| 194 | *Merremia tridentata* (L.) Hall<br>*Ipomoea angustifolia* N. J. Jacq. | | ALX/CY(E)R/1246 | Convolvulaceae |
| 195 | *Microstemma tuberosum* R. Br. | | ALX/CY(E)R/46 | Asclepiadaceae |
| *189* | *Mimusops browniana* Benth. | | | — |
| 196 | *Mimusops elengi* L.<br>*M. parvifolia* R. Br. | | CY(E)F | Sapotaceae |
| *189*[34] | *M. kauki* L. or R. Br. | | | — |
| *196* | *M. parvifolia* R. Br. | | | — |
| 197 | **Morinda citrifolia* L. | | AL(4)F(D)/CY(E)F/MG[35](M/T)/146 | Rubiaceae |
| 198 | *Musa* sp. | Banana | CY(E)F | Musaceae |
| *199* | *Myoporum dampieri* A. Cunn. ex A. DC. | | | — |
| 199 | *Myoporum montanum* R. Br.<br>*M. dampieri* A. Cunn. ex A. DC. | Native myrtle; native daphne; dogwood | CA F | Myoporaceae |
| *231* | *Myrsine crassifolia* R. Br. | | | — |

| Nos. | Names and synonyms | Common names | Codes | Families |
|---|---|---|---|---|
| 200 | *Nauclea orientalis* L.<br>*Sarcocephalus cordatus* (Roxb.) Miq. | Leichhardt's tree; Indian mulberry | ALX/CY(E)F/123456 | Rubiaceae |
| *201* | *Nelumbium speciosum* Willd. | | | — |
| 201 | *Nelumbo nucifera* Gaertn.<br>*Nelumbium speciosum* Willd. | Pink water-lily; pink lotus-lily | AL(2)R(D1)/CY(E)S/ MG[36]/3467 | Nymphaeaceae |
| *202* | *Nymphaea caerulea* Savigny | | | — |
| 202 | *Nymphaea gigantea* Hook.<br>*N. caerulea* Savigny | Blue water-lily | AL(3)R/F/G(D1)/CY (E/W)R(E)S/G/123467 | Nymphaeaceae |
| 202a | *N. gigantea* Hook. var. *violacea* (Lehm.) H. S. Conard | | AL(3)R/F/G(D1)/246 | |
| 203 | *Nymphoides geminatum* (R. Br.) Kuntze<br>*Limnanthemum geminatum* (R. Br.) Griseb. | A water plant | CY(W)R | Gentianaceae |
| 204 | *Opilia amentacea* Roxb. | | AL(5)F(W)/CYX/MX/ 12346 | Opiliaceae |
| *205* | *Oryza fatua* Koen. ex Trin. | | | — |
| 205[37] | **Oryza rufipogon* Griff.<br>*O. fatua* Koen. ex Trin.<br>*O. sativa* auct. non L. | Wild rice | AL(2)S(D1)/CY(E)S/ MS/123456[38] | Graminae grasses |
| *205* | *O. sativa* auct. non L. | | | — |
| 206 | *Owenia reticulata* F. Muell. | | ALX/CAS | Meliaceae |
| 207 | *Palaquium galactoxylum* (F. Muell.) H. J. Lam<br>*Lucuma galactoxyla* Benth. & Hook. | Pencil cedar | CY(E)F | Sapotaceae |
| 208 | *Pandanus spiralis* R. Br. | Pandanus palm; screw palm | AL(3)S(D2)/13456 | Pandanaceae |
| 209[24] | *P.* sp. | | CY(E)S(E/W)F | |
| 210 | *Panicum decompositum* R. Br. | Native millet; barley grass | ALX/CYX/CAS | Graminae grasses |
| *47* | *P. distachyum* [non L.] Benth. | | | — |
| *163* | *P. pauciflorum* R. Br. | | | — |
| *109* | *Pappaphorum avenaceum* Lindl. | | | — |

| Nos. | Names and synonyms | Common names | Codes | Families |
|---|---|---|---|---|
| 211 | *Parinari nonda* F. Muell. ex Benth. *Parinarium nonda* F. Muell. ex Benth. | | CY(E/W)F | Rosaceae |
| *211* | *Parinarium nonda* F. Muell. ex Benth. | | | — |
| 212 | *Pentatropis kempeana* F. Muell. | | CA G | Asclepiadaceae |
| 213 | *Persoonia falcata* R. Br. | Geebung | AL(5)F(W)/CY(E)F/ 123456 | Proteaceae |
| 214 | *Physalis minima* L. | Cape gooseberry | AL(2)F(D2)/CYX/ MF[39]/23467 | Solanaceae |
| 215 | *Planchonella arnhemica* (F. Muell. ex Benth.) van Royen | | AL(6/7)F(W/D)/CYX/ 246 | Sapotaceae |
| 216 | *P. brownlessiana* (F. Muell.) van Royen *Sideroxylon brownlessianum* F. Muell. | | CY(E)F | |
| 217 | *Planchonia careya* (F. Muell.) R. Knuth *Barringtonia careya* F. Muell *Careya australis* F. Muell. | Billy-goat plum; broad-leafed apple | AL(5)F(W)/CY(E)F/ 12346 | Barringtoniaceae |
| *63* | *Plectronia barbata* auct. non (Seem.) Benth. & Hook. f. ex Hemsl. | | | — |
| *64* | *P. latifolia* (F. Muell. ex Benth.) Black | | | — |
| *65* | *P. odorata* F. Muell. | | | — |
| *218* | *Pleiogynium solandri* Engl. | | | — |
| 218 | *Pleiogynium timorense* (DC.) Leenh. *P. solandri* Engl. | Burdekin plum; tulip plum | CY(E)F | Anacardiaceae |
| 219 | *Podocarpus elatus* R. Br. | | CY(E)F | Podocarpaceae |
| 220 | *Polyalthia nitidissima* (Dun.) Benth. | | CY(W)F | Annonaceae |
| 221 | *Polygonum plebeium* R. Br. | | CA S | Polygonaceae |
| 222 | *Portulaca australis* Endl. | Purslane | CY(W)R | Portulacaceae |
| 222a[40] | *P. ballonnensis* | | | — |
| 223 | *P. intraterranea* J. M. Black *P. oleracea* L. var. (?) *grandiflora* Benth. | Pigweed | CAR/G | Portulacaceae |

| Nos. | Names and synonyms | Common names | Codes | Families |
|---|---|---|---|---|
| 224 | *P. oleracea* L. | Pigweed; purslane | ALX/CAR/S | |
| *223* | *P. oleracea* L. var. (?) *grandiflora* Benth. | | | — |
| 225[41] | *Pothos longipes* Schott. *P. loureirii* auct. non Hook. & Arn. | | CY(E)F | Araceae |
| *225* | *P. loureirii* auct. non Hook. & Arn. | | | — |
| 226 | *Pouteria sericea* (Ait.) Baehni *Lucuma sericea* (Ait.) F. Muell. | | AL(4/5)F(D2)/ CY(E/W)F/12456 | Sapotaceae |
| 227 | *Psoralea badocana* (Blanco) Benth. | | ALX/CY(E)R/1346[42] | Leguminosae |
| 228 | *Ptychosperma elegans* (R. Br.) Blume | | AL(4) central shoot (W/D)/CYX/46 | Palmae palms |
| 229 | *Pygeum turnerianum* F. M. Bail. | | CY(E)S | Rosaceae |
| 230 | *Randia fitzalani* (F. Muell.) F. Muell. ex Benth. | | CY(E)F | Rubiacae |
| 231 | *Rapanea crassifolia* (R. Br.) Mez *Myrsine crassifolia* R. Br. | | CY(E)F | Myrsinaceae |
| 232 | *Rhodomyrtus macrocarpa* Benth. | Native loquat | CY(E)F | Myrtaceae |
| 233 | *Rumex crystallinus* Lange | Dock | CA S | Polygonaceae |
| 234 | *Santalum acuminatum* (R. Br.) A. DC. *Eucarya acuminata* (R. Br.) Spr. & Summerh. | Quandong; native peach | CA F | Santalaceae |
| 235 | *S. lanceolatum* R. Br. | Plum bush; native plum | ALX/CYX/CAF | |
| *200* | *Sarcocephalus cordatus* (Roxb.) Miq. | | | — |
| *21* | *Schmidelia serrata* DC. | | | — |
| 236 | *Scirpus littoralis* Schrad. | | ALX/CY(E/W)R/ 124678 | Cyperaceae sedges |
| 237 | *Securinega obovata* Muell.-Arg. | | AL(4)F(D2)/CAX/ MF[43]/124510 | Euphorbiaceae |
| 238[44] | *S. virosa* (Roxb. ex Willd.) Pax & Hoffm. *Fluggea obovata* (Willd.) Wall. ex F.-Vill. | | CY(E/W)F | |

| Nos. | Names and synonyms | Common names | Codes | Families |
|---|---|---|---|---|
| 239 | *Semecarpus australiensis* Engl. | Australian cashew nut; marking nut | AL(4)S(D2)/CY(E)F/456 | Anacardiaceae |
| *240* | *Sida inclusa* Benth. | | | — |
| 240 | *Sida platycalyx* F. Muell. ex Benth.<br>*S. inclusa* Benth. | | CA F | Malvaceae |
| *216* | *Sideroxylon brownlessianum* F. Muell. | | | — |
| 241 | *Siphonodon pendulum* F. M. Bail. | | CY(E)F | Siphonodontaceae |
| 242[45] | *Solanum centrale* J. M. Black<br>*S. nemophilum* F. Muell. | | CA F | Solanaceae |
| 243 | *S. coactiliferum* J. M. Black | | CA F | |
| 244 | *S. ellipticum* R. Br. | Desert raisin; wild gooseberry | CA F | |
| 245 | *S. eremophilum* F. Muell. | Wild gooseberry | CA F | |
| 246 | *S. esuriale* Lindl. | Tomato plant | CA F | |
| 247 | *S. melanospermum* F. Muell. | | CA F | |
| *242* | *S. nemophilum* F. Muell. | | | — |
| 248 | *S. orbiculatum* Dun. | | CA F | Solanaceae |
| 249 | *S. petrophilum* F. Muell. | | CA F | |
| 250 | *S. phlomoides* A. Cunn. | | CA F | |
| 251 | *Stenopetalum velutinum* F. Muell. | | CA G | Cruciferae |
| *49* | *Sterculia caudata* Heward ex Benth. | | | — |
| 252 | *Sterculia quadrifida* R. Br. | Kurrajong | AL(4/5)S/F(D2)/CY(W)R(E/W)S/13467 | Sterculiaceae |
| 253 | *Stictocardia tiliifolia* (Desv.) Hall. f.<br>*Ipomoea grandiflora* (L. f.) Lam. | | CY(W)R | Convolvulaceae |
| 254 | **Tacca leontopetaloides* (L.) Kuntze<br>*T. pinnatifida* Forst. | Polynesian arrowroot | AL(6)R/F(W2)/CY(E/W)R/MR/12346 | Taccaceae |
| *254* | *T. pinnatifida* Forst. | | | — |
| 255 | **Terminalia catappa* L. | Indian or country almond | CY(E/W)S/MS | Combretaceae |

| Nos. | Names and synonyms | Common names | Codes | Families |
|---|---|---|---|---|
| 256 | *T. grandiflora* Benth. | | AL(5)S(W/D)/CAX/ 12410 | |
| *257* | *T. microcarpa* auct. non Decne. | | | — |
| 257[46] | *T. muelleri* Benth. *T. microcarpa* auct. non Decne. | | CY(W)S | Combretacea |
| 258 | *T. seriocarpa* F. Muell. | | ALX/CY(E)F/46 | |
| 259[47] | *Thysanotus exiliflorus sens.* J. M. Black | Fringed violet | CA R | Liliaceae |
| 260 | *T. patersonii* R. Br. | Fringed violet | CA R | |
| 261 | *Tragus australianus* S. T. Blake | Prickle grass; bur grass | CA S | Graminae grasses |
| 262 | *Tribulus solandri* F. Muell. | | CY(W)R | Zygophyllaceae |
| 263 | *Triglochin procera* R. Br. | Arrow grass | AL(3)R(D1)/CYX/ CAX/4678910 | Juncaginaceae |
| 263a | *T. procera* R. Br. var. *dubia* Benth. | Small arrow grass | AL(3)R(D1)/2456 | |
| 264 | *Trigonella suavissima* Lindl. | Native clover; Australian shamrock | CA G | Leguminosae |
| 265 | *Typha angustifolia* L. | Bulrush | ALX/CYX/CAR/G | Typhaceae |
| 266 | *Typhonium angustilobium* F. Muell. | | AL(3)R(W/D)/CY(W) R/456[48] | Araceae |
| 267 | *T. brownii* Schott. | | AL(3)R(W/D)/CY (E/W)R/46 | |
| 268 | *Vandasia retusa* (Benth.) Domin *Hardenbergia retusa* Benth. | | CY(E/W)R | Leguminosae |
| 269 | *Vigna canescens* C. T. White | | CY(W)R | Leguminosae |
| 270 | *V. lanceolata* Benth. | Yam | CYX/CAR | |
| 270a | *V. lanceolata* Benth. var. *filiformis* Benth. | | AL X | |
| *271* | *V. lutea* (Swartz) A. Gray | | | — |
| 271 | *V. marina* (Burm. f.) Merr. *V. lutea* (Swartz) A. Gray | | CY(E)R | Leguminosae |
| 272 | *V. vexillata* (L.) Benth. | | CY(W)R/MX/2345678[49] | |
| 272a | *V. vexillata* (L.) Benth. var. *youngiana* F. M. Bail. | | AL(3)R(D)/46 | |

| Nos. | Names and synonyms | Common names | Codes | Families |
|---|---|---|---|---|
| 273 | *Vitex glabrata* R. Br. | Chaste tree | AL(4/5)F(W)/CY(E/W) F/MT[50]/1234567 | Verbenaceae |
| *28* | *Vitis acetosa* (F. Muell.) Benth. | | | — |
| *78* | *V. clematidea* F. Muell. | | | — |
| *84* | *V. opaca* F. Muell. | | | — |
| *79* | *V. trifolia* L. | | | — |
| *274*[51] | *Wallrothia* | | | — |
| 274[51] | *Waltheria indica* L. | | ALX/CY(E)F/CAX/ 1234610 | Sterculiaceae |
| *275* | *Xanthorrhoea arborea* auct. non R. Br. | | | — |
| 275[52] | *Xanthorrhoea johnsonii* A. Lee<br>*X. arborea* auct. non R. Br. | | CY(E) base of young leaves | Xanthorrhoeaceae grass trees |
| 276 | *X. thorntonii* Tate | | CA base of inner leaves | |
| 277 | *Ximenia americana* L. | Seaside plum; yellow plum | CY(E)F | Olacaceae |

*Notes:*

[1] Black (1943-57:**2**, 412) has both *Acacia frumentacea* Tate and *A. murrayana* F. Muell. ex Benth. as synonyms of *A. leptopetala* Benth. Eichler (1965:172) doubts the conspecificity of *A. murrayana* and *A. leptopetala.* Burbidge (pers. comm.) agrees, but suggests treating *A. frumentacea* as synonymous with *A. murrayana.*

[2] Roth (1901:9) records *Acrostichum aureum* L. for eastern Cape York. Pedley (pers. comm.) says that Queensland collections all appear to be *A. speciosum* Willd. *A. aureum* L. is in New Guinea but has not yet been found in Queensland. Roth's specimen having been misidentified, I have adopted *A. speciosum* Willd. as the equivalent of *A. aureum* auct. non L.

[3] Distribution as interpreted from Tindale (1958:180).

[4] Of *Avicennia tomentosa* R. Br. (Roth 1901:9) Pedley (pers. comm.) says that Robert Brown was not the author of the name but misidentified Australian plants as *A. tomentosa* Jacq. Of *A. officinalis* L. (Roth 1901:9) Black (1943-57:**4**, 726) says that it does not occur in Australia, having been misidentified for *A. marina* (Forsk.) Vierh., which therefore = both *A. officinalis* auct. non L. and *A. tomentosa* auct. non Jacq.

[5] Specht's reference (1958d:499) to use at Borroloola is outside Arnhem Land proper.

[6] 8 and 9 are not in Specht's distribution but Black (1943-57:**4**, 726) says that the plant occurs all round the Australian coast, except for Tasmania.

[7] Pedley (pers. comm.) says that this is the Australian plant and that *B. racemosa* Gaudich. as recorded by Roth (1901:10) is a later homonym for a distinct non-Australian species, now treated as a synonym of *B. samoense* A. Gray. Thus *B. racemosa* (L.) Spreng. = *B. racemosa* auct. non Gaudich.

[8] Distribution as interpreted from Tindale (1958:180-1).

[9] Leaves India, roots Fiji, both as famine foods.

[10] Pedley (pers. comm.) has *Buchanania arborescens* (Blume) Blume.

[11] There is discussion in Black (1943-57:**2**, 348) and especially in Eichler (1965:138) as to whether in South Australia *Calandrinia volubilis* Benth. has not been wrongly applied for *C. eremaea* Ewart. On Burbidge's advice (pers. comm.) I have retained *C. volubilis* because the relevant record is for the Northern Territory (Lawrence 1968:82).

[12] Lawrence (1968:208) lists *Candelia*, a mangrove, recorded as a food source in Cape York during the voyage of H.M.S. *Rattlesnake*. The identification offered here seems the only possible one.

[13] *Plectronia barbata* = *Canthium barbatum* Seem. is recorded by Roth (1901:15). Pedley (pers. comm.) says that this is a Pacific Islands species and that some early collections of the Australian *C. coprosmoides* F. Muell. were misidentified as such: that is, *C. coprosmoides* F. Muell. = *C. barbatum* auct. non Seem. and *Plectronia barbata* auct. non (Seem.) Benth. & Hook. f. ex Hemsl.

[14] Pedley (pers. comm.) accepts the synonymy of *Canthium lucidum* Hook. & Arn. with *C. odoratum* (Forst. f.) Seem. but cannot trace the name *Plectronia odorata* F. Muell. Maiden (1889:392) has *P. odorata* and *Canthium odoratum* as synonyms of *C. lucidum*, which is also Specht's form.

[15] Pedley (pers. comm.) says that *Capparis nobilis* F. Muell., as recorded by Roth (1901:10) for Cape York, is restricted to Lord Howe Island, the Queensland species intended being *C. arborea* (F. Muell.) Maiden. Because of the misidentification *C. arborea* (F. Muell.) Maiden = *C. nobilis* auct. non F. Muell.

[16] Single reference, Malaya, which may be suspect.

[17] *Carissa*, not being a member of the specifically South Australian flora, does not appear in Black (1943-57). I therefore follow Ewart and Davies (1917) on the synonymy. Interior Australia (10) is not in Specht's distribution for *C. lanceolata* but is for *C. ovata* R. Br. (distribution 246710). Maiden (1889:14) has *C. brownii* as a synonym of *C. ovata*. *Index Kewensis* also comments on the taxonomic problem.

[18] Here I follow Eichler (1965:291).

[19] Pedley (pers. comm.) says *Dioscorea bulbifera* L. is the Australian plant = *D. sativa* auct. non L.: that is it has been misidentified as *D. sativa* L.

[20] Specht gives no interior distribution (10) for *Dioscorea* but Meggitt (1957:144) talks about its use amongst the Walbiri (*jala*) and Sweeney (1947:291, 295-6) mentions the importance of *yala* to the same people. Strangely, Lawrence omits all mention of *Dioscorea* in his central Australian table (1968: Table 2). Though unidentified as to species, the central Australian plant must be either *D. bulbifera* or *D. transversa* and Table 15:5 is calculated on this basis.

[21] The unidentified species is included because it provides additional information about the range of use of *Dioscorea*.

[22] *Erythrina australis*, recorded for central Australia by Cleland and Tindale (1954:84; cf. Lawrence 1968:83) under the name of bean tree, I can find in no botanical list. I presume it is an error for *E. vespertilio* Benth.

[23] I omit this from my calculations as being possibly only a source of nectar (cf. Lawrence 1968:83), though Lawrence includes it in his main food lists.

[24] I include these unidentified species of *Eucalyptus*, *Ipomoea*, and *Pandanus* because they provide extra information about the food uses of the three genera.

[25] Eichler (1965:94) has *Ficus platypoda* (Miq.) A. Cunn. ex Miq. Interior Australia (10) is not in Specht's distribution but Black (1943-57:**2**, 260) confirms its presence there.

[26] Specht (1958a:217) has *Ficus lacor* Buch.-Ham. for what is evidently the same plant. Pedley (pers. comm.) says that at the time of Specht's paper *F. virens* Ait. was being misidentified as *F. lacor* Buch.-Ham.: in other words *F. virens* Ait. = *F. lacor* auct. non Buch.-Ham.

[27] *Gahnia psittacorum* Labill., recorded by Roth (1901:13) for eastern Cape York is, according to Pedley (pers. comm.), a Tasmanian and Victorian species. The north Queensland plant is *G. sieberana* Kunth, which because of the misidentification = *G. psittacorum* auct. non Labill.

[28] Presence or absence uncertain for 12357.

[29] Not yet recorded between New Guinea and India.

[30] On the assumption that Burkill's *H. radiatus* Willd., whose leaves are sometimes eaten, is the same plant.

[31] *Lepidium ruderale* is recorded by Stirling (1896:57) with no botanical author cited (cf. Lawrence 1968:83). Black (1943-57:**2**, 382-3) divides *L. ruderale* [non L.] F. Muell. amongst three species. Of these Burbidge (pers. comm.) thinks *L. fasciculatum* Thell. the most likely to be meant by Stirling.

[32] Cape York (6) does not appear in Specht's distribution.

[33] Blake (1954:127) establishes the distinctiveness of *Livistona mariae* F. Muell. from *L. humilis* R. Br.

[34] With no access to the manuscript cited by Lawrence (1968:207) for *Mimusops kauki* I do not know what, if any, botanical author is quoted. *Index Kewensis* gives *M. kauki* L. and Maiden (1889:45) gives

*M. kauki* R. Br., but both agree on synonymy with *M. browniana* Benth., which van Royen (1953) assigns to *Manilkara*.

[35] F Indo-China.

[36] R/S/F reported marginally to Malaysia and for Singapore Chinese.

[37] *Oryza sativa* L. is recorded by Roth (1901:14) for wild rice but Pedley (pers. comm.) and others say that this should be reserved for the cultigen. Because of this misapplication of the Linnaean name, the form becomes *O. sativa* auct. non L. Specht (1958a:197) has *O. fatua* Koen. ex Trin. for wild rice in Arnhem Land. Pedley says the correct name is *O. rufipogon* Griff. He also says that there are two species of wild rice in north Queensland, *O. rufipogon* and *O. australiensis* Domin, which were probably not distinguished in Roth's time.

[38] Cape York (6) does not appear in Specht's distribution for *O. fatua*.

[39] No geographical details of use in Burkill.

[40] *Portulaca ballonnensis*, recorded by Basedow (1904:18; cf. Lawrence 1968:83), exists in no botanical list. Basedow gives as the native name *munyeroo*, which is also recorded for *Calandrinia balonensis* (cf. Lawrence 1968:81, under *Claytonia*). *Portulaca* is probably a mistake for *Calandrinia* (*Claytonia*) and has been so taken.

[41] Pedley (pers. comm.) says that *Pothos loureirii* Hook. & Arn., recorded by Roth (1901:15) for eastern Cape York, is a Malaysian species. The commonest Queensland species is *P. longipes* Schott. I have taken this as the species which Roth collected but which was misidentified, i.e. *P. longipes* Schott. = *P. loureirii* auct. non Hook. & Arn.

[42] Presence or absence uncertain for 5.

[43] Reference is to Indonesia, with no details given by Burkill.

[44] On Pedley's advice (pers. comm.) I follow Merrill on the form of *Fluggea obovata*.

[45] Black (1943-57:**4**, 748) has *Solanum centrale* J. M. Black as synonymous with *S. nemophilum* F. Muell. Eichler (1965:272) doubts that the two are conspecific and says the correct name for the South Australian plant is *S. centrale* Black. Burbidge (pers. comm.) says that central Australian material labelled *S. nemophilum* should be called *S. centrale*.

[46] *Terminalia microcarpa* Decne. is recorded for western Cape York by Roth (1901:16). Pedley (pers. comm.) says that this is a Malaysian species which has never been collected from Cape York, Roth's specimen being misidentified as such. There is Roth material in the Queensland Herbarium under *T. muelleri* Benth., which is therefore accepted here as = *T. microcarpa* auct. non Decne.

[47] For this see Black (1943-57:**1**, 191) and Eichler (1965:82).

[48] Cape York (6) not in Specht's distribution.

[49] Distribution as interpreted from Specht (1958a:246).

[50] F Thailand.

[51] Lawrence (1968:208) cites a record of food use of *Wallrothia* from H.M.S. *Rattlesnake*'s visit to eastern Cape York. In MacGillivray (1852:**1**, 127), *Wallrothia* is described as providing a large yellow plum. The suggested identification seems to be the only possible one.

[52] Roth (1901:16) records *Xanthorrhoea arborea* R. Br. for eastern Cape York. Pedley (pers. comm.) says this is a New South Wales species, the north Queensland species being either *X. johnsonii* A. Lee or hybrids of this. *X. johnsonii* A. Lee is accepted as the plant meant by Roth = *X. arborea* auct. non R. Br.

## Appendix 15:2

# Generic List of the Food Plants of Arnhem Land, Cape York, and Central Australia

Besides listing, by botanical affiliation, all genera providing food plants in Arnhem Land, Cape York, and central Australia, this appendix gives information about the food or other uses in Malaysia of genera shared with that region: R rootstocks, S seeds (including nuts, kernels), F fruits, G greens, M medicine, T technology, B (personal) beautification, O ornamental, X no recorded uses.

Those genera with members accorded some degree of domestication in Malaysia are italicised: where a food-providing genus is the object of domestication, but not for food, the purpose of domestication is indicated in parentheses.

Genera providing food in Arnhem Land which are absent from central Australia are indicated by * before the name; those providing food in central Australia which are absent from Arnhem Land by † before the name.

**Australasian Genera (23)**

Amyema, Archontophoenix
†Blennodia, Bowenia, Brachychiton
Calostemma, *Cartonema[1], Castanospermum[2]
Davidsonia[3]
†Enchylaena, †Eremophila, Eupomatia[4]
Fenzlia
†Leichhardtia[5], Lysiana
Microstemma[4]
Owenia
*Persoonia[6], Pleiogynium[4]
†Stenopetalum
Triglochin[4]
Vandasia[4]
†Xanthorrhoea

*Note:*

As well as Australia found [1] Aru Islands; [2] New Hebrides and New Caledonia; [3] Fiji; [4] New Guinea; [5] New Caledonia; [6] New Zealand.

**Other Non-Malaysian Genera (14)**

Aristida[1]
Calandrinia, Carissa, †Chenopodium, †Convolvulus, Cucumis
Dactyloctenium
Enneapogon, †Erodium
†Lepidium
†Pentatropis
†Rumex
†Tragus, †Trigonella

*Note:*

[1] Ridley (1922-5: **5**, 242) has one species in peninsular Malaya, noted as rare.

**Malaysian Genera (124)**

Acacia M/T[1], Acrostichum G, *Aleurites* S, Allophylus M/T[2], *Alocasia* G[3], *Amaranthus* G, *Amomum* F, **Amorphophallus* R/F/G[4], *Ampelocissus G[5], Aneilema G, *Antidesma* F/G, Aponogeton[6] X, Atylosia X, Avicennia S/F

*Barringtonia* F/G (M), Beilschmiedia M/T, *Blechnum X, Boerhavia M[7], Brachiaria X, *Bridelia M/T, Bruguiera S/F/G, *Buchanania F[8]

Calamus F/G, Canthium F/G[9], Capparis F[10], *Carallia F[11], Casuarina M/T, *Cayratia M, *Celtis M, Ceriops M/T[12], Ceropegia X, Cissus G, *Clerodendrum* G (O), *Colocasia* R/G, *Curculigo F, *Curcuma* R, **Cycas* S/pith, Cyclosorus X, *Cymbopogon* G, Cyperus R

*Dioscorea* R, *Dolichos* S/G

*Elaeocarpus* F, Eleocharis R, Endiandra T, Enhalus F[13], Entada S/G, Eragrostis X, Eriachne X, Eriosema X, Erythrina S/G, Eucalyptus[6] X, **Eugenia* F/G, Exocarpos[6] T

*Ficus* F/G

Gahnia X, *Ganophyllum[6] T, Grevillea[6] X, Grewia F

*Hemicyclia T, *Hibiscus* G (T/O), *Hornstedtia* F

Ichnanthus X, *Ipomoea* G, Ixora F[14]

*Lepironia* T, Licuala stem bud/pith, Livistona stem bud[15]

*Maba M/T[16], *Malaisia X, *Manilkara* F, Marsilea G, Melastoma F/G, Melodinus F, Melothria F, *Merremia* R/G, Mimusops F, **Morinda* G[17] (M/T), *Musa* F, †Myoporum[6] X

Nauclea F, *Nelumbo G[18], *Nymphaea S[19], Nymphoides X

*Opilia X, **Oryza* S

Palaquium F, **Pandanus* F/G, Panicum X[20], Parinari M/T, *Physalis F[21], *Planchonella M/T, *Planchonia G[22], Podocarpus M/T, Polyalthia M/T, Polygonum G[23], Portulaca G[23], Pothos M/T, *Pouteria X, Psoralea X, *Ptychosperma X, Pygeum M/T

Randia F, Rapanea X, Rhodomyrtus F

Santalum[6] M/B, Scirpus T, Securinega F[24], *Semecarpus S/F/G, Sida M/T, Siphonodon X, *Solanum* F/G, *Sterculia S, Stictocardia X

**Tacca* R, *Terminalia* S/F, Thysanotus[6] X, Tribulus X, Typha R[25], *Typhonium X[26]

Vigna G, *Vitex M/T[27]

Waltheria M/T

Ximenia F

*Notes:*

[1] S Thailand (*A. farnesiana*).
[2] F Indo-China.
[3] R India.
[4] F/G only reported for the Philippines within Malaysia.
[5] F Indo-China.
[6] Restricted distribution within Malaysia, mainly E Malaysia.
[7] Leaves India, R Fiji, both as famine foods (*B. diffusa*).
[8] Single reference, Malayan Aborigines.
[9] One atypical reference to F, Java.
[10] Single reference, Philippines, but F also Pacific.

[11] Single reference, Malaya, which may be suspect (Burkill).

[12] F Andaman Islands (*C. tagal*).

[13] No geographical details for use in Burkill.

[14] Single reference, Malaya, which may be suspect (Burkill). Otherwise M/T.

[15] Single reference, Malaya, which may be suspect (Burkill). Certainly Indo-China.

[16] F (and cultivated) Pacific.

[17] F Indo-China (*M. citrifolia*).

[18] R/S/F reported marginally to Malaysia and for Singapore Chinese (*N. nucifera*).

[19] Single reference, Philippines, R/F marginal to Malaysia.

[20] S (and cultivated) continental Asia.

[21] No geographical details for use in Burkill.

[22] Single reference, Malaya. For rest T.

[23] Small reference only to food use.

[24] Reference to Indonesia without details.

[25] Burkill records no uses. *Flora Malesiana* (**4**, 244) suggests R (*T. angustifolia*).

[26] R Indo-China.

[27] F Thailand (*V. glabrata*).

*Sources:* Distribution and affiliations of Australian genera Burbidge (1963), supplemented for within Australia by Black (1943-57) and Specht (1958a, c) and for outside Australia by Burkill (1935), Ridley (1922-5), and the as yet uncompleted series *Flora Malesiana* (1948-67).

Malaysian uses of shared genera Barrau (n.d.), Burkill (1935).

# Appendix 15:3

## Genera and Species Recorded as Food Sources in Malaysia, Present but Apparently Unused in Australia

### A. Genera

1. *Agathis* 3 spp. Cape York to SE Queensland.
2. *Canarium* 3 spp., together covering Northern Territory, Cape York to SE Queensland.
3. *Caryota* 1 sp. Cape York.
4. *Coleus* 1 sp. tropical north.
5. *Corchorus* 25 spp. tropical and arid north.
6. *Diospyros* 6 spp. (5 endemic) tropical north, coastal Queensland, north coast N.S.W.
7. *Gevuina* (*Kermadecia*) 1 sp. Cape York.
8. *Setaria* 5-6 spp. widely tropical north and central Australia.
9. *Uvaria* 2-3 spp. Cape York.

### B. Species

1. *Amaranthus lividus* L. (*A. viridis* L.) Cape York and coastal Queensland.
2. *Antidesma ghaesembilla* Gaertn.[1] Hamersley/Kimberley to Arnhem Land.
3. *Bruguiera parviflora* (Roxb.) W. & A.[1] Darwin/Katherine to Arnhem Land.
4. *B. sexangula* (Lour.) Poir. Arnhem Land, Cape York.
5. *Calophyllum inophyllum* L.[2] Arnhem Land, Cape York.
6. *Cardiospermum halicacabum* L. Tropical north, coastal and interior Queensland.
7. *Cordyline fruticosa* Goeppert[3] 3-4 spp. of genus in Australia, 2-3 of which are endemic.
8. *Decaspermum fruticosum* Forst.[3] 2 spp. of genus Cape York and coastal Queensland, one of which is endemic.
9. *Diospyros montana* Roxb. Kimberley to Darwin.
10. *Emilia sonchifolia* (L.) DC. Cape York.
11. *Hibiscus tiliaceus* L. Darwin/Katherine, Arnhem Land, Cape York, coastal Queensland.
12. *Melastoma polyanthum* Blume Tropical north, coastal Queensland.

13. *Nypa* (*Nipa*) *fruticans* Wurmb.[3] One species of genus Cape York.
14. *Polygonum orientale* L. Arnhem Land, Cape York, coastal Queensland, north coast N.S.W.
15. *Sterculia foetida* L.[1] Arnhem Land, coastal Queensland.
16. *Wedelia biflora* (L.) DC. Arnhem Land, Cape York, coastal Queensland, north coast N.S.W.

*Notes:*

[1] Not seen by Specht during Arnhem Land field work (1958c: 447-75).

[2] Mentioned by Maiden (1889:12) as a fruit tree at Thursday Island (Torres Strait).

[3] Not being Arnhem Land plants, these are not in Specht (1958a, c) and have not been submitted to Pedley for checking. Australian presence and nomenclature is taken from Burkill 1935.

*Sources:* Malaysian food sources: Barrau (n.d.), Burkill (1935). Australian distributions: Burbidge (1963) and Specht (1958a, c), supplemented by Burkill (1935).

## References

Barrau, J. n.d. *An ethnobotanical guide for anthropological research in Malayo-Oceania (preliminary draft)*. UNESCO Science Co-operation Office for Southeast Asia.

Basedow, H. 1904. Anthropological notes made on the South Australian Government North-West Prospecting Expedition, 1903. *Trans.R.Soc.S.Aust.* **28**:12-51.

Black, J. M. 1943-57. *Flora of South Australia.* 2nd edn., 4 vols. Adelaide: South Australian Branch of the British Science Guild.

Blake, S. T. 1954. Botanical contributions of the northern Australia regional survey II. Studies on miscellaneous northern Australian plants. *Aust.J.Bot.* **2**:99-140.

Burbidge, Nancy T. 1963. *Dictionary of Australian plant genera: gymnosperms and angiosperms*. Sydney: Angus and Robertson.

Burkill, I. H. 1935. *A dictionary of the economic products of the Malay Peninsula.* 2 vols. London: Crown Agents for the Colonies.

Chang Kwang-chih. 1967. The Yale expedition to Taiwan and the southeast Asian horticultural evolution. *Discovery, Peabody mus.nat.Hist.* **2** (2): 3-10.

Cleland, J. B. and Tindale, N. B. 1954. The ecological surroundings of the Ngalia natives in central Australia and native names and uses of plants. *Trans.R.Soc. S.Aust.* **77**: 81-6.

Eichler, H. 1965. *Supplement to J. M. Black's Flora of South Australia (second edition, 1943-1957)*. Adelaide: South Australian Branch of the British Science Guild, Handbooks Committee.

Ewart, A. J. and Davies, Olive B. 1917. *The flora of the Northern Territory.* Melbourne: McCarron, Bird and Co.

*Flora Malesiana.* Ser. 1, Vols. 1, 4, 5, 6 (1-5), 1948-67. Djakarta: Noordhoff-Kolff N. V.

Golson, J. 1970. Both sides of the Wallace line: New Guinea, Australia, island Melanesia and Asian prehistory. In *Early Chinese art and its possible influence in the Pacific basin: a symposium* (ed. N. Barnard), in press. New York: Intercultural Arts Press.

Gorman, C. F. 1969. Hoabinhian: a pebble-tool complex with early plant associations in Southeast Asia. *Science, N.Y.* **163**: 671-3.

*Index Kewensis. An enumeration of the genera and species of flowering plants.* 2 vols., 1895. Oxford: Clarendon Press.

Lawrence, R. 1968. *Aboriginal habitat and economy.* Canberra: Australian National University, School of General Studies, Department of Geography Occasional Papers No. 6.

MacGillivray, J. 1852. *Narrative of the voyage of H.M.S.* Rattlesnake. . . . 2 vols. London: T. and W. Boone.

Maiden, J. H. 1889. *The useful native plants of Australia (including Tasmania).* London: Trubner and Co.

Meggitt, M. J. 1957. Notes on the vegetable

foods of the Walbiri of central Australia. *Oceania* **28**:143-5.
——1964. Aboriginal food-gatherers of tropical Australia. In *The ecology of man in the tropical environment.* International Union for the Conservation of Nature and Natural Resources, Ninth Technical Meeting, Proceedings and Papers, pp. 30-7. Morges: IUCN (Publ. Ser. 4).
Ridley, H. N. 1922-5. *The flora of the Malay Peninsula.* 5 vols. London: L. Reeve and Co.
Roth, W. E. 1901. Food, its search, capture and preparation. *N.Qd Ethnogr.* **3**. Brisbane: Government Printer.
Specht, R. L. 1958a. The gymnospermae and angiospermae collected on the Arnhem Land expedition. In *Rec.Am.-Aust.scient. Exped.Arnhem Ld* **3** (Botany and Plant Ecology) (eds. R. L. Specht and C. P. Mountford), pp. 185-317. Melbourne: Melbourne University Press.
——1958b. The climate, geology, soils and plant ecology of the northern portion of Arnhem Land. In *Rec.Am.-Aust.scient.Exped. Arnhem Ld* **3** (Botany and Plant Ecology) (eds. R. L. Specht and C. P. Mountford), pp. 333-413. Melbourne: Melbourne University Press.
——1958c. The geographical relationships of the flora of Arnhem Land. In *Rec.Am.-Aust. scient.Exped.Arnhem Ld* **3** (Botany and Plant Ecology) (eds. R. L. Specht and C. P. Mountford), pp. 415-78. Melbourne: Melbourne University Press.
——1958d. An introduction to the ethnobotany of Arnhem Land. In *Rec.Am.-Aust. scient.Exped.Arnhem Ld* **3** (Botany and Plant Ecology) (eds. R. L. Specht and C. P. Mountford), pp. 479-503. Melbourne: Melbourne University Press.
Stirling, E. C. 1896. *Report on the work of the Horn Scientific Expedition to central Australia.* Part 4 (Anthropology) (ed. B. Spencer). London: Dulau and Co.
Sweeney, G. 1947. Food supplies of a desert tribe. *Oceania* **17**:289-99.
Tindale, Mary D. 1958. The Pteridophyta of Arnhem Land. In *Rec.Am.-Aust.scient.Exped. Arnhem Ld* **3** (Botany and Plant Ecology) (eds. R. L. Specht and C. P. Mountford), pp. 171-84. Melbourne: Melbourne University Press.
van Royen, P. 1953. Revision of the Sapotaceae in the Malaysian area in a wider sense V. *Manilkara* Adanson em. Gilly in the Far East. *Blumea* **7**:401-12.

# 16 Open Sites and the Ethnographic Approach to the Archaeology of Hunter-Gatherers

Nicolas Peterson

The ethnographic approach to the archaeology of hunter-gatherer societies, ethnoarchaeology, is the study of modern hunter-gatherers with an eye to creating analogical models for use in the interpretation of ancient remains. As a conscious approach it is new and has not as yet developed a methodology nor established its value or limitations. But wherever archaeologists are seeking the explication and explanation of assemblage differences and similarities (cf. Binford 1962: 217) in more than strictly limited terms, ethnographic analogy has an important role, for analogy implicit or explicit lies behind all descriptive reconstructions of the past.

Though several ethnoarchaeological studies have been made in Australia, archaeologists have so far made little use of them (but see Jones 1966, 1968; White and Peterson 1969). This is partly because most of the studies are recent, but it is also because there are some strong criticisms of the use of ethnographic analogy in archaeological interpretation (Binford 1968; Freeman 1968). The purpose of this paper is to discuss these criticisms and to argue that it is ethnographic data on Aboriginal ecology that has most relevance to Australian hunter-gatherer archaeology.

## Ethnographic Analogy in Archaeology

As a form of inductive argument, analogy can only indicate probable or probably true relationships. There are, however, criteria for establishing the degree to which the conclusions reached by analogy are more or less probable.

Chang (1967:229), following Willey (1953:181) and Ascher (1961), has distinguished two types of analogy in archaeology.

I am indebted in particular to Rhys Jones, but also to Ian Glover, Betty Hiatt, and Carmel White for many influential discussions. They are not, of course, responsible for the contents.

**1** The general comparative analogy: this is based on correlations between artifacts and the behaviour associated with them on a cross-cultural basis. It can be expanded to include any cross-cultural analogy of single traits to the generalised model of hunter-gatherer society.
**2** The direct historical analogy: this is based on a demonstrable continuity from the prehistoric to the ethnographic. Two types can be distinguished: the area ethnographic model, that is one based on recent field work where re-study remains possible, and the area historical model based on historical records and descriptions. Either of these models may be for a single trait, a complex of traits, or for the whole culture of an area.

In the sense that these analogies can only express probable relationships they are all 'limiting models'. Analogies of the first type generally suffer from being statements of the apparently obvious. This leads to a lack of rigour in their formulation, which often means that they are based on very superficial knowledge of the ethnographic literature, or worse, on common sense alone. The latter is inherently dangerous since common sense and common knowledge generally lag well behind the specialist's knowledge and therefore tend to have already invalidated preconceptions built into them. The more generalised this type of model becomes the looser the 'fit' with the past in any particular area and the smaller the likelihood of the analogy providing any real insight.

The tightest analogies are direct historical, especially where they are ethnographic and involve a correlation between artifact and a limited range of behaviour. The advantage of the area ethnographic model over the area historical model is that the latter may be based only on a limited number of observations with no record of contrary examples. The area historical model, however, often has the advantage that the observations were made among people virtually unaffected by alien cultures.

It is only recently that much care has been spent on creating well researched analogies for Australian archaeology (Hiatt 1967-8; Gould 1968a; Lawrence 1968; Peterson 1968; see also Allen 1968 and Bickford 1966). In the past the tendency has been to raid the ethnographic literature for the odd fact and to generalise on the strength of it. A methodology for constructing analogies would avoid much of this problem and ensure that common sense, which has an important role in assessing interpretations, is based more firmly in a hunter-gatherer framework.

### THE CRITERIA OF ANALOGY

There are six criteria by which an analogy in logical argument can be assessed as more or less probable. These criteria provide the first steps towards a methodological framework within which to construct ethno-archaeological analogies. The criteria (after Copi 1961: 343-8) can best be stated by an example.

Ethnographically there is a direct relationship between the number of pairs of pestles and mortars in an Arnhem Land wet season camp and the number of married women living there. The number of pairs of pestles and mortars in such camps can therefore give a good idea of band size. I have argued (1968) that for as long as the climate has remained unchanged and therefore the vegetable diet similar, pestles and mortars found archaeologically can in certain situations give a good indication of the size of bands occupying wet season camps in the past. This is a strong analogy on the basis of the six criteria:
**1** *The structural relevance of the analogy*. The analogy is directly relevant. It is dealing with the same objects in the same surroundings and there is a causal relationship

between them. Pestles and mortars are necessary in food preparation now, and presumably were necessary for preparing the same foods in the past. Women in all Australian Aboriginal societies are associated with vegetable food preparation now and it is reasonable to assume that they were in the past when the same foods were available.

**2** *The number of entities for which the analogy is valid.* The basic behaviour of pounding or grinding is associated with all pestles and mortars, so the analogy is valid for them all.

**3** *Invalid entities.* The number of uses associated with the pestle and mortar which do not conform with the analogy, such as by men, is small and can be shown to be subsidiary to the basic behaviour associated with them.

**4** *The number of respects in which the things are said to be analogous.* The uses to which women now put the pestle and mortar are said to be analogous in all respects with the uses they put them to in the past. Though today the men sometimes use them as anvil and hammer for fashioning metal spears, they probably used them for making some stone tools in the past. If they did not, it does not affect the analogy since this is concerned with the use of the pestle and mortar by women.

**5** *The number of dissimilar instances in the premise.* The analogy is true for all pestles and mortars, irrespective of shape or size, and for all active married women, since once women are married they become food preparers and remain so virtually until the end of their days.

**6** *The strength of the conclusion relative to the premise.* Since the association between the pestle and mortar and its use by women is tight, a slightly looser statement about its uses in the past is likely to have a high probability of being probably true.

Such a methodology is most useful for analogies concerned with particular artifacts. Its main purpose is to ensure a proper investigation of the literature before an analogy is used so that assumptions do not creep in unconsciously. However, as I will show below, it is the concentration on ethnographic analogy, largely as an aid to identifying the possible function of artifacts in the past, that often is responsible for the ethnographic approach to archaeological interpretation proving of little help.

### OBJECTIONS TO THE USE OF ETHNOGRAPHIC ANALOGY IN ARCHAEOLOGICAL INTERPRETATION

There are three principal objections raised against the use of ethnographic analogy in archaeological interpretation. It is said that it introduces the danger of anthropological frames of reference being imposed on the archaeologist; that the units observed ethnographically have no existence archaeologically; and that archaeological and anthropological time scales are so different that it is dangerous to transpose modern analogies to periods in the distant past. An important distinction to make with regard to all three criticisms is between those analogies to do only with artifact function and those which have wider sociological implications.

Freeman (1968:262) recently examined the first objection; 'the most serious failings in the present models for interpreting archaeological evidence are directly related to the fact that they incorporate numerous analogies with modern groups [demanding] that prehistorians adopt the frames of reference of anthropologists'. This is most likely with sociological analogies. However, in Australia, by far the commonest analogies are concerned with artifact function. Since the analogies are based on descriptive observation, the possibility of anthropological theory being built into them is small. The danger remains, however, as long as archaeologists fail to treat the construction of analogies seriously, and continue to make *ad hoc* use of ethnography. As regards artifacts, when models are constructed for or by archaeologists to answer archaeological questions, this danger will be greatly reduced. Where the anthropologist is interested in cultural evolution his theoretical framework has a direct and logical relevance to the archaeologist, and vice versa (see Service 1964). However, even when this is the case there are problems which are raised by the following objections to the ethnographic approach generally.

The second objection is certainly valid. Since ethnoarchaeology has been a conscious approach, pristine Aboriginal societies have ceased to exist. Not even in the remotest parts of the continent are there functioning societies using a purely stone technology, though there are some men who have grown up in such societies and who have provided valuable information (see Gould 1968b). Direct observation therefore cannot provide much data on the sorts of artifacts that are most commonly found in archaeological deposits. There are some historical data but not as many as one might expect, since metal seems to have replaced stone for most uses immediately it became available to the Aborigines.

Sociologically it is clear from the work of Lee and DeVore with the Bushmen (reported by Deetz 1968:285), Gould in central Australia (1968a:119), and my own work in Arnhem Land that individual hunter-gatherer bands are below the threshold of archaeological visibility. Within a few years of abandonment, Arnhem Land camp sites are largely obliterated by wind, sun, rain, fire, and, most important of all, site re-use. If a single occupation were to survive in a place such as a cave where it might be preserved for many years, the rates of deposition are so slow that a second use of the cave even several hundred years later would result in the mixing of artifacts. At the Nawamoyn shelter in Arnhem Land the dates 21,450 ± 600 BP and 7,110 ± 160 BP lie less than 40 cm apart vertically (White, C. 1967:151; see also ch. 12, Fig. 12:2).

The density of artifacts at this site is 28 per $m^3$, or, more significantly, only six artifacts were deposited every one hundred years (White and Peterson 1969). Though there are richer sites, this is not atypical (see Mulvaney and Joyce 1965:172).

The third objection concerns the propriety of transposing analogies from the present to periods in the more distant past without in some way allowing for the cumulative effects of change. This is a major problem with sociological analogies, since anthropologists have a small enough understanding of change over tens of years, much less over thousands. Despite past ideas on stability, there is ample evidence for cultural changes taking place both at the time of European contact (e.g. Meggitt 1962:168; Warner 1958:451-71) and long before (e.g. McCarthy 1939; Hiatt pers. comm.; Davidson 1934).

As far as material culture is concerned, archaeology itself provides the evidence for long periods of unchanging stone technology. Here one could be confident of analogies based on modern observation were the stone artifacts still in use. Unfortunately those stone artifacts whose function is most problematical are the ones that dropped out of use even before European contact.

These three objections therefore have a substantial degree of validity and could be taken as sufficient reason for making little use of ethnographic analogy. However, they largely affect analogy used in a reconstruction of the details of a particular regional culture, in the anthropological sense. Once it is recognised that there is little direct relationship between the stone assemblages excavated by the archaeologist of hunter-gatherer sites and the particular local group that produced them, these objections fall into their correct perspective.

### The Relationship between Stone Assemblages and the Cultures that Produced Them

Traditionally there has been a tendency to assume that there is necessarily a close relationship between a culture and its technology, allowing the former to be identified by the latter. In an ethnographic situation, when the full range of material culture is present, this is likely to be true, but in an archaeological situation it is not.

New Guinea presents the nearest and most dramatic example of this situation, where the gradual though major change from hunting to horticulture in the Highlands is not reflected in the stone technology (White, J. P. 1967). In Australia, although over 500 linguistic groups existed at the time of European arrival, they are not reflected by a comparable number of typologically distinguishable stone assemblages. Though there are some regional variations, the remarkable feature is their uniformity. Indeed, it is a striking characteristic of hunter-gatherer stone artifact assemblages the world over that many of the artifacts are typologically similar. For example, most of the tool types in the European Palaeolithic are found in Australia.

Apart from their many world-wide typological similarities, the other common feature of hunter-gatherer stone technologies is their extreme degree of conservatism. The Lower Palaeolithic industries of Europe provide the classic example, but even the Australian time scales are impressive. In Arnhem Land, 16,000 years passed without any recognisable variation in the stone technology (White, C. 1967; see also ch. 12).

It is inconceivable that any culture could remain totally unchanging over such long periods and still survive. Survival requires flexibility and adaptability, which must introduce some change, however small. The cumulative effects of hundreds of years of coping with natural disasters and accommodating different personalities alone would ensure this. Certainly the burden of proof that this is not so lies with those who wish to maintain that it is; they are often archaeologists whose main evidence is the stone technology.

These world-wide typological similarities and the technological conservatism have a simple explanation. The tool types involved form a basic woodworking and boneworking kit. They are largely intended for the making of the tools used in the food quest, and other items of material culture, but are rarely employed in the quest itself. This has been noticed before—for example, Binford and Binford (1966: 291) distinguish 'maintenance' tool types from 'extractive' tool types—but it has not been stressed. The fact that it has not been stressed has allowed archaeologists to write of tens of thousands of years of cultural stability. There is no reason, however, to expect change in the generalised wood/boneworking kit. Changes to accommodate introduced ideas, new food sources, or bright inventions would have taken place in the specialised tools made for specific tasks. These perishable and specialised tools are also the ones that may have carried the signs of local group variation, which the generalised wood/boneworking kit items rarely do.

Some stone tools were used directly in the food quest, and therefore are more specialised. Projectile points are the most obvious example, but it is of interest to note that in Australia these appear after the giant marsupials died out. If, as Jones (1968) has recently suggested, the death of the Australian

Pleistocene fauna was due largely to fire and overkill, then the hunters must have been using wooden spears. Therefore, if this interpretation is valid, it must be inferred that even stone projectile points were not a necessary part of the equipment for exploiting the environment.

If the technology excavated is several removes from reflecting most temporal and spatial variations between individual regional cultures, the archaeologist's typological groups approximate to the ethnologist's culture areas rather than to the ethnographer's tribes or bands. As ethnographic analogy can do little to break down these groups into individual cultural units similar to the tribe or band, the archaeologist is left with the major regularities of behaviour displayed in the deposits as a central fact of his data. The causes of these regularities, and their local variations and changes, must therefore be a central focus of interest. Though the technology alone is too generalised to provide deep insight into such causes, its regional pattern of distribution can help, for it indicates the pattern of living—the human ecology of the past.

Since the hunter-gatherer is parasitic on his environment, and since the environment is the stage of existence independent of cultural or linguistic idiosyncracies, it is clearly a key and relatively constant factor, influencing his pattern of living. This is not to say that the environment is the sole determinant, but that because the hunter-gatherer's culture is adapted to the environment at many points, the environment tends to hold it fast in its basic regularities. Archaeologists have both the evidence and the means to study the way that the environment holds culture fast.

## Ecological Ethnoarchaeology

There are several reasons why archaeologists have rarely formulated ethnographic analogies to give precedence to ecological factors. One is the slow progress in the reconstruction of past climates and environments. Another, until recently at least, was the anthropologist's lack of interest in ecology. The underlying cause may be an over-reaction by both the archaeologist and the anthropologist to the classical environmental deterministic position, which has led to suspicion of any correlations between behaviour and environment. The most that anthropologists seem prepared to say is 'environment limits but does not determine culture'. The main argument used against tighter correlations is that, though the hypothesis that like environmental stimuli produce like cultural responses is true in the broadest of senses, it is false in detail. (See Freeman 1968: 263.) For the archaeologist of hunter-gatherer societies at least, this criticism is largely irrelevant, as his concern is with major regularities and variations of human behaviour rather than the details of the social organisation of a particular past culture.

Though the purely natural environment and the purely cultural environment are only philosophical concepts, at the end of a continuum which has human expression in utilitarian behaviour as opposed to non-utilitarian behaviour, it can be said that hunter-gatherer man lives in the least modified of all human environments. Therefore his behaviour is more likely to be causally related to it than in those societies more technologically advanced. The vast majority of tasks essential for living can be done in many ways but since the archaeologist writes at a generalised level about behaviour of hunter-gatherers, the underlying determinants are the most important. At a theoretical level, the value of an ecological approach to the study of man is that it does not tie the investigator to any particular theoretical framework, though it does ensure that recognition is given to the physical surroundings his subject lives in and to his demography and subsistence pattern.

Though anthropologists have been slow to adopt the ecological approach, first formulated for them by Steward, it has informed a number of recent studies of hunter-gatherers which are relevant to the archaeologist, in particular Lee's work with the Bushmen (1965) and Watanabe's work on the northern people of the world (e.g. 1968). The main foci of these studies have been the way hunter-gatherers move about the landscape, their subsistence economy, and the nature and determinants of their work and dietary patterns. This has led to a much more sophisticated understanding of man's relationship with his surroundings and in particular it has pointed to the necessity of making fine distinctions when speaking of environment.

Too often archaeologists and anthropologists tend to treat the ecology of a region as an undifferentiated mélange of its various plant components. The result is to create a picture of past economies as highly localised and adjusted to extremely narrow ecological niches. This is the reverse of the truth. Because hunter-gatherers are parasitic on the environment they have to be as flexible as possible and be able to exploit all resources about them.

Just as the professional ecologist builds up a complete picture of the ecology of a region by first considering its basic units, so too statements about a food-gathering culture's subsistence economy and mode of living have to be built up from an examination of its relationship with each of the same basic units. These basic units are the plant communities, the visible floral divisions of the landscape. By distinguishing the plant and animal resources associated with each plant community and their seasonal availability it becomes clear that no hunter-gatherer group could be dependent on the resources of a single community. If they experienced a bad year, disastrous fire, or some other catastrophe, it would decimate the population.

Seasonal change is responsible for variation in the food supply in most regions of the world and therefore for fundamental regularities of behaviour in parasitic economies. In some regions such as those within the monsoonal belt its effects are dramatic; but throughout the world yearly variations in temperature, day length, and precipitation control growth, reproductive seasons, migration, and hibernation. The territories of hunter-gatherers must therefore include a full range of the local plant communities in order that they may have access to all resources as they become available. Jones (pers. comm.) has suggested that human economic units tend to orientate themselves at right angles to ecological boundaries. A study of the ethnogeography of the Pomo Indians of California makes the point:

> One of the most striking features of the divisions of the Pomo linguistic area is the manner in which they extend east and west across the different environmental provinces . . . thus controlling sections of all types of Pomo territory . . . The Southwestern dialect, for example, possesses only land within the coastal province, which is actually divided into the ocean beach, the wind-swept grassy coastal shelf, the redwood and tan-oak forested hills . . . All Tribes control means of obtaining fish, game, grass seeds, acorns, fruits and berries. (Stewart 1943:55)

Some American archaeologists such as Taylor (1966), Coe and Flannery (1966), Laughlin and Reeder (1961), and Davis (1963), and Jones in Australia, have already taken note of the archaeological implications.

The most obvious influence concerns site variation. A culture cannot be typed from a single site, or more importantly from a number of sites, if they are within the same plant community. Variation between adjacent sites, or between sites that fall within an area which population densities combined with band territory size indicate as falling possibly within a single subsistence range in the past, must be treated with care. This is especially true in coastal regions where there often seems to be a tendency to create sharp dichotomies. Taylor (1966:119) has recently attacked the classic inland/coastal Eskimo culture dichotomy on just these grounds and writes, 'The Eskimo hunter was rarely a neat specialist hemmed into a murderously narrow ecological niche; rather he lived by a versatile exploiting of all available niches'. Closer to home we have Lampert (1966) overstressing the maritime aspects of south coast New South Wales economies on the basis of a single site. An assessment as to whether the variation is 'cultural' or merely a seasonal change of residence and diet is important in Australia, where the knowledge that there were over 500 linguistically differentiated groups encourages archaeologists to treat site variation as automatically reflecting different groups of people.

If projects are framed in terms of areas rather than sites, this problem will be largely avoided. Recognition should be given to the fact that Aborigines are nomadic and that the vast majority of them must have lived all their lives in the open, since much of the continent is devoid of caves. While stratified cave and midden deposits will remain the key to archaeological excavation, each such excavation should include the systematic searching of a wide area for open sites. These surveys must cover areas greater than recorded band size territory for the region. The sites should be plotted, their location with respect to topographical features and local resources noted, their artifacts systematically collected and the pestles and mortars counted. This would greatly refine understanding of human adaptation to the environment as reflected in residence and diet variation, the effect of the major climatic changes on these patterns, the coincidence of cultural and natural areas, correlations between sites and activities as reflected in tool use, the regional dispersal of materials from quarries, and trade routes and natural lines of movement.

A typological distinction between the generalised tools of the wood/boneworking kit and more specialised types would be a first step towards clarifying the distinction between culture as a typological and as a social grouping.

Though the inability to pursue the historical reconstruction of the past life of hunter-gatherer groups in detail appears at first as a weakness, it is a source of strength. The archaeologist who is concerned only with simple-minded reconstruction and the use of ethnographic data merely to draw parallels with the present, is casting himself in the role of an extremely handicapped ethnographer, with little to say about the cultural development of man. If he accepts that time largely destroys the idiosyncracies of regional sub-cultures, by removing most of the superorganic aspects of cultural variation which are the preoccupation of the anthropologist, he is left with a body of data that tells the history of man's relationship with his environment, and this is unique to archaeology. Carefully constructed ethnographic analogies will be an important tool, though not the only one, in the elucidation of these data. They can provide a modern geographical template of man's adaptation to his surroundings, be used to create hypotheses to test by excavation and be drawn upon for the reconstruction of some aspects of the past. It is not that our knowledge of the past is only as good as our knowledge of the present, but the better our knowledge of the present the better our knowledge of the past.

**References**


Allen, H. R. 1968. Western plain and eastern hill: a reconstruction of the subsistence activities of the Aboriginal inhabitants of central eastern Australia. B.A.(Hons) thesis, University of Sydney (unpublished).

Ascher, R. 1961. Analogy in archaeological interpretation. *SWest.J.Anthrop.* **17**: 317-25.

Bickford, R. Anne. 1966. Traditional economy of the Aborigines of the Murray Valley. B.A.(Hons) thesis, University of Sydney (unpublished).

Binford, L. R. 1962. Archaeology as anthropology. *Am.Antiq.* **28**: 217-25.

——1968. Methodological considerations of the archaeological use of ethnographic data. In *Man the hunter* (eds. R. B. Lee and I. DeVore), pp. 268-73. Chicago: Aldine Publishing Co.

——and Binford, Sally R. 1966. A preliminary analysis of functional variability in the Mousterian of Levallois Facies. *Am. Anthrop.* **68** (2) (Special publ.): 238-95.

Chang, K. C. 1967. Major aspects of the interrelationship of archaeology and ethnology. *Curr.Anthrop.* **8**: 227-43.

Coe, M. D. and Flannery, K. V. 1966. Microenvironments and Mesoamerican prehistory. In *New roads to yesterday: essays in archaeology* (ed. J. R. Caldwell), pp. 348-59. New York: Basic Books.

Copi, I. M. 1961. *Introduction to logic.* 2nd edn. New York: Macmillan.

Davidson, D. S. 1934. Australian speartraits and their derivations. *J.Polynes.Soc.* **43**: 41-72.

Davis, Emma Lou. 1963. The desert culture of the Western Great Basin: a lifeway of seasonal transhumance. *Am.Antiq.* **29**: 202-12.

Deetz, J. 1968. Hunters in archeological perspective. In *Man the hunter* (eds. R. B. Lee and I. DeVore), pp. 281-5. Chicago: Aldine Publishing Co.

Freeman, L. G. 1968. A theoretical framework for interpreting archeological materials. In *Man the hunter* (eds. R. B. Lee and I. DeVore), pp. 262-7. Chicago: Aldine Publishing Co.

Gould, R. A. 1968a. Living archeology: the Ngatatjara of Western Australia. *SWest. J.Anthrop.* **24**:101-22.

——1968b. Chipping stones in the outback. *Nat.Hist., N.Y.* **77** (2): 42-8.

Hiatt, Betty, 1967-8. The food quest and the economy of the Tasmanian Aborigines. *Oceania* **38**: 99-133, 190-219.

Jones, R. 1966. A speculative archaeological sequence for north-west Tasmania. *Rec. Queen Vict.Mus.* No. 25.

——1968. The geographical background to the arrival of man in Australia and Tasmania. *Archaeol. & phys.Anthrop.Oceania* **3**: 186-215.

Lampert, R. J. 1966. An excavation at Durras North, New South Wales. *Archaeol. & phys.Anthrop.Oceania* **1**: 83-118.

Laughlin, W. S. and Reeder, W. G. 1961. Rationale for the collaborative investigation of Aleut-Konyag prehistory and ecology. *Arct.Anthrop.* **1**:104-8.

Lawrence, R. 1968. *Aboriginal habitat and economy.* Canberra: Australian National University, School of General Studies, Department of Geography Occasional Papers No. 6.

Lee, R. B. 1965. Subsistence ecology of !Kung Bushmen. Ph.D. thesis, University of California, Berkeley (unpublished).

McCarthy, F. D. 1939. 'Trade' in Aboriginal Australia, and 'trade' relationships with Torres Strait, New Guinea and Malaya. *Oceania* **9**: 405-38; **10**: 80-104, 171-95.

Meggitt, M. J. 1962. *Desert people. A study of the Walbiri Aborigines of central Australia.* Sydney: Angus and Robertson.

Mulvaney, D. J. and Joyce, E. B. 1965. Archaeological and geomorphological investigations on Mt. Moffatt Station, Queensland, Australia. *Proc.prehist.Soc.* **31**: 147-212.

Peterson, N. 1968. The pestle and mortar: an ethnographic analogy for archaeology in Arnhem Land. *Mankind* **6**: 567-70.

Service, E. R. 1964. Archaeological theory and ethnological fact. In *Process and pattern in culture* (ed. R. A. Manners), pp. 364-75. Chicago: Aldine Publishing Co.

Stewart, O. C. 1943. Notes on Pomo ethnogeography. *Univ.Calif.Publs Am.Archaeol. Ethnol.* **40**: 29-62.

Taylor, W. E. 1966. An archaeological perspective on Eskimo economy. *Antiquity* **40**: 114-20.

Warner, W. L. 1958. *A black civilization.* (Rev. ed.) Harper: New York.

Watanabe, H. 1968. Subsistence and ecology of northern food gatherers with special reference to the Ainu. In *Man the hunter* (eds. R. B. Lee and I. DeVore), pp. 69-77. Chicago: Aldine Publishing Co.

White, Carmel. 1967. Early stone axes in Arnhem Land. *Antiquity* **41**: 149-52.

——and Peterson, N. 1969. Ethnographic interpretations of the prehistory of western Arnhem Land. *SWest.J.Anthrop.* **25**: 45-67.

White, J. P. 1967. Taim bilong bipo: investigations towards a prehistory of the Papua-New Guinea Highlands. 2 vols. Ph.D. thesis, Australian National University, Canberra (unpublished).

Willey, G. R. 1953. Archeological theories and interpretations: new world. In *Anthropology today* (ed. A. L. Kroeber), pp. 361-85. Chicago: Chicago University Press.

# 17 Habitat and Economy: A Historical Perspective

Roger Lawrence

The aim of this paper is to assess the role of environmental adaptation in differentiating Aboriginal economies in mainland Australia. The subject is treated generally here, but a more detailed assessment has been given elsewhere (Lawrence 1968).

It might be expected that the habitat would impinge on many aspects of the life of Aboriginal hunter-gatherers. Even the way in which an Aboriginal perceives *his* way of life and *his* tribal area in relation to that of the groups around him will be, to a very large extent, a reflection of the effect of the habitat on his economic and social life. At a very basic level the habitat, through determining the range of foodstuffs available, will affect the diet, and by extension, the type and design of equipment used in getting a living. In respect of territoriality the habitat will influence the size of the local group at any one time, its mobility, the population density, the size of the local area, and possibly the place of warfare in the lives of the people. In Aboriginal religious life totemism, increase ceremonies, myths, legends, and song cycles also indicate the pervasive nature of habitat in the cycle of life.

Other papers in this series stress the changes in environmental conditions which must have taken place during the Aborigines' occupation of Australia; and, indeed, the role of the Aborigines in effecting some changes in the character of parts of the environment. These changes may have called for significant modifications and adaptations in their way of life and in the areal spread of their activities. However, this paper is not concerned with these changes through time, but rather with the range of economic activities present in Australia at, or as close as possible to, the period of European contact.

This raises several problems of approach. Obviously, 'time of contact' varied from place to place. On parts of the eastern coast it occurred in 1770, and more effectively after 1788, whereas in central Australia initial contact did not occur until well into the 1850s, with effective contact at a very much later date. Because of the varying chronology of European contact in the area, the quality of source material varies considerably. For example, in reconstructing the activities of the Aborigines around Sydney, unsophisticated accounts of the First Fleeters and settlers of the period are the sole evidence. In the more remote regions, late occupation ensured that trained anthropologists were able to study Aboriginal groups in the field. To what extent these more remote environments changed between the arrival of Europeans in Australia generally and their arrival in the area concerned may never be known. It is highly probable that some non-indigenous plants and animals preceded settlers into the more remote areas, but the changes wrought by these would have been small in comparison with those which followed permanent European settlement in the area. The rapidity with which these later changes occurred makes it imperative to use material recorded at the beginning of the contact period in any region.

For the most part it is a matter of trying to reconstruct a picture from diverse pieces. Very few workers have looked at the problem in the field. The project by McCarthy and McArthur (1960) remains perhaps the only attempt to approach the problem in a scientific, fieldwork-oriented manner. Gould's project (1967) dates from a later period, and while having a different emphasis, produced some interesting results.

Even where Aborigines are still living in any numbers in or near their tribal territories, they differ from their forebears in that they have lived for the whole or part of their lives on mission stations, government reserves, or on cattle stations. Where hunting and gathering is still practised, it is generally only for short periods and involves the use of a somewhat modified technology. Its efficiency may have improved, but this is by no means certain. At the same time the environment has changed considerably. With the concentration of people on mission stations and reserves population pressure on the land has decreased and consequently some animal species have multiplied, while new species have been introduced. Rabbits, feral cats, dogs, buffaloes, camels, and cattle all provide new sources of food in themselves, and upset the existing ecological balance by competing with the existing animals for food. The effect of these animals, particularly rabbits, on the native vegetation is also considerable.

**Problems of Sources**

The sources available for ethnohistorical reconstruction fall into three main groups: the large body of written material, the information held by people with varying degrees of contact with Aborigines in their traditional life, and the increasing body of information derived from archaeological excavations.

Unfortunately, over much of Australia the quantity of information obtainable from informants is limited, and, even where available, its quantity and usefulness is often restricted. Information provided by archaeological excavations is limited both by the interpretative problems the excavated material poses, and also by the differential survival of food remains.

The literature provides by far the richest source of information. At a basic level this consists of writings which report direct observations or long-term impressions, and those which Betty Hiatt (1967-8:110) aptly calls 'gossip accounts'. The latter have no direct basis in observation, and their accuracy varies greatly.

The remaining literature falls into four main categories:

**1** The chiefly incidental observations of the explorers and travellers.

**2** The memoirs, reports, and reminiscences of settlers, squatters, missionaries, government officials, escaped convicts, and shipwrecked sailors who had contact with the Aborigines over longer periods of time.

**3** Edited and annotated collections of information which attempt to synthesise and generalise about the Aborigines. At their best, these are distinguished from 'gossip' works in that they are based on observations attributable to particular persons, and the collators often had knowledge of, and interest in, the Aborigines.

**4** The products of individual research, including most of the anthropological and ethnographic texts.

Each of these categories poses problems in respect of the interests and capabilities of the writers, the nature of the contact they had with the Aborigines and hence the usefulness of their observations, and the effect of current social attitudes towards the Aborigines on their writings.

#### EXPLORERS' AND TRAVELLERS' WRITINGS

Many of these journals arose from government sponsored expeditions and the personnel accompanying them frequently had some scientific training, or an interest in natural history. In addition to this, the expeditions came as the vanguard to European entry into the areas concerned. A large proportion of the material included in these journals is descriptive and related to a particular area. The writers were not concerned with making generalisations about the Aborigines over large areas. The very nature of a journal as a day-to-day account means that an observation can be dated and put into a broad seasonal context which gives it great value. Yet at the same time, the period of observation was frequently very short and it is difficult to place such comments in the context of the total life of the Aborigines. It is possible to say that at such and such a time the Aborigines ate a particular food, but this gives no indication of the relative importance of particular food types in the total diet of the people.

Even where it might be possible to build up a collection of observations about one group from several sources, it is not possible to equate this composite picture with the total diet of the people concerned. There are several reasons for this. The observers probably knew and recognised the larger game, such as kangaroos and wallabies, more readily than the smaller, less common or less obvious animals, and would record them more frequently. The larger animals were also brought back to the camp for division and distribution. A male traveller would be more likely to see men's, rather than women's, hunting activities, and thus one would expect vegetable foods and small game to be recorded less frequently than the larger game. Finally, such observers rarely travelled with the Aborigines during their food quest and so did not see those foods collected and eaten out of camp. McCarthy and McArthur (1960) found that these formed a reasonably important part of the food intake during the day, particularly in the women's diet, but would not be recorded in observations made at the camp. Thus, from this type of information it is very difficult to come to any meaningful conclusions about the importance of different food types, except that such foods were eaten.

#### MEMOIRS, REPORTS AND REMINISCENCES

Because of the greater contact such writers had with the Aborigines, and the correspondingly greater opportunities they had to observe their everyday life, their writings might be expected to provide valuable data, but frequently this is not so. From the middle nineteenth century, when most of these works were published, attitudes expressed towards the Aborigines reflect prejudice and contempt, while particular topics of interest to the nineteenth-century mind figure prominently. In particular, their 'curious' customs were featured, and whether they believed in a supreme being, their intelligence, and the possibility of civilising and Christianising them. While there are some very useful accounts belonging to this group, particularly those of Eyre (1845) and Beveridge (1883), compared with the volume of material published they are disappointingly rare.

THE COLLATED MATERIAL

These volumes, of which the works of Brough Smyth (Smyth 1878) and Curr (1886-7) are prime examples, appeared from the last quarter of the nineteenth century. Their publication was motivated by the belief that the Aborigines were dying out, and that there was a need to collect information quickly. The questionnaire technique was commonly used in data collection, and this restricted the nature of the information collected to that in which the questioning editor was interested. Even at the later stage of compiling the returns, frequently only the observations deemed interesting, useful, or unusual were published.

PRODUCTS OF INDIVIDUAL RESEARCH

While this material shows a focusing of interest and a narrowing of the field of inquiry, it also shows an increasing interest in social organisation. Later nineteenth- and early twentieth-century literature shows a marked decrease in the amount of description of material culture and economic activities. This specialisation tends to limit the value of such work to this study.

**The Regions Studied**

The source limitations partly dictated the areas studied—central Australia, coastal and inland riverine southeastern Australia—but these represent a considerable range of habitat.

Aridity is a dominating and unifying condition in central Australia, whether in the rocky ranges or in the contrasting sandhill and sandplain country. These differences are reflected in their floral and faunal assemblages. As surface water is scarce, animal populations and human activities concentrate around the available water source. The growth patterns of permanent and ephemeral plant species reflect effective falls of rain.

The climate in the coastal areas of Cape York is strongly seasonal, with a marked wet season from December to March. The environment on either side of the Cape is markedly different. The east coast is coral-fringed, allowing exploitation over the wider area of the reef and offshore islands. The hinterland is backed by relatively high mountain ranges, often covered with dense rainforest. There is no coral on the western coast, but again the offshore profile is a very gentle one, while the hinterland is also extensive, with large areas subject to flooding during the wet season.

Seasonality is less marked in the coastal areas further south. The coastline of southeastern Australia is a high wave energy coastline without fringing reefs and with a rapidly falling offshore profile. The coastline itself shows considerable diversity, with lakes, inlets, and larger, more regular embayments, with a coastal plain generally narrow and backed by mountain ranges.

The riverine area is one grading from semi-aridity to aridity, but with permanent rivers as a basic environmental feature. These have their sources outside the area and their fluctuations are a reflection of conditions in the source areas. The Murray River reaches peak volume in December-January as a result of snowmelt in the Eastern Highlands. The Darling also reaches its peak in December, but this results from tropical rain in Queensland. Under natural conditions widespread flooding was common at such time.

## The Aborigine and His Environment

The Aborigines were aware of even small differences in their environment and recognised the effect these had on their economic life. Several instances are recorded where the Aborigines used distinctive names associating particular groups with a particular environment or food source. These names are not to be confused with tribal names: they referred to sections, local groups, or in some cases to constellations of tribes. The practice appears to have been widespread, but varied in character. R. M. and C. H. Berndt (1942:325) record that the Ooldea tribes classified people into groups of *'jabu* people and *'bila* people. The first name was applied to the people living in the rocky country of the Musgrave, Everard, and Petermann Ranges, while the *'bila* people came from the sand and spinifex areas. These were blanket terms and applied to several tribes. Along the coast of Princess Charlotte Bay in Queensland, Thomson (1934b:237) records that a group of tribes were known as *Ka:wadji*, or the 'people of the east'. They called themselves the *Pämä Mälnkäna*, or the 'people of the sandbeach'. The inland people in the area were known to these tribes as *Kanidji*, meaning 'bushmen' or 'inlanders', and were to some extent feared. To the south of this area Hale and Tindale (1933-4:69) note that within one tribe a clan was given the distinguishing name *alei*, meaning 'salt-water', showing that particular clan's association with the coastal environment. At Sydney, Collins (1798-1802:558) records that the Hawkesbury River tribe was known as the *Cah-bro-gal* because the Cah-bro worm formed an important part of its diet.

These names imply that the Aborigines recognised the effect of environmental differences on their lives, and that this was of sufficient importance to warrant the use of distinguishing names.

### THE ENVIRONMENT AND ECONOMIC LIFE

The environment, through its effect on food supply, had a very obvious effect on human diet. Because of this it also affected the range of hunting techniques employed, and the equipment used, both in hunting and collecting and also in preparation. Variations in food supply were reflected in the movements of people around their tribal territories.

The impression gained by observers in central Australia was that the Aborigines ate most of the species of birds, animals, and reptiles occurring in their tribal territory, and utilised a significant number of the plant species. In better seasons when food was more plentiful preferences were shown for particular foods.

The division between spinifex and sand country and the range country was reflected in the food supply of the people and the hunting methods they employed. The flesh diet of the spinifex people consisted mainly of small mammals and birds hunted with spears and throwing sticks, aided by the use of fire; lizards and rodents were dug out of the dunes with digging sticks. There is little evidence to suggest that large-scale drives were organised by the spinifex tribes, such as the Bindubi.[1]

[1] Spelling of all Aboriginal tribal names in this paper follows that of the Australian Institute of Aboriginal Studies.

This diet contrasts with that of the inhabitants of the ranges and adjacent areas, where in better seasons large groups came together to carry out organised drives for large game, including emus, kangaroos, wallabies, and euros. In hunting, they used spears and throwing sticks, in association with firing of the vegetation, brush fences, pitfalls, and beaters. Hides, decoys, and narcotics were also used to aid the hunter. In poorer seasons game was hunted in smaller parties. In this type of country the range of hunting techniques appears to have been greater than that employed in the sandhill country, and this may be a reflection of the differences in the floras and faunas of the areas, and also the ease with which larger gatherings could be supported. The methods used also show adaptations to capitalise on the peculiar habits of the game being hunted (details in Lawrence 1968: Table I).

The contrast between range and spinifex country can also be seen in the vegetable foods eaten, though perhaps not to the same extent. At least 178 plant species were utilised for food in central Australia. Meggitt (1964:33) claims that vegetable foods composed between 70 and 80 per cent of the total food supplies, and other observations support this for some seasons at least. Seeds of several acacias and grasses, together with various rootstocks, appear to have formed staples, while fruits, leaves, and stems provided a more shortlived and less certain source of food. Large groupings of people came together when plant foods were plentiful and surface water was still prevalent.

Riverine groups are rather difficult to define. Perhaps the best basis of definition recognises them as those who spent at least some seasons of the year in continuous settlement on rivers, and whose economy at that time showed some specialisation in the collection of aquatic foods. It is evident from many of the early reports that the population visible from the rivers increased downstream, and that the country downstream from the junction of the Murray and Darling Rivers was much more heavily populated.

The exploitation of the riverine environment appears to have been markedly seasonal, both in the fishing methods used and in the numbers able to be supported by these activities. Fish and crayfish were outstandingly important in the diet of large groups of people in the area at certain seasons of the year, while at other seasons numbers living on the river thinned noticeably. The faunal remains from the excavations of Mulvaney (1960) and Hale and Tindale (1930) show, however, that even when the Aborigines camped on the rivers they hunted land animals and brought them back to the shelters.

With the annual flooding of the river the swamps filled and the river flats became inundated. Large groups of people lived by netting fish and crayfish, spearing fish, and trapping waterfowl. The rootstocks of aquatic plants were also important sources of food. Large populations remained at the rivers while the water level dropped and the billabongs flowed back into the mainstream. At such times, fish were trapped in weirs built across the channels and yields were high. Incidentally, this method was also used in the reverse situation, to trap the fish leaving the mainstream to enter the billabongs on the rising floodwaters. Once the river level had subsided different fishing methods were employed. Fish were speared, often underwater rather than from canoes, women scoured the shallow pools with movable fish weirs, and poisons were used in the smaller water holes. These methods alone, however, could not support such large population concentrations and in addition land animals were sought, including kangaroos, wallabies, emus, large birds, dingoes, platypus, wombats, koalas, echidnas, and possums. These techniques therefore show adaptation to the prevailing seasonal conditions, the size of the game being hunted, its habits, and the number of people able to be employed in a hunting or fishing expedition.

The marine environment of coastal areas is a potential source of food in addition to that supplied by the land. The marine bias of the Aborigines around Sydney was immediately obvious to the First Fleeters. Their activities consisted mainly of collecting shellfish, spearing fish, and line-fishing from canoes. The midden contents support this to some extent. Lampert's excavations at Durras North (Lampert 1966) and Megaw's at Gymea Bay (J. V. S. Megaw pers. comm.) show that while the Aborigines were living at these camps their food was predominantly marine. The excavation at Curracurrang (Megaw pers. comm.) shows far more variety in the faunal remains, with land mammal, including dingo, remains being present in the midden, as well as fish, seal, bird, and whale. These variations probably reflect the basic differences in the character of the coastal environment of these areas, the differences in the faunal assemblages present, or perhaps their occupational function.

With longer acquaintance, the First Fleet journal keepers noticed that there was considerable seasonal variation in the pattern of life in the Sydney area. Large groups of people came together in the spring and early summer, coinciding with the arrival of the school fish. In winter, when fish were scarce and extreme hunger not unusual, the groups dispersed, either going inland to hunt land animals or breaking up into smaller groups and spreading themselves more thinly along the coast. There is little direct evidence adding weight to one or other of these conclusions.

Fish were speared and caught with hooks and fish weirs and poisons were also used. There is no clear reference to the use of nets in fishing, although the people possessed nets and used them as carrying receptacles. The method of hunting land animals appears to have differed little from that used in the areas discussed previously, except that nets were not employed. Little prominence is given in the sources to vegetable foods, but whether this indicates that they were of little importance in the diet, or were seen and not recorded is not clear.

The dependence of the Aborigines of eastern Cape York on marine sources of food was even more marked. At the northern end of Princess Charlotte Bay there was no seasonal movement away from the coast and the local groups appear to have been remarkably stable. At Port Stewart the only moves made by the camp were those in the interests of hygiene and a movement on to the exposed dunes during the wet season. In other areas there may have been seasonal movements to the offshore islands. The economy showed the same concentration on fishing and shellfish collection, with very little attention given to the hunting of land animals and the collection of vegetable foods. This bias was so marked that in some areas vegetable foods were considered a luxury (Hale and Tindale 1933-4:69).

The most distinctive feature of the marine economies of some groups in this area was the importance of turtle and dugong hunting. During the whole time Thomson spent with the Yintingga at Stewart River, the camp was not without dugong meat for any length of time. The periods involved were from May to July and from November to December (Thomson 1934b:239). The equipment used in hunting turtle and dugong included outrigger canoes and harpoons, both of which have obvious cultural affinities with areas to the north. The methods used in hunting turtle varied with the species being taken and with the seasonal changes in the behaviour of the turtles. On land, the hunting activities do not appear to have differed greatly from those described elsewhere.

Of the tribes on the western coast of Cape York, the Munggan figure pre-eminently in the literature. Despite this prominence it is not clear whether they can be properly labelled as a coastal group. They might be designated more appropriately as an estuarine group. From McConnel's observations (1930), it seems reasonably clear that they lived on the estuary of the Archer River and not on the seashore, as is suggested in Thomson's work (1939). In the latter's description of the Munggan he includes the people living at Cape Keer-Weer, but according to McConnel (1930:191), the coastal strip was settled by the Ngadanja, not the Munggan, whose tribal area did not extend fully to the coast (McConnel 1930:97).

Regardless of this confusion, the Munggan provide an interesting variation from the pattern evident in the other coastal areas, as they did not specialise in the exploitation of marine resources throughout the year. During the wet season they lived on the estuary and ate fish, shellfish, crabs, turtle, dugong, tortoises, crocodile, and waterfowl. At this season vegetable foods were not very important in the diet. Mobility was restricted by flooding and the dense, seasonal plant growth. Nomadic life began at the end of the wet season when movement was possible and hunting and burning could begin. Vegetable foods, freshwater fish, wallabies, and honey became important sources of food. However, the vegetable harvest decreased with the progression of the dry season and mobility was restricted because of the scarcity of water. Possums, bandicoots, goannas, and native cats became more important in the diet. Despite the great differences in the patterns of activities in the area, the actual hunting equipment and methods used were comparable.

The question of the Aborigine's relation to his environment is a very complex one. In the broadest sense the environment did not produce any momentous changes in the character or structure of Aboriginal economic life. All economies remained at a hunting and gathering level, and the structure remained broadly the same. There was little evidence of craft specialisation and, with minor variations, the division of labour based on sex was universal. But these observations are rather general.

### ENVIRONMENT AND EVERYDAY LIFE

By studying the seasonal activities of one small group it could be demonstrated that their diet and their hunting and gathering activities altered with the seasonal cycles of the plants and animals in the region, and that their activities can be understood only against this background.

Yet this is an unsatisfactory conclusion because inherent in it is the assumption that environmental stimulus *produced* the economic machinery with which the environment was exploited. The evidence does not support such a conclusion. The habitat provides a wide range of edible species, and which of these are used depends largely on the preferences and technological abilities of the people who exploit it.

At this point it is impossible to determine what effect preference had on the character of Aboriginal diet. The harsh environment of central Australia may have put severe limitations on the diet of tribes in the area, but even here, in better seasons, preferences for particular foods were shown. In addition, there is little information on the efficiency with which the environment was exploited.

This factor is very important to the analysis of the activities in areas where alternative sources of food were available. Not all tribes in coastal areas had what could be called a 'coastal' economy. This was the case at the southern end of Princess Charlotte Bay. The Lama-lama inhabited the mangrove lined estuaries of the area and had an economy more like that of an inland tribe (Hale and Tindale 1933-4:69). This may also have applied to some extent to the tribes around Port Phillip Bay, where the lack of emphasis on the exploitation of marine resources may have been a reflection of the character of the marine environment in the local area.

Even over small distances variations occurred in the degree to which certain species were exploited, and it is doubtful whether these can be explained solely on environmental grounds. Dugong is widely distributed in the northern waters of Australia, but apparently it was a major food source for only a small group of tribes. This may have been because of the possession of a more advanced technology, in the form of harpoons and outrigger canoes, which enabled the Ka:wadji tribes to capitalise on this resource. This appears an attractive and useful conclusion, but care is necessary, as the information on these tribes comes from one source only, Thomson (1934b). Hale and Tindale (1933-4), who refer briefly to these people in their paper, give no detailed account of their dugong hunting activities. There were other tribes in Cape York with an identical technology yet, as far as it is known, they did not emphasise the hunting of the dugong to the same extent.

### THE ENVIRONMENT AND MATERIAL CULTURE

Looking more specifically at material culture and hunting techniques, there appears to have been a basic range of equipment common to all tribes. This included spears, spearthrowers, throwing sticks, digging sticks, brush fence traps, and several other implements, including fist grasped stone core tools or hafted axes.

The Bindubi had simple spears, spearthrowers, throwing sticks, boomerangs which were regarded almost as ceremonial objects, stone axes, digging sticks, shallow wooden dishes, grindstones, and bark sandals. The people of the central Australian ranges possessed a slightly greater range of equipment including spears, spearthrowers of several designs, throwing sticks, boomerangs, clubs, shields, adzes, stone knives, axes, flakes, digging sticks, grindstones, fur twine, grass baskets, skin wallets, and water bags. The hunting devices they used included hides, brush fences, pitfalls, and poisons.

On the lower Murray and Darling Rivers fish spears, nets, canoes, brush fence fish weirs, stone fish traps, hollow log traps, and fish poisons were used in fishing. The equipment used in hunting land animals included spears, spearthrowers, clubs, boomerangs, and nets. Other implements in use included stone axes, flake knives, shell, bone, and reed knives, stone and wooden chisels, bone awls and needles, grindstones, digging sticks, wooden spades, bark and gnarl troughs, skin water bags, grass baskets, and net bags. The hunting devices included decoys, hides, brush fences, pitfalls, nets, and snares.

The tribes at Sydney used canoes, spears, fish hooks, log traps, fish poisons, and possibly nets in fishing. They also possessed spears, spearthrowers, clubs, boomerangs, wooden swords, parrying sticks and shields, digging sticks, hafted and unhafted stone axes, shell scrapers, net bags, folded bark troughs, palm spathe baskets, wooden and bark dishes. The hunting devices employed included pitfalls, brush fences, brush fence traps, and food decoys.

In the Cape York area the tool kit was

enriched by the addition of outrigger canoes, harpoons and harpoon floats, fishing nets, bamboo knives, pleated bark containers and baler shell cooking vessels. Additional hunting methods included the use of purse nets, snares, flails, switches, and bird lime.

These inventories reveal surprisingly little significant variation in the range of weapons and techniques employed by the Aborigines. It seems that a basic range of equipment was used in hunting and that the greatest proliferation of design occurred in fighting weapons, in ceremonial objects, and perhaps in the equipment used in transporting and preparing food. The standard hunting equipment consisted of spears, clubs, and throwing sticks, together with the use of hides, decoys, brush fences, pitfalls, and poisons. It is interesting to note that, with some modifications, spears, nets, brush fences, and poisons were used both in hunting and fishing.

The most noteworthy additions to Aboriginal material culture can be attributed to diffusion from outside cultures, and would include outrigger canoes (Davidson 1935; Thomson 1952), fish hooks (Massola 1956) and netting (Davidson 1933). The diffusion which may explain the adoption of this equipment did not result from passive acceptance on the part of the Aborigines. They accepted those innovatory elements which could be accommodated within their way of life, and rejected others (Thomson 1933, 1934a).

In discussing the relationship between local design modification and adaptation, there is a very great danger of attributing particular significance to a design modification when the use of the implement is not adequately documented or understood. Most writers are quite willing to accept the multi-pronged spear as being a 'fish' spear, but it was also used in hunting jabiru, emus, wallabies, and kangaroos in western Cape York (McConnel 1953:25-6), and also in central Australia (Spencer and Gillen 1938: 575-8). The generalised design of the leaf spearthrower in central Australia is explained as being an adaptation to the rigorous, highly mobile life in the harsh conditions of the area (McCarthy 1957:94). It was used as a spearthrower, a parrying stick, an adze handle, a vessel, a musical instrument, and a fire saw. However, this design was also present in eastern Australia where conditions were not so rigorous and the need for a multi-purpose implement was presumably not as great. If the design advantages were so important, it is surprising that it did not displace the stick spearthrower of the Warramanga of northern central Australia, a device which had none of the above advantages. It is difficult to assign functional relevance to many of the design modifications observable in Aboriginal implements. The opinion of the Aborigines themselves might differ, however, and this is a technological problem still worth pursuing.

The relationships between local design variations and the influence of differing raw materials pose similar problems of interpretation. The substitution of shell and bone for stone in coastal areas comes immediately to mind. This probably does not represent a straight substitution of bone and shell for stone. The alternative implements are of a legitimate and recognisable design and were manufactured in some areas where stone was also available. Trade would also tend to lessen the importance of local variations resulting from differences in raw materials.

Thus material culture consists of a basic range of equipment enriched by diffusion, and possibly modified by the effects of local materials on manufacture, and affected by a certain amount of design elaboration. The latter may have reflected environmental adaptation in the first instance, but the modified designs frequently spread outside the area of development. It is difficult, therefore, to see the design of the material culture of any one group simply as being a rational and distinctive response to life in a particular environment.

This conclusion relates only to the *origin* of particular cultural traits. It does not explain the employment of the equipment by Aborigines living in a particular area. The pattern of activities of any group of Aborigines, and their decision to employ particular hunting techniques, cannot be fully comprehended without viewing them against the background of the habitat.

### THE ENVIRONMENT AND POPULATION NUMBERS

Finally, there is the question of the relationship between the habitat and population numbers and densities. The information for this is sparse in the extreme, and consists mainly of very disparate observations of the numbers present at any one place. Other information is contained in the estimate of tribal size by people with longer contact with the Aborigines, or from reconstruction from tribal genealogies.

It is reasonably clear that the usual hunting group in all areas studied, except Stewart River, was the family or extended family. This group was generally mobile through the year and its composition fluctuated with changing food supplies. In seasons when food was plentiful, larger groupings formed, their stability being conditioned by the availability of food and water. These larger congregations ranged from about forty-two among the Bindibu (Thomson 1962:154, 269)—although this was possibly an unusually large grouping attracted by the presence of Thomson himself—to assemblies probably as large as 240 among the Pitjanjadjara (Tindale and Hackett 1933: 102). Groups of about 600 were recorded at Lake Victoria on the Murray River (Sturt 1834:2, 107). Meggitt's reconstruction suggests that these gatherings during good seasons may have allowed the whole community to come together, and this would have represented about one-quarter of the Walbjiri tribe (Meggitt 1962:49, 244). Among the Ka:wadji the hunting groups appear to have been larger and more sedentary. This may have been a reflection of dugong hunting as a co-operative activity.

Estimates of population density are probably rather meaningless because the conditions to which they relate are uncertain. In central Australia Meggitt gives densities of one person to 35 square miles (90 $km^2$) for the Walbjiri and one person to 12·5 square miles (32·4 $km^2$) for the Aranda. This may reflect the differences in quality of the spinifex and range environments. In the riverine area we have two estimates for the Ba:gundji tribe. Bonney (1883:123) gives a density of one person to 20 square miles (52 $km^2$) and Teulon (in Curr 1886-7:**2**, 186) one of one person to 3 square miles (7·8 $km^2$). Eyre's (1845:**2**, 372) estimate of the permanent population at Moorunde on the Murray as being 3 to 4 people per mile of river (2 to 2·5 km) may be more meaningful, and it is well within the range of similar estimates in coastal areas. These latter estimates are very variable, ranging from 12 per square mile (4·6 per $km^2$) in northern Cape York (based on Jardine 1866:83 and Tindale 1940), to 5 per square mile (1·9 per $km^2$) and one to 15·6 square miles (40·5 $km^2$) at Sydney (Phillip 1788 and Arnold 1810 respectively). The number of people per kilometre of coast could be estimated at about 2·5 for Sydney and 21·5 for Cape York (4 and 34·5 respectively per mile).

These figures hardly warrant any conclusion. It can safely be said that densities were higher on the coast and in riverine areas than in the arid areas, but this is not very illuminating. If Meggitt's figure for the Aranda is correct, then Radcliffe-Brown's (1930:696) estimate of 300,000 for the whole continent would demand an overall density of population no greater than that in the more favourable parts of central Australia.

The relationship between the Aborigines and their environment was not a clear cut one. The environment does not appear to have stimulated any radical change in the nature of the Aboriginal economy, nor does it appear to have encouraged any great amount of inventiveness in Aboriginal material culture. But once the functioning of individual economies is investigated the interplay between the habitat and the choice of hunting techniques and equipment is striking. Frequently their habitat was such that all techniques known to the Aborigines were not used uniformly. The decision to employ particular methods and make specific adaptations depended on the Aborigines, and although it was made in reference to the conditions prevailing in the local environment, the element of human choice is a striking phenomenon of 'ethnohistoric' Australia.

## References

Arnold, J. 1810. Letter to his brother, 18 March, from Sydney Cove. Arnold Letters, A1849-2, pp. 1-16. Mitchell Library, Sydney.

Berndt, R. M. and Berndt, Catherine H. 1942. A preliminary report of field work in the Ooldea region, western South Australia. *Oceania* **12**:305-30.

Beveridge, P. 1883. Of the Aborigines inhabiting the Great Lacustrine and Riverine depression of the lower Murray, lower Murrumbidgee, lower Lachlan, and lower Darling. *J.Proc.R.Soc.N.S.W.* **17**: 19-74.

Bonney, F. 1883. On some customs of the Aborigines of the River Darling, New South Wales. *Jl R.anthrop.Inst.* **13**:122-36.

Collins, D. 1798-1802. *An account of the English colony in New South Wales.* 2 vols. London: Cadell and Davies.

Curr, E. M. 1886-7. *The Australian race.* 4 vols. Melbourne: John Ferres.

Davidson, D. S. 1933. Australian netting and basketry techniques. *J.Polynes.Soc.* **42**: 257-99.

——1935. The chronology of Australian watercraft. *J.Polynes.Soc.* **44**:1-16, 69-84, 137-52, 193-207.

Eyre, E. J. 1845. *Journals of expeditions of discovery into central Australia, and overland from Adelaide to King George's Sound, in the years 1840-1.* 2 vols. London: Boone.

Gould, R. A. 1967. Notes on hunting, butchering, and sharing of game among the Ngatatjara and their neighbors in the West Australian Desert. *Pap.Kroeber anthrop. Soc.* **36**:41-66.

Hale, H. M. and Tindale, N. B. 1930. Notes on some human remains in the lower Murray valley, South Australia. *Rec.S.Aust. Mus.* **4**:145-218.

—— and —— 1933-4. Aborigines of Princess Charlotte Bay, north Queensland. *Rec. S.Aust.Mus.* **5** (1):63-116, (2):117-72.

Hiatt, Betty. 1967-8. The food quest and the economy of the Tasmanian Aborigines. *Oceania* **38**:99-133, 190-219.

Jardine, J. 1866. Description of the neighbourhood of Somerset, Cape York, Australia. *Journal of the Royal Geographical Society* **36**:76-85.

Lampert, R. J. 1966. An excavation at Durras North, New South Wales. *Archaeol. & phys.Anthrop.Oceania* **1**:83-118.

Lawrence, R. 1968. *Aboriginal habitat and economy.* Canberra: Australian National University, School of General Studies, Department of Geography Occasional Papers No. 6.

McCarthy, F. D. 1957. Habitat, economy and equipment of the Australian Aborigines. *Aust.J.Sci.* **19**:88-97.

—— and McArthur, Margaret. 1960. The food quest and the time factor in Aboriginal economic life. In *Rec.Am.-Aust.scient.Exped. Arnhem Ld* **2** (Anthropology and Nutrition) (ed. C. P. Mountford), pp. 145-94. Melbourne: Melbourne University Press.

McConnel, Ursula. 1930. The Wik-Munkan tribe of Cape York Peninsula. *Oceania* **1**:97-104, 181-205.

——1953. Native arts and industries on the Archer, Kendall and Holroyd Rivers, Cape York Peninsula, north Queensland. *Rec. S.Aust.Mus.* **11**:1-42.

Massola, A. 1956. Australian fish hooks and their distribution. *Mem.natn.Mus.Vict.* **22**: 1-16.

Meggitt, M. J. 1962. *Desert people. A study of the Walbiri Aborigines of central Australia.* Sydney: Angus and Robertson.

——1964. Aboriginal food-gatherers of tropical Australia. In *The ecology of man in the tropical environment.* International Union for the Conservation of Nature and Natural Resources, Ninth Technical Meeting, Proceedings and Papers. pp. 30-7. Morges: IUCN (Publ.Ser.4).

Mulvaney, D. J. 1960. Archaeological excavations at Fromm's Landing on the lower Murray River, South Australia. *Proc.R.Soc. Vict.* **72**:53-85.

Phillip, A. 1788. Letter to Sir Joseph Banks, 2 July, from Sydney Cove. Phillip Letters, C213, pp. 8-22. Mitchell Library, Sydney.

Radcliffe-Brown, A. R. 1930. Former numbers and distribution of the Australian Aborigines. In *Official year book of the Commonwealth of Australia.* No. 23, pp. 687-96. Melbourne: Government Printer.

Smyth, R. B. 1878. *The Aborigines of Victoria.* 2 vols. Melbourne: Government Printer.

Spencer, B. and Gillen, F. J. 1938. *The native tribes of central Australia.* London: Macmillan.

Sturt, C. 1834. *Two expeditions into the interior of southern Australia. . . .* 2 vols. London: Smith, Elder and Co.

Thomson, D. F. 1933. The hero cult, initiation and totemism on Cape York. *Jl R. anthrop.Inst.* **63**:453-537.

——1934a. Notes on a hero cult from the Gulf of Carpentaria, north Queensland. *Jl R. anthrop.Inst.* **64**:217-35.

——1934b. The dugong hunters of Cape York. *Jl R.anthrop.Inst.* **64**:237-63.

——1939. The seasonal factor in human culture. Illustrated from the life of a contemporary nomadic group. *Proc.prehist. Soc.* **5**:209-21.

——1952. Notes on some primitive watercraft in northern Australia. *Man* **52**:1-5.

——1962. The Bindibu Expedition: exploration among the desert Aborigines of Western Australia. *Geogrl J.* **128**:1-14, 143-57, 262-78.

Tindale, N. B. 1940. Distribution of Australian Aboriginal tribes: a field survey. *Trans.R.Soc.S.Aust.* **64**:140-231.

—— and Hackett, C. J. 1933. Preliminary report on field work among the Aborigines of the north-west of South Australia, May 31st to July 30th, 1933. *Oceania* **4**: 101-5.

# 18 Arid Region Aborigines: The Pintubi

J. P. M. Long

The area near Lake Mackay in western central Australia, along with neighbouring areas to the west and south, was among the last parts of the continent where Aborigines maintained their traditional way of life, largely undisturbed by contact with non-Aborigines. This paper is based on observations made on a number of journeys into the area between 1957 and 1964 before the last of the inhabitants left in 1966 (Evans and Long 1965; Long 1964). Most encounters with Aborigines occurred within a rectangular area 270 km long and 110 km wide, west of the border of Western Australia between Lake Mackay and Lake Macdonald, through which a road was made in 1960 by the Weapons Research Establishment for mapping purposes. Some information was also obtained during a patrol in 1957 to the area around the north of Lake Mackay. All these journeys were made with the object of assessing the numbers, condition, and needs of the people in the area, and no detailed ethnographic work was undertaken.

I also refer to published reports of travels and of anthropological work in the area, notably Carnegie's account of his remarkable journeys north and south across the desert in 1896 and 1897, and Thomson's several articles dealing mainly with the people living in 1957 in the area north of Lake Mackay.

The former inhabitants of the area explored on these journeys have conveniently been referred to as Pintubi (Pintupi) or Bindibu.[1] The evidence suggests that at least three distinct dialects of what has been labelled the Western Desert language were spoken in this area, and that the inhabitants distinguished at least as many regional groups of people within the area.

[1] Bindubi in the orthography of the Australian Institute of Aboriginal Studies.

By 1957 the evacuation of the area was well advanced. People had moved north to the Balgo Hills mission, southwest to the Warburton Range mission and east to Haasts Bluff settlement. The people remaining in the desert were separated from their relatives who had gone to Haasts Bluff by an uninhabited area about 250 km wide, of which they knew little, and they knew nothing of the fate of those who had left. At least one or two of the men living in the extreme west of the area, not far from Carnegie's Family Well (Carnegie 1898:240-5, 249), had visited wells on the Canning Stock Route and one of these men had certainly been with parties which killed cattle on or near the stock route. At least one man in the desert had visited the Warburton Range mission and others had visited the Giles Meteorological Station. Some of those who remained in the desert in 1957 had travelled some of the distance with parties walking towards Balgo Hills and Haasts Bluff. Some living in the Lake Mackay area in 1957 had made journeys to the edge of the occupied pastoral country (Mount Doreen Station). In 1957 groups in the area still had occasional contact with groups farther to the south and west, but depopulation had certainly reduced the range and frequency of their social contacts.

Some fragments of clothing and some metal items had been brought into the area but the evidence does not suggest that all or even most of the people possessed such items. Some metal had been in the area for many years, since Carnegie (1898:244) found metal objects with groups he met in 1896. The making of the road through the area in 1960 was the means of introducing many more such objects. Though the road making party, perhaps not surprisingly, met no Aborigines, the mapping parties in 1961 met several families, and by 1962 most groups were probably using cans for carrying water and food.

## The Environment

### TOPOGRAPHY

The whole country is flat or undulating at levels between 400 and 460 m above sea level, with a few scattered hills rising up to 90 m above the plain. The area generally is lower than the Macdonnell Range country to the east. Sandstone outcrops and low hills are characteristic of the southern part of the area. Sand ridges and dunes cover much of the area and nowhere is one more than about 8 km from the nearest sand ridge. Two forms of sandhill country can be distinguished: the areas through which run long ridges, some of which are 60 or 80 km long, separated by 'valleys' up to several hundred metres wide, and the areas of confused and dense ridges and dunes. The long, east-west ridges predominate in the northern parts of the area and the confused dune country in the south and west. Claypans occur, and there are 'lakes' of mud with a salt crust in large depressions. After exceptionally heavy rain in the late summer of 1967 quite extensive shallow pools were formed in places between and alongside sand ridges. To the west and southwest of the area lies Carnegie's 'Great Undulating Desert of Gravel' (Carnegie 1898:208), the *rira* country of the Aborigines.

### VEGETATION

The so-called spinifex (*Triodia* spp.) is the dominant ground cover on hills, sand plains, and in sand ridge and dune country. The Aborigines distinguish at least two main varieties here: one inedible variety with a spiny awn, and an edible one with a wheat-like head. Mulga (*Acacia aneura*) and grass country occurs in only small patches near hills and in belts between sand ridges. Areas of desert oak (*Casuarina decaisneana*) occur, often in the confused dune country and often in association with limestone outcrops and with the depressions of the salt lakes.

CLIMATE

Rainfall is unreliable and the area probably has an average annual rainfall of less than 25 cm, possibly less than 20 cm. Heavy falls tend to occur most often in the late summer and early winter months, if at all. Evaporation can be assumed to be between 280 and 305 cm a year. From November till March days are extremely hot and nights warm to hot, and in winter the nights are cold and the days generally warm.

FAUNA

Broadly the range of birds and animals found elsewhere in central Australia appears to be present, at least sometimes. In the dry years up to 1964 few kangaroos and no signs of emus were seen, but in 1967, when after two good summer seasons surface water was abundant, there were emu tracks in many areas. Thomson (1964:404) also observed emu and kangaroo tracks near a claypan east of Lake Hazlett after heavy rains in 1957. Tracks of wild camels were not seen in the area until 1967 and it may be only since the road was made that camels have come so far west. Feral cats appear to be widely distributed at least in the southern sand dune country: Carnegie found a woman who had caught one west of Lake Macdonald in 1897, and I met a man with one in the same general area in 1962. Although Thomson and Hosmer (1963:236) suggest that feral cats seek refuge from Aboriginal hunters in rock outcrop areas, they are certainly not restricted to such areas. Some signs of rabbits have been seen and they inhabit places to the east and north, as well as to the south, as do foxes. The absence of reliable surface waters no doubt affects the distribution of some animals, and the southern parts of the area, where waters seem generally more accessible, may be better for some species.

## Demographic and Subsistence Patterns

POPULATION DENSITY

There were about 125 people living in this area of about 52,000 km$^2$ in 1962, or one person to 415 km$^2$. There had already been substantial emigration by then and only informed guesses can be made of the density of population before any permanent emigration occurred. Present sketchy knowledge of the origins of the people who have emigrated suggests that it is unlikely that the population of this area would have been less than 200 and improbable that it could have been much over 300, so that the probable density would have been something between 130 and 260 km$^2$ per person. I am inclined to think that a ratio of one person to 200 km$^2$ would be a reasonable estimate, but one to 260 km$^2$ is possible.

This estimate is in line with Radcliffe-Brown's estimate of one person to 80 square miles (210 km$^2$) for the arid parts of South Australia (Radcliffe-Brown 1930:690). He does not discuss the probable population of the desert areas in any detail, indicating merely that the population of the one million square miles (2·6 million km$^2$) of arid country in Western Australia, South Australia, and central Australia must have been very sparse (1930:687 and map at p. 672).

Early accounts suggest that the population was sparse. Carnegie (1898) met three separate groups of people, numbering probably about thirty in all, in the course of a month's journey over about 320 km through the west of the area in 1896 and four groups (about twenty-five people) on the equivalent section of his return journey through the eastern part the following year. He saw signs of a few other groups but nothing to suggest that the population was anything but extremely sparse in the whole region. His party was constantly and anxiously looking for recent tracks and 'smokes' and he probably noted all such evidence. It is unlikely that any groups within 15 km either side of his line of march would have been unobserved. Tietkens in 1889 met no groups during a month's travel into the eastern part of the area but found indications of relatively intensive occupation south of Lake Macdonald and saw smokes in the area southeast of the lake (Tietkens 1890).

By all accounts the area which a man or a group might range over in the course of a few years, or even months, could be considerable. Few if any of the adults living within these 52,000 km$^2$ would not from time to time have lived outside that area. Any calculations of population densities must therefore be somewhat arbitrary and imprecise. It may yet be possible to gather enough information about water supplies of the area, the normal range of families or groups of families, the extent of individuals' travels over a period and the local origins of the Pintubi now living at settlements, to make rather more confident guesses about the pre-contact population of the area. On the basis of the apparent resources of different areas, I would be inclined to assume that the western gravel downs supported the sparsest population and that the sandstone country to the south, where there is generally more mulga and grass country, would have been more densely populated than other parts of the area.

### SIZE AND COMPOSITION OF GROUPS

Groups have been met with which ranged in size from two people (a man and a boy) to twenty-two, the term 'group' being here used to mean a number of people camping overnight at a single waterhole. The group of twenty-two was found at a claypan west of Lake Mackay in 1957 shortly after heavy rain. It consisted of three men, their wives and children and three single women (widows), and a group of three single males was reported to be camped some kilometres away to the west.

Generally, however, groups have consisted of one or two families, often with one or more adolescent males and/or elderly or widowed females. Such 'family groups' ranged in size from three to twelve people. The period 1957-64 was one of generally average or below average rainfall but, since good seasons are the exception and drought the rule, it is safe to assume that the people of this area lived in scattered groups of this size for most of the time (cf. Gould 1968:103).

It seems that one can think in terms of loose 'associations' of families, which commonly foraged independently but often within a day's march of each other, and some or all of which came together at a single camp from time to time. Operating from separate but relatively close waterholes the individual families would be aware of each other's presence and movements by observing hunting and camp fire smokes as well as by occasional meetings. The families and single males of such associations might have come together not only after rain, but in the hot dry summer months, when it seems that groups may have tended to congregate at the more reliable wells (*inta* or *tjila*), where there might be good shade as well as water. After rain, when both water and food were relatively abundant, member groups of more than one association might foregather for ceremonies. In the winter of 1962, for example, groups were living in the Dovers Hills area because rains had recently filled the rockholes in the hills. They were visited by two single males, emissaries from groups living in the sand ridge country to the south-west, who invited them to foregather for initiation ceremonies. It seems likely that gatherings of more than fifty people would have been extremely uncommon in this area.

On three occasions small groups of young single males were met, or reported, living apart from family groups at least temporarily. On two other occasions married men have been found or reported moving with one or two adolescent males, and on another occasion an elderly widower was met camped with one of his sons and another youth. It seems that where an association included a number of unmarried males these would commonly forage independently as a group as did the individual families of the association. Young men seem to have played an important function in communications between groups (arranging ceremonial gatherings for example), but often a married man would leave his family to travel with single males on such journeys.

Husbands and wives seem to have parted company quite frequently. On one occasion a man's three wives (all sisters) and children were found with the parents of the wives while their husband and adolescent brother were away with another group. On another occasion a man arrived with two youths at a group we had just met which included his older wife (and her mother and brothers), having left his younger wife with her parents some 60 or 80 km away. Several other instances of temporary separation of husbands and wives could be cited.

#### LOCAL TIES

Groups have been found living in areas a long way from the 'clan country' of the male members of the groups. When, in 1957, we first met people around the north of Lake Mackay, we assumed that they were all in their traditional territory. Later inquiries showed that at least four of six adult males were up to 160 or 250 km from their country. They had apparently been living in the area for some years. Indeed Thomson's account of one of these men reciting the names of waterholes in the area 'mapped' on the backs of spearthrowers is perhaps suggestive of relatively long residence in, or at least of an intimate knowledge of, the Lake Hazlett area (Thomson 1962:274). An old man met near Jupiter Well in the extreme west of the area was said to have lived for many years there, some 100 km from his own country, apparently because he preferred the company of the group there.

No evidence was seen of any marked preference for family residence in the husband's country nor that the so-called 'mother-in-law' avoidance rule had the effect of separating married daughters from their parents. Three groups consisted of parents, married daughter or daughters and son-in-law, and two other groups included a couple and the wife's mother. One man made a journey of several days to meet us and travelled with his wife and widowed (classificatory) 'mother-in-law'. The rule of avoidance was observed—the 'mother-in-law' sat some 40 m from the man in camp—but it did not prevent her travelling and living with her 'son-in-law' and 'daughter'.

#### WATER SUPPLIES

There are no permanent surface waters in the area. The large, shallow, transient pools formed by heavy rain, the claypans and the rock reservoirs in the hills that might be filled by lighter falls could have supplied groups for only a small part of the time in any period of years. There are few sandy creek beds and no large ones, and only one creek 'soakage' well has been seen in use. The sources probably most used were the 'wells' in the sand, or in rock between the sand ridges. By no means all of these are permanent or reliable, and one man was reported to have perished between waters in about 1961 or 1962 (his wife survived). It is hard to imagine that the original occupation of the area could have occurred at a time when the water supplies were so sparse, unreliable and difficult to find as they are now. Carnegie, who dug out a number of 'wells' and found them to be buried rockholes, observed that in the southern part of the area there were more rockholes, and speculated that in the sandy northern areas similar surface rockholes had been gradually buried and 'transformed into "Native Wells"' (Carnegie 1898:233, 409).

Permanent wells seem often to be found in association with tea-tree scrub, with limestone occurrences, with desert oak forests and with the depressions around salt lakes. Some of these supplies are probably saline: Jupiter Well, sunk to a depth of over 4 m in 1961 by a National Mapping party in tea-tree scrub among desert oaks, has been undrinkable on each of my visits. It is claimed that the old 'native well' 400 m away is fresh but this was not used after 1961, while Jupiter Well was much used by Aborigines.

Carnegie noted (1898:274) that the best waters seemed to be least used. The limitations of food resources and firewood supplies in any locality would be an incentive to use the permanent waters as little as possible and to make maximum use of ephemeral supplies while they lasted. Carnegie's observation (1898:274) that 'first the small native wells are used, and only when these are exhausted are the more permanent waters resorted to' is supported by Gould's recent account (1968:104-5) of site use in the Warburton Range area to the southwest.

As previously indicated, there is some evidence that groups may sometimes have congregated at permanent waters in hot dry summers. But the most intensively used sites may have been around water sources that could provide generous supplies of water for short periods after rain and give access to areas where good supplies of vegetable foods would be available at such times (cf. Gould 1968:119).

As Gould (1968:118) has noted, 'there is always a distinct separation between the water hole and habitation area at each campsite'. Groups have been found camped a good 400 m from the water supply where this was in rocky country and up to 100 m away in sand country (cf. Gould 1968:118). In this environment rock shelters would be occupied in the daytime for their shade and for shelter from rain, as children's drawings on the walls of such shelters elsewhere in the desert attest, but none of some forty waters visited in this area had any rock shelters nearby. Camp debris from the recent past is therefore likely to be scattered over a considerable area at each water.

#### FOOD SUPPLIES

Because no extended studies of any of the groups have been possible I can offer only a few observations on the diet of the people of the area. Only two of the twenty-two groups were found to be using grass seeds for flour making. Mortars for grinding seeds were noticed at some sites in the area and Thomson (1964:403) states that he found such stones at most camps in the Lake Mackay area. Aborigines have indicated that one species of spinifex has edible seeds but none of the groups we met was harvesting these seeds. Other seeds and fruit were more often available than grass seeds, the most used being probably *Solanum* fruits and the quandong (*Santalum acuminatum*).

On the one occasion when the harvesting of grass seeds was observed, the whole plant was torn up and threshed in piles by beating with sticks on hard flat termite beds. Seed was collected and taken to camp for winnowing and grinding of the grain, before being mixed with a little water and baked into a cake. All these operations were carried out by women but men have been seen gathering solanum and other fruits in quantity, as well as lizards and rodents.

It seems reasonable to assume that the main staple in the area was lizard meat (cf. Thomson 1964:402). Virtually every group met had been gathering lizards; in most instances this was the only food brought back to camp. Small lizards (*linga*) seem to have provided the staple for children and to some extent for women. Of the larger lizards probably the 'blue tongues' (*lungkata*; *wana*) were the most important source of food. One doctor who examined blood samples from a group was surprised at the fat levels, for which presumably a diet of lizards was responsible. It seems probable that lizards were similarly the reliable, all-seasons staple in the sandy deserts to the south (and probably on the Nullarbor Plain) where the mean annual rainfall is even lower than in this area but where insects and lizards are similarly numerous.

Thomson and Hosmer (1963) give an account of the lizards of the Lake Mackay area. They note that the Bindibu (Pintubi) exploited the agamids which Aborigines elsewhere do not eat but indicate that the mountain devil (*Moloch horridus*) was not used for food (1963:232). But a man of a particularly ill-nourished group met in 1963 at Jupiter Well brought in and ate one of these lizards.

Other game which has been observed 'in hand' include a feral cat, small desert rodents (unidentified), and a rabbit. Carnegie's narrative suggests that desert rats were relatively important in the diet. When camping in hill country our party secured euros but there was no indication that the group there had been hunting euros (not a highly regarded meat). Groups have reported recent meals of kangaroos and 'porcupines'. It appears that nowhere in this area have rabbits been the staple food they were farther south, for example in the Petermann Ranges.

### TOOLS AND WEAPONS

The most important weapons and implements in daily use were made of wood and, certainly for women and probably for men too, the indispensable implement is a tool for digging. Thomson (1964) in his detailed account of the implements of the groups near Lake Mackay stresses the importance of the women's pointed digging sticks and the care with which these tools are maintained. Numbers of such sticks have been seen with metal blades attached. Thomson does not mention a special digging stick among the implements carried by men and, except for one boy who carried a kind of short spear shaft, I do not recall seeing men carrying specialised digging tools. Thomson and Hosmer (1963:233) report seeing a man using his spearthrower as a scoop for digging and Thomson (1964:408) observed a man using his wife's digging stick. (I have observed men using women's wooden dishes for scooping out sand from wells). Certainly most of the food men have been seen to bring into camp would have been gathered by digging. Possibly the blunt ends of spears and unshaped sticks gathered as needed and discarded are used to supplement the spearthrower in digging but it appears that the cultural 'rule' that men hunt game while women dig and gather may have inhibited the adoption by males of a digging tool as efficient and useful in these arid conditions as the female's digging stick.

Stone implements, other than the scraper

usually attached with spinifex resin to spear-throwers, were rarely seen, and certainly the main use of stone here was for shaping and maintaining wooden tools and weapons. A hand axe was found at one unoccupied site but neither stone axes nor stone knives were actually seen, though Thomson (1964:405) reports one man as having a 'bi-faced axe of quartzite'. Men may have had them wrapped in what Carnegie termed their 'portmanteaus', but it is interesting that the only stone tools in the 'portmanteau' of which he gives an inventory were 'fragments of quartz suitable for spear and chisel heads' (Carnegie 1898:391). One major 'quarry' area for stone knives, west of Pollock Hills, where fragmented chalcedony littered the surface over several thousand square metres, might have supplied the people of the whole area with knives and scrapers.

Men were always equipped with spears and spearthrowers. Spears were made with minimum effort from *mulyati*, an *Acacia* growing in a wide belt of country in the western part of the area. The acacia is erect in form and has few branches, and this may possibly be the only area where it grows in this form. The roots are broken by bending the stem to and fro, and the bark is removed by burning. What straightening is necessary is done by bending the warm stem, using the teeth as a vice, and the shaft is scraped (and sometimes decorated with incised rings) and the blade shaped with a stone or metal scraper. Thomson (1964:410) reports the making of spears from acacia roots in the area immediately to the south, near the Rawlinson Range. Several forms of spear have been seen: a barbless type; one with one or sometimes two barbs tied on with sinew; and a shaft with from one to four barbs cut in the shaft wood. A heavy, broad-bladed jabbing spear used only in combat is made usually from mulga wood.

No boomerangs have been seen in this area, but Carnegie reported boomerangs 'of several shapes', as well as throwing sticks, in his inventory of the implements of the group he met at Family Well immediately to the west (Carnegie 1898:244). Thomson (1964:410) reports that boomerangs were owned by men near Lake Mackay but these had evidently been acquired by trade and seemed to have prestige value and ritual rather than practical use.

Women were all equipped with at least one wooden dish, and smaller, rough bark dishes were also much used.

### THE DINGO

Not all groups had a dog: I recall only three groups with dogs and another that had recently taken a litter of pups from their mother. Where a dingo was in camp any bones that reached him whole—and the larger bones (of euro and kangaroo for example) are crushed for the marrow—would be thoroughly worked over (cf. Gould 1968: 119). One man, who reported recently killing two kangaroos, indicated that his dog had run them down, but I have not actually seen how dogs are used in foraging operations. All camp dogs were conspicuously less well nourished than dogs seen at large. The camp dogs are named and there is evidently a strong emotional tie between owners and their dogs. There is some evidence that dingoes may have a more restricted range than their masters, so that they may from time to time change masters or may be only intermittently attached to camps.

### TIME AND THE FOOD QUEST

Virtually no observations of daily foraging routines were made. In one attempt to gain an idea of what might be the fruits of a normal day's foraging, three young men were induced to spend most of a day hunting and they returned with one metre-long snake, a blue tongue lizard, and a goanna. One group met with in early November was resting in camp at midday and it can be assumed that it is normal to rest in the shade in the middle of the day in the hot months. At another camp (in winter) all of the group except two small girls were away foraging in the late morning and did not begin to return until mid-afternoon—mother and daughter first, then the man—and by late afternoon a youth and a boy had not returned. On two occasions men returned to camp after sundown and in one instance smokes indicated that a man had been at least 16 km from camp during the day.

### Research Possibilities

Most of the area was, for practical purposes, inaccessible for study until the roads were made in 1960, and travel through the sand ridges is so difficult as to make the examination of even a limited area a time consuming and expensive job. When the area was inhabited the difficult terrain, aridity, and sparse food supplies would have made extended studies of the daily life of the groups immensely difficult. But there is still scope for further detailed studies, of the kind Gould (1968) has made, of representative camp sites and their use, and of the animal and plant resources of the surrounding country. There are hundreds of people who have experienced life in the desert and it is still possible by interviewing such people to learn more, for example, of local organisation and the movements of the people, their classifications of types of habitat and waters, and of the ritual and mythological associations of sites and areas.

### References

Carnegie, D. W. 1898. *Spinifex and sand.* London: C. Arthur Pearson.

Evans, E. C. and Long, J. P. M. 1965. The Aborigines of western central Australia. *Geogrl J.* **131**:318-29.

Gould, R. A. 1968. Living archeology: the Ngatatjara of Western Australia. *SWest.J. Anthrop.* **24**:101-22.

Long, J. P. M. 1964. The Pintubi patrols: welfare work with desert Aborigines. *Aust. Territories* **4** (5):43-8; (6):24-35.

Radcliffe-Brown, A. R. 1930. Former numbers and distribution of the Australian Aborigines. *Official year book of the Commonwealth of Australia* No. 23, pp. 687-96. Melbourne: Government Printer.

Thomson, D. F. 1960. A bark sandal from the desert of central Western Australia. *Man* **60**:177-9.

——1961. A narcotic from *Nicotiana ingulba*, used by the desert Bindibu. *Man* **61**:5-8.

——1962. The Bindibu expedition: exploration among the desert aborigines of Western Australia. *Geogrl J.* **128**:1-14, 143-57, 262-78.

——1964. Some wood and stone implements of the Bindibu tribe of central Western Australia. *Proc.preh.Soc.* **30**:400-22.

——and Hosmer, W. 1963. A preliminary account of the herpetology of the Great Sandy Desert of central Western Australia. *Proc.R.Soc.Vict.* **77**:217-37.

Tietkens, W. H. 1890. Journal of Mr W. H. Tietkens' central Australian exploring expedition. *South Australian Parliamentary Papers.* 1890, Vol. 4, No. 111.

# 19 The Demography of Hunters and Farmers in Tasmania

Rhys Jones

The cultural change which is focal in our view of the history of man is the Neolithic Revolution. This is what separates the hunter from the farmer, the savage from the barbarian. If we look back at the history of the last 10,000 years, the main thread of the story is the invention, development, and spread of agriculture throughout the entire inhabited world.

The interface between Palaeolithic and Neolithic fascinates us, and, when we think of it, we automatically cast our minds back thousands of years ago to the end of the last Ice Age and to the beginnings of the post-glacial period. Yet in Australia, the Neolithic Revolution is happening all around us, and it will be one of the fascinating problems of Australian archaeology and history to combine our disciplines to study it. We are fortunate, for we have a detailed historical documentation of the entire process; the exploration, economy, and race relations. It happened such a short time ago in most places that the countryside is essentially the same now as it was during the first impact, and we can look at the land and see our own effect on it and compare it with that of the hunters, unhampered by 10,000 years of previous farming (Tindale 1959:42-6; Jones 1968:201-11, 1969; Merrilees 1968:4-5, 16-20). There are still trees standing which were already old when the first squatter walked by and ring-barked them; and there are people living, both farmers and hunters, who have experienced the change. Indeed, in some parts of Australia, the process is going on at the present day.

In this brief essay, I hope to begin an attack on the problem by considering the relative sizes of the human populations of Tasmania under both hunting and farming economic régimes. I have chosen population, as it is one of the most important parameters used by biologists when seeking to establish

*Fig. 19:1 Tasmanian localities referred to in text*

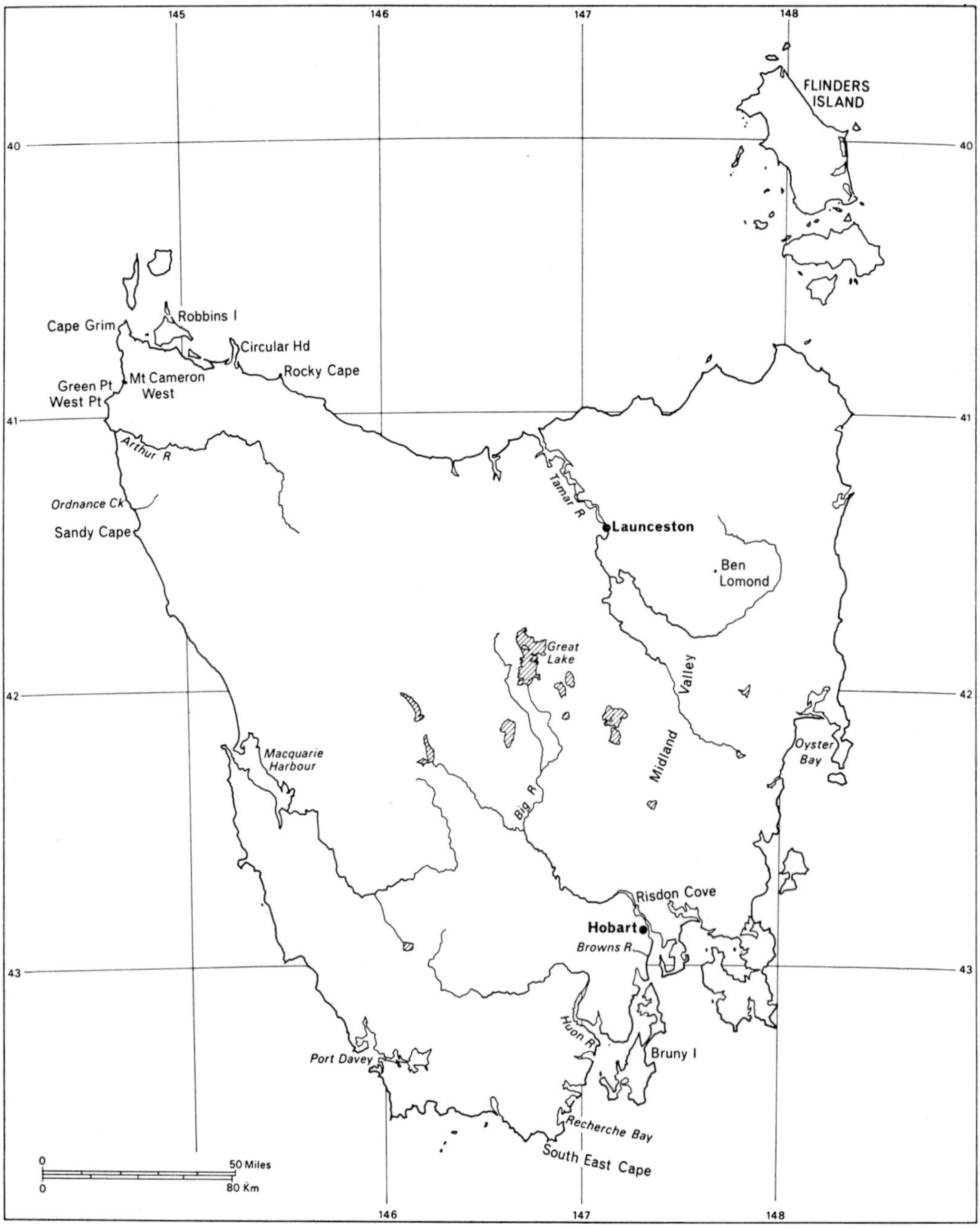

the success or failure of a species. Tasmania is convenient because, being an island, it is circumscribed (Fig. 19:1). The ethnographic accounts are surprisingly good, considering that they were all written before the science of anthropology had begun (Hiatt 1967-8), and recently they have been vastly augmented by the publication of the field journals of G. A. Robinson from 1829 to 1834 (ed. Plomley 1966). The wealth of archaeological sites on the island (Jones 1965, 1966; Lourandos 1968) gives us some hope that one day it will be possible to extend this picture into the past in some detail. In addition the geography and history of the European colony there have been well documented.

A comparison between the populations of hunters and farmers in Tasmania brings out the revolutionary nature of the changes in human ecology which have occurred there in the past two hundred years.

## The Hunters

The Aboriginal population level is difficult to assess, and because all the Aborigines are now dead it is impossible to test our conclusions in the field. However, some attempt must be made, and a careful analysis of the historical sources allows us to reach figures of a similar order of magnitude by several methods.

Contemporary estimates quoted by Roth (1899:163-5) range from Melville's 20,000 to G. W. Walker's 500. We can begin with a base line, for Robinson mentions the names of 281 Aborigines in his journals (1829-34), most of whom he actually met in his travels (Plomley 1966:977-88). A few Aborigines met or alluded to are not named, and seven more were captured in 1842 (West 1852: **2**, 65). It would be possible to make a diligent search of Robinson's and other journals, official reports, police records, and newspaper accounts—particularly in the period 1820 to 1830—of all the Aborigines captured, killed, died through disease, found as corpses, or abducted. We might thus obtain a minimum number for the population at that time. I have not done this systematically, but my guess is that the figure of approximately 300 from Robinson could easily be doubled. These records pertain to the period 1825 to 1835, fully twenty to thirty years after the original British settlement, and must be regarded very much as minimum estimates for the original Tasmanian population. The groups seen by Robinson were only remnants of former numbers, some 'tribes' being represented by as little as one individual (Robinson 3 November 1831, in Plomley 1966:500-1) and others had totally disappeared. Large areas of Tasmania had been entirely depopulated. I think that we can safely say that the original population was considerably in excess of 600.

### TASMANIAN TRIBAL ORGANISATION

The most convincing estimates of the precontact Aboriginal population have been based on studies of Tasmanian social organisation, into tribes and language groups, coupled with contemporary estimates as to the size of these groups (Milligan 1890:6; Walker 1898; Roth 1899:163-71; Radcliffe-Brown 1930a:695). These, combined with Schmidt's linguistic work (1952), produced a coherent picture of the pattern of Tasmanian tribal organisation at a general level. These workers, however, did not have the benefit of Robinson's field notes, nor the results of archaeological work, and the new data can be used to test hypotheses already in the literature. I differ slightly from the above authors in my final estimate of the size of the Tasmanian population, but it is heartening to find that such a re-analysis brings out a broad consensus of opinion as to the distribution, order of magnitude, and basic social structure of the Tasmanian Aboriginal population.

First, there is a question of nomenclature. All writers on Tasmanian social organisation have referred to the basic units as 'tribes', but the word has had different meanings for different people. Berndt (1966:26-33, 56) has shown that for Australianists, the term 'tribe' has a wide range of meanings from the broadest linguistic units to small-scale aggregations of local groups. The names 'horde', 'clan', and 'community' have been defined in terms of sociological attributes so specific that they are outside the scope of the Tasmanian historical evidence. The colonial term 'mob' is not synonymous with 'tribe', since it referred to Aboriginal groups ranging from foraging units of a few families to large seasonal congregations of several hundred people. Given the flexible nature of the term 'tribe', the precedence established by its use in the early historical literature, and the lack of an obvious alternative, it is useful at least for the time being to retain the word in discussions on Tasmanian social organisation. To avoid any confusion, I propose the term *Robinson tribe* to describe those units called 'tribes' or 'nations' by Robinson, and *Walker tribes* for the units called 'tribes' by Walker (1898:178-87), following a common practice in the natural sciences, e.g. Beckmann thermometer, Gaussian curve.

*Robinson tribes.* Throughout his journals, Robinson talked about the Aborigines being organised into 'tribes' or 'nations'. He was particularly interested in their composition and location, because the information was vital to him in finding them in the first place, and in making sure that all surviving members of such groups had been contacted. His journals contain a wealth of information on the subject, gathered from his own observations, and from interviews with Aborigines, both as to the contemporary situation and what it used to be in the past.

A Robinson tribe consisted of a group of people who called themselves by a particular name, and were known by that name to other Aborigines. Many had two names (Plomley 1966:969-70), the reasons behind this not being known, but a linguistic analysis of the meaning of the Aboriginal names for these tribes would be most illuminating, if it could be done (e.g. see Plomley 1966:970). Each tribe was related to a particular locality, the core areas of which were well known; thus people talked of the 'Sandy Cape Tribe', 'Port Davey Tribe' and so on. From Robinson's journals, Plomley (1966:970-6) has counted and mapped forty-six of these tribes (see Fig. 19:2). This distribution is similar to that of archaeological remains and direct ethnographic observations (Fig. 19:3), indicating that a reasonable coverage of the island's tribes has been achieved. There are, however, some important gaps where we have no evidence of tribal names in areas known to have once supported large Aboriginal populations. The most important of these areas are in the Midland Valley and the country abound Hobart and Launceston, all occupied early by the British colonists. Some named locations on the map refer to what were once several tribes, though there is no extra information to allow us to map them separately, and Plomley (1966:969) thinks that as little as half of the original tribes have been recorded. I think that, taking into account these gaps in our knowledge, the figure of forty-six should be increased by 50 per cent, so let us take a figure of seventy as being a reasonable, conservative estimate for the number of Robinson tribes in Tasmania before European contact.

Along the coasts, where we have the most

*Fig. 19:2 Tribal and linguistic map of Tasmania*

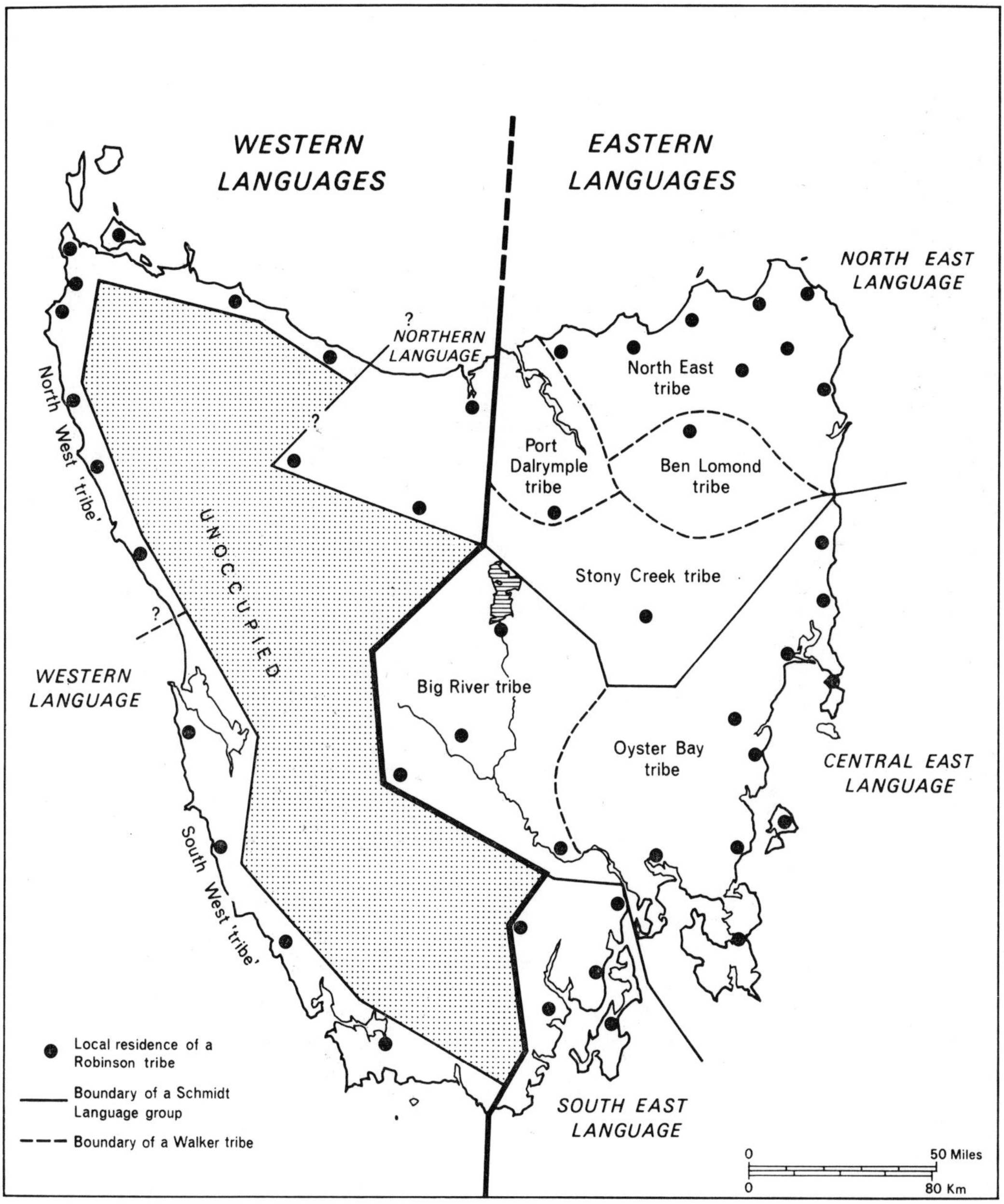

*Fig. 19: 3 Distribution of Aboriginal man in Tasmania. A, archaeological remains (after Jones 1968: 209); E, direct ethnographic observations (after Hiatt 1967-8: 192).*

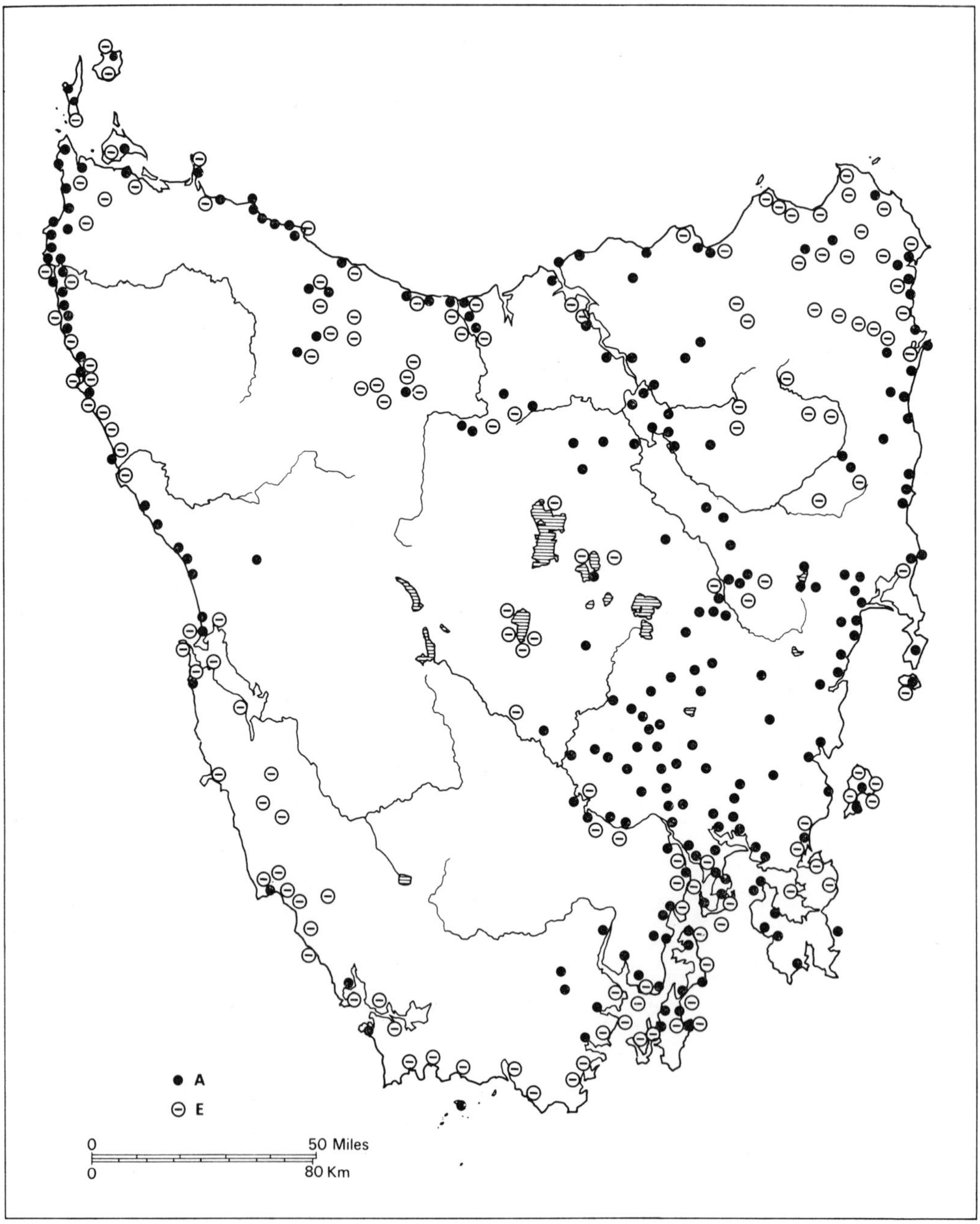

complete evidence, we see that the Robinson tribes were regularly spaced, each one occupying about 24 to 32 km (15 to 20 miles), and in the southwest, with its lower food resources, this distance was approximately doubled (Plomley 1966:969). Aborigines knew when they were in the country belonging to a certain tribe, and some evidence on the native geography of part of the southeast coast suggests that natural features such as rivers or hills sometimes formed boundaries (Robinson 11 January 1831, in Plomley 1966:312). Although a tribe was often to be found in or near its recognised territory, the annual seasonal movements of most groups took them 150 km or more away from their core areas and across the territories of neighbouring groups (Walker 1898; Kemp 1963:243; Hiatt 1967-8:190-205; Plomley 1966:969). Ideally, the tribes were exogamous (Robinson 21 June 1834, in Plomley 1966:888), the wife usually moving to her husband's tribe, and analysis of the data from northwest Tasmania shows that marriages took place with neighbouring tribes.

There are difficulties in trying to calculate the size of these tribes. Robinson, our best source as to their composition and size, was working when the Aboriginal population had collapsed in most places and consisted of the ageing remnants of former numbers, whereas the late eighteenth-century maritime explorers, who were seeing an intact population, only stayed a short while in Tasmania and did not venture far from their ships. In southeastern Tasmania, on 12 February 1793, Labillardière (1800:308-9) saw a group of forty-eight people, consisting of ten men, fourteen women, and twenty-four children, eating their food around seven fires; elsewhere he saw forty-two people, consisting of seven men, eight women, and twenty-seven adolescents and children. On 31 January 1802 Péron (1809:195-6) saw a party of twenty women returning to their husbands after shell fishing, implying a total population of sixty to eighty people including men and children. Later, at Oyster Bay, he saw twenty-five to thirty people from his ship. In 1804, Knopwood reported that an exploring party led by Collins found a native village at the Huon River consisting of twenty families (Plomley 1966:18), and Kelly in 1816 saw what he thought were at least fifty people on Robbins Island on the northwest coast (Bowden 1964:30). In 1824 a tribe, probably from Oyster Bay, consisting of sixty-four people, visited Hobart, and the following year a tribe of fifty people visited the town also. A despatch to Governor Arthur in 1828 relates how a party of about fifty Aborigines used regularly to visit Bruny Island and Recherche Bay (West 1852: **2**, 15; Plomley 1966:49, 100 n.3).

In 1830 Robinson met his first tribe in the bush, namely the Port Davey tribe (Robinson 18 March 1830, in Plomley 1966: 132). This consisted of twenty-six people, but of these twenty were adults, two were adolescents, and only four were children or babies. To be a viable breeding unit under hunter-gatherer conditions, there would have had to have been as many children as adults, giving us a population of forty. Although this was Robinson's first encounter with them, the tribe itself was by no means intact. One year previously, at least nine people from this tribe had died in a pulmonary epidemic at Bruny Island (Plomley 1966:76-7), and we have no means of knowing how many more had died in the bush from similar causes. The remnants of the West Point tribe met in 1832 consisted of twenty-three people, of whom ten were men, five women, four adolescents, and only four children (Robinson 17 July 1832, in Plomley 1966:633). Given sex and age ratios similar to those observed by the first French explorers, such a group would have consisted of ten families giving about forty people.

Each tribe apparently consisted of a number of *hearth groups*, a hearth group approximating to a single family of man, wife, children, and sometimes dependants and friends. In the tense evening before the attack on Robinson at the Arthur River, he describes the Aborigines taking up their positions for eating and sleeping according to their tribal affiliation, the remnants of each tribe in family hearth groups close to one another, and away from those of other tribes (Robinson 3 September 1832, in Plomley 1966:649). Labillardière, in the observation referred to above, saw forty-eight people in seven hearth groups, and elsewhere, in February 1793, he saw nineteen people in three groups (1800:303) (Plate VII). In the bush, in May 1792, he saw fourteen fireplaces in one spot, the people being absent (1800: 127), but a population of between fifty and eighty people is implied.

Robinson and other observers saw many groups of huts clustered together in what they called villages. On the west coast these huts were dome-shaped, about eight to ten feet (2·5 to 3 m) in diameter and height, and they contained anything from about six to twelve people. I suggest that on the west coast each hut may have corresponded to a hearth group, and the village to the local residence of part or all of a Robinson tribe. Observers of these villages usually refer to 'several' huts or 'some' huts, suggesting figures from between two and ten. Near the Great Lake Robinson saw a village consisting of five good bark huts (Robinson 10 November 1831, in Plomley 1966:512), and on the west coast he saw a village of four huts, from which he deduced that it belonged to a tribe of forty people (Robinson 6 June 1830, in Plomley 1966:168-70). These villages, and through them the tribes, are probably represented in the archaeological record. Near Ordnance Creek, on the west coast, Robinson (9 March 1834, in Plomley 1966:858) described a large shell midden with concave holes ten to twenty feet (3 to 6 m) wide and from three to five feet (1 to 1·5 m) deep in the surface of the ground. These were for Aboriginal habitations and they had large piles of shells beside them. Near the same place six months earlier, he saw native habitations in hollows dug out of the side of a sandhill, and his field sketch shows five huts (Robinson 4 September 1833, in Plomley 1966:790).

Identical circular depressions have been found on shell middens on the west coast (e.g. Jones, J. F., 1947). Practically every large stable and well-grassed midden that I have seen on the coast between Mount Cameron West and the Arthur River has these well-defined and standard shaped depressions on their surfaces, and I have seen no such features elsewhere on the coast. I think that they were hollows made for huts, or at least for shelter from the wind. A midden (W.P.2) 1 km south of the West Point lighthouse has five of these circular depressions on its surface; the midden (W.P.1; Jones 1966: 6-8) 1 km north of the lighthouse has seven depressions (Plate VIII), and another midden at Green Point (G.P.1) some 4 km north of the lighthouse has five. If my arguments are valid, this gives us a picture of these large shell middens as local residences of one or more Robinson tribes, a tribe itself numbering from about thirty to fifty people. The huge size of the middens (e.g. Jones 1966) and their proximity to elephant seal breeding grounds makes this ecologically feasible. The period referred to is from 1,000 to 2,000 years ago.

To sum up this section, I think that in traditional times the extreme population range of a Robinson tribe was between 30 and 80 people and that the majority lay between 40 and 70 people; that is between 10 to 20 families. In Tasmania there were of the order of seventy such social units.

*Walker tribes and Schmidt language groups.* Tasmanian society was also organised into larger units than the Robinson tribes. From an analysis of the accounts of colonial settlers, official reports and the writings of learned visitors to the Flinders Isle settlement, J. Backhouse Walker (1898) divided the Aboriginal population of Tasmania into four regional groupings: 1. Southern tribes; 2. Western and northwestern tribes; 3. Central tribes; 4. Northern and northeastern tribes.

Schmidt (1952) carried out an exhaustive analysis of all the Aboriginal vocabularies available to him and published his work without having had the opportunity of seeing Robinson's journals. He defined the existence of five distinct languages or dialects (1952: 54-6), a conclusion which finds general support from Capell (1953:315, 1968:7). These could be divided into two major groups; the eastern and the western languages (Fig. 19:2). In the eastern half of Tasmania, there were three languages or dialects, those of the northeast, central east, and southeast. Along the west and northwest coast one language, the western language, was spoken and another related one called the northern language was situated in northern Tasmania, somewhere between the Circular Head District and the mouth of the Tamar. These linguistic units were identified with Walker's general tribal divisions, and I have mapped their position on Fig. 19:2.

At one time or another, Robinson had with him Aborigines from most parts of Tasmania. By cross checking who could or could not talk to whom and by tabulating all the observations Robinson himself made on linguistic differences across the island, it is possible to build up a parallel picture which conforms well with the Walker-Schmidt pattern. The distribution of some elements of material culture such as huts and watercraft types together with non-ecologically based cultural traits such as myths, hair styles, body cicatrice patterns, uses of body ochre, and food tabus, etc., parallels the linguistic pattern. So also do details of traditional enmities and friendships, who was afraid and who was pleased to go to certain parts of the island and meet certain other Aborigines. This broad configuration of linguistic and cultural groups makes excellent sense ecologically and receives support from archaeological work. Let us call these major units *Schmidt language groups*.

There were also social units in order of magnitude somewhere between these language groups and the Robinson tribes, and I will call these *Walker tribes* (after Walker 1898). Within the central eastern language group there were two Walker tribes, the Big River and the Oyster Bay. Each of these acted together as a self-contained political unit on some occasions. The members of each tribe met periodically, probably on occasions dictated by the seasonal fluctuations in the abundance of resources. Although the Big River and Oyster Bay peoples spoke the same language and sometimes co-operated, especially in their last desperate counter-attack against their European oppressors, yet the diet, seasonal movements, and exploitation of mineral resources of each tribe had a decidedly different pattern. Both the European settlers and the Aborigines themselves recognised the separate identity of the two groups.

Within the northeastern language group, there were four Walker tribes, some of which were almost entirely independent from some of the others. There may indeed have been some slight dialectical and cultural differences between them. Walker (1898:184, 186) cites evidence that the people of the remote Ben Lomond tribe spoke a different dialect from some of the others in the same language group. Such minor dialectical differences between Walker tribes within a single language group may be what Clark, the catechist at Flinders Island, was referring to when he said that on his arrival at the settlement there were about eight to ten different dialects being spoken by the 200 Aborigines there (Walker 1898:179). In some cases the language group and the Walker tribe coincided. Such was the case with the northern and possibly the southern languages. On the west coast, I think that from a reading of Robinson's journals we are justified in defining at least two cultural units of the same status and order of size as Walker tribes of the east, namely one in the northwest and one in the southwest. I have entered these on Fig. 19:2 as 'tribes'.

Large congregations of Aborigines consisting of several hundred people were sometimes seen by the early settlers. Thus Knopwood (1948:99) met between 250 and 300 people near Browns River, south Tasmania, in 1807 and 200 Aborigines were seen near Launceston in 1824 (West 1852:**2**, 16). Many other examples have been given by West (1852:**2**, 6), Walker (1898:177-8), and Roth (1899:164). Whereas in some cases the actual counts may be suspect, we have no reason to disbelieve all of these reports outright, and I think that they represented the seasonal aggregations of the bands of people who made up part or all of a Walker tribe. These reports range from 200 to 500 people and I think that this was the order of size of a Walker tribe.

To sum up, we may define a Walker tribe as the largest social unit which periodically and systematically met together for reasons of food collecting, leisure, politics, marriage, or ritual. All members would have met all other members at one time and place on several occasions, probably at least annually, and the Walker tribe was probably the biggest unit to which any single person felt himself totally part of, though he was aware of neighbouring groups who shared some of his culture. In some language groups there were several such tribes, in others the linguistic and tribal groups coincided. Walker tribes probably consisted of between 200 and 500 people, and we have evidence of about ten such tribes in Tasmania.

*Table 19:1 Calculations of Tasmanian Aboriginal population*

| Social unit | No. of people in each unit | No. of units in Tasmania | Total Aboriginal population |
|---|---|---|---|
| Robinson tribe | 40-70 | 70 | 3,000-5,000 |
| Walker tribe | 300 (200-500) | 10 | 3,000 (2,000-5,000) |
| Schmidt language group | ?600 (100-1,500) | 5 | ?3,000 |

### DISCUSSION AND CALCULATION OF ABORIGINAL POPULATION

The Aboriginal population of Tasmania can be divided into a hierarchy of social/cultural units. There was one culture (Kemp 1963: 243) divided into five language groups, into ten Walker or 'true' tribes and into seventy Robinson or 'local' tribes. In the areas where we have the best information such as the northeast coast and the west coast, we see from Fig. 19:2 that there could be from four or five up to more than eight or nine Robinson tribes within a Walker tribe, and between one and four Walker tribes within a full language group. With some idea of the size of each of these social units, we are in a position to try to calculate the original total Aboriginal population of Tasmania, and I have done this in Table 19:1.

We have no direct evidence of the size of the Tasmanian language units, but for the mainland of Australia Elkin (1964:11) says that membership of a linguistic tribe ranged from about 100 to 1,500, averaging around 500 to 600, and Birdsell (1968:230) thinks that such units averaged 500 people.

We see that there is a fair degree of correlation between the results, and I feel that a figure of between 3,000 and 5,000 for the Aboriginal population of Tasmania is a reasonable approximation to the truth. This figure is higher than the 2,000 of Milligan (1890:6) and Walker (1898:178) or the 2,000 to 3,000 of Radcliffe-Brown (1930a: 695), but they did not have the data available to us today, particularly from the west coast. Plomley (1966:18), who may have done a similar calculation to mine, arrives at a figure of 4,000, though he gives no details.

Tasmania has an area of 67,000 $km^2$ (26,000 square miles), giving a population density of between one person per 13 $km^2$ (5 square miles) and one per 23 $km^2$ (9 square miles). The densities on the ground would have been slightly higher because a quarter of the area of Tasmania was not occupied by the Aborigines. All of the Aboriginal groups had access to the coast or coastal estuaries at some time during the year. Tasmania has about 1,600 km (1,000 miles) of coastline, giving a population density of 2 to 3 people per km of coast (3 to 5 per mile). If we include offshore islands recorded in the literature of early contact as being economically exploited by Tasmanian Aborigines, these figures are somewhat reduced (Jones n.d.). They are still high, however, compared with those of most hunter-gatherers, but they fit well into the range of the highest populations living in the rich coastal and riverine regions of Australia (Meggitt 1966:59-62), and North America (Kroeber 1963:131-81). The archaeological record supports this picture of high coastal populations in Tasmania, for the west coast between Cape Grim and Sandy Cape probably has a higher density of archaeological remains on it than any other coastline so far recorded in Australia.

I have deliberately tried not to draw analogies from the Australian mainland, for fear of confusing the issue in Tasmania. It is interesting to note, however, that the tribal/linguistic pattern set up here for Tasmania shows a strong structural similarity to that on the mainland. The Robinson tribe shares many of the attributes of Radcliffe-Brown's 'horde' (1930b) or L. R. Hiatt's 'community' (1965:24-7), and it was probably similar to them. The hierarchy of social units from hearth groups through to linguistic and cultural units is also paralleled in detail on the mainland (e.g. Radcliffe-Brown 1930a:688; Berndt 1966:53; Meggitt 1966:68-9). Whether these similarities are due to an ancient shared cultural tradition, or whether they are similar because of the action of powerful ecological mechanisms, is a question worth pursuing further.

## Death of the Hunters

Man has lived continuously in Tasmania for at least 8,000 years (Reber 1965; Jones 1968: 197-201). During that period, archaeological evidence points to the survival of a single cultural tradition with only small changes in some traits. Put in other words, it is likely that the direct ancestors of the ethnographically observed Aborigines had been living on their island in an unbroken line back to the beginning of the post-glacial period. They had probably been there much longer. Carbon dates on the southeast mainland of Australia show that man was occupying the lands adjacent to the ice-locked peninsula of Tasmania at least twenty-five thousand years ago (Jones 1968:186-91). Having

*Table 19:2 Decline of Tasmanian Aboriginal population*

| 1770-1800 | 1830 | 1833 | 1834 | 1836 | 1838 | 1847 | 1854 | 1859 | 1863 | 1869 | 1877 |
|---|---|---|---|---|---|---|---|---|---|---|---|
| 3,000-5,000 | 280 | 151 | 134 | 116 | 82 | 45 | 16 | 14 | 6 | 1 | 0 |

allowed for all the adjustments which had to be made in the occupation of the new lands made free by the melting of the ice, and in the reorganisation of society forced by the inundation of Bass Strait, there were still 8,000 years of relatively stable conditions in Tasmania for its human population to adjust itself to the environment.

We have two models to consider. Either the population fluctuated widely about some mean, so that in good years it increased rapidly only to be drastically reduced when population exceeded food supply; or having reached equilibrium, the population maintained itself at about that level over a long period of time. If the latter, then this equilibrium position would be somewhat below the maximum number of people the environment could support at any one time, so that the population could be buffered against fluctuations in the abundance of natural resources. There are many ways in which a hunting and gathering society is able to exercise some control over its numbers, for example through infanticide (N. Peterson pers. comm.).

The structure of Tasmanian society was in harmony with the environment, and it closely matched other hunting and gathering societies in Australia and elsewhere both in form and in population density. My own feeling is that the numbers and distribution of Aboriginal man in Tasmania at the time of European contact were the result of a long established equilibrium between population and environment which had probably been set up over thousands of years.

This ancient balance was completely shattered on the arrival of the Europeans. The fate of the Tasmanians constitutes one of the few examples in written history where an entire people has become totally extinct. Whether it occurred through design or negligence of the Europeans, or whether it was ecologically inevitable, it is the example *par excellence* of genocide. It was an old story in 1803, and it was to be re-enacted many times afterwards on the savannah plains of Australia and in the jungles of Brazil. Savage and barbarian met face to face, and the savage died.

From records kept at various Aboriginal settlements we have fairly detailed figures documenting the final decline of the Aboriginal population from 1830 onwards. There are some minor discrepancies in the published collations of the actual counts (West 1852: **2**, 71-2; Roth 1899:164-5; Hormann 1949: 188; Turnbull 1965:163-4; Plomley 1966: 977-88), due to fluctuations resulting from new arrivals and a high death rate, but the figures tabulated below are probably close to being accurate.

I have tried to plot this population collapse in Fig. 19:4, using an approximation to a model logistic curve proposed by Allee *et al.* (1949) to describe the extinction of animal populations. I have taken 1800 to represent the date at which the process began, though Plomley (1966:964) has reminded us that exotic diseases could have been introduced by maritime explorers in the thirty years after 1770. Two model curves are shown, using the top and bottom of the range estimated above for the original Aboriginal population. If the population did in fact decline as shown, then the greatest rate of decrease would have been in the period

*Fig. 19:4 Presumed decline and collapse of Tasmanian Aboriginal population. Dots refer to actual counts.*

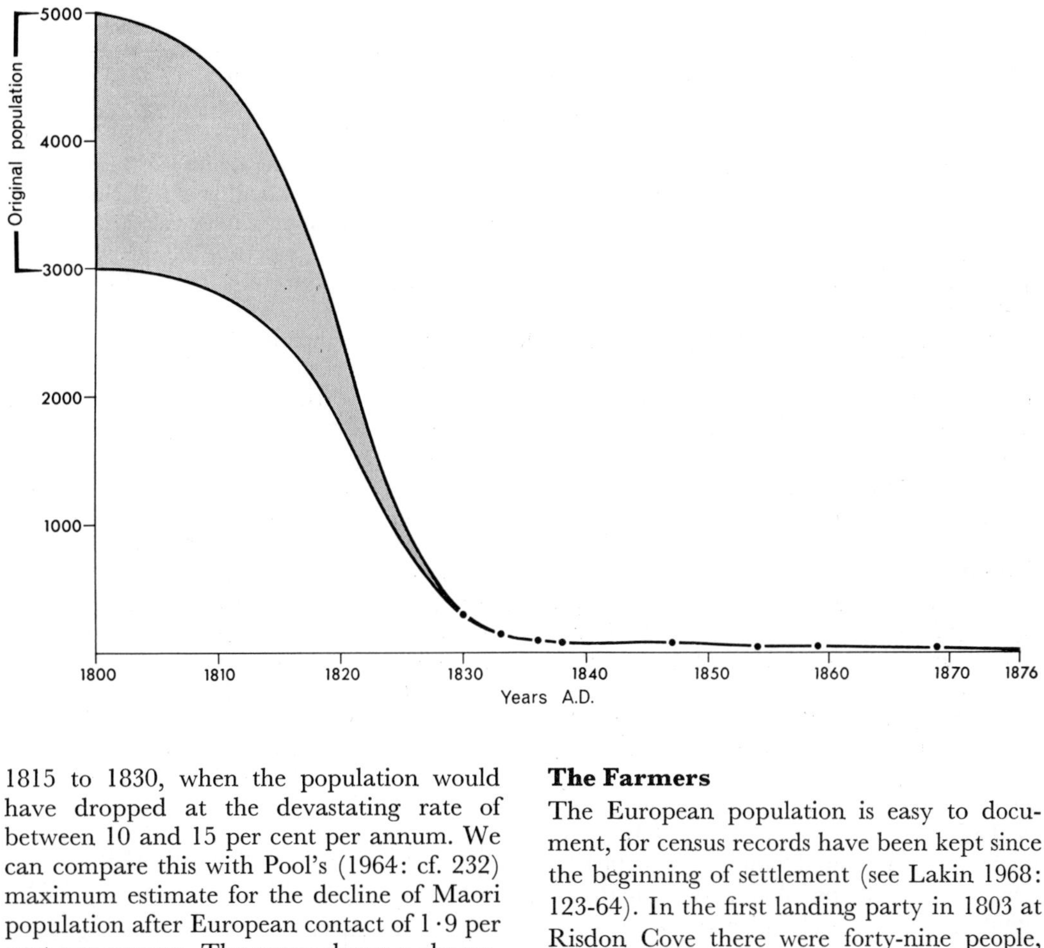


1815 to 1830, when the population would have dropped at the devastating rate of between 10 and 15 per cent per annum. We can compare this with Pool's (1964: cf. 232) maximum estimate for the decline of Maori population after European contact of 1·9 per cent per annum. The curve shows a characteristic long tail on the right hand side, indicating the survival of the remnants of an ageing adult population. The last fullblood Tasmanian died in 1876, seventy years after the establishment of European settlement on the island. Tindale (1937, 1953) and Birdsell (1958:50-2) have documented the establishment and growth of a hybrid Tasmanian-European population on some of the Bass Strait islands and on Kangaroo Island. These populations exist today.

**The Farmers**

The European population is easy to document, for census records have been kept since the beginning of settlement (see Lakin 1968: 123-64). In the first landing party in 1803 at Risdon Cove there were forty-nine people, and in the following year this was augmented to more than 400 people. Table 19:3 shows the subsequent growth of the European population.

In the early decades most of this increase resulted from massive immigration of convicts, military units, and free settlers. Throughout this period, until about the 1850s, men outnumbered women by two to one, and it is only then that natural increase began to provide the major increment to

*Table 19:3 European population of Tasmania*

| 1804 | 1816 | 1820 | 1830 | 1850 | 1900 | 1967 |
|---|---|---|---|---|---|---|
| 400 | 1,500 | 5,400 | 24,000 | 70,000 | 173,000 | 375,000 |

population growth (West 1852: **1**, 31, **2**, 356; Clark 1962:237, 383; Lakin 1968:124).

In Fig. 19:5 I have plotted the Aboriginal and European populations in Tasmania between 1780 and 1850. We see that in twenty years from first settlement the agricultural population equalled that of the original hunting-gathering one; in fifty years it was fifteen times as great; and nowadays, one hundred and sixty years later, the population is between sixty and a hundred times as great as it was before Risdon Cove was occupied.

**Conclusions**

The lesson from Tasmania is clear. A population of hunters and gatherers, well adapted to their environment and established on their land for thousands of years, collapsed almost instantaneously at their first contact with farmers. The story was the same on mainland Australia (e.g. Campbell 1939:33-5; Corris 1968). Some of the reasons for the extinction in Tasmania have been analysed by Hormann (1949) and Plomley (1966:964-7). The Aboriginal population was high by hunter-gatherer standards, being of the same order of size as some of the highest coastal populations on mainland Australia and North America, yet only a hundred and fifty years after the arrival of the Europeans the agriculturally based population is some fifty to a hundred times as great as that of the hunters. The entire Aboriginal population of A.D. 1800 would fit into a small country town of modern Tasmania.

Yet there are some curiosities. In spite of the population explosion large parts of Tasmania have become almost entirely depopulated since the demise of the hunters. If we exclude the small town of Strahan in Macquarie Harbour, which owes its existence to inland mining operations, the west coast of Tasmania is virtually uninhabited today (Farmer 1965:47). From Macquarie Harbour north to the Arthur River there are two farms, some stockmen, fishermen, and prospectors, while south of Macquarie Harbour to South East Cape there are a family of tin miners and some lighthouse keepers. Much of the coast has barely been walked on since Robinson's day, yet when the Aborigines held sway, the west coast supported some five hundred to a thousand people. Hunters and farmers use their land in different ways, but in Tasmania the farmers decided that every hunter had to be removed from the face of the earth to make way for the new order.

Charles Darwin, observing a similar situation in New South Wales in 1836, wrote (1965:230):

> Wherever the European has trod, death seems to pursue the aboriginal. We may look to the wide extent of the Americas, Polynesia, the Cape of Good Hope, and Australia, and we shall find the same result. . . . The varieties of man may seem to act on each other; in the same way as different species of animals—the stronger always exterminating the weaker.

We are beginning to look at the historical events in Tasmania in terms of theories of cultural evolution. It is interesting to speculate what effect the events themselves had on the minds of men like Darwin, Tylor, and Lubbock in the original formulation of these theories. Have we come full circle?

*Fig. 19:5 European population growth in Tasmania compared with Aboriginal population decline. Dots refer to actual counts.*

## References

Allee, W. C., Park, O., Emerson, A. E., Park, T., and Schmidt, K. P. 1949. *Principles of animal ecology*. Philadelphia, London: W. B. Saunders.

Berndt, R. M. 1966. The concept of 'the tribe' in the Western Desert of Australia. In *Readings in Australian and Pacific anthropology* (eds. I. Hogbin and L. R. Hiatt), pp. 26-56. Melbourne: Melbourne University Press.

Birdsell, J. B. 1958. Some population problems involving Pleistocene man. *Cold Spring Harb.Symp.quant.Biol.* **22**:47-69.

——1968. Some predictions for the Pleistocene based on equilibrium systems among Recent hunter-gatherers. In *Man the hunter* (eds. R. B. Lee and I. DeVore), pp. 229-40. Chicago: Aldine Publishing Co.

Bowden, K. M. 1964. *Captain James Kelly of Hobart Town*. Melbourne: Melbourne University Press.

Campbell, T. D. 1939. Notes on the Aborigines of the south-east of South Australia, part II. *Trans.R.Soc.S.Aust.* **63**:27-35.

Capell, A. 1953. Review of *Die tasmanischen Sprachen* by W. Schmidt. *Oceania* **23**:315-16.

——1968. What do we know of Tasmanian language? *Rec.Queen Vict.Mus.* No. 30.

Clark, C. M. H. 1962. *A history of Australia*. Vol. 1. Melbourne: Melbourne University Press.

Corris, P. 1968. *Aborigines and Europeans in western Victoria*. Occasional Papers in Aboriginal Studies No. 12 (Ethnohistory Ser.No. 1). Canberra: Australian Institute of Aboriginal Studies.

Darwin, C. 1965. Journey across the Blue Mountains to Bathurst in January, 1836. In *Fourteen journeys over the Blue Mountains of New South Wales* (coll. and ed. G. Mackaness) pp. 226-36. Sydney, Melbourne, London: Horwitz-Grahame.

Elkin, A. P. 1964. *The Australian Aborigines: how to understand them*. 4th edn. Sydney: Angus and Robertson.

Farmer, R. S. J. 1965. Population. In *Atlas of Tasmania* (ed. J. L. Davies), pp. 47-52. Hobart: Lands and Surveys Department.

Hiatt, Betty. 1967-8. The food quest and the economy of the Tasmanian Aborigines. *Oceania* **38**:99-133, 190-219.

Hiatt, L. R. 1965. *Kinship and conflict*. Canberra: Australian National University Press.

Hormann, B. L. 1949. Extinction and survival: a study of the reaction of aboriginal populations to European expansion. Ph.D. thesis, University of Chicago (unpublished).

Jones, J. F. 1947. Huts of Tasmanian Aborigines. *Pap.Proc.R.Soc.Tasm. for 1946*: 133.

Jones, R. 1965. Archaeological reconnaissance in Tasmania, summer 1963/1964. *Oceania* **35**:191-201.

——1966. A speculative archaeological sequence for north-west Tasmania. *Rec. Queen Vict.Mus.* No. 25.

——1968. The geographical background to the arrival of man in Australia and Tasmania. *Archaeol. & phys.Anthrop.Oceania* **3**: 186-215.

——1969. Fire-stick farming. *Aust.nat.Hist.* **16**:224-8.

Kemp, Thérèse B. 1963. The prehistory of the Tasmanian Aborigines. *Aust.nat.Hist.* **14**: 242-7.

Knopwood, R. 1947. The diary of the Rev. Robert Knopwood, 1805-1808 (eds. W. H. Hudspeth and S. Angel). *Pap.Proc.R.Soc. Tasm.for 1946*:51-124.

——1948. The diary of the Rev. Robert Knopwood, 1805-1808 (eds. W. H. Hudspeth and S. Angel). *Pap.Proc.R.Soc.Tasm. for 1947*:69-126.

Kroeber, A. L. 1963. *Cultural and natural areas of native North America.* Berkeley: University of California Press.

Labillardière, J. J. H. de. 1799-1800. *Rélation du voyage à la recherche de La Pérouse . . . .* 2 vols., atlas. Paris: Jansen.

——1800. *Voyage in search of La Pérouse . . . during the years 1791, 1792, 1793 and 1794.* 2 vols. London: John Stockdale.

Lakin, R. 1968. *Tasmanian year book.* No. 2. Hobart: Commonwealth Bureau of Census and Statistics.

Lourandos, H. 1968. Dispersal of activities—the east Tasmanian Aboriginal sites. *Pap. Proc.R.Soc.Tasm.* **102**:41-6.

Meggitt, M. J. 1966. Indigenous forms of government among the Australian Aborigines. In *Readings in Australian and Pacific anthropology* (eds. I. Hogbin and L. R. Hiatt), pp. 57-74. Melbourne: Melbourne University Press.

Merrilees, D. 1968. Man the destroyer: Late Quaternary changes in the Australian marsupial fauna. *J.Proc.R.Soc.West.Aust.* **51**:1-24.

Milligan, J. 1890. *Vocabulary of the dialects of some of the Aboriginal tribes of Tasmania.* Hobart: Government Printer.

Péron, F. A. 1809. *A voyage of discovery to the Southern Hemisphere, performed by order of the Emperor Napoleon, during the years 1801, 1802, 1803, and 1804.* Translated from the French. In *A collection of modern and contemporary voyages and travels.* III. Analyses of new works of voyages and travels, lately published in London. London: Richard Phillips.

Plomley, N. J. B. (ed.) 1966. *Friendly mission. The Tasmanian journals and papers of George Augustus Robinson 1829-1834.* Hobart: Tasmanian Historical Research Association.

Pool, D. I. 1964. The Maori population of New Zealand. Ph.D. thesis, Australian National University, Canberra (unpublished).

Radcliffe-Brown, A. R. 1930a. Former numbers and distribution of the Australian Aborigines. In *Official year book of the Commonwealth of Australia.* No. 23, pp. 687-96. Melbourne: Government Printer.

——1930b. The social organisation of Australian tribes. *Oceania* **1**:34-63, 322-41, 426-56.

Reber, G. 1965. Aboriginal carbon dates from Tasmania. *Mankind* **6**:264-8.

Roth, H. L. 1899. *The Aborigines of Tasmania.* 2nd edn. Halifax: F. King. (Facsimile edn. Hobart: Fullers Bookshop, 1968.)

Schmidt, W. 1952. *Die tasmanischen Sprachen.* Comité International de Linguistes. Publications de la Commission d'Enquête Linguistique **5**. Utrecht: Comité International de Linguistes.

Tindale, N. B. 1937. Relationship of the extinct Kangaroo Island culture with

cultures of Australia Tasmania and Malaya. *Rec.S.Aust.Mus.* **6**:39-60.
——1953. Growth of a people: formation and development of a hybrid Aboriginal and white stock on the islands of Bass Strait, Tasmania, 1815-1949. *Rec.Queen Vict.Mus.* No. 2.
——1959. Ecology of primitive Aboriginal man in Australia. In *Biogeography and ecology in Australia* (eds. A. Keast, R. L. Crocker, and C. S. Christian), pp. 36-51. Monographiae biol. **8**. The Hague: W. Junk.
Turnbull, C. 1965. *Black war. The extermination of the Tasmanian Aborigines.* Melbourne: Cheshire-Lansdowne.
Walker, J. B. 1898. Some notes on the tribal divisions of the Aborigines of Tasmania. *Pap.Proc.R.Soc.Tasm.for 1897*: 176-87.
West, J. 1852. *The history of Tasmania.* 2 vols. Launceston: Henry Dowling.

# 20 Changes in the Aboriginal Population of Victoria, 1863-1966

Diane E. Barwick

Early estimates of the original native population of Victoria range from 5,000 to 15,000. The anthropologist Radcliffe-Brown (1930: 690-3) calculated, on the basis of estimated population densities, that the population just before direct contact was at least 11,500, and had already been reduced by one or several smallpox epidemics which spread westward from tribe to tribe between 1789 and 1835.

Fewer than 2,000 remained by 1863. All historical sources agree that there was an extraordinary decline in numbers during the 1840s and 1850s as a consequence of European settlement. The causes of this excessive mortality are almost clichés: increased intertribal warfare, the violence of settlers and the punitive expeditions of police, intemperance and, above all, the introduction of alien diseases.

## Distribution and Government Policy

Under the Port Phillip Protectorate, which endured from 1839 to 1849, occasional rations were issued to induce some natives to settle at the four depots in the Loddon Valley, the Western District, the Goulburn Valley, and near Melbourne. Salaries consumed most of the funds allowed and little remained for food, clothing, and medicines. Three church mission stations established later, near Colac and on the Murray, also lacked money to settle the Aborigines and provide adequate food and medical care. The Guardian of the Aborigines was able to give some food to the tribes near Melbourne during the 1850s, and those in the Loddon Valley also received some government aid, but elsewhere the natives were ignored.

A Select Committee (Legislative Council, Victoria 1858-9) inquiry was held in 1858 and the many settlers consulted agreed that the Aboriginal population had been almost halved in a decade, that certain tribes were already extinct, and that few children had been born since 1850. The Select Committee recommended that the government should reserve land for the various tribes in their own territories and grant funds for the regular distribution of food and clothing.

A Board for the Protection of the Aborigines was established in June 1860 to oversee this expenditure, in co-operation with a number of sympathetic settlers appointed as Local Guardians who maintained depots where occasional supplies were issued to the aged, sick, and orphaned children. It was the Board's policy, buttressed by the government's unwillingness to grant funds adequate for the care of all natives, that the able-bodied should support themselves and their families by working for settlers.

The depot system could not provide adequate protection from exploitation by unscrupulous employers, prostitution, and the illicit supply of alcohol, and many Aborigines wanted land to farm for themselves, so the Board sought large reserves suitable for supervised Aboriginal stations. For many reasons, not least the anticipated cost of salaries, it was thought best to encourage missionary endeavour rather than establish secular government stations.

Moravian missionaries were already active in the Wimmera, and Anglicans on the Murray. The petition of the Goulburn Valley tribes for land on the Acheron River had been granted in 1859, but the protests of settlers forced the government to choose a new site, Coranderrk, in 1863. This, the Board's first secular station, was designed to serve as a central asylum for children and young people from the Goulburn, Loddon, and Murray Rivers. The Gippsland tribes were cared for by Anglican and Presbyterian-Moravian stations established in 1861 and 1862. The Anglican mission committee, in response to local petitions ignored by the Board, opened the first Western District station, Framlingham, in 1865. The committee's attempt in 1867 to resettle these tribal remnants farther west, at the Lake Condah Mission, failed, and the Board had to rebuild Framlingham as a second secular station in 1869. Other small reserves, of 2 to 64 acres (0·8 to 26 hectares), were set aside for certain tribal remnants who would not leave their home territories. Other groups, mainly along the Murray, camped on the properties of friendly settlers.

Fig. 20:1 shows the locations of the supervised stations and the smaller reserves, together with the adjacent reserves and stations in New South Wales which have contributed to the present-day Victorian Aboriginal population.

Genetic criteria were used by the Victorian Board to define the number of persons entitled to assistance. At first the Board aided all persons of Aboriginal ancestry, but in 1886 the Aborigines Act was amended to implement the Board's decision to 'merge into the general population all half castes capable of earning their living'. By 1886 half of the estimated 844 surviving natives were of mixed ancestry. All those under thirty-four years of age who were not genetically 'full-blood' were told they were now 'legally white' and must find homes and jobs away from the Aboriginal stations. Unfortunately the serious economic depression which lasted through the 1890s wrecked the Board's experiment in assimilation, and discrimination was a further hindrance in obtaining jobs and housing. The Board did have power until 1893 to license needy half castes and their dependants to return temporarily to the stations for aid, but the Board's funds were so reduced that it could assist only the most desperate cases.

The Board's 'absorption' policy was rigidly enforced for three decades. Many children were transferred to institutions to be 'reared as whites'. Needy half castes

*Fig. 20:1 Locations of Victorian Stations and Reserves, 1860-1966*

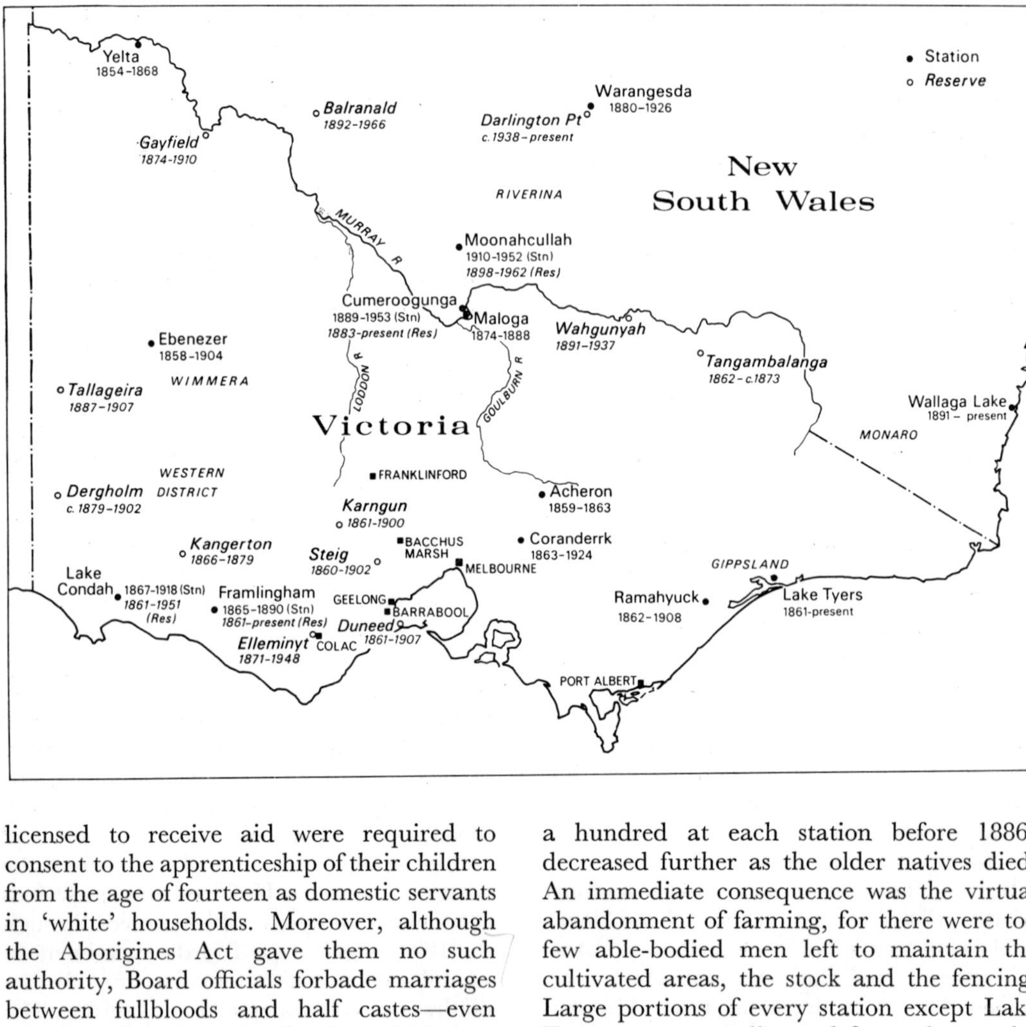


licensed to receive aid were required to consent to the apprenticeship of their children from the age of fourteen as domestic servants in 'white' households. Moreover, although the Aborigines Act gave them no such authority, Board officials forbade marriages between fullbloods and half castes—even using another clause in the Act which prescribed a fine or gaol term for any person who 'harboured' an Aboriginal without the Board's permission, to prosecute half castes who eloped with and attempted to marry fullbloods.

The size of the station communities had been halved by the 1886 policy decision, and the numbers there (an average of seventy to a hundred at each station before 1886) decreased further as the older natives died. An immediate consequence was the virtual abandonment of farming, for there were too few able-bodied men left to maintain the cultivated areas, the stock and the fencing. Large portions of every station except Lake Tyers were soon alienated for settlement by white farmers, despite pleas for occupation rights by the dispossessed half castes camping nearby. The Board then decided to reduce expenses by amalgamating the formerly separate regional groups. Five of the six stations were closed between 1904 and 1924 and the Aborigines still eligible for care were transferred to the most remote station, Lake

Tyers in eastern Gippsland. The Board used various means of persuasion to resettle those unwilling to leave their homes: the withdrawal of rations, the seizure of children and, if necessary, the forcible transfer of families and individual adults under police escort after obtaining an Order-in-Council prescribing their new place of residence. Yet a large number resisted, forgoing the aid to which they were entitled in order to camp near the ruins of their demolished homes with their half caste kinsmen and friends.

The 1886 legislation was amended in 1910 to empower the Board, at its discretion, to give half castes (which term was again defined to include 'all other persons whatever of mixed Aboriginal blood') the same assistance as fullbloods: that is, to reside at any reserve or station and receive rations and some clothing in return for their work. But from 1926 to 1956 officials interpreted this to mean only that needy persons of *not less than half caste standard* could be settled temporarily at Lake Tyers. The growing number camped elsewhere were ignored. Welfare organisations campaigned for reform, but government spokesmen continued to insist that the Board had no power to assist persons of part-Aboriginal ancestry, who were 'legally white'. Adults today bitterly recall the hopelessness of their situation: 'We were too black to get work, or State Relief, and too white to get help from the Board'. The government declined to take part in the important Native Welfare Conference convened at Canberra in 1951, explaining that there was 'no native problem in Victoria'. But by 1955 the fiction that people of Aboriginal descent had equal opportunities and were successfully assimilated could no longer be maintained. An official investigation led to the establishment in 1957 of an Aborigines Welfare Board, empowered to assist 'any person of Aboriginal descent'.

The policy decisions of the past are responsible for the disappearance of Aborigines from some districts of Victoria and the concentration of their descendants in the areas where the six Aboriginal stations were located. The regional populations—Gippsland, the Western District, the Wimmera, the Lower Murray, and the Goulburn Valley—are geographically and historically separate. Each group is closely knit by ties of kinship and common experience, and workers seldom travel far outside the regional boundaries. In 1962 the Western District group formed only one-sixth of the Victorian Aboriginal population, and the tiny Wimmera group and the (now rapidly increasing) Lower Murray people another one-sixth. The Goulburn Valley people (former residents of Coranderrk and Cumeroogunga Stations, and their descendants) formed more than one-third of the total, and the Gippsland people (including descendants of those transferred forcibly from other districts between 1904 and 1924) just less than one-third.

Before presenting a detailed discussion of the characteristics of the Victorian Aboriginal population at the time of the 1863 and 1877 enumerations and subsequently, I should describe the sources used.

### Records of the Victorian Aboriginal Population

Obviously one of the first tasks for the Board, on its appointment in 1860, was to ascertain the number of Aborigines in the colony and their locations, in order to budget for the distribution of supplies. Most of the settlers consulted in 1858 had only guessed at the numbers surviving. The Board immediately sent questionnaires to all Local Guardians and other interested settlers and missionaries, but the totals given in the Reports for 1861 and 1862 were only approximate, for not all had supplied adequate information.

*Table 20:1 Enumeration of Aborigines in Victoria, 1863*

| | Enumeration of Aborigines in Victoria (Compiled by BPA, 25 September 1863) (Totals only published in 3rd Annual Report 1863) | | | | | Numbers of half castes 1863 (Data not given in published census) | | | | |
|---|---|---|---|---|---|---|---|---|---|---|
| **Locality** | **Adults** | | **Children** | | **Total** | **Adults** | | **Children** | | **Total** |
| | **M** | **F** | **M** | **F** | | **M** | **F** | **M** | **F** | |
| Southern District (Melbourne) | | | | | | | | | | |
| Woiwurrung or Yarra Tribe | 9 | 8 | 3 | 2 | 22 | — | — | 1 | — | |
| Bunurong or Coast Tribe | 8 | 3 | — | — | 11 | — | — | — | — | |
| | 17 | 11 | 3 | 2 | 33 | — | — | 1 | — | 1 |
| Northern District (Goulburn Valley) | | | | | | | | | | |
| Goulburn and Delatite River Tribes | 40 | 28 | 14 | 13 | 95 | 1 | 2 | 4 | 4 | |
| (Murchison) | (3) | (6) | (3) | (1) | (13)[a] | — | — | 1 | 1 | |
| Echuca and Campaspe River | 30 | 17 | 16 | 11 | 74 | — | — | 11 | 4 | |
| | 73 | 51 | 33 | 25 | 182 (169) | 1 | 2 | 16 | 9 | 28 |
| North-eastern District (Upper Murray) Cobram, Wyuna, Barnawartha, Chiltern, Wangaratta, Wahgunyah, Tangambalanga, Wodonga, etc. | 51 | 24 | 14 | 21 | 110 | 2 | 3 | 4 | 5 | 14 |
| North-western District (Lower Murray) | | | | | | | | | | |
| Gunbower, Gannawarra | 38 | 22 | 5 | 7 | 72 | — | 5 | 4 | 1 | |
| Lower Loddon, Boort | 36 | 21 | 5 | 3 | 65 | 4 | 6 | 3 | 3 | |
| Swan Hill, Piangil | 88 | 39 | 23 | 21 | 171 | 3 | 2 | 5 | 5 | |
| Yaacko Yaacko Tribe (Yungera to Euston) | 30 | 21 | 10 | 4 | 65 | — | — | 2 | 1 | |
| Kulkyne | 25 | 12 | 8 | 5 | 50 | — | — | 2 | — | |
| Yarre Yarre Tribe (Mildura) | 23 | 13 | 1 | 2 | 39 | — | — | — | — | |
| Kamink Tribe (Ned's Corner) | 17 | 9 | 1 | — | 27 | — | — | 1 | — | |
| | 257 | 137 | 53 | 42 | 489 | 7 | 13 | 17 | 10 | 47 |
| Wimmera | | | | | | | | | | |
| Lake Hindmarsh and vicinity | 57 | 31 | 17 | 7 | 112 | — | 2 | 7 | 1 | |
| Horsham | 20 | 11 | — | — | 31 | 1 | 2 | 2 | 2 | |
| Richardson River and Morton Plains | 27 | 22 | 3 | — | 52 | — | 1 | — | — | |
| Glenelg and Mt Talbot | 32 | 13 | — | — | 45 | 1 | — | — | — | |
| | 136 | 77 | 20 | 7 | 240 | 2 | 5 | 9 | 3 | 19 |
| South-western District (Western District) | | | | | | | | | | |
| Casterton and Balmoral | 62 | 19 | 9 | 8 | 98 | 1 | — | 3 | 4 | |
| Hamilton | 33 | 14 | 8 | 3 | 58 | 1 | 2 | 4 | 1 | |
| Portland[b] | 49 | 25 | 13 | 13 | 100 | 1 | 4 | 5 | 6 | |
| Belfast and Port Fairy | 13 | 4 | — | — | 17 | — | — | — | — | |
| Warrnambool | 21 | 15 | 6 | 9 | 51 | — | 1 | 3 | 4 | |
| Mortlake[c] | 25 | 10 | 5 | 3 | 43 | 1 | 1 | 1 | 2 | |
| Wickliffe, Mt Rouse, Hexham | 32 | 18 | 13 | 7 | 70 | 4 | 2 | 8 | 7 | |
| Camperdown | 25 | 10 | 1 | 4 | 40 | — | — | — | 2 | |
| Colac and Geelong | 18 | 9 | 1 | — | 28 | 1 | 1 | — | — | |
| Bacchus Marsh | 24 | 8 | 1 | — | 33 | — | 2 | 1 | — | |
| Carngham (Mt Emu) and Ballarat | 37 | 29 | 2 | 1 | 69 | — | 2 | — | — | |
| Franklinford and vicinity | 19 | 12 | 5 | 2 | 38 | 2 | 1 | 3 | 1 | |
| | 358 | 173 | 64 | 50 | 645 | 11 | 16 | 28 | 27 | 82 |

| Locality | Enumeration of Aborigines in Victoria (Compiled by BPA, 25 September 1863) (Totals only published in 3rd Annual Report 1863) | | | | | Numbers of half castes 1863 (Data not given in published census) | | | | |
|---|---|---|---|---|---|---|---|---|---|---|
| | Adults | | Children | | Total | Adults | | Children | | Total |
| | M | F | M | F | | M | F | M | F | |
| South-eastern District (Gippsland) | | | | | | | | | | |
| Port Albert | *9* | *5* | *2* | *2* | *18* | — | — | — | — | |
| La Trobe and Rosedale | *20* | *18* | *8* | *4* | *50* | — | — | *2* | *1* | |
| Maffra, Macalister, Upper Mitchell, Omeo, etc. | *28* | *13* | *4* | *7* | *52* | — | — | — | *2* | |
| Swan Reach, Mitchell, Lake Tyers [d] | *31* | *19* | *4* | *5* | *59* | — | — | — | *2* | |
| Snowy River, Buchan, etc. [d] | *23* | *9* | *6* | *4* | *42* | — | — | *1* | *1* | |
| | *111* | *64* | *24* | *22* | *221* | — | — | *3* | *6* | *9* |
| Totals | 1,003 | 537 | 211 | 169 | 1,920[a] (*1,907*) | 23 | 39 | 78 | 60 | 200 |

[a] Additional number from 1866 newspaper report; the 1863 A.R. says 'there is reason to believe that some Aborigines in the central part of Victoria are not included'.
[b] Names of 6 men and 6 women unknown.
[c] Names of 14 men unknown.
[d] Bulmer's more complete figures used in lieu of Hagenauer's.

*Sources:* Figures in italics appear in surviving documents used for this compilation. Other figures for men, women, boys, and girls reconstructed from data in surviving papers of BPA; geneaologies; and attendance rolls 1874-86 of Maloga Mission N.S.W., where many Murray River and Goulburn Valley people settled.

The Board's Report for 1863 gave the first nearly complete and correct enumeration, of 1,907 persons, based on these returns. The next accurate enumeration, in 1877, showed that that Aboriginal population had declined by almost half, to only 1,067 persons.

Table 20:1 presents my reconstruction of the Aboriginal population in 1863. Only the totals for each locality were published in the 1863 Annual Report, with the comment that the Board now possessed the names and other particulars of 1,788 of the 1,907 counted.

I have been able to find the names of all but twenty men and six women by collating data from the surviving papers of the Board, the inward correspondence of the Chief Secretary's Department, and various unpublished records concerning the Aborigines of the Riverina region of New South Wales.[1] Twenty-four of the original returns, giving names and numbers of men and women, boys and girls, and specifying those who were half castes, have survived, but the notebooks of the Board's Inspector, naming those in another nine localities, could not be found. The Board's unpublished papers include attendance rolls and correspondence naming station residents and wanderers. The published Reports name every Aboriginal tried and sentenced between August 1861 and December 1875, those treated by medical officers at the stations, and give the names, ages, and physical descriptions of all men

[1] The archives of the Board for the Protection of the Aborigines were in the care of the Victorian Aborigines Welfare Board, now the Ministry of Aboriginal Affairs. The Chief Secretary's Papers, to 1900, are held by the La Trobe Library, Melbourne. The Maloga Mission Reports and Daniel Matthews's Papers are lodged in the Mitchell Library, Sydney. I must thank these authorities for permission to consult the documents.

*Table 20:2 Enumeration of Aborigines in Victoria, 15 March 1877* [a]

| Locality | Fullblood | | | | | Half caste | | | | | Total |
|---|---|---|---|---|---|---|---|---|---|---|---|
| | Adults | | Children | | Total | Adults | | Children | | Total | |
| | M | F | M | F | | M | F | M | F | | |
| Central (Goulburn Valley) | | | | | | | | | | | |
| Melbourne and Western Port | 1 | 1 | | | 2 | | 1 | | | 1 | 3 |
| *Coranderrk Aboriginal Station | 22 | 16 | 6 | 7 | 51 | 15 | 24 | 19 | 26 | 84 | 135 |
| Avenel, Alexandra | 4 | 6 | | | 10 | 1 | | | | 1 | 11 |
| Wyuna | 10 | 5 | | | 15 | 2 | 1 | 2 | 2 | 7 | 22 |
| Ulupna | 19 | 13 | | 5 | 37 | 5 | 1 | 5 | 5 | 16 | 53 |
| Romsey, Corop | 3 | | | | 3 | | | | | | 3 |
| | 59 | 41 | 6 | 12 | 118 | 23 | 27 | 26 | 33 | 109 | 227 |
| North-eastern (Upper Murray) | | | | | | | | | | | |
| Wangaratta | 10 | 5 | | 1 | 16 | | 1 | | 1 | 2 | 18 |
| Tangambalanga, Wharparilla, Myrtleford | 3 | 2 | 3 | | 8 | | | | | | 8 |
| | 13 | 7 | 3 | 1 | 24 | | 1 | | 1 | 2 | 26 |
| North-western (Lower Murray) | | | | | | | | | | | |
| Durham Ox, Terrick-Terrick | 2 | | | | 2 | | | | | | 2 |
| Kerang | 6 | 4 | 4 | 1 | 15 | 1 | | 1 | | 2 | 17 |
| Swan Hill (and Piangil) | 72 | 39 | | | 111 | 2 | 1 | | | 3 | 114 |
| Cowana, Narung | 17 | 7 | 1 | | 25 | | | | | | 25 |
| Kulkyne | 7 | 2 | 2 | 1 | 12 | 2 | | 1 | | 3 | 15 |
| Mildura | 18 | 7 | | | 25 | | | | | | 25 |
| Ned's Corner | 17 | 4 | 1 | 2 | 24 | | | | | | 24 |
| | 139 | 63 | 8 | 4 | 214 | 5 | 1 | 2 | | 8 | 222 |
| Wimmera | | | | | | | | | | | |
| *Lake Hindmarsh Aboriginal Station | 17 | 4 | 6 | 5 | 32 | 6 | 6 | 7 | 9 | 28 | 60 |
| Towaninnie | 17 | 5 | 1 | | 23 | | 3 | 3 | | 6 | 29 |
| East Charlton | 5 | 2 | | | 7 | 1 | 1 | | | 2 | 9 |
| Carr's Plains, Banyenong | 5 | 1 | 1 | | 7 | 1 | 2 | 4 | | 7 | 14 |
| Barton, Elmhurst | 2 | | | | 2 | | | | 1 | 1 | 3 |
| St Arnaud, Navarre | 3 | 3 | | | 6 | 2 | | | | 2 | 8 |
| Horsham, Warracknabeal | 5 | 1 | | | 6 | 1 | | | | 1 | 7 |
| Edenhope, Apsley | 11 | 2 | | | 13 | | | | | | 13 |
| | 65 | 18 | 8 | 5 | 96 | 11 | 12 | 14 | 10 | 47 | 143 |
| South-western (Western District) | | | | | | | | | | | |
| Casterton, Dergholm, Nareen | 7 | 6 | 2 | 1 | 16 | 2 | | | | 2 | 18 |
| Balmoral, Coleraine, Cavendish | 9 | 4 | | 2 | 15 | 2 | 1 | | | 3 | 18 |
| Hamilton, Merino, Heywood, Portland | 4 | | | | 4 | 2 | 2 | 2 | 2 | 8 | 12 |
| *Lake Condah Aboriginal Station | 23 | 13 | 6 | 9 | 51 | 4 | 7 | 8 | 11 | 30 | 81 |
| Belfast, Warrnambool, Camperdown | 6 | 1 | | | 7 | 1 | | | | 1 | 8 |
| Wickliffe, Hexham, Dunkeld | 6 | 1 | | | 7 | 1 | | | | 1 | 8 |
| *Framlingham Aboriginal Station | 29 | 11 | 4 | 1 | 45 | 5 | 8 | 4 | 7 | 24 | 69 |
| Colac and Geelong | 5 | 2 | | | 7 | 1 | 1 | 1 | 2 | 5 | 12 |
| Beaufort, Skipton | 1 | 1 | | | 2 | 1 | 1 | | 2 | 4 | 6 |
| Castlemaine | — | 2 | | | 2 | | 1 | 1 | | 2 | 4 |
| | 90 | 41 | 12 | 13 | 156 | 19 | 21 | 16 | 24 | 80 | 236 |

| Locality | Fullblood | | | | | Half caste | | | | | Total |
|---|---|---|---|---|---|---|---|---|---|---|---|
| | Adults | | Children | | Total | Adults | | Children | | Total | |
| | M | F | M | F | | M | F | M | F | | |
| South-eastern (Gippsland) | | | | | | | | | | | |
| Sale, Toongabbie | 2 | 2 | | | 4 | 1 | | | | 1 | 5 |
| *Lake Wellington Aboriginal Station | 18 | 12 | 11 | 12 | 53 | 2 | 4 | 6 | 8 | 20 | 73 |
| Bairnsdale | 14 | 9 | 8 | 9 | 40 | | | | | | 40 |
| *Lake Tyers Aboriginal Station | 18 | 18 | 6 | 10 | 62 | | 2 | 1 | 3 | 6 | 68 |
| Bendoc, Omeo | 7 | | | | 7 | 3 | 2 | 8 | 7 | 20 | 27 |
| | 59 | 41 | 25 | 31 | 166 | 6 | 8 | 15 | 18 | 47 | 213 |
| Totals | 425 | 211 | 71 | 67 | 774 | 64 | 70 | 73 | 86 | 293 | 1,067 |

[a] Census conducted by police. Published in Thirteenth Annual Report, 1877.

* Aboriginal Stations.

obtaining work certificates from 1870 to 1875. There is also considerable material on named individuals in early histories and in newspaper files.

I am satisfied that my reconstruction of the 'base population' of 1863 is fairly adequate. The figures in italics in Table 20:1 are those appearing in the surviving returns. Other sub-totals were reconstructed.

Table 20:2 presents the census of 1877. It was published as a random listing of localities, compiled from returns obtained from every police district. For Table 20:2 the localities had to be identified with the aid of a contemporary gazetteer (Whitworth 1870), and regrouped according to region to parallel the 1863 enumeration. None of the original police returns survives in the Board's papers and we cannot know whether names were supplied. But a useful check on the numerical totals of the published census is furnished by a listing which survives in the Chief Secretary's Inward Correspondence (File 76/J9682, La Trobe Library Archives). In August 1876 the Board had ordered the station managers to compile the names of all adult residents together with those of adult wanderers who should be compelled to settle at the stations for their better protection. The lists were submitted as a mass application for an Order-in-Council specifying residence, but the request was not granted, the government perhaps considering it an infringement of the natives' liberties or, more probably, deterred by the cost of the police details which would be required to locate and escort wanderers to the stations, and keep them there.

About many of the people enumerated in 1863 we know nothing but their name, sex, and whether they were old or young. They died before 1877 and left no descendants. About others, particularly those who settled, married, bore children, and died at the stations, we know a great deal. Their ages can be verified, their incomes assessed. The illnesses they were subject to, their reputation as workers and householders, the numbers of children they bore and lost, their loves and foibles and deathbed sufferings are recorded not only in the memories of their descendants, but also in the meticulous reports of the sometimes fond, sometimes exasperated officials who cared for them. Numerous letters and petitions in their own carefully elegant script also survive to record their opinions, grateful or indignant, of the managers and missionaries, and of the Board and its policies.

The primary sources for the ensuing discussion of population changes in Victoria are the oral histories, the recounted genealogies, collected from most adult members of the present-day population. On his appointment as Superintendent in 1958, Mr P. E. Felton set out to collect such data in order to discover the number of persons for whom the newly-formed Aborigines Welfare Board was responsible. When I began field work in 1960 I found that such family histories were essential in understanding the regional affiliations and social networks of the Victorian dark people. My own findings were collated with the records for my enumeration of 1962 and again at the end of 1967, to produce an enumeration for the Welfare Board and the Centre for Research into Aboriginal Affairs, Monash University. I was also able to compare these oral histories with versions collected by Mr N. B. Tindale and Dr J. B. Birdsell during the Harvard-Adelaide Universities Expedition of 1938-9, and with data collected in 1951 by Dr Donald Tugby.

Dates have been verified by checking birth, death, and marriage records published in the Board's reports from 1876 to 1912, and with unpublished registers and correspondence in the Board's papers for the period 1860-1956. In 1878 the managers were appointed District Registrars of Vital Statistics and their records are complete thereafter until the closure of the stations. Separate rolls, naming those eligible and ineligible for aid, were maintained by the Board's secretary until the 1920s. After 1924 births and deaths outside Gippsland were only haphazardly recorded, but can be satisfactorily reconstructed from living informants. Similar data were published in the *Maloga Mission Reports*, 1874-88 (Mitchell Library), and the monthly journals *The Australian Evangel* 1929-67 and *Our Aim* 1907-61 (Aborigines Inland Mission, Sydney) for the New South Wales Riverina and parts of Victoria.

These contemporary records, and others, provide a check on the accuracy of the oral histories, and they are extraordinarily accurate. The 1938, 1951, and 1958-67 oral accounts are virtually identical, and of course the latter can be cross-checked by comparing the testimonies of numerous descendants. Almost always errors and variations turn out to be due to the ignorance of the recorder, not lapses by the informant. Stillbirths are likely to be forgotten unless there is specific questioning on this point, and infant deaths occurring several generations ago may be missed except by a close relative, but an *affair de coeur* which resulted in issue is remembered even when the offspring have been dead for fifty years. Because marriage with first or even second cousins was forbidden (and all insist this is 'the old law'),[2] accurate identification of the genitor has always been important, although the name might be forgotten, as irrelevant, in the case of an outsider from another state or a white man. In such a small population, of course, the facts of physical resemblance permit few secrets.

The whole period of European occupation of Victoria is, after all, little more than two lifetimes, and in these small communities where almost every member is descended from the dozen or twenty founding couples who settled at each station during the 1860s, and households are still composed of three and even four generations, genealogies are subject to constant discussion and little

[2] The most widespread kinship systems in Australia forbade actual first-cousin marriage, but permitted marriage between classificatory second cousins (Elkin 1948:60-4).

distortion is possible. Today the genealogical knowledge of most older adults is extraordinarily vast: they can reconstruct the complete history of their own families and, indeed, most of the cognatic stocks of the same home region, dating back to the founding couples born in the 1850s or earlier. Only a few members of the seventh descending generation have been born; most of the sixth generation are still children; and many of the fifth generation have yet to marry. The names of members of the ascending generations are unknown, unless they survived to 1863, because of the customary reluctance to speak the tribal names of the deceased. European nicknames did not replace tribal names until the late 1850s and surnames were not widely adopted until the 1870s.

Estimates of age, before births were registered, are of course imprecise, but fortunately several estimates, made years apart, are available for most individuals and the presumed birth dates usually do not differ by more than five years. In assessing ages, officials apparently made use of datable events such as the arrival of certain settlers, and markers such as the 'Black Thursday' fires of 6 February 1851.

In the last section of this paper I shall describe the base population at 1863, the changes which had occurred by 1877, and certain geographical, historical, physiological, and cultural factors which influenced population size in subsequent decades.

## An Analysis of Population Changes

Before discussing in detail characteristics of the 1863 population and subsequent changes I must define what I mean by the 'Victorian Aboriginal population'. The census lists of 1863 form a base population. The adults and children (many of them orphaned) who were counted then, together with certain accretions—immigrants from other states who 'married in'—and their descendants, compose the 'Victorian' population for which statistics are presented in the last part of this paper. I have not included immigrants who have come within the last decade and not married Victorians, except when mentioning the enumerations of 1962 and 1967. Earlier accretions are included. For example, the Moravian missionaries imported a dozen girls from Western Australia during the 1860s as brides for their Christian converts, and also welcomed a few Queensland servants who had been abandoned in Melbourne by their employers. Immigrants from New South Wales settled in Gippsland during the 1870s and 1880s and subsequently intermarried. There was also some intermarriage with men and women brought from Queensland and Sydney to the Maloga Mission (later Cumeroogunga Station) at the Murray-Goulburn junction. Many half castes from this district had been removed to Coranderrk during the 1860s, but some fifty rejoined their kinsmen and friends at Maloga between 1876 and 1885. Another sixty exiled half castes left Coranderrk for Cumeroogunga after 1886, and there was constant visiting until Coranderrk was closed. The New South Wales Aborigines Protection Board in its 1909 Act and 1910 Regulations copied (indeed, quoted verbatim) the Victorian legislation excluding half castes from care, and as a result 150 were dismissed from Cumeroogunga between 1909 and 1915. Most camped at a Victorian township directly across the Murray, and spread southward during the 1920s and 1930s.

Many of the Lower Murray natives refused to settle at the Victorian stations. By 1900 most of the aged fullbloods had died, leaving few descendants. The half castes, ineligible for aid after 1886, apparently moved across the Murray, drawn by employment opportunities and, probably, with the intention of finding spouses among the tribal remnants beginning to settle at the adjacent New South Wales reserves and Aboriginal stations. During the 1930s their descendants again migrated, drawn to the expanding fruit-growing centres in Victoria. In compiling my 1962 enumeration I had assumed, as did Victorian officials, that most of the newcomers to towns such as Swan Hill and Robinvale were New South Wales folk, as their names are identified with the reserves where most have lived for the last few generations, and I was surprised to discover, when preparing this paper, just how many of their surnames appeared in the Victorian rolls of the 1860s and 1870s.

There was one other minor source of increments: the issue of European wives of Aboriginal men. The 1877 census noted three such marriages. Of the 2,300 Aboriginal men married since 1863 only 107 are known to have married white women, and most of these unions have occurred since World War II. Of the 111 white wives (several widowers remarried) 95 produced 212 children, but the number born to the other sixteen wives is not accurately known. These women are not included in the tables showing mean family size and juvenile mortality presented at the end of the paper, but their daughters who lived to reproductive age are counted, as they are of course part-Aboriginal.

## POPULATION SIZE, 1863-1966

The total number alive at 1863 was 1,920. Settlers and missionaries then seemed to use age thirteen as the dividing line between childhood and adult status, so we can assume that 1850 was the birth year of the oldest of those classified as children. The base population of 1863 had certain peculiar characteristics. The most noticeable was the excessive masculinity. Of the total number, 52 per cent were *adult* males. There were almost twice as many adult males as adult females. Although children formed only 19 per cent of the total number, the proportion of juvenile males and females was fairly normal: 55 per cent were males. This suggests that although infanticide may have been generally practised during the 1850s there was no special emphasis on killing female children.

By 1877 there were only 1,067 alive, and 297 of these, classified as children, were presumably born after 1863. There had in fact been a loss of 1,150 persons, 60 per cent of the 1863 total, within fourteen years. There was still an excess of adult males at 1877: 489 to 281 adult females. But now there was a very slight excess of females among the juveniles: 144 males to 153 females.

The regional distribution had changed markedly by 1877 in that a majority of the children and young women had been settled at the stations. The apparent losses from some districts are illusory. The Upper Murray District numbers had seemingly declined from 110 to 26, but in fact many had been resettled at Coranderrk and some others had crossed the Murray to the nearby Maloga Mission. The Lower Murray District population had apparently decreased from 489 to 222, but a number had been resettled at Coranderrk and Ebenezer Stations. The numbers in the Goulburn Valley were artificially inflated as a result of this resettlement at Coranderrk. There were genuine losses in the Wimmera: a decline from 240 to 143 despite the resettlement of Lower Murray people at Ebenezer. The most serious decrease occurred in the Western District: the population here declined from 645 to 236 (a loss of 63 per cent) but only thirty young people were removed from the district to Coranderrk. The Board's neglect of this region, before stations were established in 1867 and 1869, may account for this substantial loss.

The apparent stability of the Gippsland population—221 in 1863, 213 in 1877—deserves special examination, for none were resettled outside this region. Because lists of names of all district residents at 1863, 1876, and 1879 have survived it is possible to calculate actual losses. By 1877, 102 members of the original population (46 per cent) had died and been replaced by 94 children and immigrants, the latter mostly from the Monaro region of New South Wales. The Lake Tyers manager recorded, in the Board's Annual Report for 1877, the arrival in November and December 1876 of 39 natives 'belonging partly to the Bendock and Delegate tribes, and a few, the last remnant, of the Bidwell tribe'.

The regional distribution of the 200 half castes counted in 1863 is shown in the right-hand columns of Table 20:1. The total in 1877 (Table 20:2) was 293. Some of those counted as 'half castes' at 1877 were, it is clear from the genealogies, actually three-quarter and one-quarter castes as well as first and second generation half castes. Indeed, at 1863 there was already a handful of these second generation hybrids, issue of half caste parents born during the first decades of contact. In ensuing generations hybridisation has grown ever more complex. Miscegenation was common during the 1840s and 1850s but settlement at the stations was an effective protection. Most of the succeeding generations counted as half castes were descended from the relatively small number alive at 1877, for unions with whites did not become common again until after World War II. By December 1967 only three 'Victorian' fullbloods, descendants of those counted at 1863, survived. The number of fullbloods has steadily diminished because some couples were childless, many lost all their offspring in infancy, and those children who did survive were as likely to marry half castes. By official fiat the ensuing generations were counted among the half castes, for the Victorian Board after 1886 classified all descendants of half castes as ineligible for aid, even when the other parent was full-blood or three-quarter caste.

The total number listed as station residents in the 1877 census was 486, or 45 per cent of the total, but the average attendance in that year in fact fluctuated between 432 and 617. Most of the wanderers were on the Murray, but a fairly high proportion of the Gippsland people were still camping away from the stations and in fact a substantial number of conservative old people continued to wander until late in the 1890s. In 1885, just before the removal of the half castes, the estimated total of surviving natives was 844, and the average attendance at the stations was 426-594: at most only 70 per cent of the Victorian Aborigines were ever settled at the stations.

Figures given for the Aboriginal population in the Victorian Year Books after 1877, whether based on colonial and later Commonwealth censuses, or on enumerations and estimates by the Board, are of dubious value. The accuracy of the colonial censuses may be judged by an exchange of correspondence in June 1882. The Government Statist had forwarded his total of 780 and asked for 'the correct number'. The Board's secretary replied (Letter Book 1880-3):

> Judging by my returns there are about *870* but it is hard to get information on the numbers at Swan Hill and that district as the Aborigines consider the river no boundary so far as they are concerned.

The accuracy of the official totals after 1877 is doubtful for two reasons: the continuous migration between Victoria and New South Wales, and changes in the official definition of 'Aborigines'. Migration to and from South Australia has been negligible, but large numbers have always travelled across the Murray and between Gippsland and the south coast of New South Wales for seasonal employment, and there have also been mass migrations as a consequence of the two governments' changes in Aboriginal policy.

In the early censuses the sub-enumerators' judgment of 'caste'—according to physical appearance—was used, rather than the respondents' self-definition, and this may have affected the totals given in official census returns. The Board's enumerations, too, omitted those considered 'legally white'. In any case the censuses and the Board's figures vary wildly. The Commonwealth census for 1921 gave a total of 586, yet the returns for a special Board enumeration in May 1921 include only 402 names, and of these only 197 were eligible for aid. In 1933 the census total was 1,034 but the Board's estimate was 1,229. The 1947 census total was 1,277, yet the Board's estimate for the following year was only 801.[3]

The census total for 1961 was 1,796, but my own count at March 1962 gave a total of 2,989 people actually resident in Victoria who identified themselves as Aborigines. Almost 51 per cent were then under fourteen years of age.

Only part of the 1966 census total is available: the number stating they were of less than half Aboriginal descent has not been released. Those who identified themselves as half, three-quarter, or fullblood Aborigines totalled 1,790. A survey of the Welfare Board's records in December 1967 gave a total of 4,432 persons of Aboriginal descent (Tatz pers. comm.). Several hundred reside in New South Wales and only seek seasonal work in Victoria, so a total of 4,000 permanent residents would be more accurate. There has been considerable immigration since my field work in 1960-2, perhaps 500 in all, including about 150 from distant states.

My own enumeration of March 1962, corrected by the collection of further genealogical data, omitted about a hundred persons. The difference, then, between the corrected total of 3,100 at March 1962 and the 3,500 'Victorian' population at December 1967 is a natural increase of 400 persons in almost six years. This is not unreasonable, in that the changing age structure of this population means a continuing increase in the proportion of females in the high-fertility age group 15-29 years.

Over a century there have been great changes in the Victorian Aboriginal population. Today half or more are children, the proportion of males and females is normal, and the population is likely to increase even more rapidly since most females now survive to reproductive age and (judging by intervals between births) there is still little attempt to limit family size. But a century earlier, at 1863, there was only a 'dying' population, with a small proportion of adult females, and few children.

### CHARACTERISTICS OF THE BASE POPULATION

Some peculiarities of the age and sex structure of the 1862 population have already been noted. These abnormalities were presumably a consequence of rapid population decline during the first two decades of contact. The extraordinary mortality of the 1850s was commented on by all settlers consulted during the 1858 Select Committee inquiry, but unfortunately few gave detailed population estimates. Analysis of the losses suffered in certain districts is possible, how-

[3] It is possible to chart actual population growth during this period, using genealogies of Victorian families and data on the arrival of immigrants, but I did not have time to complete the analysis for this paper.

ever, because of the publication (Legislative Council, Victoria 1853-4) of enumerations made in 1852 by Crown Lands Commissioners appointed as guardians after the abandonment of the Protectorate.

The Commissioner for the Portland Bay District in December 1852 compiled returns subdivided into localities which parallel (except for the exclusion of Geelong, Bacchus Marsh, and Franklinford) those of the South-Western District in the 1863 census. Grouped, these gave a total of 700, which had declined by 22 per cent to 546 in 1863. These 1852 figures separate adult males and females, but group the children, so it is not possible to show precisely changes in sex composition. But certain characteristics were already marked: the excess of adult males (367 to 240 adult females) and the small proportion of juveniles (93 children, or 13 per cent of the total). Eleven years later there were 105 children (with a nearly normal male-female ratio of 57:48), forming 19 per cent of the total. But only 303 adult males survived at 1863 (a loss of 17·5 per cent), and only 139 adult females (a loss of 42 per cent).

Another Commissioner counted thirty in the Geelong tribe in January 1852, and this number was halved by 1863. There had been 275 in 1837, when they were first assembled by Fyans with the aid of William Buckley (Bride 1898:115), but only 118 of the 'Barrabool tribe' survived in 1842 (Bride 1898:310-11) and although the male-female ratio of adults (44:41) was still normal, the proportion of children was small (28 per cent) and there were twice as many boys as girls.

The Jajowrong (Jaara) tribe of Franklinford and the Loddon Valley had numbered 282 when Assistant Protector Parker (1841a) prepared a complete census in February 1841. The male-female ratio of the 'aged and adult' was then 95:83, and that of the 'youths and children' was 58:46. In December 1852 only fifty men, forty-two women, and fifty children survived, a decrease of 50 per cent in less than twelve years. In 1863 only thirty-one adults and seven children remained, a further decrease of 73 per cent in eleven years.

The Port Albert tribe of western Gippsland totalled thirty-two when Commissioner Tyers counted them in November 1852, but only eighteen survived in 1863. Tyers estimated there had been about three hundred in 1844, but great losses were caused by an intensification of inter-tribal fighting in 1846-7.

William Thomas kept complete records of the losses suffered by the Woiwurrung and Bunurong tribes of Melbourne and vicinity who were under his care from 1839 to 1863. He counted 207 in 1839 and estimated there had been 350 only three years earlier, before settlement began (Foxcroft 1941:53). In 1839 there was already a surplus of males (124:83) and at December 1852 only thirty-three men and twenty-six women survived. The total in 1863 was thirty-three: a loss of 86 per cent since 1839.

These figures indicate that the tribes of the Western District, Melbourne, and western Gippsland were already 'remnant' populations in 1852. Unfortunately the 1852 counts gave no detailed information on age structure. But the original returns for the 1863 enumeration gave precise ages for the individuals comprising four tribal groups and sufficient data survive for a reconstruction of the age distribution within the five tribal remnants which made up the total Gippsland population in 1863. The age and sex distributions of the nine groups are shown as population pyramids in Fig. 20:2.

Chance variation, in such small groups, may produce abnormalities, and some distortion may be caused by errors in estimating ages, but the absence of a normal pyramidal structure in these groups is primarily due to

*Fig. 20 : 2 Age structure of nine tribal remnants, 1863*

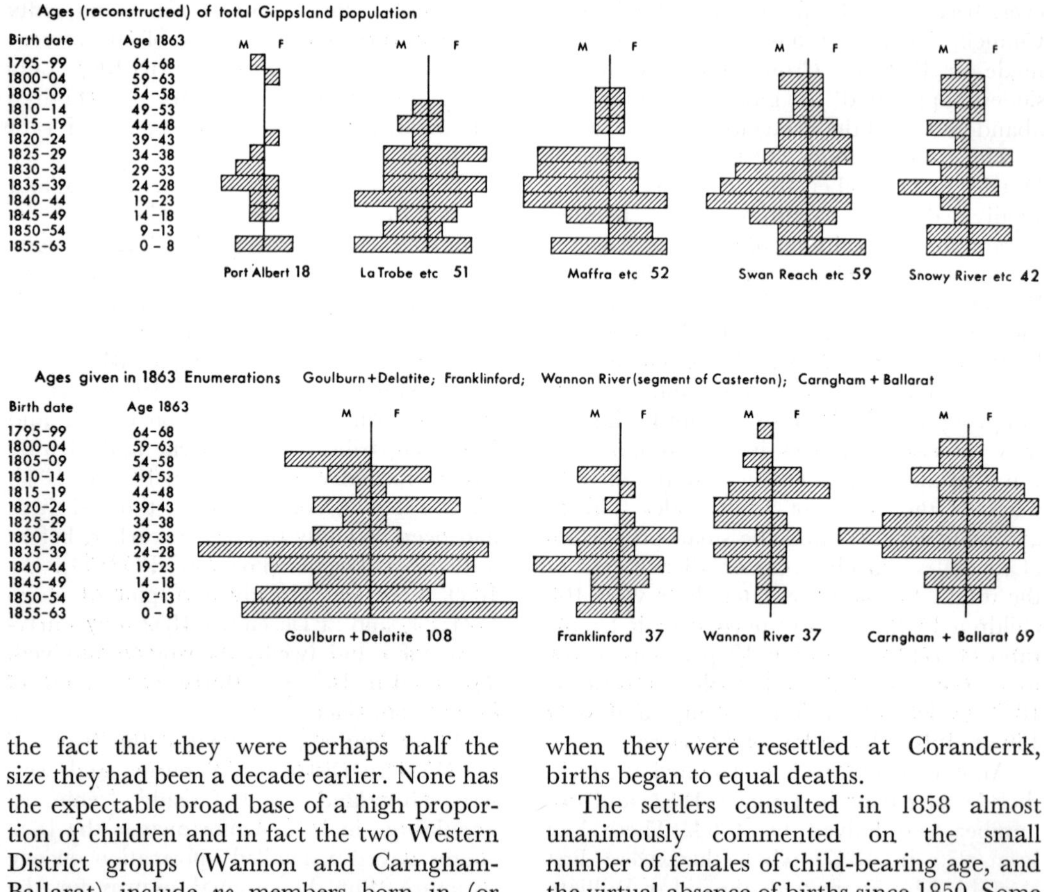


the fact that they were perhaps half the size they had been a decade earlier. None has the expectable broad base of a high proportion of children and in fact the two Western District groups (Wannon and Carngham-Ballarat) include *no* members born in (or at least surviving from) the years 1855-63.

The pyramid on the lower left of Fig. 20:2, showing the Goulburn and Delatite tribes, is of special interest in that this is the only group with a sizeable proportion of children and a fairly normal sex ratio (56:42). Thomas had counted eighty-four Taungerong (53:31), including only two children, when he toured the Upper Goulburn in 1852. They, with a related group from the Delatite River, were settled at the Acheron Station in 1859. Infanticide was ended in 1860 and venereal disease eradicated in 1862. By 1863, when they were resettled at Coranderrk, births began to equal deaths.

The settlers consulted in 1858 almost unanimously commented on the small number of females of child-bearing age, and the virtual absence of births since 1850. Some suspected that infanticide was practised in their districts, and half mentioned that venereal disease, specifically 'syphilis', was common.

Their observations are supported by Thomas's records (1861:27-9) for the Woiwurrung and Bunurong: between 1839 and 1859 there were 135 deaths and twenty-one births, only eight of these in the latter decade. He noted that 'most died before the first month'. Thomas did not specify the incidence of infanticide, but his comment that most adults had 'syphilis' before he controlled the disease in 1855 by applications of 'bluestone' (copper sulphate) suggests that infertility was largely due to post-venereal sterility.

Parker's reports and letters (1841b, 1841d, 1842) noted that 'syphilis' also spread extremely rapidly among the Jajowrong: there were five cases among 170 natives in December 1840; 'half' of the women were infected by October 1841; and 'the mothers, with one exception, have been or still are diseased' at July 1842. Parker had noted in May 1841 that infanticide was very rare in this tribe: only two women were known to have killed an infant. But as a result of the spread of venereal disease he had to report in August 1842 that births were few, and fewer still survived, for 'in a few weeks, sometimes in a few days, after their birth, the poor children become covered with a syphilitic eruption and speedily perish'.

Early sources consistently identify the venereal infection as syphilis; gonorrhoea is never mentioned, but this form is primarily responsible for sterility. The ascription of infant deaths and deformities to 'congenital syphilis' was probably correct. Almost certainly both forms were present. Consequent sterility and infant mortality were largely responsible for the distorted age structure of the 1863 population.

We do not have sufficient information to explain the excessive masculinity of the Victorian Aboriginal population at 1863 and earlier. None of the ethnographic accounts suggest that there was an emphasis on female infanticide. The exposure of the females to venereal infection may have made a difference, but the disease was soon widespread among the men. It may be that the women and children bore the brunt of inter-tribal warfare and the settlers' violence, but Corris's (1968:153-8) attempt to estimate the numbers killed in the Western District before 1860 shows only that the victims were not identified sufficiently clearly to count the females among the 159 known killed. It is likely, of course, that the perpetrators were disinclined to admit to the slaughter of women and children. The excessive masculinity may be largely the result of differences in mortality, as death records for fullbloods in the period 1876-1912 show that a substantially higher proportion of females died as children and young adults.

#### THE CAUSES OF MORTALITY

The most complete surviving records of causes of death are the annual lists published in the Board's Reports from 1876 to 1912. The death dates of 698 station residents, with cause as recorded on the death certificates, appear. The deaths of another 146 residents, and their causes, were not listed, either because of inadvertent omission, or because they occurred in hospitals or on employers' properties. The managers were always reluctant to inflate their mortality figures by listing deaths which occurred off their stations.

Age at death is known for all of these 844 residents. Of this total 658 were fullbloods. The records are skewed, of course, because most of the half castes were not station residents after 1886, and the majority belonged to later generations so their life span exceeded the cut-off date of 1912 when the Board ceased to publish Annual Reports. For Figs. 20:3 and 20:4 I have used the Board's classification: the 'fullbloods' are genetically Australoids, and the 'half caste' numbers include all other degrees of hybridisation.

In Fig. 20:3 (representing 844 deaths) the percentages of deaths at various ages among fullbloods and half castes are shown as a bar graph. The proportion of juvenile deaths is very great: 44 per cent of fullblood deaths and 59 per cent of half caste deaths were

*Fig. 20:3 Age at death, Victorian Aboriginal Station residents, 1876-1912*

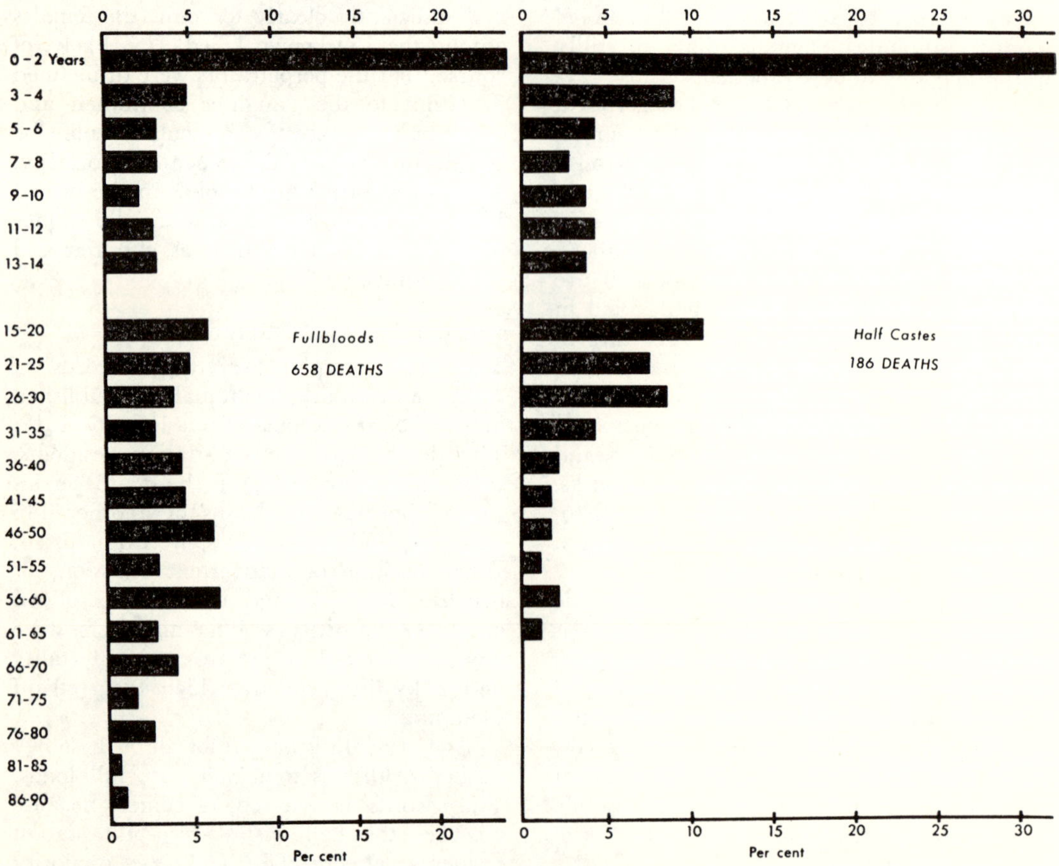


those of children aged 0-14 years. The incidence of juvenile mortality among the half castes is distorted, as a result of the 1886 removal policy: of the 186 half caste deaths recorded only fifty-two were adults, but of course hundreds reached adult ages after leaving the stations. Moreover, it was only the sickly with many dependants who were temporarily licensed to return to the stations for care.

Fig. 20:4 reclassifies these 844 deaths to show sex differences. Juvenile deaths are shown in six-month age groups for the first two years, then two-year groups to age fourteen. Adult deaths are given in five-year age groups.

Little can be said about the proportions of half caste deaths because of the known skewing, but the differences between male and female fullbloods are significant. For example, 52 per cent of the deaths of females, but only 38 per cent of the deaths of males, occurred under the age of fourteen years. Another 24 per cent of females died aged 15-40, but the male loss was only 22 per cent. This difference must have made it more

*Fig. 20:4 Age at death, classified by sex and caste, 1876-1912*

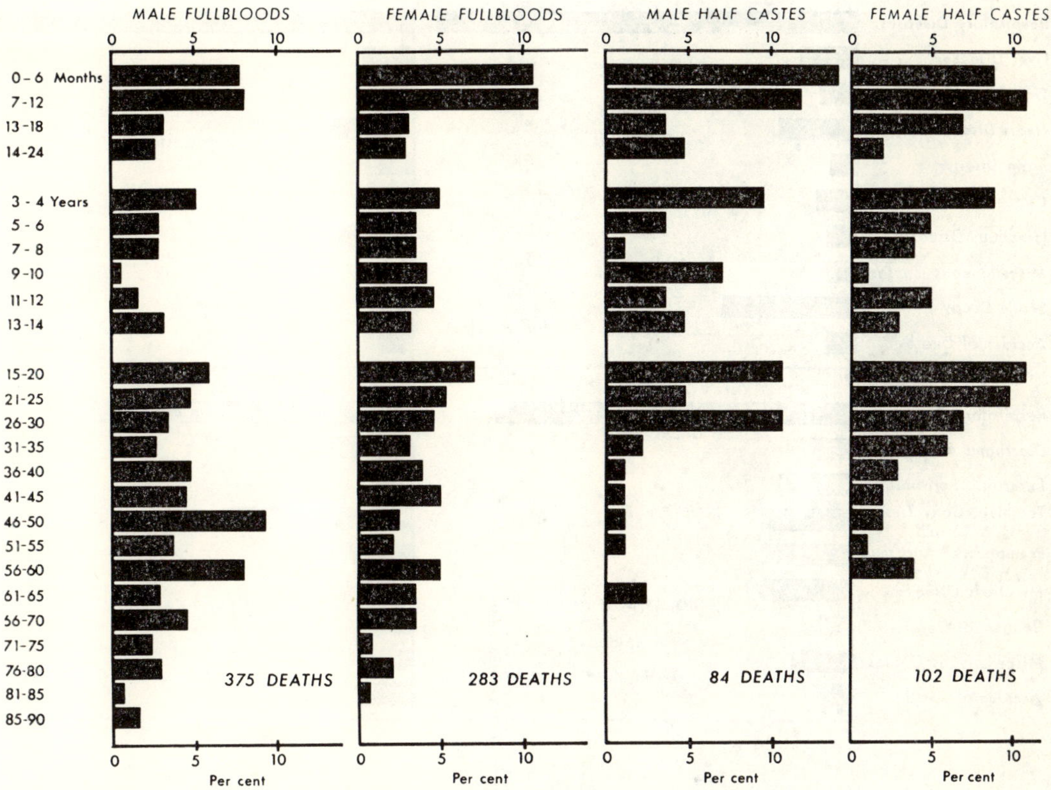


difficult for fullblood males to find spouses, especially after the Board tried to forbid marriages with half castes. Of the 658 fullblood station residents who died in this thirty-six year period, three-fifths of the males and three-quarters of the females were under the age of forty, surely an indication of the generally poor health of this population.

The 736 deaths from known causes are analysed in Fig. 20:5. This bar graph shows percentages of deaths from various diseases among juvenile (0-14 years) and adult (15-90 years) fullbloods and half castes.

Immediately obvious is the high proportion of deaths ascribed to respiratory diseases: 47 per cent of fullblood and 44 per cent of half caste adults, 41 per cent of fullblood and 36 per cent of half caste juveniles. Under this heading I have grouped tuberculosis (also called consumption, phthisis, and lung disease or inflammation) with pleuro- and broncho-pneumonia and bronchitis. Tuberculosis itself caused 38 per cent of adult deaths, but only 18 and 19 per cent of the fullblood and half caste juvenile deaths.

'Liver disease' killed twenty-five adults as well as five of the children included under 'miscellaneous causes' in Fig. 20:5. Fifteen deaths were explicitly ascribed to hydatid infection, four to hepatitis, and two to

*Fig. 20: 5 Known causes of death, Victorian Aboriginal Station residents, 1876-1912*

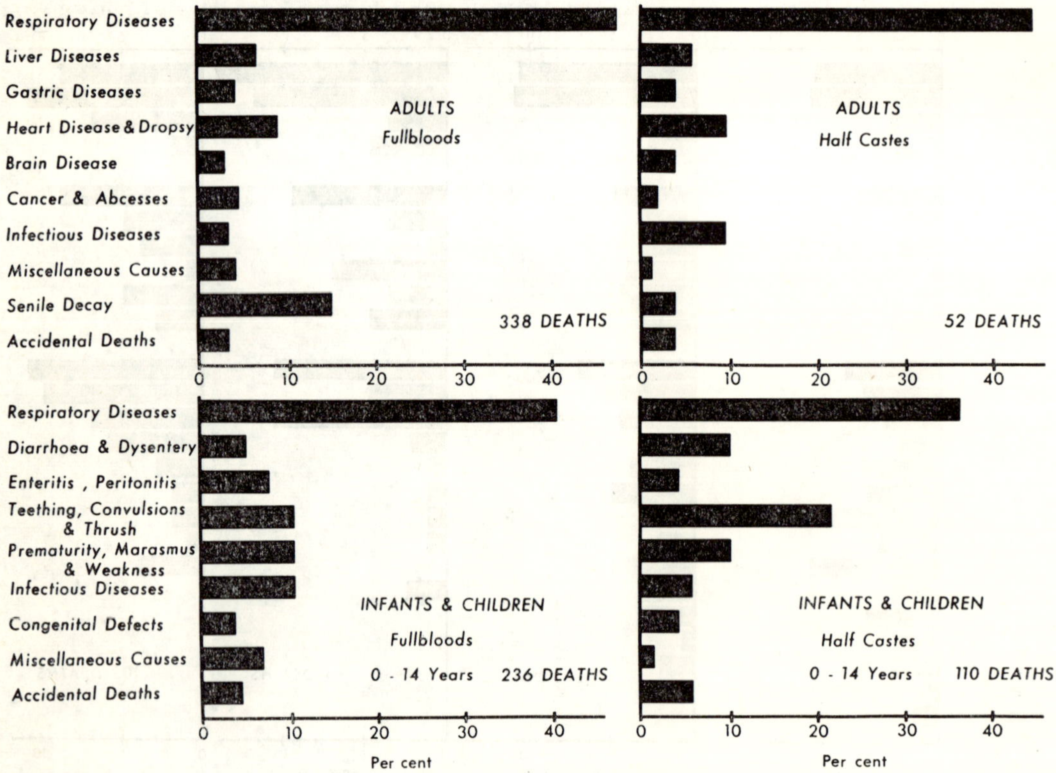


cirrhosis. The remainder, described as 'chronic liver enlargement', may have been due to hydatids also.

Gastric disease caused thirty-seven deaths, but only ten were explicitly described as 'enteritis'. The deaths of another two adults and nineteen children from 'diarrhoea' and 'chronic dysentery' may have been caused by gastroenteritis, which in more recent times has been a common cause of child deaths among Aborigines camping in unsanitary conditions.

The thirty-four infant deaths ascribed to 'teething fever' and 'convulsions' may also have resulted from the high fevers caused by gastroenteritis. Another thirty-two infant deaths resulted from 'prematurity', 'weakness', and the wasting disease marasmus, while six died of thrush, a fungus infection of the mouth and throat.

Fourteen children and thirteen adults died accidentally: eighteen by drowning, six by fire, two by suffocation and one by snakebite. Under 'miscellaneous causes' I have grouped such things as 'scrofula' (6), 'paralysis' (6), 'gangrene' (4), and 'childbirth fever' (7 cases). 'Syphilis' killed only two lads and an infant girl.

The infectious diseases measles, mumps, and scarlatina were cited as the direct cause of death for only sixteen adults during this period, but it may be that many had achieved immunity as survivors of the disastrous known epidemics before 1876. Only fourteen adult and child deaths were specifically attributed to measles, but other deaths ascribed to pleuro-pneumonia were in fact a consequence of measles epidemics among the children.

Smallpox, the scourge of the tribes of North America and many other British colonies, was apparently unknown in Victoria after 1860. Its incidence during the Protectorate period is unknown, although Curr (1883:239) saw one child ill near the Murray-Goulburn junction about 1843. A number of early settlers (Curr 1883:239; Hamilton 1913:100; Haydon 1846:23; Kerr 1872:16; Snodgrass, in Bride 1898:209) saw many adult Aborigines marked by the disease before 1841 and were told that it had spread, many years earlier, among the tribes of every region but Gippsland. The 1877 census listed the names of five pockmarked old people. The Board repeatedly authorised vaccinations by managers during the 1860s and 1870s to ensure that all residents were protected, and had the stations quarantined whenever a case of scarlatina or smallpox was suspected in the surrounding districts.

Incidentally, the one physician on the Board, Dr Thomas Embling, was the strongest opponent of the newfangled idea of vaccination: the manager of Framlingham once complained that Embling's prescription for *all* ailments there was 'a soap and sugar plaster'. Even this was better than the medical care offered in earlier decades: the missionary Goodwin noted in the Board's 1861 report that well-meaning settlers gave their employees doses of 'salts' as 'the almost universal remedy for all their complaints' and this treatment was 'often the source of much aftersuffering to them, producing haemorrhoid, etc.', while the Assistant Protectors (Parker 1841c), relying on the popular medical guides of the time, treated 'inflammation of the lungs' by 'prompt bleeding and repeated small doses of tartarised antimony' (tartar emetic).

How well do these 1876-1912 figures represent the incidence of disease in earlier decades? There are few published references to the causes of mortality during the Protectorate period, 1839-49, when medical officers held appointments at the depots. A history of the Goulburn Station, by a physician (Bossence 1965), does not record the care given by the medical officer who had charge from 1842 to 1849, possibly because there are no surviving returns. Accounts of the Loddon Station (Morrison 1967a, 1967b) only cite Parker's attempts to get medical assistance to cope with the high incidence of 'syphilis' and lung and liver inflammation. A recent study (Corris 1968:119) of Mount Rouse Station in the Western District notes that Dr John Watton, medical officer from 1842 to 1848, kept a count of illnesses treated and their nature:

> Of the 182 cases treated by Watton in 1845, 52 were syphilitic. Syphilis continued to be the major single disease suffered by the Aborigines down to 1848 when Watton reported it to be less common. Apart from venereal, respiratory diseases were the most common.

William Thomas, who had charge of the Melbourne and Westernport natives for twenty-four years, analysed the diseases of the Aborigines for the Board's 1861 Report. Among the ills to which they had been subject before contact he listed rheumatism, boils, skin eruptions, colds, dysentery, wounds, and burns, which were 'effectively' treated with lotions and decoctions of wattle bark and gum. The same treatment was used for the introduced venereal disease and achieved some cures which 'amazed' physicians, so he said. Parker (1841e), however, noted that the Jajowrong 'doctors' made no attempt to treat 'the wombi or syphilitic complaint now so prevalent' and considered this proof of its recent introduction.

An early ethnographic account by Brough Smyth (1878:**1**, xxxviii-xxxix), who was secretary of the Board from 1860 to 1875, listed as indigenous diseases ophthalmia, colds, eczematous skin infections, and 'hydatids in the liver and lung', which he ascribed to 'the imperfect cooking of food'. It is doubtful whether hydatids were in fact present in Australia before the introduction of sheep: they could have been introduced with the arrival of the dingo, for kangaroos can serve as secondary hosts, but they were almost certainly not as prevalent. The Aborigines' fondness for keeping large packs of hunting dogs was a continuing nuisance at the stations until the 1890s, and the frequency of hydatid infestation was largely due to this habit.

In the Board's 1861 Report, Thomas summarised the opinions of thirty settlers (including one doctor) consulted in 1858. Twenty stated that respiratory complaints, specifically 'consumption', were the primary cause of the rapid population decline, and were encouraged by intemperance and consequent exposure. Thirteen noted that 'syphilis' was a frequent cause of deaths, four said 'liver disease' was common, and eight mentioned that influenza epidemics had caused serious losses. Thomas himself attributed 'eight-tenths of the mortality' to intemperance, which brought on 'pulmonary disorders, pleurisy, pneumonia, disorders of the chest, consumption, etc.' In the same report Goodwin, a missionary since 1855, attributed the extraordinary mortality on the Murray to introduced 'inflammation of the lungs' which was 'almost invariably fatal'. Influenza epidemics, heart disease, dropsy, and paralysis also caused many deaths. He blamed their poor diet for the very common chronic diarrhoea and dysentery and diagnosed their chronic skin infection as a form of scabies.

The Board appointed a visiting medical officer for Coranderrk in 1865, and others for the western stations in 1871. The two Gippsland missionaries, before physicians settled in nearby towns about 1877, had no recourse but their chests of 'homeopathic medicines' and, one presumes, the grace of God. The ailments diagnosed by visiting doctors before 1875 had apparently the same incidence as given in the death records for 1876-1912. Tuberculosis, pneumonia, and bronchitis were by far the most common, but hydatids and stomach complaints were also frequent. In addition, there were repeated epidemics of 'intermittent fever', 'low fever', influenza, whooping cough, and measles. In 1875, when the last major epidemic of measles occurred, thirty-one of approximately 150 Coranderrk residents died of measles or subsequent pleuro-pneumonia.

The doctors' reports suggest that venereal disease had been effectively controlled, at least among the station residents, by the end of the 1860s, for only a dozen cases of 'syphilis' were noted, invariably among newcomers to the station. For the 1877 Royal Commission inquiry (1877:81) the causes of fifty-one deaths occurring between 1863 and 1874 at Coranderrk, the station where most wanderers were resettled, were summarised by the manager. He was trying to counter the Board's plans to sell the reserve and to rebut evidence that the climate 'caused' a higher incidence of tuberculosis, so he perhaps over-emphasised the importance of venereal disease. He noted that three died accidentally, four of old age, and twenty-one of 'pleuropneumonia, lung disease and low fever', while fifteen new arrivals died of 'pleuropneumonia and syphilis' and eight children of 'syphilitic parents' died of 'congenital syphilis' and pleuro-pneumonia.

In 1879 the Board asked the government to authorise an investigation of mortality at the stations, explaining that Aborigines who came young or were born there mostly died under the age of twenty-five and the usual cause was 'a disease of the lungs peculiar to the natives which ends fatally in every case'. When the colony's chief medical officer merely reiterated his comments of 1876 on Coranderrk, blaming the climate and inadequate diet, clothing, and medical care, the Board retained a Melbourne surgeon, William Thomson, to investigate the causes of 'tubercular phthisis'. His report was finally presented at an official inquiry in 1881 (Coranderrk Aboriginal Station 1881:35-7). Thomson concluded, from information he had collected on deaths at stations in Victoria and Queensland, that climate had no bearing on the incidence of the disease, and speculated that the cause might be 'contagion':

> That is the theory of phthisis that is rapidly growing among pathologists at home. They suppose that there are some germs or organisms, whether animal or vegetable I do not exactly know, that, multiplying in the lungs, destroy the tissue.

Thomson deplored the Board's failure to provide isolated quarters for the sick, suggesting that their infectious spittle impregnated the cottages and could be 'breathed' by others. Another physician corroborated Thomson's opinion that the Aborigines were specially susceptible to an extraordinarily rapid and fatal 'tubercular consumption', but gave no theory of its causation. The colony's medical officers ignored the theory of 'contagion' and so, it seems, did the Board, for the sick were not isolated until the first station hospital was built at Lake Tyers during the 1920s.

The high incidence of tuberculosis among Victorian Aborigines, and their apparent lack of resistance to respiratory infections, was frequently noted by officials in ensuing decades. The 1910 amendment of legislation to enable the Board to give needy half castes the same aid as fullbloods was introduced for this reason. To solicit additional government funds, the Board's secretary explained on 8 June 1910 (Letter Book 1909-11) that:

> such half castes are practically Aborigines and the Board find that owing to their predisposition to consumption and other diseases it is just as essential to assist them as the pure blacks.

## FERTILITY AND POPULATION GROWTH

The number of females of child-bearing age is the essential limiting factor governing rates of population growth, so juvenile mortality, as well as absolute fertility, conditions population increase.

It is important to remember that there were five separate regional groups, each virtually endogamous, during this century. The size of the groups in the Western District, the Goulburn Valley, the Wimmera, the Lower Murray, and Gippsland regions changed over time as a consequence of government policy decisions stimulating migration or transfer from station to station. The absolute size and the age structure of each separate regional group obviously conditioned the availability of marriage partners. The often gross age differences of husbands and wives, apparent during the 1860s and 1870s, were a consequence of this limitation of choice and also, possibly, a continuation of traditional preferences.

Cultural preferences regarding age at marriage and opportunities for remarriage also affect population growth. In tribal times girls married soon after reaching puberty, and the verified ages of mothers in the genealogies suggest that before settlement at the stations they did begin child-bearing when very young. The managers discouraged very early marriage, so the usual age was seventeen to nineteen years. In Gippsland, where there are few employment opportunities, this is still the norm, but many of the Western District and Goulburn Valley girls now obtain work in Melbourne and tend to delay marriage until the middle twenties. Because of the shortage of females during the first decades of settlement at the stations every young widow remarried very soon and almost no women remained unmarried. Those listed as single in the ensuing discussion of all women living to reproductive age from 1863 to 1966 were (except for those in the most recent cohorts) mostly females who died between the ages of fifteen and twenty-one, many of them so ill with tuberculosis that marriage was impossible.

Population growth may also be influenced by poor general health, and a high incidence of sterility. Venereal disease obviously caused infertility and a high rate of infant mortality during the early days, but I have no information on the extent to which latent tuberculosis (prevalent among Victorian Aborigines until the 1950s) may affect fertility. Mr Jeremy Long (pers. comm.) informs me that there is some evidence, from his surveys of Aboriginal communities in the Northern Territory, that such things as influenza epidemics may result in 'wasted' pregnancies—that is, in miscarriages and stillbirths.

Artificial controls on fertility, such as the spacing of children by prohibitions on intercourse or contraceptive practices, seem (judging by intervals between births) not to have been important in limiting the Victorian Aboriginal population, and infanticide was apparently abandoned about 1863.

The last two graphs present information on the children born to those women who belonged, by descent, to the Victorian Aboriginal population of 1863-1966. Fig. 20:6 shows the average size of their families and Fig. 20:7 shows the proportion of their offspring who died before reaching adulthood.

The basic data come from a compilation of the issue of the 2,084 women who lived to reproductive age (fifteen years) in the five regional groups, from the time of the 1863 census to the present day. Some women were excluded from Figs. 20:6 and 20:7 because their apparent childlessness cannot be verified: 237 born before 1863, who mostly died before 1877, are not counted because we cannot know whether they had lost children before the 1863 census, while members of later cohorts (fifteen born 1872-1926, and sixty-five born 1927-51) are excluded because insufficient information has been collected—in fact, half are or were living in other states. Adequate data are available for 1,767 females (1,415 mothers, 149 childless wives, and 203 single girls). These women are grouped, by date of birth, in five-year cohorts to coincide with the groups of the 1966 Commonwealth census: that is, the youngest, born 1947-51, were aged 15-19 years at June 1966. The earliest cohorts, born 1802-6 and so on, were the survivors counted at the 1863 enumerations.

Fig. 20:6 shows the total number of mothers, the numbers of children they bore, and the average family size for each cohort. The bars on the left show the number of mothers born in a five-year period, while the bars for their issue show the number of live births throughout their child-bearing span. This span usually begins eighteen to twenty years after the mother's birth date, and may last twenty years or more. For example, women born 1922-6 began their child-bearing careers between 1941 and 1953, and finished

*Fig. 20:6 Mean family size, Victorian Aboriginal women, 1863-1966*

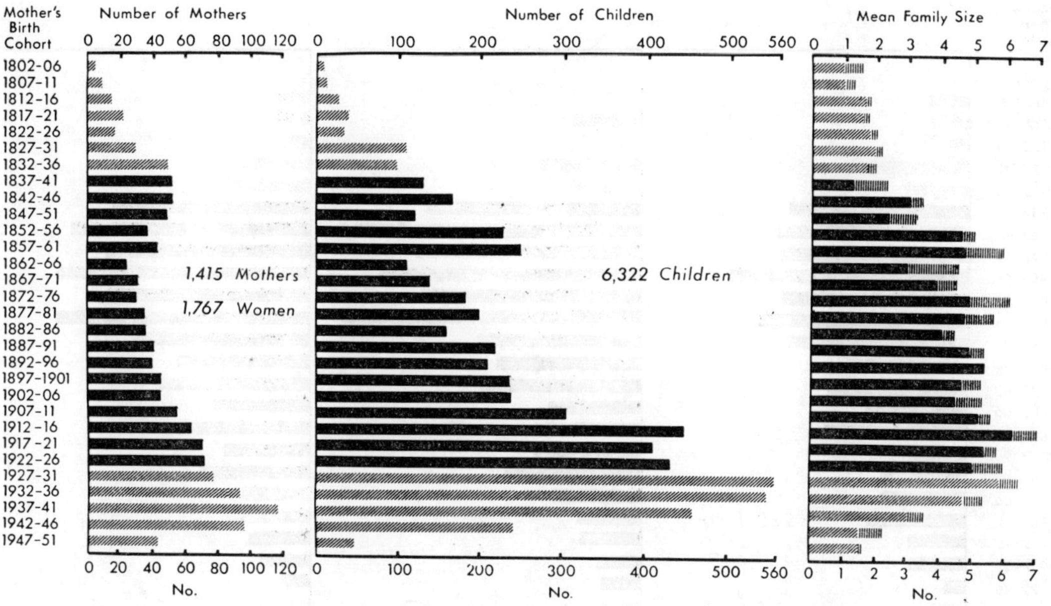


between 1941 (one child) and 1962 (twelve children). Obviously the child-bearing spans of members of adjacent cohorts overlap considerably.

Fig. 20:6 is divided horizontally into three segments. The upper shaded segment, representing women born 1802-36 who had almost completed their child-bearing careers by 1863, must be considered incomplete: the average family size is small and we cannot know whether this was due to infertility or infant mortality before 1863. The lower shaded segment is demarcated because these women (aged fifteen to thirty-four at June 1966) have not completed their spans. To show the extent of incompleteness for these five cohorts at June 1966, when 131 were still single and some of the newly-married were childless, I should mention that by December 1967 another seventy-three children had been born and another fifteen women had become mothers.

Mean family size is shown for all women (the point indicated by shading) and for mothers only (the total length of the bar) in the right-hand column of Fig. 20:6. The apparent fluctuations in family size had a negligible effect in fact, since the birth spans of adjacent cohorts overlapped so greatly.

Details of the 1966 census were not available at the time of writing, but roughly comparable figures for family size, the average issue of all *wives*, from the 1961 census have been published (Commonwealth Bureau of Census and Statistics 1967:381). Australian wives in the cohorts aged 15-34 averaged 0·62 to 2·50 children while the averages for Victorian Aborigines of equivalent age (shown in the lower segment of Fig. 20:6) were 1·58 to 5·26 for mothers only, and 0·34 to 4·88 for all women. The Australian averages for wives aged 35-69 were stabilised at about 2·50, while the Aboriginal cohorts born 1892-1931

*Fig. 20:7 Infant and child mortality, issue of women, 1863-1966*

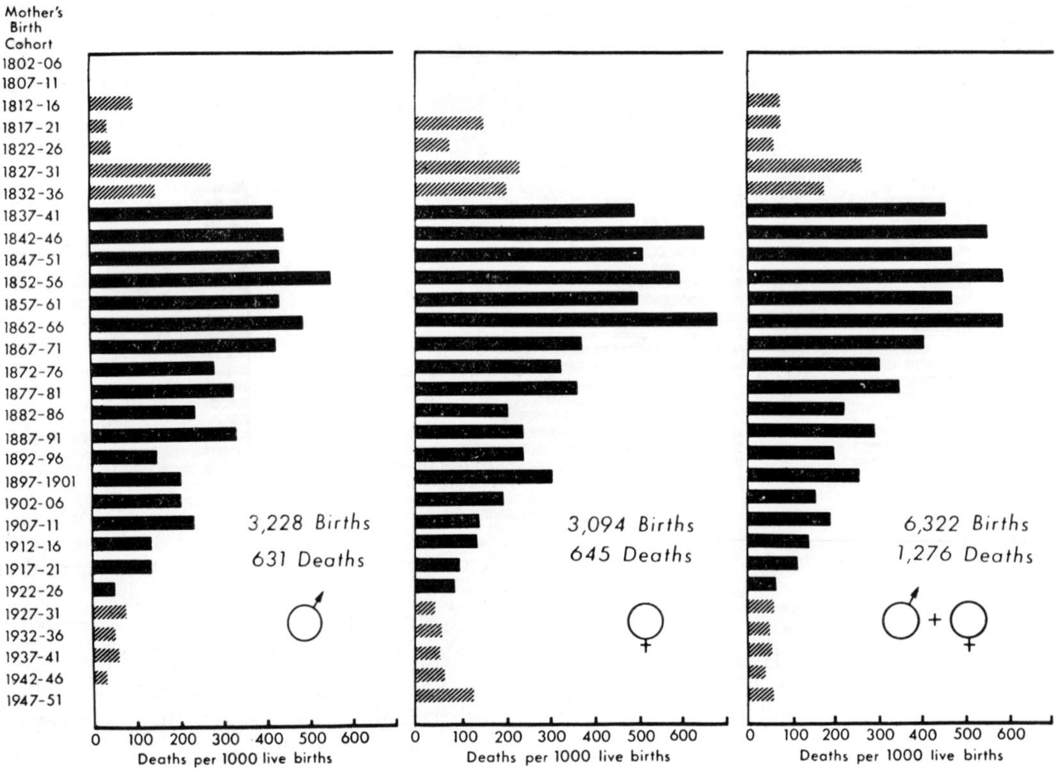


had averages of 5·24 to 7·00 for mothers only, and 4·50 to 6·31 for all women. The average size of the completed families of these Aboriginal women has remained fairly stable, but the Australian averages have declined markedly over the last forty years: Australian wives aged 35-69 at the time of the 1921 census had averages of about 3·7 to 5·8 (Current Affairs Bulletin 1966) while the Aborigines of equivalent cohorts (1852-86) had averages of 4·45 to 6·07 for mothers only, and 2·95 to 4·92 for all women.

Although mean family size in the Victorian Aboriginal population has long been well above the Australian norm, population growth has not been as rapid and as great as the above figures suggest, for 1,276 of the 6,322 children born to these mothers died before reaching their fifteenth year.

Fig. 20:7 shows the extent of infant and child mortality among the children born to the various cohorts. I have not distinguished 'infant' deaths under the age of one year, for the significant figure for population growth is the total who survive to reproduce themselves. The total number of juvenile deaths must be considered.

This graph, too, is divided horizontally into an upper shaded segment of cohorts whose losses before 1863 can never be precisely known, and a lower shaded segment which does not provide a comparable total of juvenile deaths because the children (born 1942-66) of these cohorts are mostly very young. Indeed, virtually all of the losses here are infant deaths.

Mortality rates are calculated as proportions per thousand live births, as in the Australian census. They are not strictly comparable, in that the census figures are based on a one-year calculation period: for example, the infant mortality rate for 1961 is the 'number of deaths under one year of age per thousand live births registered' in that year.

Fig. 20:7 shows that the proportions of juveniles lost were astronomical. Women of the six cohorts born 1837-66, whose child-bearing spans extended from 1857 to 1904, lost more than half the children they bore (453 to 578 per thousand). Women of the next three cohorts, bearing children from 1885 to 1925, lost more than a third (302 to 398 per thousand). The women born 1882-1901 lost a quarter (195 to 287 per thousand) of the children they bore between 1902 and 1944. The two other cohorts (1902-11) whose children were all aged fourteen or older at 1966, have lost less than a fifth (140 to 191 per thousand). The child-bearing spans of the three cohorts born 1912-26 began between 1927 and 1941 and ended between 1958 and 1962; their families are complete, but many of their children are young, and the mortality rates (now 80 to 133 per thousand) may yet increase should more die before reaching the age of fifteen.

Women in the last five cohorts (born 1927-51) were still bearing children in 1966 and 1967. Their losses, by June 1966, were 41·2 to 61·2 per thousand. These figures do include some deaths over the age of one year, but can be roughly compared with the Australian infant mortality rates for 1961 (19·5 infant deaths per thousand live births) and for 1951 (25·2 per thousand). Infant mortality among Victorian Aborigines, who now live mostly in the towns and cities, is still approximately twice the Australian norm.

Virtually maximum fecundity has been assured in the Victorian Aboriginal population of 1863-1966 by several cultural characteristics: a positive desire for children, the lack until recent years of effective contraceptive devices and a continuing disinterest in their use, together with relatively early marriage and almost universal remarriage when unions dissolve. The only limiting factors were general poor health and perhaps some difficulty, within the smaller regional groups, in finding spouses. Child-bearing has always begun relatively early, and continued long, and today it is still usual for the elder daughters to begin bearing children when their mothers' families are incomplete. Families have always been large, but population growth has been slowed by an extraordinarily high incidence of infant and child mortality and, in earlier decades, by an apparently abnormal incidence of respiratory disease which caused the deaths of a majority of adults before the age of forty.

## References

Board for the protection of the Aborigines, Victoria 1861-1912. Annual Reports Nos. 1-48. *Papers presented to both Houses of Parliament, Victoria.*

Bossence, William H. 1965. *Murchison. The J. G. Kenny memorial history.* Melbourne: The Hawthorn Press.

Bride, Thomas F. (ed.) 1898. *Letters from Victorian pioneers.* Melbourne: Government Printer (for the Trustees of the Public Library of Victoria).

Commonwealth Bureau of Census and Statistics. 1967. *Statistician's report, census of the Commonwealth of Australia, 30th June, 1961.* Vol. 8. Canberra: Government Printer.

Coranderrk Aboriginal Station. 1881. Report of the Board to enquire into, and report upon, the present condition and management of Coranderrk Aboriginal Station. *Papers presented to both Houses of Parliament, Victoria.* 1882-3, Vol. 2, No. 5.

Corris, Peter. 1968. *Aborigines and Europeans in western Victoria.* Occasional Papers in Aboriginal Studies No. 12 (Ethnohistory Ser.No. 1). Canberra: Australian Institute of Aboriginal Studies.

Curr, Edward M. 1883. *Recollections of squatting in Victoria.* Melbourne: George Robertson.

Current Affairs Bulletin. 1966. *Australia's birth rate.* **39** (2). Sydney: Department of Adult Education, University of Sydney.

Elkin, A. P. 1948. *The Australian Aborigines: how to understand them.* 2nd edn. Sydney: Angus and Robertson.

Foxcroft, Edmund J. B. 1941. *Australian native policy.* Melbourne: Melbourne University Press.

Hamilton, J. C. 1913. *Pioneering days in western Victoria.* Melbourne: Exchange Press Pty Ltd.

Haydon, G. H. 1846. *Five years experience in Australia felix.* London: Hamilton, Adams and Co.

Kerr, J. H. 1872. *Glimpses of life in Victoria by a resident.* Edinburgh: Edmonston and Douglas.

Legislative Council, Victoria. 1853-4. Aborigines, return to address, Mr Parker, 21st October, 1853. *Votes and Proceedings of the Legislative Council, Victoria.* 1853-4.

—— 1858-9. Report of the select committee on the Aborigines. *Votes and Proceedings of the Legislative Council, Victoria.* 1858-9.

Morrison, Edgar. 1967a. *Early days in the Loddon Valley, memoirs of Edward Stone Parker, 1802-1865.* 2nd edn. Daylesford: 'Advocate' Press.

——1967b. *Frontier life in the Loddon Protectorate. Episodes from early days, 1837-1842.* Daylesford: 'Advocate' Press.

Parker, Edward Stone. 1841a. Census of the Jajowrong tribe or nation, 23 February 1841. Papers of the Port Phillip Protectorate. La Trobe Library Archives, Melbourne.

——1841b. Quarterly journal, 1st March-31st May 1841. Papers of the Port Phillip Protectorate. La Trobe Library Archives, Melbourne.

——1841c. Quarterly journal, 1st June-31st August 1841. Papers of the Port Phillip Protectorate. La Trobe Library Archives, Melbourne.

——1841d. Letter to Chief Protector Robinson, 15th October 1841. Papers of the Port Phillip Protectorate. La Trobe Library Archives, Melbourne.

——1841e. Quarterly journal, 1st September-20th November 1841. Papers of the Port Phillip Protectorate. La Trobe Library Archives, Melbourne.

——1842. Quarterly journal, 1st June-31st August 1842. Papers of the Port Phillip Protectorate. La Trobe Library Archives, Melbourne.

Radcliffe-Brown, A. R. 1930. Former numbers and distribution of the Australian Aborigines. In *Official year book of the Commonwealth of Australia.* No. 23, pp. 687-96. Melbourne: Government Printer.

Royal Commission on the Aborigines, Victoria. 1877. Report of the Commissioners appointed to enquire into the present condition of the Aborigines of this Colony, and to advise as to the best means of caring for and dealing with them in the future. *Papers presented to both Houses of Parliament, Victoria.* 1877-8, Vol. 3, No. 76.

Smyth, R. Brough. 1878. *The Aborigines of Victoria.* 2 vols. Melbourne: Government Printer.

Thomas, William. 1861. Report on the diseases of the Aborigines. First report of the Central Board . . . App. III, pp. 27-9. *Papers presented to both Houses of Parliament, Victoria*. 1861-2, Vol. 3, No. 39.

Victorian Government Statist. 1875-1966. *Victorian year books*. Melbourne: Government Printer.

Whitworth, R. P. (comp.) 1870. *Victorian gazetteer and road guide*. 2nd edn. Melbourne: F. F. Bailliere.

# 21 The Racial Affinities and Origins of the Australian Aborigines

A. G. Thorne

Any contribution to the problem of the racial affinities and derivation of the Australian Aborigines increasingly involves a discussion of two theories—hybridity or homogeneity. As in other parts of the world it is inevitable that variation in a population, particularly a widely-spread and partly extinct one, will be interpreted as either internally-developed variation, or variation based on the mixture of distinct racial elements.

Proponents of hybridity begin with Topinard (1872), Davis (1874), and Lesson (1880) and include Hrdlička (1928), Fenner (1939), Wunderly (1943), Hooton (1947), Gates (1960), Tindale and Birdsell (1941), Birdsell (1949, 1950, 1967) and Morrison (1967). The most specific theorist in this group, Birdsell, postulates a tri-hybrid origin for the Australians. According to Birsdell three waves of people, Negritos, Murrayians, and Carpentarians, entered the continent and the population at the time of European contact was a mixture of all three. He believes all three elements can be traced back to the Asian mainland and that the three groups entered Australia during the fourth glacial period, when Australia was joined to New Guinea. He further states (1949:120) that aboriginal New Guinea and Melanesia contain 'the same racial ingredients as the tri-hybrid Australians, with small additions of Mongoloid genes present in some areas'.

Those who maintain a single racial basis for the Australians include Klaatsch (1908), Wood-Jones (1934), Campbell *et al.* (1936), Howells (1937), Abbie (1951, 1960, 1963, 1966, 1968a, 1968b) and Macintosh (1963). Abbie (1966:42) has recently asserted that 'all the evidence points to the fact that the Aborigines, everywhere, are one people'. In a comment on the status of the Cairns rainforest people, described by Tindale and Birdsell, and Gates, Abbie suggests (1963: 102) that this information has to be 'set against a wealth of physical data which very strongly suggest . . . Aborigines widely dispersed through the rest of the continent are practically uniformly homogeneous'.

It is worthwhile examining the bases of these two theories, because of their relevance to current work in other areas of Aboriginal research. In doing so I will concentrate mainly on the work of Birdsell and Abbie, on hybridity and homogeneity respectively, because their publications are the most modern and detailed approaches to the problem.

Birdsell sees the Oceanic Negritos as the first wave of people entering Australia. The greatest evidence of this wave he sees in Tasmania and in the Cairns rainforest area of northeast Queensland. He finds evidence of the same group in highland New Guinea, in the interior of the Gazelle Peninsula of New Britain, and in the northern interior of New Caledonia (1949:120). Though there are no longer living people to be studied in Tasmania, Birdsell predicted that archaeological work in Tasmania, and in fact in Australia, would show that the earliest inhabitants were pure negritic in type.

As early extinction has precluded a thorough examination of living pure Tasmanians, comparative studies have been limited almost entirely for the last ninety years to skeletal material, especially cranial. Three of the five known whole skeletons were destroyed in World War II and the number of listed crania has never exceeded 200. Thus statistical and morphological analyses have been based on quite small samples and perhaps even more important, the sources and documentation of the cranial collections are in most cases extremely poor. The fact that large numbers of Melanesian, Polynesian, and especially Australian Aborigines were taken to Tasmania soon after contact raises the question of authenticity. Macintosh (Macintosh and Barker 1965:43) believes that because the major anatomical studies of the Tasmanian-labelled crania are in general agreement we can accept at least 90 per cent as genuine. However, this agreement is no proof of authenticity. The fact that different workers come to similar conclusions, sometimes based on overlapping collections of crania, is no validation of the labels on the specimens, particularly when a major aim is to test the extent or lack of relationship of the Tasmanians and mainland Australians.

However, if we assume that at least a reasonable proportion of Tasmanian crania are genuine, a picture of the Tasmanian cranium does emerge. There is general agreement among osteologists about the similarities between Australians and Tasmanians and a similar agreement about certain differences. As Macintosh (Macintosh and Barker 1965:49) has pointed out, there is no conflict about the observed data—only conflict about their interpretation.

Birdsell (1949:108) indicates three basic features of the Tasmanians which demonstrate negritic affiliation—marked development of parietal bosses, high cranial breadth, and tightly helically curled hair. In these characteristics the Tasmanians, he says, resemble the Andamanese, regarded by Birdsell as the least mixed of the existing Oceanic Negrito populations. He supports and extends the belief that the Tasmanians are a hybrid group and says the negritic element is a minority one, submerged by a predominantly Murrayian morphology.

New material from Tasmania is the only check on the status of these people. New finds derived from formal archaeological excavation, and with a chronological guarantee that they are pre-contact, are the most valuable of all; the osteological material recovered by Rhys Jones (1966) falls into this category. The skeletal material from the site at West Point, though limited, is unique in that by predating European contact by more than 1,000 years there can be no suggestion of dubeity. The West Point bones, except perhaps for the skeleton of Truganini, are the only fully-authenticated specimens of Tasmanian Aborigines. A molar tooth from this site has been described by Barker (Macintosh and Barker 1965). Other material, largely calcined fragments from cremation deposits, has now been analysed (Thorne 1967) and a report prepared for publication. Briefly, the main conclusions of this report are as follows: bits and pieces of about eight individuals are represented. The bones are badly fragmented and although no single individual skeleton, or even skull, can be reconstructed, several isolated characters, mostly non-metrical, are valuable for comparison with other skeletal collections of Tasmanians and Australians.

Parietal bossing has been referred to as prominent in Tasmanians by Turner (1908), Wunderly and Wood-Jones (1933); and Birdsell, as mentioned above, says that bossing is a distinguishing characteristic of the negritic element in Tasmanians. In three of the West Point specimens the areas of possible bossing are present but in none of them is there any sign of this feature. Another feature noted for Tasmanians is that the sagittal suture lies in a longitudinal depression (Wunderly and Wood-Jones 1933:585; Wunderly 1938:323). In the West Point fragments which include the suture there is no depression. Thirdly, longitudinal parasagittal grooving has been stressed in Tasmanians, and some anatomists have regarded this feature as an exclusive trademark of the islanders. It has been noted in South Australia, and Birdsell suggests that this trait is a genetic consequence of the hybridisation of markedly different cranial types—one with marked sagittal keeling and one with marked development of parietal bosses (1949:117). Three of the West Point fragments exhibit no trace of parasagittal grooving, the bone being quite rounded in this region.

As far as teeth are concerned, fourteen specimens were examined metrically. The results agree with Barker's (Macintosh and Barker 1965) conclusions about another tooth from this site, that the Tasmanian teeth fall within the ranges for the Australians and that there is no reason to suspect any dichotomy other than a sexual one.

The morphological features of the West Point individuals are curious. On the evidence, assuming for the moment one was not aware of the source of the material or its radiocarbon age, there would be no reason to identify it as Tasmanian. Indeed, there would be strong grounds for saying the morphology was very consistent with individuals from coastal New South Wales in the recent past. The West Point material, however, *is* Tasmanian and yet it lacks the allegedly-identifying features of the Tasmanians.

There is another feature in the material which also seems non-Tasmanian. This is a maxilla with a healed or resorbed central incisor socket. There is no evidence of pathology and the specimen is quite consistent with the possibility of tooth avulsion. However, Macintosh (Macintosh and Barker 1965:22) concluded that tooth avulsion in a specimen labelled Tasmanian 'would quite strongly impugn its authenticity'. If tooth avulsion did exist in Tasmania, and Labillardière (1800:320) records it quite specifically, then the rejection of crania because of possible tooth avulsion unjustifiably reduces the range of specimens available for study.

I believe that anatomical conclusions about the Tasmanians have tended to stress their differences from mainland groups. Some workers have simply assumed the Tasmanians were very different or completely unrelated to mainland Australians because of early ethnographic reports. Others, because of the limited material available, have been forced to see differences, just as I am forced to see strong similarities because of the West Point finds. Then there are those whose comparative Australian material is limited and unrepresentative. It is probable that some French collections of Australian crania, for example, are from northern Australia, particularly from around Port Essington and Darwin. Comparisons of Northern Territory crania with those of Tasmania show differences, but then so do Northern Territory comparisons with Victorian crania.

My present attitude is to regard the Tasmanian Aborigines as local variants of a southern Australian population, based on a morphology existing about the time of Tasmania's connection and subsequent separation from the mainland. The degree of individuality developed by the islanders has been heightened by environmental stimulus and accelerated genetic effects due to population size. (Differences in hair form, skin colour, body proportions, and cranial module, among other features, could be explained in this way.) Changes in mainland morphology since separation (either internal or introduced) would also widen any apparent gap.

The Barrineans are the second major remnant population contributing to Birdsell's hypothesis that Oceanic Negritos were the initial occupants of Australia. The Barrinean groups occupy a rainforest area in coastal northeast Queensland. Birdsell believes the negritic element in the Barrineans is greater than in the Tasmanians (1949:116). He has recently published (1967) some of the data on which his racial separations are based and the majority of the observations are from ninety-five male subjects from twelve contiguous tribes. Many of the morphological characters listed are somatological features. I will confine my comments to osteological aspects of the cranium, using as a comparison data from Larnach and Macintosh's current study of a series of Queensland crania, which include twelve from the rainforest area. I thank them for permitting me to quote from their results.

For brow ridge shape Birdsell notes an incidence of 38·3 per cent of the continuous type in Barrineans. In the series now under study there is one rainforest specimen with continuous brow ridges, 8·3 per cent. A wider discrepancy is revealed in percentages for Birdsell's comparative group of forty-four Murrayians from southeast Australia. He notes 45·5 per cent with continuous brow ridges. In a survey of New South Wales coastal crania, from within Birdsell's Murrayian area, Larnach and Macintosh (1966: 14) earlier found only 1·7 per cent continuous and if their continuous and pseudo-torus types are combined the total is only 20·7 per cent. As far as divided brow ridge type is concerned Birdsell notes 50·0 per cent for his Murrayians. For this type Larnach and Macintosh (1966:14) note 74·1 per cent and since their cases of pseudo-torus indicate traces of division, both should be combined as a total of 93·1 per cent of divided brow ridges. Furthermore, Birsdell notes an incidence of 21·8 per cent of continuous brow ridge type in 'whites'. Yet no true continuous brow ridges have been found in Europeans. In fact the absence of this brow ridge form is one diagnostic character distinguishing modern Europeans from western Neanderthals (see Leakey 1934; Keith 1925).

Birdsell notes a series of dental morphological traits which distinguish Barrineans from other Aboriginal groups. The most striking comparison from Birdsell's data is that Barrineans have a high incidence (57·3 per cent) of mesodont molars as opposed to nil in Murrayians. Yet in the twelve rainforest crania now under study, molar size can be matched in Murray Valley and New South Wales coastal specimens, at both extremes of the molar size range. Birdsell also notes 46·03 per cent of four-cusped mandibular first molars. In Larnach and Macintosh's twelve specimens there are no four-cusped examples.

A third point concerns the presence of parasagittal grooving, which Birdsell sees in the Tasmanian as an indication of a hybridised negritoid bossed cranium with a Murrayian keeled cranium. If the Barrineans 'owe an approximate equal indebtedness to Negrito and Murrayian contributions to their gene pool' (1967:151) and therefore retain more negritic elements than the Tasmanians, it is to be expected that the rainforest group will exhibit parasagittal grooving to at least the same extent. In his latest statement Birdsell makes no mention of the feature. In the series being examined by Larnach and Macintosh none of the twelve rainforest crania shows any trace of parasagittal grooving. It should be mentioned, however, that the rainforest crania do display a higher incidence of parietal bossing (33·3 per cent) than either New South Wales (9·3 per cent) or Queensland (12·1 per cent) crania.

The Murrayian element of the tri-hybrid hypothesis is allegedly the dominant one across Australia. Birdsell regards this population as basically an archaic white or Caucasoid group, having affinities with the Ainu of Japan. Howells (1937) and Hooton (1947) have also suggested an Australian-Ainoid relationship. Birdsell (1967:140) notes twenty-two characteristics in which Ainu and Murrayian both deviate, in the same direction, from the Australian population as a whole. Yamaguchi has recently surveyed an Australia-wide series of crania, specifically to determine whether there was any similarity between the Ainu and the Australian Aborigines. Although he notes similarities in a preliminary note (1966) he concludes in a final report (1967:32) that his results, after statistical analysis, are 'generally unfavourable for assumption of any relationship'.

The Carpentarians form the third and most recent element in the tri-hybrid theory. Birdsell considers the ancestral Carpentarian stock as a fourth major human racial group, having its evolutionary centre in India. In Australia the area where the Carpentarian component is most evident is 'the northern coastal regions between Cape York and the northern Kimberleys' (1967:146). This area, however, almost exactly coincides with the limits of contact in relatively recent times of the Indonesian trepang fishermen and the trading voyages of the Papuans. Macknight's survey (Macknight and Thorne 1968) of Macassan sites shows intensive contact in the area between Melville Island and about the west Queensland border, with another branch of the trepang industry along the Kimberley coast, south from Cape Londonderry. Although Macknight believes the regular visits by Macassans date from only the beginning of the eighteenth century, for the next 200 years as many as 1,500 men came each year to northern Australia. In at least some areas, contact with Aborigines was more than fleeting and hybridisation is to be expected. There are hybrid crania in the Australian Museum's collection from Queensland which are probably Macassan-Australian crosses. The results of Papuan contact have produced a similar hybridisation in Cape York. Finally, Birdsell states (1949:111) that there is no archaeological evidence for the early presence of Carpentarians in Australia. The cultural material recovered archaeologically in the Northern Territory from the Ingaladdi site, stretching back some 7,000 years (Mulvaney 1969:147), as well as from Arnhem Land, demonstrating occupation at least by 20,000 years BP (White, see ch. 12), requires at least some modification of this view.

I believe that definition of the degree of homogeneity or hybridity of Aborigines from northern Australia should be deferred until at least some of the factors produced by known contact in recent times are more clearly understood.

Those who maintain the homogeneity of the Aborigines are aided in their use of genetic drift and adaptation by estimations of a relatively low continental population. They would appear to be supported by linguistic and serological evidence. Linguistic evidence of homogeneity is probably of limited value, however. Given the known length of occupation at this time, say 20,000 years, linguistic evolution, aided by sociological factors, could well smooth differences or even submerge and extinguish languages overlaid by other languages for some time. Serological data might well be subject to similar sorts of pressures.

Abbie has long maintained a homogeneous origin for the mainland Australians. He sees the Tasmanians as a separate entity: 'all the evidence available indicates that the Tasmanians were of Melanesian origin and never were on the Australian continent' (1968a:23).

As far as mainland Aborigines are concerned, Abbie's physical data, mostly somatological, have been gathered from a relatively small area, a slice of the middle third of the continent, from Maningrida in the north to Yalata in the south. The possibility of significant variability in this area is reduced by linguistic and ethnographic evidence that some southward movement of people has occurred in relatively recent times (Birdsell 1950; Kirk, see ch. 22).

In his most recent statement, Abbie (1968b) uses six computed criteria to screen for racial homogeneity, to permit him to pool all his data. Despite the fact that further work in this area may support his concepts, expansion of 'homogeneity' to continent-wide status is surely premature and would presumably be criticised by the proponents of hybridity.

I would like briefly to refer to the Australian 'fossil' crania. The number of recognised specimens in this category makes a very short list indeed and for that reason any deductions based on their morphology are at present of limited value. Like certain European fossil specimens, more time seems to have been spent commenting on the Australian fossils than describing them. Of the specimens considered to have significant antiquity, the Keilor cranium is the only specimen which has been dated, although Macintosh's intensive efforts with the Talgai cranium promise an analysis of chronology in this case. Of other relics, the Cohuna cranium remains undated, while an ultimately satisfactory statement about the age of the Mossgeil skeleton seems possible. Analysis is proceeding on the new find from Kow Swamp. A few fragments of this very rugged and mineralised skeleton were found during a survey of the skeletal collections held by the National Museum of Victoria. Subsequent investigations and excavations in early 1968 established its origin to be from a lunette near the northeast shore of Kow Swamp in northern Victoria. A considerable proportion of the total skeleton is now in hand, including most of the cranium and mandible. This individual was interred less than 2 km from the site of the Cohuna cranium and it is hoped that an age assessment of Kow Swamp will aid a similar assessment of Cohuna. At the present stage of examination it seems the Kow Swamp cranium is a somewhat larger and more rugged version of Cohuna. The brow ridges and the palate are larger. The degrees of frontal recession and post-orbital constriction are probably at the extreme for Australia. The Kow Swamp mandible is very large indeed, as is indicated by the fact that its metrics probably exceed those of the Mauer mandible in all but two dimensions—thickness of the body at the first molar and the anteroposterior width of the ramus.

There are a few features of the Australian fossils which seem to hint at a morphological pattern. Talgai, Cohuna, Mossgeil, and Kow Swamp all share marked frontal retreat and postorbital constriction, large brow ridges, large palates, and considerable rugosity. The teeth are large, especially the Talgai canine. The 8,000-12,000-year-old incisor from Western Australia described by Davies (1968) is also at the upper limit for Australia. These features suggest that ruggedness and primitive characteristics of the frontal and palatal areas were features of Aborigines in the past. The Keilor and Green Gully specimens, on the other hand, are almost the reverse, orthognaphic, with well-rounded frontal areas, minimal postorbital constriction, and horseshoe-shaped palates. At present the total relic collection can be viewed either as examples of different types which are in a general sense co-existent, or as variants within a continuum.

Speculatively, it is possible also to regard the majority of the relics as examples of an early Australoid morphology which differs in several important respects from the recent continent-wide morphology. These characteristics include frontal recession and dental size particularly. It is difficult to account for differences between early and late morphology in terms of internal change, either randomly or selectively. On the other hand migration could well account for it, though not in migrational terms so far proposed.

I believe it is worth considering Australian Aboriginal physical history in two stages—early and late Australoid. Whether this suggested dichotomy proves to be the result of internal change or successive migration remains to be seen, though there is some archaeological evidence to support a migrational explanation. Several workers have noted a typological division of early and late industries, from widely scattered sites in Australia. The early tradition covers a time-span from at least 20,000 years BP to 6-7,000 years BP, when it is replaced by a late stone tradition. Mulvaney (1966; Mulvaney and Joyce 1965) has correlated the change with the introduction of hafting techniques, based on material from his Kenniff excavations. Whether this explanation proves valid or not, industries from the Oenpelli area, Ingaladdi, Seelands, Curracurrang, Laura, Capertee, and Noola (see endpaper map and Fig. 10:1) all show a distinctive typological change from industries characterised by large artifacts to those which include smaller points and microlithic tools.

A second source of evidence for migration is the dingo. Basically, the ecological competitors for this importation lie with *Sarcophilus* (Tasmanian devil) and *Thylacinus* (Tasmanian tiger) rather than with man. The late extinction of the devil and the tiger on the Australian mainland and their continued success in Tasmania points to the introduction of the dingo after the formation of Bass Strait. Given that the earliest known specimens of dingo in Australia are older than 3,000 years it may well be that the 6-7,000-year-old typological changes and the introduction of the dingo are related.

What I have sought to point out in this paper is the very incomplete state of our knowledge of the physical makeup of the Aborigines and the dangers of detailed theorising on the derivation of this morpholgy.

Those who believe that the Aborigines are a homogeneous or uni-racial population are at an advantage, relative to those who propose a more complex origin for these people. In a sense, workers who see more than one migration, of different stocks, have a greater obligation to prove their case. The uni-racialists tend to fall back on the arguments of environmentally-produced differences, genetic drift, and social pressures to explain observed variation.

Much more basic analysis, particularly of osteological material, is necessary before detailed constructs will permit a clear picture of the peopling of the continent. To date there are very few detailed studies of regional groups to determine internal variation of such aspects as discontinuous morphological features and metrical characteristics, or dental morphology and metrics.

As far as osteology is concerned, I see the problem in two stages—detailed regional studies of existing skeletal collections to determine morphological features in contact and immediately pre-contact populations, to form a basis of comparison with specimens and populations of some antiquity. The start made by Larnach, Macintosh, and Freedman, using the cranial collections from coastal areas of New South Wales, is a most important move in this direction. Larnach and Macintosh are at present analysing the Queensland material in a similar fashion and a parallel study has begun on the National Museum collection of Victorian crania. Studies of extremely localised populations of known age, such as the Broadbeach burial ground material being studied by Dr Wood, and the collections resulting from the National Museum project in the Chowilla Dam site area, also provide valuable populations which have a spatial and chronological integrity not found in other study collections.

It is important that excavations of burial grounds be encouraged, especially rescue operations to secure large samples in areas threatened with disturbance or destruction. The solution of the Tasmanian problem also requires a collection of some size, especially if cremation has had a long history on the island. Midden excavation, particularly on the west coast, promises to supply the necessary bone material. Finally, fossil skeletons are urgently needed to establish a base of comparison with recent Aborigines. Projects specifically designed to secure such specimens require some priority

## Addendum

Since the 1968 symposium at which this paper was originally given, a number of significant skeletal discoveries have been made. At Kow Swamp, a minimum of eight new specimens has been located. Most, but not all, are rugged, with thick cranial vault bone and archaic features of the face and forehead. Although absolute dates are not yet available it is likely the Kow Swamp series is late Pleistocene in age. The form and size of several characteristics (the frontal area particularly) overlap the ranges for modern Australian Aborigines. The Mungo I specimen from western New South Wales (Bowler *et al.* 1970) has been dated at 25-32,000 years BP. At the present stage of reconstruction the Mungo cranium appears to be essentially modern in form.

## References

Abbie, A. A. 1951. The Australian Aborigine. *Oceania* **22**:91-100.

——1958. The original Australians. *Leech, Johannesb.* **28**:120-30.

——1960. Doctor Ruggles Gates and the aboriginal Australian. *Nature, Lond.* **187**: 375-6.

——1963. Physical characteristics of Australian Aborigines. In *Australian Aboriginal studies* (ed. Helen Shiels), pp. 89-107. Melbourne: Oxford University Press.

——1966. Physical characteristics. In *Aboriginal man in South and central Australia.* Part 1 (ed. B. C. Cotton), pp. 9-45. Adelaide: Government Printer.

——1968a. Australian Aborigines. In *Rassengeschichte der Menschheit* (eds. A. A. Abbie *et al.*), 1 Lfg., pp. 7-24. Munich: R. Oldenbourg.

——1968b. The homogeneity of Australian Aborigines. *Archaeol.& phys.Anthrop.Oceania* **3**:223-31.

Birdsell, J. B. 1949. The racial origin of the extinct Tasmanians. *Rec.Queen Vict.Mus.* **2**:105-22.

——1950. Some implications of the genetical concept of race in terms of spatial analysis. *Cold Spring Harb.Symp.quant.Biol.* **15**:259-311.

——1967. Preliminary data on the trihybrid origin of the Australian Aborigines. *Archaeol.& phys.Anthrop.Oceania* **2**:100-55.

Bowler, J. M., Jones, R., Allen, H., and Thorne, A. G. 1970. Pleistocene human remains from Australia: a living site and human cremation from Lake Mungo, western New South Wales. *World Archaeology* **2**: 39-60.

Campbell, T. D., Gray, J. H., and Hackett, C. J. 1936. Physical anthropology of the Aborigines of central Australia. *Oceania* **7**:106-39, 246-61.

Davies, P. L. 1968. An 8,000 to 12,000 years old human tooth from Western Australia. *Archaeol.& phys.Anthrop.Oceania* **3**:33-40.

Davis, J. B. 1874. On the osteology and peculiarities of the Tasmanians, a race of man recently become extinct. *Natuurk. Verh.holland.Maatsch.Wet.* 3 de Verz., **2** (4).

Fenner, F. J. 1939. The Australian Aboriginal skull: its non-metrical morphological characters. *Trans.R.Soc.S.Aust.* **63**:248-306.

Gates, R. R. 1960. Racial elements in the Aborigines of Queensland, Australia. *Z. Morph.Anthrop.* **50**:150-66.

Hooton, E. A. 1947. *Up from the ape.* 2nd edn. New York: Macmillan.

Howells, W. W. 1937. Anthropometry of the natives of Arnhem Land and the Australian race problem. *Pap.Peabody Mus.* **16**:1-97.

Hrdlička, A. 1928. Catalogue of human crania in the United States National Museum collections. (Australians, Tasmanians, South African Bushmen, Hottentots, and Negro). *Proc.U.S.natn.Mus.* **71** (2696) article 24:1-140.

Jones, R. 1966. A speculative archaeological sequence for north-west Tasmania. *Rec. Queen Vict.Mus.* No. 25.

Keith, Sir A. 1925. *The antiquity of man.* 2nd edn. London: Williams and Norgate.

Klaatsch, H. 1908. The skull of the Australian Aboriginal. *Rep.path.Lab.Lunacy Dep.N.S.W.* **1** (3):43-167.

Labillardière. J. J. H. de. 1800. *Voyage in search of La Pérouse . . . during the years 1791, 1792, 1793, and 1794.* 2 vols. London: John Stockdale.

Larnach, S. L. and Freedman, L. 1964. Sex determination of Aboriginal crania from coastal New South Wales, Australia. *Rec.Aust.Mus.* **26**:295-308.

—— and Macintosh, N. W. G. 1966. The craniology of the Aborigines of coastal New South Wales. *Oceania Monogr.* No. 13.

Leakey, L. S. B. 1934. *Adam's ancestors.* London: Methuen.

Lesson, A. 1880. *Les Polynésiens.* Vol. 1 Paris: Libraire de la Société Asiatique de Paris.

Macintosh, N. W. G. 1963. Origin and physical differentiation of the Australian Aborigines. *Aust.nat.Hist.* **14**:248-52.

—— and Barker, B. C. W. 1965. The osteology of Aboriginal man in Tasmania. *Oceania Monogr.* No. 12.

Macknight, C. C. and Thorne, A. G. 1968.

Two Macassan burials in Arnhem Land. *Archaeol.& phys.Anthrop.Oceania* **3**:216-22.

Morrison, J. 1967. The biracial origin of the Australian Aborigines. *Med.J.Aust.* **2**: 1054-6.

Mulvaney, D. J. 1966. The prehistory of the Australian Aborigine. *Scient.Am.* **214** (3): 84-93.

——1969. *The prehistory of Australia.* London: Thames and Hudson.

—— and Joyce, E. B. 1965. Archaeological and geomorphological investigations on Mt. Moffatt Station, Queensland, Australia. *Proc.prehist.Soc.* **31**:147-212.

Thorne, A. G. 1967. The racial affinities of the Tasmanian Aborigines—some new skeletal evidence. M.A. thesis, University of Sydney (unpublished).

Tindale, N. B. and Birdsell, J. B. 1941. Tasmanoid tribes in north Queensland. *Rec. S.Aust.Mus.* **7**:1-9.

Topinard, P. 1872. Étude sur les Tasmaniens. *Mém.Soc.Anthrop.Paris* Sér.1, **3**:307-29.

Turner, W. 1908. The craniology, racial affinities and descent of the Aborigines of Tasmania. *Trans.R.Soc.Edinb.* **46**:365-403.

Wood-Jones, F. 1934. *Australia's vanishing race.* Sydney: Angus and Robertson.

Wunderly, J. 1938. The cranial and other skeletal remains of Tasmanians in collections in the Commonwealth of Australia. *Biometrika* **30**:305-37.

——1943. The Keilor fossil skull: anatomical description. *Mem.natn.Mus.Vict.* **13**:57-69.

—— and Wood-Jones, F. 1933. The nonmetrical morphological characters of the Tasmanian skull. *J.Anat.* **67**:583-95.

Yamaguchi, B. 1966. A comparative osteological study of Ainu and Australian Aboriginal material: a preliminary report. *Aust.Inst.Aborig.Stud.Newsl.* **2** (3):16-19.

——1967. *A comparative osteological study of the Ainu and the Australian Aborigines.* Occasional Papers in Aboriginal Studies No. 10 (Human Biology Ser. No. 2). Canberra: Australian Institute of Aboriginal Studies.

# 22 Genetic Evidence and its Implications for Aboriginal Prehistory

R. L. Kirk

Various theories have been propounded to explain the prehistoric peopling of Australia. They have relied heavily on comparisons of anthroposcopic characters in living races, sometimes supplemented with reference to the scanty skeletal material from prehistoric sites in Australia, nearby islands, and other parts of the world. It is not my intention here to review these theories in detail, but only to survey the evidence on the distribution of various genetic markers in Aboriginal populations and compare this with similar information for selected groups outside Australia, concluding with an estimate of the length of time required to achieve the present genetic status.

The review will be restricted to those genetic markers which can be detected by tests performed on samples of blood, thus excluding characters such as ability to taste phenylthiocarbamide (PTC) and red-green colour blindness. Despite these exclusions the number of genetic-marker systems potentially valuable in anthropological studies is already formidable and is continuing to grow. They can be divided broadly into three categories: (a) blood group antigens (Table 22:1) (b) serum protein groups and (c) red cell and serum enzyme groups (Table 22:2). Not all these systems have been explored in Aborigines: indeed, some of them have been described relatively recently and we know their distribution in only a few populations in the world. However, there is a sufficient body of information already in existence for Aborigines on many of the genetic marker systems to warrant detailed consideration here.

## Genetic Markers

### BLOOD GROUP ANTIGENS

The distribution of blood group antigens in Aboriginal populations has been the subject of investigation for more than forty years and

*Table 22:1 Human blood group systems*

| | | | | | |
|---|---|---|---|---|---|
| 1 | ABO | | 9 | Kidd | (Jk) |
| 2 | MNSs | | 10 | Diego | (Di) |
| 3 | P | | 11 | I | |
| 4 | Rhesus | (Rh) | 12 | Auberger | (Au) |
| 5 | Lutheran | (Lu) | 13 | Xg | |
| 6 | Kell & Sutter | (K & Js) | 14 | Yt | |
| 7 | Lewis | (Le) | 15 | Dombrock | (Do) |
| 8 | Duffy | (Fy) | | | |

has been summarised by Kirk (1965) and Simmons (1966). Little has been added during the last few years to alter the pattern of distribution described by these authors. Looking at the continent as a whole (Table 22:3) Aboriginal populations are characterised by the following patterns:

A complete absence of the $A_2$ gene in the ABO blood group system. In the same system the B gene is also absent over the greater part of the continent: its presence in Arnhem Land, along the southern shores of the Gulf of Carpentaria, and in parts of Queensland will be discussed more fully below. In the MNS system, the frequency of gene N is everywhere high, reaching its highest value in the world in the Western Desert. More interesting is the complete absence of the large S gene. The Rhesus blood group system is also of interest, Aboriginal populations having no rh-ve (*cde*) genes, and variable frequencies of $R_1$, $R_2$, $R_0$ and $R_z$. $D^u$ variants occur sporadically, and a unique variant, the $E^T$ antigen, discovered by Vos and Kirk (1962) is found in many places, and its distribution is referred to more fully below.

Among the other blood group systems the $P_1$ antigen frequency is variable, but tends to be lower than in other parts of the world, and all persons are Fy(a+), Di(a−), Lu(a−), Js(a−) and K(−). A small sample of Bentinck and Mornington Islanders studied by Simmons *et al.* (1962) showed just over 60 per cent to be Jk(a+), and Scott and Scott (1966) for a small series at Wiluna found 88 per cent Jk(a+). For the Lewis blood group system the Le(a+) frequency is low, as also is the number of persons who are secretors of ABH substance in their saliva. Vos and Comley (1967) have recently made a valuable contribution to the study of the Lewis system in Aborigines, revealing a much greater range of variability in expression than is found in Caucasians. Finally, recent unpublished results by Simmons (pers. comm.) show that there are a few persons who give positive reactions with the Duffy antibody anti-Fy(b) and also that there is a high frequency of Xg(a+) reactions. Further studies on these markers may be important.

The general picture of the blood group distributions given above is sufficient to distinguish Aboriginal populations from all others in the world. At the same time, however, this general picture conceals some striking regional differences to which it is necessary also to draw attention, and two examples must suffice for this purpose.

The antigenic variant, $E^T$, in the Rhesus blood group system was first detected in the Laverton area of Western Australia. Fig. 22:1 shows the proportion of persons in various parts of Australia who carry the $E^T$ antigen. It varies from 50 per cent in the Western Desert to 100 per cent in north Queensland. Though insufficient populations have been tested throughout the whole range, the pattern of distribution suggests the spreading of a mutation, occurring originally as a deficiency change in the normal E antigen in the Western Desert. The distribution of this mutation from the centre of origin has reached the Kimberleys but has not so far reached the Queensland area.

The second example concerns the ABO blood group distribution. Over the greater part of the continent group B is absent, but the frequency of the A gene varies from values below 10 per cent in some localities in Cape York, southern Gulf of Carpentaria,

*Table 22:2 Polymorphic serum protein and enzyme systems in man*

| **Serum** | | **Cells** | |
|---|---|---|---|
| 1 Transferrin | (Tf) | 1 Haemoglobin | (Hb) |
| 2 Haptoglobin | (Hp) | 2 Glucose-6-phosphate dehydrogenase | (G6PD) |
| 3 Group Specific Component | (Gc) | 3 6-Phosphogluconate dehydrogenase | (6PGD) |
| 4 Albumin | (Alb) | 4 Phosphoglucomutase | (PGM) |
| 5 Gamma globulin | (Gm & Inv) | 5 Adenylate kinase | (AK) |
| 6 $\beta$-Lipoprotein | (Ag & Lp) | 6 Peptidases | |
| 7 Cholinesterase | | 7 Acid phosphatase | |
| 8 Caeruloplasmin | | 8 Phospho-hexose-isomerase | |
| 9 Alkaline phosphatase | | 9 Adenosine deaminase | |
| 10 Prealbumins | | 10 Carbonic anhydrase | |
| 11 c′3 ($\beta_{1_c}$-globulin) | | 11 Placental alkaline phosphatase | |
| | | 12 Phosphoglycetate kinase | |
| | | 13 Glutathione reductase | |

*Table 22:3 Range of gene frequencies and phenotype frequencies for blood group systems by geographical regions in Australia*

| | **Gene frequencies (per cent)** | | | | | | | |
|---|---|---|---|---|---|---|---|---|
| **Locality** | **p** | **q** | **r** | **n** | **$R_1$** | **$R_2$** | **$R_0$** | **$R_Z$** |
| Cape York | 4-22 | 0-14 | 74-96 | 66-88 | 66-79 | 6-8 | 4-25 | 1-11 |
| Southeast Queensland | 18-25 | 2-4 | 71-80 | 72-79 | 66-67 | 11-12 | 13-16 | 6-8 |
| Southern Gulf of Carpentaria | 0-10 | 0-24 | 76-93 | 49-69 | 52-67 | 5-24 | 9-43 | 0-1 |
| Northern Australia[a] | 8-23 | ? | 77-92 | 66-73 | 64-69 | 7-25 | 6-29 | 0-2 |
| Central Australia | 22-46 | 0 | 54-78 | 61-87 | 56-57 | 25-41 | 1-12 | 0-7 |
| Western Desert | 31-50 | 0 | 50-69 | 90-100 | 51-57 | 38-44 | 2-3 | 2-3 |
| Kimberleys | 11-23 | 0 | 77-89 | 59-78 | 55-76 | 13-28 | 4-14 | 2-9 |
| Western central | 35-46 | 0 | 54-65 | 68-78 | 57-62 | 27-29 | 0-9 | 2-13 |

| | **Phenotypes (per cent)** | | | | | | | | |
|---|---|---|---|---|---|---|---|---|---|
| **Locality** | **Le(a+)** | **$P_1$(+)** | **Fy(a+)** | **Lu(a+)** | **Lu(b+)** | **K(+)** | **Di(a+)** | **Js(a+)** | **Jk(a+)** |
| Cape York | 8-21 | 29-32 | 100 | — | — | — | 0 | — | — |
| Southeast Queensland | — | — | — | 0 | — | — | — | — | — |
| Southern Gulf of Carpentaria | 0-9 | 9-27 | 100 | — | — | — | — | — | 61-69 |
| Northern Australia[a] | — | — | — | — | — | — | — | — | — |
| Central Australia | 0-7 | 18-65 | 0 | — | — | 0 | 0 | — | — |
| Western Desert | 5 | 44-73 | 100 | 0 | 100 | 0 | 0 | — | 88 |
| Kimberleys | 0-19 | 24-61 | 100 | 0 | — | 0 | 0 | — | — |
| Western central | 14-21 | 26-28 | — | — | — | — | — | — | — |

[a] Including Bathurst and Melville Islands.

*Fig. 22:1 Area distribution in Australia of Rh-E positive persons reacting with a naturally occurring anti-$E^T$ antibody*

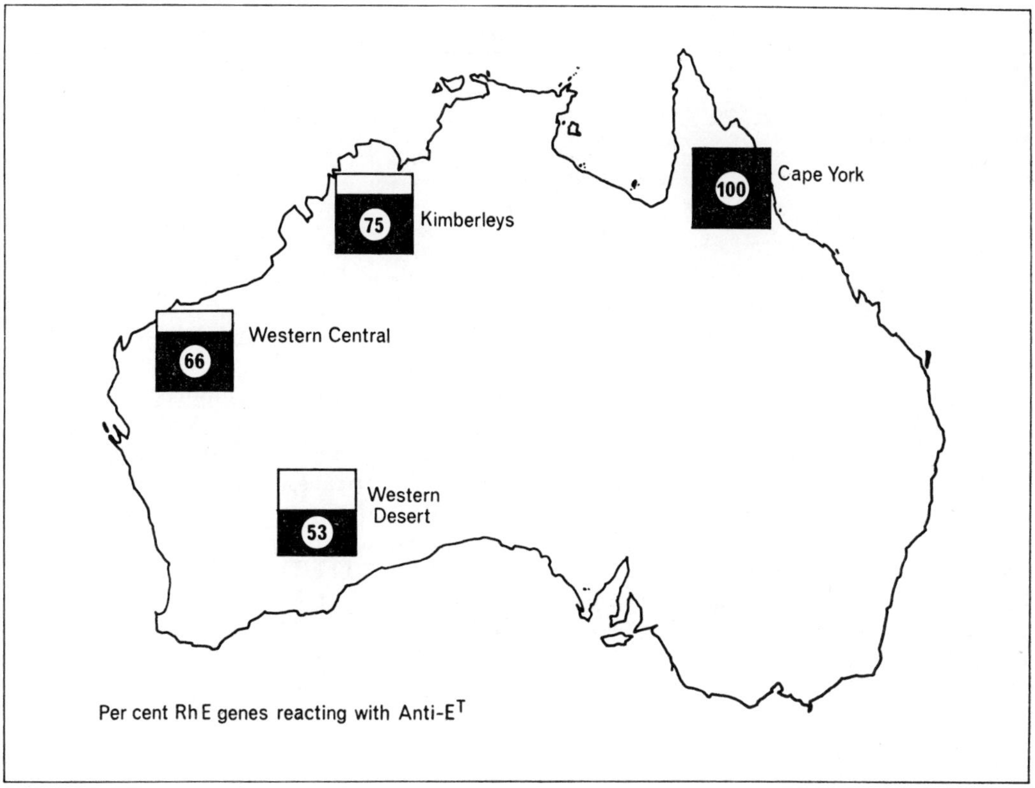


and Arnhem Land to 50 per cent in the Western Desert. The blood group O gene shows an inverse relationship. The general trend is clinal, populations in the centre having high values for the A gene, with lower values on the coast, particularly in the north. There are, however, some differences in frequency from locality to locality, and also, where information is available, from tribe to tribe. In addition to these variations in the distribution of the A and O genes the presence of group B in parts of Queensland, the Barkly Tableland, and Arnhem Land presents another interesting problem. Earlier surveys in Queensland (Birdsell 1950; Simmons *et al.* 1958) suggested that the group B was introduced from New Guinea via a Torres Strait migratory route. Mapping of the distribution of group B in Cape York, however, reveals that the higher frequencies of the B gene are in the south, and this information has been supplemented more recently by the studies of Simmons and his colleagues (Simmons *et al.* 1962; Simmons *et al.* 1964) showing the widespread distribution of the B gene in the Barkly Tableland area. They found also an unusual distribution among the small population of Bentinck Island, where 40 per cent of the inhabitants are group B and the remainder group O, the A gene having disappeared completely.

Simmons and his colleagues (1964) believe

*Table 22:4 Range of gene frequencies for serum protein groups by geographical regions in Australia*

| Locality | $Hp^1$ | $Tf^D$ | $Tf^B$ | $Gc^1$ | $Gc^{Ab}$ | $Gm^1$ | $Gm^{1,2}$ | $Gm^{1,5}$ | Inv(1+) |
|---|---|---|---|---|---|---|---|---|---|
| Cape York | 10-21 | 2-11 | 6 | 72-90 | 2-7 | 62-85 | 14-34 | 0-5 | 39-61 |
| Northern Australia | — | — | — | 87 | 2 | — | — | — | — |
| Central Australia | 13-29 | 5-10 | 0 | 86-94 | 1-5 | 62-70 | 29-37 | 0-1 | 37-47 |
| Western Desert | 15 | 20 | 0 | 92-98 | 0-1 | 73 | 27 | 0 | 49 |
| Kimberleys | 22-32 | 7-15 | 0 | 81-90 | 1 | 58 | 25 | 17 | 45 |
| Western central | 25-39 | 10-16 | 0 | 91-93 | 2-3 | — | — | — | — |

that the Karawa tribe has been the main focus for the spread of the B gene to other mainland tribes in the Gulf area, and suggest that the introduction of the B gene into groups such as the Kaiadilt, Lardiil, and Karawa, together with some other noted peculiarities of their blood group distribution, could be explained by separate admixture (and probably at different periods) with different non-Australian races to the north. Contact with peoples from the north on the basis of archaeological and cultural evidence is known to have taken place in Arnhem Land, and many claim to detect a Malayan or Macassan influence in the physical appearance of Aborigines living at various places along the north coast. The presence of the blood group B in these populations is one possible consequence of this contact, and further genetic evidence which may bear on this problem will be discussed below.

SERUM PROTEIN SYSTEMS

The distribution of a number of serum protein systems in Aboriginal populations was reviewed by Kirk (1965, 1966). The results have been extended recently by extensive studies in the Northern Territory, and the overall picture at the present time is summarised in Table 22:4. As in the case of several of the blood group systems, marked regional differences occur in all four of the serum protein systems which have been examined in detail.

In the haptoglobin system, the frequency of the gene $Hp^2$ is relatively high by world standards, but it falls markedly in the Kimberleys and around Port Hedland and Marble Bar in Western Australia. The frequency of hypohaptoglobinaemia (HpO) is low (ranging from 0-3 per cent) and the Hp2-1 mod. phenotype, which is found commonly in Africans, does not occur in Australia.

The distribution of genetic variants of the iron binding protein, transferrin, is especially interesting. The $D_1$ variant is present in all populations, the $Tf^{D_1}$ gene frequency approximating 10 per cent overall, but showing a clinal variation similar to that for the blood group A gene. $Tf^{D_1}$ values are lower in the north and reach their highest values of more than 20 per cent in the Western Desert area.

Kirk *et al.* (1964) have tested critically the transferrin variant in Aborigines against D variants from other parts of the world. They failed to detect any difference between that found in New Guinean or African populations, though it can be discriminated from the $D_{Chi}$ variant found in Mongoloid populations. More recently Wang *et al.* (1967) have confirmed the identity of the African and Aboriginal $D_1$, showing that the molecule in each case has an identical amino acid substitution when compared with the common C transferrin type.

Another transferrin variant, with the B-type electrophoretic mobility, has been detected at Edward River, in Cape York (Kirk *et al.* 1962). It has a mobility intermediate between that of $B_1$ and $B_2$, but does not appear to be identical with the $B_{1-2}$ found in Venezuela (Arends *et al.* 1962). It has a completely restricted distribution and so far has not been detected elsewhere in the world.

The Group Specific Component system (Gc) has been studied extensively in Australia. Of the two universally distributed alleles, $Gc^1$ and $Gc^2$, the $Gc^1$ frequency among Aborigines is everywhere high, exceeding 80 per cent, except in Cape York where the $Gc^1$ values lie between 70 and 80 per cent (excluding a very small sample from Weipa). The Cape York area provides us, however, with another interesting variant which we have recently been studying in more detail. The allele controlling this variant $Gc^{Aborigine}$ (Cleve *et al.* 1963) has its highest frequency in Australia in the Cape York area, with lower values in the Kimberleys and the central coastal areas of Western Australia. In the Western Desert it is virtually absent, but significantly higher frequencies of $Gc^{Ab}$ occur in some populations in central Australia, one of the highest values, 5 per cent, being recorded from Amoonguna. When the results from central Australia are analysed on a tribal basis, the highest $Gc^{Ab}$ value, although on a small sample of less than 50 persons, is found among the Aranda, where the gene frequency is 7 per cent (Nicholls *et al.* 1965).

The $Gc^{Ab}$ gene, however, is of interest also because it is widespread in Melanesian populations, both in New Guinea (Kirk *et al.* 1963; Kenrick 1967) and the New Hebrides (Cleve *et al.* 1967) although it has not been detected elsewhere in the world. Kitchin and Bearn (1966) state that they find a variant which appears to be identical with $Gc^{Ab}$ in Africans, but this important observation has not so far been confirmed.

Another important serum protein system, the Gm and Inv groups, is associated with the gamma globulins. This system is increasing rapidly in complexity, and studies in Australia have so far been restricted to only some of the possible allelic combinations. The Inv(1) frequency shows little consistent variation in different parts of the continent, but in contrast the Gm alleles vary widely and significantly. $Gm^1$ is present in all populations studied, ranging in frequency from 58 to 85 per cent. Similarly the $Gm^{1,2}$ allele varies in frequency from 14 to 37 per cent. The distribution of the $Gm^{1,5}$ allele, however, is of the greatest interest for although it is present in low frequency in Cape York and occurs with a frequency of 17 per cent in the Kimberleys it is completely absent in the Western Desert. Despite intensive study no single Gm(5) reaction has been detected in these latter populations, suggesting the complete absence of gene flow from the northwest into the desert areas to the southeast. Even in central Australia the $Gm^{1,5}$ allele is virtually absent and two cases only have been observed.

For the other serum protein systems known to be polymorphic in at least some human populations, studies in Australia have so far been scanty. The Ag(a) lipoprotein reaction was found to be high (84 per cent) in a small number of samples from Western and central Australia, but the Lp and Xm systems have not been tested. Recent work in our laboratory has failed to detect any variation in serum albumins, but no tests have been carried out for the pre-albumins, the $\beta_{1c}$-globulin or caeruloplasmin variants. Future study of these systems may add further information of value in understanding the genetic heterogeneity of Aborigines.

### ENZYME GROUPS

During the last few years considerable advances have been made in our understanding of the genetic control of variations

*Table 22:5 The distribution of 6PGD types in northern Australia*

| Locality | No. tested | Phenotypes AA | AC | CC | Elcho | Gene frequencies $PGD^{A}$ | $PGD^{C}$ | $PGD^{Elcho}$ |
|---|---|---|---|---|---|---|---|---|
| Elcho Island | 621 | 549 | 47 | 1 | 21 | ·944 | ·039 | ·017 |
| Delissaville | 94 | 87 | 7 | 0 | 0 | ·963 | ·037 | 0 |
| Angurugu | 90 | 74 | 5 | 0 | 11 | ·911 | ·029 | ·061 |
| Roper River | 100 | 90 | 9 | 0 | 1 | ·950 | ·045 | ·005 |
| Maningrida | 102 | 73 | 23 | 2 | 4 | ·848 | ·132 | ·020 |
| Millingimbi | 102 | 92 | 9 | 1 | 0 | ·946 | ·054 | 0 |
| Victoria River Downs | 103 | 95 | 6 | 0 | 2 | ·961 | ·029 | ·010 |
| Robinson River | 36 | 30 | 6 | 0 | 0 | ·917 | ·083 | 0 |
| Amoonguna | 104 | 92 | 12 | 0 | 0 | ·942 | ·058 | 0 |
| Hermannsburg | 61 | 57 | 4 | 0 | 0 | ·967 | ·033 | 0 |
| Yuendumu | 98 | 92 | 6 | 0 | 0 | ·969 | ·031 | 0 |
| Areyonga | 100 | 88 | 12 | 0 | 0 | ·940 | ·060 | 0 |
| Maryvale | 99 | 94 | 4 | 1 | 0 | ·970 | ·030 | 0 |
| Aileron | 70 | 67 | 2 | 1 | 0 | ·971 | ·029 | 0 |

in enzyme proteins, particularly those present either in serum or in haemolysates prepared from red blood cells. Table 22:2 lists the systems known to be polymorphic in the populations studied initially, and information is accumulating on the distribution of variants in a limited number of other populations in various parts of the world.

In Australia progress has been slow, but in the last few months my own laboratory has been able to begin accumulating data on a number of these enzyme systems. Earlier work (Horsfall *et al.* 1963; Kirk 1966) had shown a very low frequency of $C_5$ esterase variants in Aboriginal serum. More recently we have been studying the distribution of variants of red cell acid phosphatase, phosphoglucomutase, adenylate kinase, 6-phosphogluconate dehydrogenase, lactic dehydrogenase and malic dehydrogenase. Lai (1968) has shown that the red cell acid phosphatase types in series from the Northern Territory and Queensland was predominantly of type B, with about 10 per cent AB. Our own results from north Australia are approximately the same, suggesting that there is little variation for the acid phosphatase types in Aboriginal populations.

The adenylate kinase system has two alleles, $AK^1$ and $AK^2$, in Caucasian populations, but $AK^2$ appears to be absent in black Africans. We have tested nearly 2,000 samples in northern Australia and find only the $AK^1$ gene present. For the LDH and MDH systems no variants have so far been detected in Australia. They are rare in most parts of the world, though there is some evidence that the frequency of LDH variants may be slightly higher in Africans than in other racial groups. The PGM and 6PGD systems, on the other hand, appear to be potentially important. So far we have only been able to investigate the distribution of PGM and 6PGD variants in populations in the Northern Territory. In particular, we have delineated a new 6PGD allele, $PGD^{Elcho}$ (Blake and Kirk 1969) after the discovery of an unusual variant in a sample from Elcho Island. This allele is present in many populations in Arnhem Land, but it has not been detected in populations from central Australia (Table 22:5). Since its distribution, superficially at least, is similar to that for blood group B and since we have not been able to detect $PGD^{Elcho}$ in several hundred samples from the New Guinea Highlands, it

*Fig. 22:2 Localities mentioned in Table 22:5*

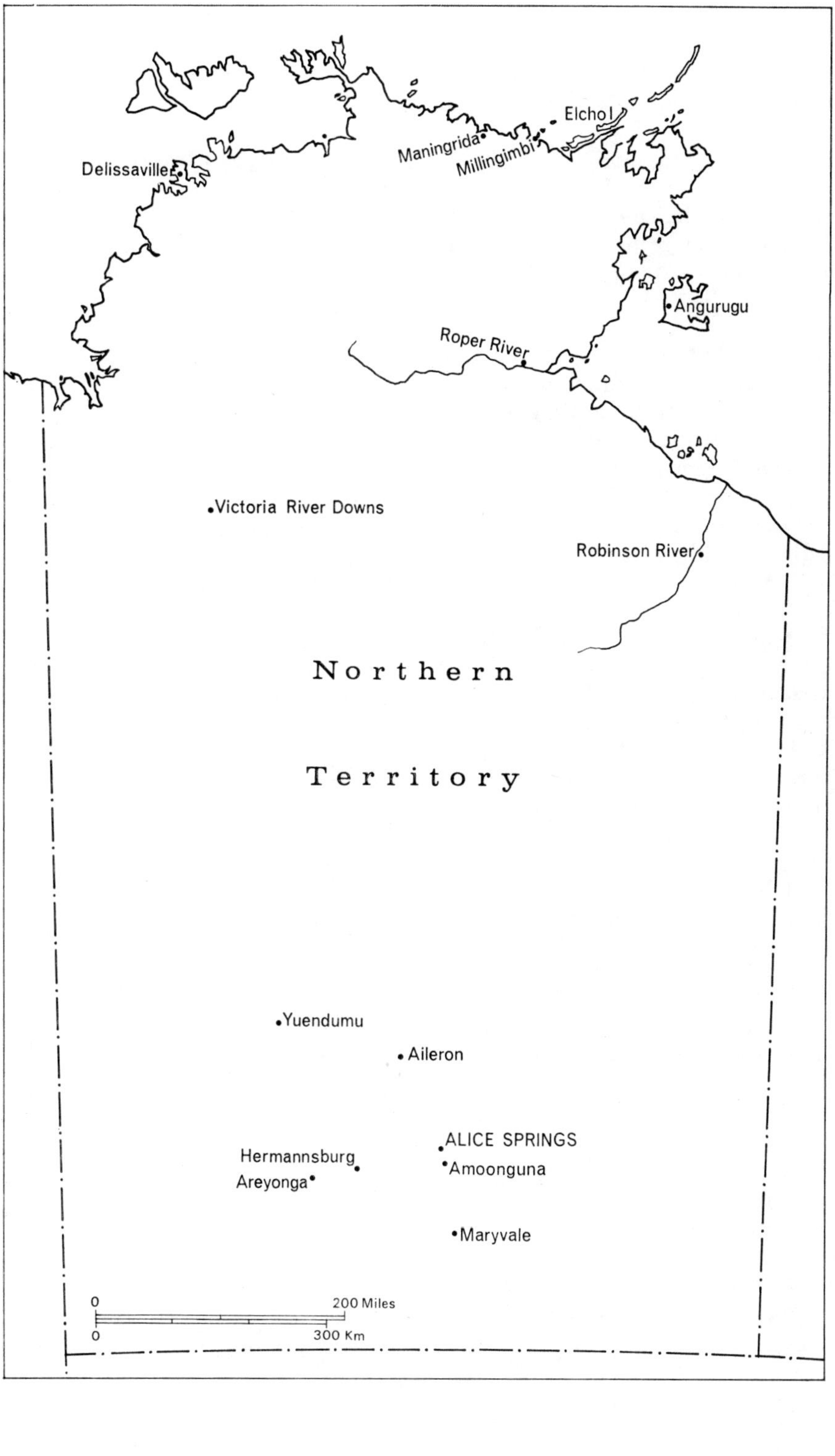

would be interesting to know if $PGD^{Elcho}$ occurs in Indonesian or Malay populations. If this proves to be the case a more specific marker for the influence of outside populations on Australian Aborigines would be available.

OTHER GENETIC SYSTEMS

So far I have left out of consideration several genetically controlled variations in blood, all of which probably have some significant selective advantage for populations living in holoendemic malarious regions. I refer to the abnormal haemoglobins, the thalassaemias and glucose-6-phosphate dehydrogenase deficiency.

In New Guinea, coastal populations show high frequencies of G6PD deficiency and of elevated $A_2$ levels, the latter being indicative of thalassaemia: abnormal haemoglobins, although present, are rare. Repeated tests in Australia, however, have failed to demonstrate elevated $A_2$ levels, G6PD deficiency, or the presence of abnormal haemoglobins (Horsfall and Lehmann 1953, 1956; Simmons 1958; Budtz-Olsen 1958; Budtz-Olsen and Kidson 1961; and Kirk unpublished).

The failure to detect variants for abnormal haemoglobins, G6PD deficiency, and thalassaemia in Aboriginal populations suggests that, in the absence of selection due to holoendemic malaria (which does not appear to have been present in this continent), the genes for these traits, even if present in the original populations or introduced at a later date, would have been lost. Livingstone (1964) and Lisker and Motulsky (1967) have shown that gene frequencies for these traits can change relatively quickly under the action of selective forces and it is not unreasonable to suppose that in the absence of malaria a frequency of an abnormal haemoglobin gene would fall from 10 per cent to zero in 50 generations, i.e. about 1,000 years.

## Anthropological Implications

The results set out above can be examined for the light they throw on three problems: the genetic diversity of Aboriginal populations within Australia; the origins of Aborigines; and the length of time needed for Aboriginal populations to achieve their present status.

GENETIC DIVERSITY WITHIN AUSTRALIA

Birdsell (1950) first drew attention to marked differences of some genes between tribal groups. For example, in considering the distribution of the A gene for the ABO blood group system and the N gene in the MN system he constructed contours along supposed isogenic lines, showing high values for gene A among the Pitjandjara with lower values to the southwest among the Ngadadjara, and a very steep decline to the east among the Aranda. The distribution of gene N shows similar differences, the Aranda having the lowest value, but the changes in frequency were not as sharp as for the A gene. Birdsell argued that these differences reflect in part differences in the origins of the Aborigines. He wrote (1950:263):

> Broadly, the northern portion of the Continent, in which the Carpentaria racial element predominates, shows low frequency for the gene $I^A$. Southern Australia and the marginal coastal districts, in what is essentially Murrayian territory, are characterised by higher frequencies. These general trends indicate that the Murrayians introduced relatively high frequency for the gene $I^A$ into Australia, and were followed by Carpentarians who showed considerably reduced values. This seemingly simple situation is complicated by the occurrence of the highest frequencies for this allele in the desert regions. There the populations can be described taxonomically only as a hybrid group in which the Murrayian and Carpentarian genetic contributions are roughly equivalent.

Birdsell continues, however, by stating 'The high frequencies observed in and around the peak area and even more widely within the contour line of ·35 value cannot be attributed to hybridization between Murrayian and Carpentarian elements. Hybridization may be excluded as a force responsible for the Pitjandjara peak'. He goes on to consider that such a peak may have been the result of selection, random genetic drift, and hybridisation. Birdsell considered that the lower values of gene A and N among the Aranda tribe were capable of a different interpretation. This is that the Aranda migrated southwards to their present domain at a relatively recent date, and he claims that linguistic and other cultural data for the Aranda tend to substantiate this view. That there are significant differences in gene frequency for a number of separate alleles between the Aranda and some of their neighbours now seems clear (Nicholls *et al.* 1965). What is clear also, however, is that when single systems are considered alone very striking differences in frequency between neighbouring tribes can sometimes occur due to the operation of random processes.

Whether the Aranda are genetically closer to populations to the north than they are to their neighbours to the west is still an intriguing question which must depend on the application of a quantitative measurement of genetic distance, which combines results for all genetic marker systems into a single measure. In collaboration with Dr L. D. Sanghvi we have carried out a preliminary analysis of this type and the results are of great interest. The Aboriginal populations sampled fall into two broad clusters widely separated from one another. One cluster includes all populations in central Australia, the Western Desert, and Port Hedland, whilst the other cluster includes the Kimberleys, Arnhem Land, and north Queensland. This division into two clusters is now being analysed in more detail, but it suggests a basic differentiation of Aboriginal populations corresponding to at least part of the major linguistic division to which attention has been drawn by Tryon (see ch. 23).

When one looks beyond the tribal boundaries to regional groupings there is no doubt about the genetic heterogeneity of Aboriginal populations. Though certain markers, such as transferrin $D_1$, are found in all populations, others show either a restricted or a clinal distribution. Among the latter are the $Gc^{Ab}$ gene, the $E^T$ rhesus blood group antigen, and the $Gm^{1,5}$ allele. Among more restricted markers is blood group B and possibly the $PGD^{Elcho}$ enzyme variant.

The genetic structure of the Western Desert population, together with several tribes from central Australia, is particularly striking. In this area are found extreme values for the blood group genes $A$, $N$, and $P_1$ and the serum protein genes $Hp^1$, $Tf^{D_1}$, $Gc^1$, and $Gm^{1,5}$, for all of which the population is approaching, or has reached, either loss or fixation of the gene concerned. This situation implies a long period of isolation without the benefit of fresh additions from outside. Indeed, the evidence from the gamma globulin group system indicates the complete absence of gene flow from the coast in toward the desert regions of the interior, though the distribution of the $E^T$ antigen suggests that gene flow from the centre towards the coast has taken place.

The occurrence of regionally restricted distributions, such as blood group B and possibly $PGD^{Elcho}$, as well as the clinal distribution of $Gc^{Ab}$ from a focus in Cape York, is testimony to the introduction of genetic material probably not present in the original Aborigines. The wide distribution of $Gc^{Ab}$ in Melanesian populations would

*Fig. 22:3 Frequency ranges for various alleles and phenotypes in Aboriginal populations in Australia and in New Guinea*

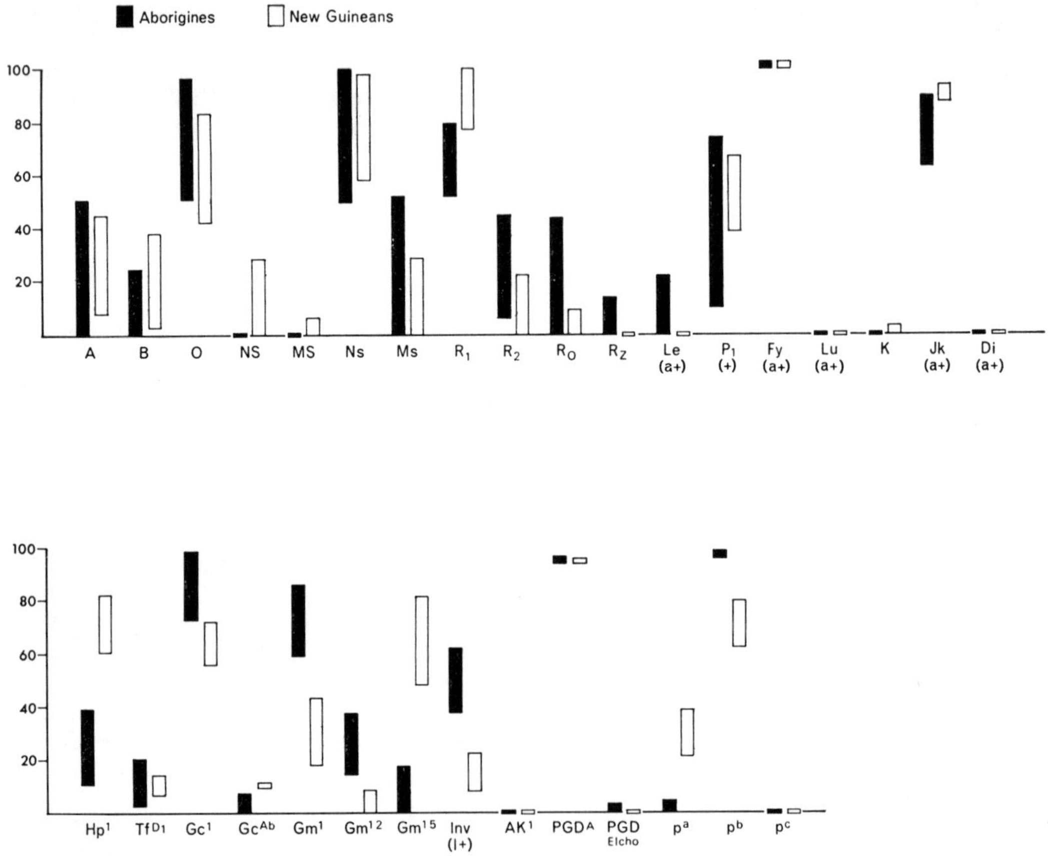


suggest its introduction into Australia from this source. By contrast the distribution of blood group B and $PGD^{Elcho}$ would indicate a non-Melanesian source for this gene, and one is tempted to accept the hypothesis that contact along the Arnhem Land coast with Malay fishing fleets was the primary source.

### THE ORIGIN OF AUSTRALIAN ABORIGINES

Any attempt to use genetic markers to identify the possible place or places of origin of Aborigines must take into account the fact that evolution will have continued in both the parent population and its offshoot since the separation took place. The possible evolutionary changes include the effects of random drift, of random accumulation of new mutants, of selection on existing alleles, and of gene flow from other groups into either the descendants of the parent population or its offshoot. If the time of separation was long ago one might expect the summation of all these effects would produce such a gross divergence in two populations that a comparison today would make difficult the identification of a common ancestor.

*Fig. 22: 4 Frequency ranges for various alleles and phenotypes in Aboriginal populations in Australia and the Ainu of Japan*

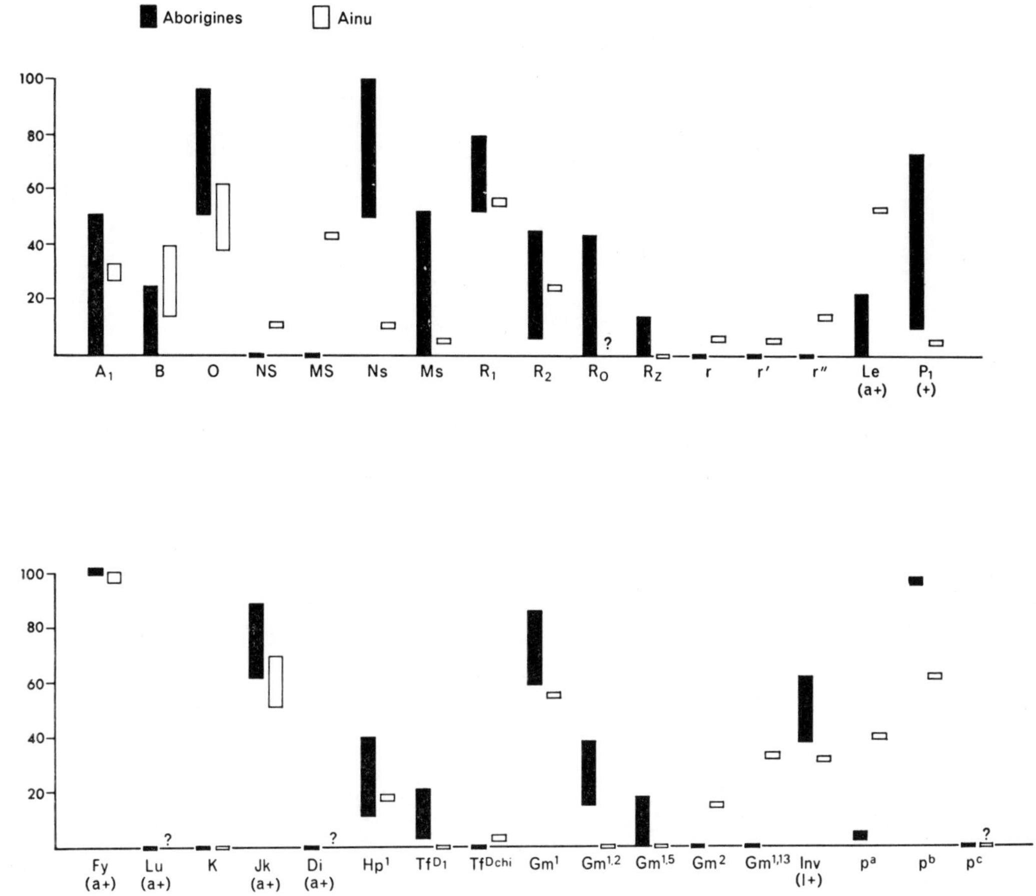


Despite the apparent difficulty two approaches can be made to the problem, one qualitative and the other quantitative. In the qualitative approach we may ask what special genetic characters present in Aboriginal populations are found elsewhere. This is comparable to the procedure in physical anthropology where comparisons are based on anthroposcopic characters such as skin colour, hair form, tooth structure, etc.

In Fig. 22:3 I have compared the gene frequency ranges for thirty alleles in Australian and New Guinea populations. Those alleles known to be responsive to malarial selection have been omitted. For the majority of these alleles the ranges of variation in New Guinea and Australia overlap. The most striking divergences, in which there is no overlap, occur in the frequencies of the haptoglobin genes and alleles of the Gm and Inv systems. The Gm system is known to show marked differentiation in gene frequency from population to population, so that the divergence between Australia and

*Fig. 22:5 Frequency ranges for various alleles and phenotypes in Aboriginal populations in Australia and Bushmen in Africa*

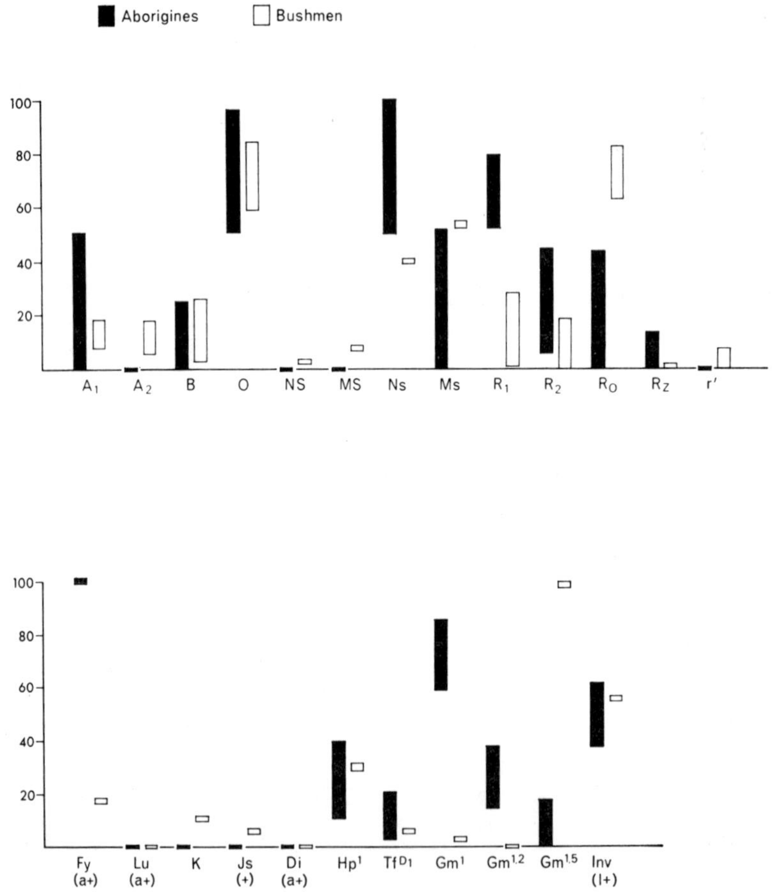


New Guinea for the Gm alleles is perhaps not too surprising.

In Figs. 22:4-6 similar comparisons have been made for a more restricted range of markers between Aboriginal populations and the Ainu, Bushmen, and a Veddoid population in south India and Ceylon. The degree of divergence shown in all three comparisons is clearly greater than that between Australians and New Guineans. It can be argued that this situation merely reflects a longer time of separation between these groups. What is needed, therefore, is some precise way of quantifying their relationship.

Recently, Cavalli-Sforza and Edwards (1965, 1967) have been exploring a quantitative approach to this problem. In the first instance they selected fifteen populations for which reasonably adequate information on the frequencies of a number of genetic markers was available. They then measured the divergence in mean gene frequency, the frequencies being summed over all alleles using an angular transformation.

The difference in frequencies between pairs of populations was then subjected to a method of cluster-analysis which separated the groups of populations into smaller and smaller clusters by a succession of binary splits. The populations were then joined by a network in the shape of a branching tree, the best net being selected on the principle of 'minimum evolution', i.e. the net which involved the least number of edges to connect all points. It is assumed that evolution is a random process, each separate population evolving independently.

In their original study, data for five blood group systems with a total of eighteen alleles were used to determine the 'minimum evolutionary' tree for the fifteen populations. The base of the tree was chosen arbitrarily, but its position does not alter the relative lengths of any of the branches, this function being proportional to the quantity of evolution. The tree shows that populations for the same continent are most closely related, a conclusion which accords with a common-sense appraisal. The greatest separation is between Australians and New Guineans at one extreme and Bantu and Ghanaian at the other. In a subsequent paper (Cavalli-Sforza 1967) the same type of analysis has been repeated using a greater number of alleles, a total of thirty-eight. The number of populations, because of inadequate data, had to be reduced to seven, but the results are not too dissimilar from those obtained from the tree constructed for the fifteen populations.

Both these trees suggest that the first split occurred between populations of east Asia (a group including populations in the Pacific area as well as aboriginal populations in America and Australia and New Guinea) and the Caucasoids. African populations separated somewhat later.

Cavalli-Sforza and Edwards admit that their evolutionary tree model is only preliminary and that improvement in the amount of information on gene frequencies in different parts of the world could lead to more accurate models. Further they point out that the model does not take into account hybridisation between populations, a phenomenon which almost certainly played an important part in the evolution of human races, nor does it allow for convergence and parallelism, that is for similar selective response to similar environmental stimuli in different populations. I have discussed this problem in more detail elsewhere (Kirk 1969). Despite these limitations, however, Cavalli-Sforza and Edwards have developed an approach which, if it is refined to allow for the other variables, may yield useful indicators of population origins.

#### TIME-SPAN

A final problem remains: is it possible on the basis of genetic data obtained from living populations to estimate the length of time required to achieve the degree of diversity exhibited by man in different parts of the world? I think it is obvious that no great degree of accuracy will ever be achieved in making such estimates, for each of the number of assumptions made is subject to an uncertainty which cannot be estimated accurately in times before the present. However, it is of interest to examine the estimates which have been made, bearing in mind the limitation mentioned.

Cavalli-Sforza (1966) has approached this problem by examining the consequences of random genetic drift as a function of time. It is assumed that the rates of evolutionary change due to drift and to random changes in selective values are essentially the same in all groups. The rates of change of the standardised variance of gene frequency $\sigma^2/pq$ has been computed for various effective

*Fig. 22:6 Frequency ranges for various alleles and phenotypes in Aboriginal populations in Australia and the Veddoids in Ceylon and south India*

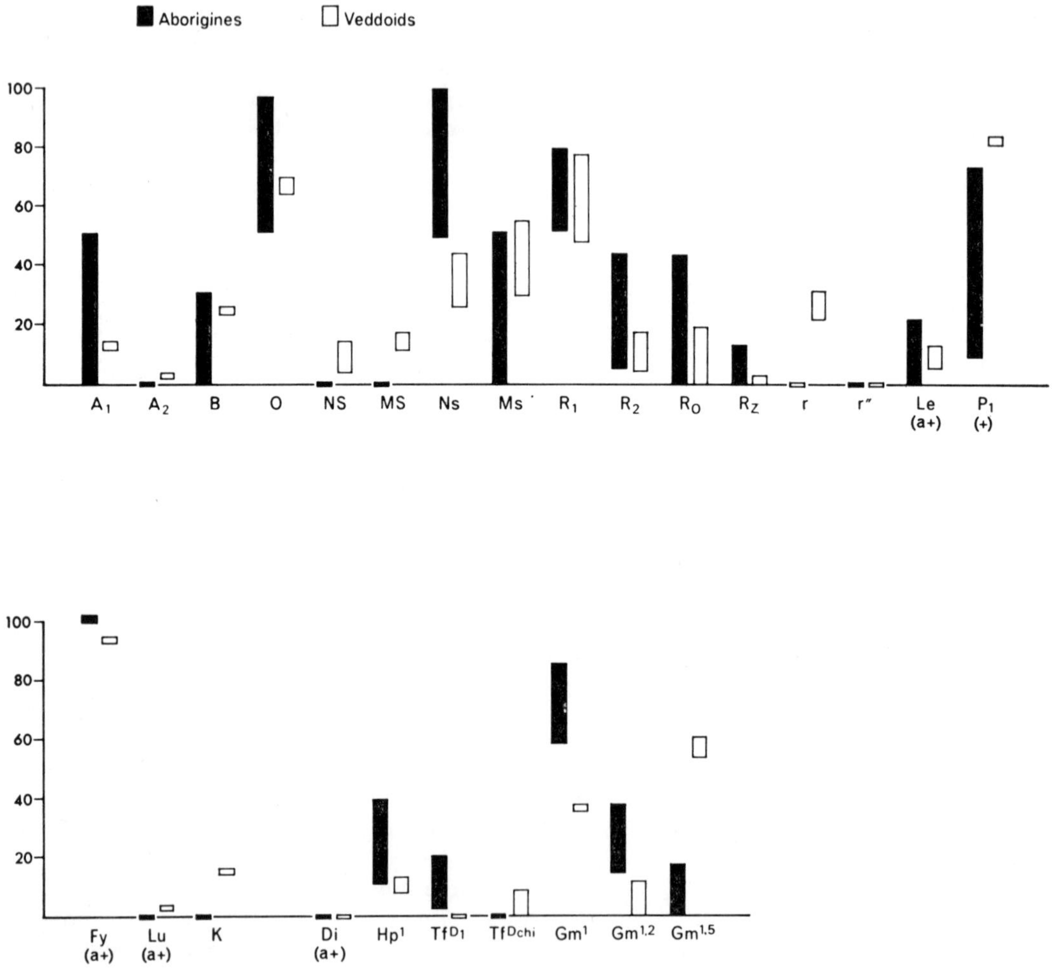


population sizes. Assuming an effective population size of 1,000, the observed world differences in standardised variance of the majority of the alleles in the ABO, MNS, and Rh blood group systems could have been achieved as a result of drift alone in just over 11,000 years. Since the divergences in gene frequencies in Australia are no greater than for the world as a whole, and since also the effective population size in Aboriginal groups is never likely to have been greater than 1,000, we may assume that an originally genetically homogeneous population split into non-interbreeding sub-populations could have achieved the degree of diversity in genetic structure observed today in at most 10,000 years. If effective population size was less than 1,000 the rate of divergence would have been greater. Population movements and hybridisation may have complicated the pattern of divergence, but it seems that random genetic drift could readily explain the variation in gene frequencies shown by present Aboriginal populations.

For an estimate of the time of separation of the ancestors of Aborigines from their parent population another approach may be made (Kirk 1969). Certain alleles are restricted to populations in one geographical region of the world, or if dispersed more widely, are present only in populations which on other grounds we assume to have been derived from the same parental population. An example of such an allele is that controlling the Di(a) blood group antigen. This is found in the majority of all Mongoloid populations, but only very rarely in other racial groups. Also present in Mongoloid populations is a transferrin variant $D_{Chi}$, widely distributed in South American Indian populations, throughout east Asia, among the Lapps in Finland, and also in some tribal populations in northeast India as well as in the Ainu. $D_{Chi}$ therefore, like the Di(a) antigen, can be considered a Mongoloid marker, and its pattern of distribution suggests that it arose as a new mutant before the oriental migrations from Asia to America took place. Since at present there is no evidence that the transferrin polymorphism is of the balanced type, let us assume that the $D_{Chi}$ mutant is gradually increasing in frequency, and that over the greater period of time required for this to occur the process has been approximately linear. On the basis of evidence for the rate of complete amino acid substitution in protein molecules in various species of mammals it is possible to calculate that the length of time required for the $D_{Chi}$ mutant to achieve its present observed frequency would lie in the range of 30,000 to 80,000 years.

A similar but quite distinct transferrin mutant $D_1$, as noted above, is widely distributed in Australian, Melanesian, and black African populations. If it is assumed that the $D_1$ variant in Africans was derived from the same source as that in Australians and Melanesians, then a similar calculation to that used for $D_{Chi}$ would suggest that the $D_1$ mutation had its origin in a population ancestral to Australians, Melanesians, and black Africans but already distinct from populations ancestral to other racial groups, roughly 100,000 years ago.

Australian Aborigines are a genetically heterogeneous collection of populations. The heterogeneity arises most probably from three sources: primarily, random genetic drift operating on small isolated breeding populations, the occurrence and spread of new mutants, and the mixing of gene pools from different groups. Selection by environmental factors, such as the conditions of a desert environment, may also have played a part, but evidence for this is lacking at present.

A comparison of the constellation of genetic characters in Aborigines shows them to be related most closely to but not identical with populations in New Guinea. Close affinity with populations elsewhere in the world cannot be shown in a simple manner, but it is possible that Aborigines, Melanesians, and Africans shared a common ancestry perhaps as long ago as 100,000 years. Separation over this length of time is enough, however, to have allowed maximum divergence in frequencies of the majority of alleles so that quantitative estimates of the rates of such divergence place Africans and the Australian-Melanesian complex at the extremes of the range for the diversification of living populations. Similar estimates for the rates of divergence in gene frequencies due to the operation of genetic drift alone suggest that the heterogeneity observed within Australia itself could have been achieved in at least 10,000 years.

## References


Arends, T., Gallango, M. L., Parker, W. C., and Bearn, A. G. 1962. A new variant of human transferrin in a Venezuelan family. *Nature, Lond.* **196**:477-8.

Birdsell, J. B. 1950. Some implications of the genetical concept of race in terms of spatial analysis. *Cold Spring.Harb.Symp. quant.Biol.* **15**:259-311.

Blake, N. M. and Kirk, R. L. 1969. New genetic variant of 6-phosphogluconate dehydrogenase in Australian Aborigines. *Nature, Lond.* **221**:278.

Budtz-Olsen, O. E. 1958. Haptoglobins and haemoglobins in Australian Aborigines, with a simple method for the estimation of haptoglobins. *Med.J.Aust.* **2**:689-93.

—— and Kidson, C. 1961. Absence of red cell enzyme deficiency in Australian Aborigines. *Nature, Lond.* **192**:765.

Cavalli-Sforza, L. L. 1966. Population structure and human evolution. *Proc.R.Soc.* Ser.B, **164**:362-79.

——1967. Human populations. In *Heritage from Mendel* (ed. R. A. Brink), pp. 309-31. Proceedings Mendel Centennial Symposium, Fort Collins, Colorado, 1965. Madison: University of Wisconsin Press.

—— and Edwards, A. W. F. 1965. Analysis of human evolution. In *Genetics today* (ed. S. J. Geerts), Vol. 3, pp. 923-33. Proceedings of the XI International Congress of Genetics, The Hague, 1963. London: Pergamon.

—— and —— 1967. Phylogenetic analysis: models and estimation procedures. *Evolution, Lancaster, Pa.* **21**:550-70.

Cleve, H., Kirk, R. L., Gajdusek, D. C., and Guiart, J. 1967. On the distribution of the Gc variant Gc Aborigine in Melanesian populations. *Acta genet.Statist.med.* **17**: 511-17.

—— *et al.* 1963. Two genetic variants of the group-specific component of human serum: Gc Chippewa and Gc Aborigine. *Am.J.hum.Genet.* **15**:368-79.

Horsfall, W. R. and Lehmann, H. 1953. Absence of the sickle cell trait in seventy-two Australian Aborigines. *Nature, Lond.* **172**:638.

—— and —— 1956. Absence of abnormal haemoglobins in some Australian Aboriginals. *Nature, Lond.* **177**:41-2.

——, ——, and Davies, Daphne. 1963. Incidence of pseudo cholinesterase variants in Australian Aborigines. *Nature, Lond.* **199**:1115.

Kenrick, K. G. 1967. Gc-Aborigine in a New Guinea population. *Acta genet.Statist.med.* **17**:222-5.

Kirk, R. L. 1965. *The distribution of genetic markers in Australian Aborigines.* Occasional Papers in Aboriginal Studies No. 4. Canberra: Australian Institute of Aboriginal Studies.

——1966. Population genetic studies in Australia and New Guinea. In *The biology of human adaptability* (eds. P. T. Baker and J. S. Weiner), pp. 395-430. Oxford: Clarendon Press.

——1969. Biochemical polymorphism and the evolution of human races. In *Proc. 8th Int.Congr.anthrop.ethnol.Sci.* Tokyo and Kyoto, 1968, Vol. 1., pp. 371-6. Tokyo.

——, Lai, L. Y. C., and Horsfall, W. R. 1962. The haptoglobin and transferrin groups among Australian Aborigines from north Queensland. *Aust.J.Sci.* **24**:486-8.

——, Cleve, H., and Bearn, A. G. 1963. The distribution of the group specific component (Gc) in selected populations in south and south east Asia and Oceania. *Acta genet.Statist.med.* **13**:140-9.

——, Parker, W. C., and Bearn, A. G. 1964. The distribution of the transferrin variants $D_1$ and $D_{Chi}$ in various populations. *Acta genet.Statist.med.* **14**:41-51.

Kitchin, F. D. and Bearn, A. G. 1966. The electrophoretic patterns of normal and variant phenotypes of the group specific (Gc) components in human serum. *Am.J. hum.Genet.* **18**:201-14.

Lai, L. Y. C. 1968. Unusual distribution of red cell acid phosphatase among Aborigines of Australia. *Nature, Lond.* **217**:1186.

Lisker, R. and Motulsky, A. G. 1967. Computer simulation of evolutionary trends in an X linked trait: application to glucose-6-phosphate dehydrogenase deficiency in man. *Acta genet.Statist.med.* **17**: 465-74.

Livingstone, F. B. 1964. Aspects of the population dynamics of the abnormal hemoglobin and glucose-6-phosphate dehydrogenase deficiency genes. *Am.J.hum. Genet.* **16**:435-50.

Nicholls, E. M., Lewis, H. B. M., Cooper, D. W. and Bennett, J. H. 1965. Blood group and serum protein differences in some central Australian Aborigines. *Am.J. hum.Genet.* **17**:293-307.

Scott, I. D. and Scott, M. K. 1966. Blood groups of some Australian Aborigines of the Western Desert. *Nature, Lond.* **212**: 545.

Simmons, R. T. 1958. A review of blood group gene frequencies in Aborigines of the various Australian States. In *Proceedings of the seventh congress of the International Society of Blood Transfusion.* Rome 1958. pp. 287-92. Basle: S. Karger.

——1966. The introduction of 'new' blood genes to the Australian Aborigines. *Archaeol. & phys.Anthrop.Oceania* **1**:67-73.

——, Graydon, J. J., and Gajdusek, D. C. 1958. A blood group genetical survey in Australian Aboriginal children of the Cape York Peninsula. *Am.J.phys.Anthrop.* **16**: 59-77.

——, Tindale, N. B., and Birdsell, J. B. 1962. A blood group genetical survey in Australian Aborigines of Bentinck, Mornington and Forsyth Islands, Gulf of Carpentaria. *Am.J.phys.Anthrop.* **20**:303-20.

——, Graydon, J. J., and Tindale, N. B. 1964. Further blood group genetical studies on Australian Aborigines of Bentinck, Mornington and Forsyth Islands and the mainland, Gulf of Carpentaria, together with frequencies for natives of the Western Desert, Western Australia. *Oceania* **35**: 66-80.

Vos, G. H. and Kirk, R. L. 1962. A 'naturally-occurring' anti-E which distinguishes a variant of the E antigen in Australian Aborigines. *Vox Sang.* **7**:22-32.

—— and Comley, P. 1967. Red cell and saliva studies for the evaluation of ABH and Lewis factors among the Caucasian and Aboriginal populations of Western Australia. *Acta genet.Statist.med.* **17**:495-510.

Wang, An-Chuan, Sutton, H. E., and Scott, I. D. 1967. Transferrin $D_1$: identity in Australian Aborigines and American Negroes. *Science, N.Y.* **156**:936-7.

# 23 Linguistic Evidence and Aboriginal Origins

D. T. Tryon

The purpose of this paper is to present an outline account of the development of knowledge of the Australian Aboriginal languages and the theories concerning Aboriginal origins and relationships that have evolved as more and more linguistic information has come to hand. To this end, the theorists and their theories will be discussed in terms of the state of knowledge at the time.

It seems most suitable, therefore, to divide the history of linguistic theories of Aboriginal origins into three periods, 1886-1930, 1930-1956, 1956-1968.

## Curr to Capell: 1886-1930

It is interesting to note that some of the theories of this early period have returned to favour in some quarters even in the 1960s. This period was one which could perhaps be described as that of 'unscientific linguistics'. None of the theorists represented here was a trained linguist. Even more hampering was the fact of the extreme paucity of knowledge of the Aboriginal languages right through the period.

The first to draw linguistic conclusions about the origins of the Aborigines in this period was E. M. Curr (1820-89). His linguistic knowledge was not great, but his theory that the first settlement in Australia was in the northwest and that the distribution of population was effected by the original stream of people crossing to the south of Australia in three separate bands is of interest, as his suggestion (1886-7:**3**, 600 ff.) as to the point of entry of the Aboriginals is not far removed from the latest linguistic theory, based on considerably more evidence. Curr believed that the Australians and the Tasmanians were distinct offshoots of 'the African race' and held that they had never been in contact since the first one of the branches was severed from the parent Negro stock.

The Rev. J. Mathew (1899:5-6) disagreed completely with Curr's theory. He rejected the idea of genetic unity for the Australian race and states that

> Australia was first occupied by a people, a branch of the Papuan family . . . They came from the north, in all likelihood from New Guinea . . . the [un-Papuan] constituents being Dravidian and Malay . . . Of these the Dravidian was the first to arrive . . . Mainly from linguistic evidence I incline to think that the people, who for convenience may be called Dravidians, first touched on the north-east coast of Queensland.

Mathew's linguistic knowledge was not extensive, his theories being more hypothetical than supported by solid evidence. This theory of the origin of the Aborigines reappears in modified form in the work of other writers, as will be seen below.

In 1907 S. H. Ray stated (1907:516):

> There is a tacit supposition in all the foregoing theories that the Australians are immigrants from some unknown place into the lands which they now occupy. Why it should be necessary to prove such an assumption is not evident to the present writer. There seems to be no more difficulty in assigning a distinctive character and local origin to the languages of the Australian aborigines, than there is in assigning a special character to the fauna and flora of the land they dwell in.

This theory was discounted because of the very limited amount of information on which it is based, and because no real proof was offered.

In 1919 the monumental work of Father Schmidt appeared in collected form. His materials, although fairly extensive, have been shown to be rather unreliable. He placed the point of entry into Australia at Cape York and links Australian with Papuan languages and suggests further links with the Dravidian languages (1919a:5, 20-2). He divided Australian languages into north and south, stating that the southern languages arrived first. Schmidt divides the north into three sections, but on 'formal' grounds only, that is, according to word shapes. Such a division is linguistically untenable, and it should be noted that his survey did not include the northwest of the continent, which has become an extremely significant area in terms of modern linguistic theory. As far as the Tasmanians are concerned, Schmidt (1919a:21) placed them on the oldest level of Australian and suggested that Tasmanian and Australian languages would ultimately be shown to be related.

Kroeber (1923) heralds the approach of a scientific and objective formulation of a theory of Aboriginal origins based on linguistic evidence. His thoughts are based mainly on the linguistic materials of Curr, plus a further twenty vocabularies to which he had access. He disagrees with Schmidt's divisions into different strata and opts for a theory of genetic unity, first because, when he plotted the occurrence of certain noun roots on his maps, he found that Schmidt's boundaries were meaningless, and secondly and more importantly because he felt that at that time too little was known about Aboriginal languages for any conclusions to be drawn (Kroeber 1923:115).

This first period, therefore, could be described as one of theorising without any real linguistic evidence. So little was known at this time that any theory could be only considered to be pure conjecture. Only with the arrival of the second period and a scientific linguistic method was the nature of the Australian languages first perceived.

### Capell: 1930-56

In 1930 Capell began working in the field of Australian linguistics. He concentrated on north and northwestern Australia, from the Kimberleys across Arnhem Land to Groote Eylandt. With this research, general comparative linguistics for the first time was launched on a scientific basis in Australia.

In 1937 he concluded (Capell 1937:27) that Australian languages exhibited considerable differences among themselves, yet that their general structure was very much the same. He observed that the Kimberley group forms a group apart in Australian languages and so posited two divisions as follows:

(a) Prefixing languages
Those languages in which noun classes play a part, combined with such phenomena as incorporation of pronoun objects (Kimberley group and northern Northern Territory).

(b) Suffixing languages
Those languages in which such features do not occur (the rest of Australia).

Capell was of the opinion that these divisions were probably historical as well as geographical.

On the basis of his own extensive knowledge and that of earlier students of Australian languages, Capell (1937) divided the languages of Australia into the following groups:

(a) Southeast Australia, east of Adelaide.
(b) The northern Kimberleys.
(c) Central Australia from Mt Margaret to southwest Queensland.
(d) The non-homogeneous north and central Queensland area.
(e) The New South Wales area (non-homogeneous).

This represents the first scientific attempt to classify Australian languages, although Capell was only too well aware of the fact that embarrassingly little was still known about them. In his survey paper (1937:60) he concluded that there was no doubt that all Australian languages belong to one family. At this time he thought that India was a likely source of Aboriginal migrants, although he did not enlarge on this idea. At the same time, he compared Australian languages with the languages of Oceania and New Guinea and concluded that Oceanic languages bore no resemblance either in general or in detail to Australian languages. He found that there was more resemblance between Australian languages and the non-Austronesian languages of New Guinea, because of their complex verb systems, although the actual forms were very different.

In 1940 Capell published 'The Classification of Languages in North and North-West Australia'. This follows the same line as his 1937 work in dividing Australian languages into two main groups, prefixing and suffixing. This division was a departure from the classifications of the writers of the pre-1930 period. He found much grammatical, but little lexical similarity between the two groups, and asked himself the question (1940:242): 'Does it mean immigration on a large scale by people speaking one language, or one type of language, or is there another explanation to be found?'

At this stage, Capell was puzzled by the fact that in spite of the differences of these northern languages from other Australian languages, there were also not inconsiderable agreements. He found that there was a definite element in north Australian languages common to the south as well, in spite of the dichotomy which he had set up. He felt that this common element was very archaic and possibly represented a substratum. He gives examples of such lexical agreements:

| | | |
|---|---|---|
| blood: | guli, gulu<br>gurug | Kimberleys<br>Narrinyeri (Adelaide) |
| | gur(u)g | Kulin (South Australia-western Victoria) |
| | gərig | East Buandik (south-east South Australia) |

He concludes (1940:432) that it may be possible that the multiple classifying languages of north Australia represent the remains of a very early language group, swamped by later comers. He felt that this would account for the constant use of the same or very similar prefixes and suffixes throughout the languages, combined with a very heterogeneous vocabulary, as grammar tends to change less easily than vocabulary under the impact of an outside influence. Capell later (1956) rejected his own theory of the dichotomy of Australian languages and perhaps peoples, as will be explained below.

In 1942 Capell continued his survey work right across Arnhem Land, as far as Groote Eylandt. His 1940 theories were not changed, but he did discover a group of languages in the northeast corner of Arnhem Land that were not prefixing. He posited that these languages bore evidence of having been brought in from more central regions of the continent. At this stage Capell decided that it was impossible to state the chronological relationship existing between the various groups and tribes. Being dissatisfied with the state of linguistic knowledge, he continued to amass information for a further fifteen years, at the end of which crystallised *A New Approach to Australian Linguistics* (1956), a completely new and integrated theory of Australian linguistics, which was to usher in and stimulate the period of intense interest and modern techniques in the classification of Australian languages.

Before examining this third period (1956-68), it should be mentioned that in 1941 Tindale and Birdsell (1941) reported the existence of a group of twelve small tribes of Tasmanoid or pygmoid people on the Atherton Tableland region of north Queensland, speaking Tasmanian-type languages. However, they supplied little linguistic information. The existence of these rain-forest people is discussed by Capell (1956), but a linguistic analysis of the languages of the area was not undertaken until 1963-4, by Dixon. Further mention of this will be made below.

**Recent Period: 1956-68**

Capell's *A New Approach to Australian Linguistics* discusses the comparative study of Australian languages from both the structural and lexical points of view. At this time he was in possession of a vast amount of material, added to which is the fact that linguistic analytical techniques had become much more effective. Capell rejected the theories which he had held previously concerning the Australian languages. It is worthwhile discussing his new theories and the reasoning underlying them as they had a marked effect on later linguistic research work.

Capell states (1956:2) that the total failure to link Australian languages with any other family suggests that they may have developed *in situ*, and that this theory is supported by theories in social anthropology.

He states that Australian languages differ widely both in structure and vocabulary among themselves, yet that a basic similarity exists in certain structural elements and a small but obstinate basic vocabulary. Prefixing and suffixing can now be explained as a development within the continent, given a sufficiently long period of time. This is Capell's new theory which has not yet been disproved. It still allows for the possibility of some foreign influences in north and northwest Australia.

Capell (1956:9-10) maintains that the development of language in Australia has been a gradual process of fixing an originally elastic word-order in the utterance, so that certain elements have come to fall into fixed positions in it. He observes that the most flexible of the present-day languages are the Western Desert languages. In these it is possible to see the original freedom of arrangement still preserved, at least in part. This freedom, as with prefixing and suffixing to the verb stem, is seen to be gradually limited in these languages, so that different morphemic elements of the verbal system begin to assume fixed positions, in different orders in different languages. Languages in other parts of the continent show what has happened in centuries past, and much of the development can be worked out by a comparison of the Western Desert languages with those in other parts of Australia.

Capell then shows that the Western Desert languages, on the grounds of typological features, still represent a very primitive stage in the development of Australian languages and compares them with the remainder of Australia, which he shows to be more evolved linguistically. From this he goes on to demonstrate how prefixing and suffixing languages evolve from the limiting process described above. He posits that as the morphological type over the larger part of the continent is suffixing, this may be presumed to be the original type. He claims that it is unnecessary to look outside Australia to explain the development of the prefixing languages, as they contain far too much 'Common Australian' to make that theory reasonable. Capell, in fact, was proved correct in 1959 when Hale discovered a language family on the Barkly Tableland which used as suffixes morphemes identical to those used as prefixes in a neighbouring language (Capell 1962a:9). So, from a typological viewpoint it would seem that Capell's claim of structural homogeneity for Australian languages was proved to be correct.

The fact that the prefixing languages are limited to the Kimberley area and Arnhem Land is unexplained. However, the fact that the languages of the northeast corner of Arnhem Land are not prefixing leads Capell to the supposition that the prefixing languages did not originate in their present home. Synchronic data, which is all that is available at present, do not allow us to say where they originated.

Also in *A New Approach to Australian Linguistics* Capell (1956:3) sets up a 'Common Australian' morphology and lexicon. By 'Common Australian' he means 'a certain group of words and constructions that are found throughout Australia, with varying frequency and clarity in different parts'. He claims that this Common Australian may also be Original Australian, a view which he rejects in subsequent publications. He states that, given an original language, subsequent migration and resultant isolation of tribes will account for the loss of many of the elements in different parts of the continent, and the development of new words to replace them. He is not surprised that the common element is small, considering the long habitation of Australia. Capell was disturbed, however, that the Arandic languages of central Australia missed most of the Common Australian roots which he set up.

Although Capell accepted the homogeneity of Australian languages, he also discovered 'regional vocabularies', words particular to certain geographical areas. 'Some, at least', he states (1956:79), 'must surely be accepted as old words which have survived regionally but not universally'. He found geographically scattered agreements which could not be explained as coincidental. In a few instances, agreements over long distances are very striking. It is easier, in Capell's opinion, to consider such examples as the tenuous remains of former linkages, rather than simply coincidences.

He quotes, for example:

| | | | |
|---|---|---|---|
| fish: | guja | Dieri | (Lake Eyre, S.A.) |
| | gjab | Maung | (northwest Arnhem Land) |
| | guj | Gogobera | (Cape York Peninsula) |
| | gi'jou | Djauan | (Sleisbeck area, N.T.) |
| | gwija | Gubabwingu | (northeast Arnhem Land) |

These cases are striking, and raise the question of common origin. They could have been part of the Common Australian word-store which was swamped or lost, or perhaps of an Original Australian sub-stratum. Capell (1962a) takes up the question of regional vocabularies (Arandic, Dampier Land, Victoria, Cape York, and the Queensland rainforest) in *Some Linguistic Types in Australia*. He draws three conclusions from his work of 1956:

(a) Australian languages are one.

(b) The two divisions previously set up are genera within the family.

(c) There are numerous subdivisions.

In 1962 Capell further condensed and crystallised his 1956 material. He re-affirmed (1962a:1) that Australian languages cannot be linked outside Australia, nor can it be demonstrated that the Tasmanian languages had any connection with the Australian mainland languages. He states that it is necessary to postulate a lengthy period of sojourn in the continent, at least 12,000 to 15,000 years to account for the development of the great diversities existing at the present day between languages of different parts of the continent.

Capell (1962a:12) further discusses the regional vocabularies mentioned above and the agreements of words found over widely scattered regions. He states that in terms of areal linguistics such occurrences would naturally be treated as remnants of a body of very early speech, pushed out to the margins of the continent by later comers. These later comers could be the speakers of Common Australian, which would not then be Original Australian. 'This [marginal] vocabulary is not to be necessarily included as CA [Common Australian]—its range is different'. In some cases Common Australian and Original Australian present different words for the same object, and it cannot be assumed without proof that these were synonyms occurring in one and the same language. Capell, therefore, sets up an Original Australian sub-stratum which was overrun by Common Australian, which he thinks could have originated in the north or could have come from outside Australia. Even if Common Australian developed within Australia, it imposed itself on the earlier language or languages so thoroughly that the language type of the earlier period can no longer be recognised. Common Australian would then set the type for modern Australian languages.

Capell's work, especially *A New Approach*

*to Australian Linguistics* (1956), stimulated much interest by both Australian and overseas research workers. This renewal of interest brought with it newer linguistic analytical techniques in the form of lexico-statistics and reconstruction of proto-languages, in an attempt to clarify the position of Australian languages themselves and their ultimate history and possible linkages outside Australia.

Hale, O'Grady, Wurm, and C. F. and F. M. Voegelin have worked in collaboration since this time towards a classification of Australian languages and the establishment of a clearer view of Australian linguistic history.

In 1959-60 Hale undertook extensive field work in Australia and discovered a new language family in the Barkly Tableland area of the Northern Territory which lent support to Capell's theory of the development of Australian languages. He also applied lexico-statistical techniques to the Arandic languages of central Australia, which had previously been considered extremely aberrant. He ultimately showed that the Arandic languages are typically Australian, obscured by unusual sound changes; they are in fact a sub-group of the Western Desert languages.

In 1963-4 two unusual and unacceptable theories of Aboriginal origins and linkages were published as the classificatory work of the above-mentioned workers continued. First, Holmer (1963:99) suggested that Australia may have been peopled from the east or northeast and was possibly occupied first by a branch of the Papuan family. There are two serious objections to this, the first being that he presents no concrete evidence, the second that it has not been established that there is any such thing as 'the Papuan family'. There is no linguistic evidence of Papuan presence in Australia except, possibly, their influence on the languages of Cape York Peninsula. On the other hand, Laycock (pers. comm.) has found a number of linguistic-cultural indications of possible contact between Cape York and the Sepik region of New Guinea.

Murdock (1964:123) cites Greenberg as formulating a Macro-phylum which includes Australians and Tasmanians, the Andamanese, the Papuans of New Guinea, and the non-Austronesian-speaking peoples in the Solomon Islands, and in Halmahera and Timor in the Moluccas. There is no evidence to support the establishment of such a Macro-phylum.

In 1964 Hale published the results of his study of the northern Paman languages of Cape York Peninsula. In this he proved that the languages of this area, previously considered un-Australian because of their aberrant phonology, are in fact Australian, again obscured by extensive sound changes. He also posited (1964:251) that the centre of dispersal of Australian languages is somewhere in the area west of the Gulf of Carpentaria. This theory follows the accepted theories of Dyen (1956 and 1962), of setting up the linguistically most diverse area as the centre of dispersal.

Hale, O'Grady, Wurm, and C. F. and F. M. Voegelin completely reassessed the linguistic materials available in 1966, and brought out a new classification of the languages of Australia. Previous classifications had been made largely on typological criteria. Theirs was the first classification based on lexico-statistics or cognate densities. However, this work did not attempt to concatenate the typology-based observations with cognate counts, as no satisfactory means of doing so has as yet been devised.

O'Grady, Wurm, and Hale (1966) published the first linguistic map of Australia since that of Tindale (1940). This map is the companion-piece to the preliminary classification of O'Grady and C. F. and F. M. Voegelin (1966).

O'Grady and C. F. and F. M. Voegelin (1966) introduce two new terms to Australian linguistics in an attempt to give a clearer picture of the situation. First, they state that Australia is made up of a large number of 'Family-like Languages'. In this they use a new criterion, that of neighbour intelligibility rather than mutual intelligibility in their division between languages/dialects. This is set up in order to account for and classify the 'dialect chains' peculiar to Australia, where the speakers of dialect/language A understand the speakers of dialect/language B, but not those of dialect/language C, while speakers of dialect/language C understand dialect/language B, but not A. On European criteria these would not represent one language at all. These Family-like Languages form Phylic Families (based on low cognate density but a high degree of homogeneity of structure). All of these Phylic Families, except Mbarbaram (Qld) and Aniwan (N.S.W.), belong to what they have called the Australian Macro-phylum. Thus they state that in 1966 they find that there are twenty-nine Phylic Families in Australia, made up of 502 Speech Communities and 228 discrete languages (O'Grady and C. F. and F. M. Voegelin 1966:28). They also find that one of the families (Pama-Nyungan) covers seven-eighths of the continent, while the other twenty-eight occupy only one-eighth, in the Kimberley/Arnhem Land area. They also remark that no one has proved a definite link between the languages of Australia and Tasmania.

The main problem of classification in Australia has proved to be the extreme dearth of cognates between Phylic Families. This hampers both lexico-statistical and reconstructive work. However, in spite of this problem an attempt is being made to arrive at a satisfactory classification also, on the basis of reconstructing the proto-languages of each Phylic Family and the lower members of its hierarchy as a check or supplement to the lexico-statistical methods employed by O'Grady and C. F. and F. M. Voegelin (1966). To this end, O'Grady (1966) and Hale (1966) each published proto-forms of sub-groups of the Pama-Nyungan Phylic Family. It is also hoped that ultimately a proto-Australian will be arrived at, which could show links with other language groups of the world. Dixon is also working towards a proto-Australian, although it will be some time before anything definitive is produced.

Dixon published the results of his investigation of the rainforest languages of Queensland, and is satisfied (1966a:114) that Mbabaram, allegedly a Tasmanian-type language, is really an Australian language, even though it appears to be an isolate. This does not eliminate the possibility that these people could be Tasmanoids who have abandoned their language, as is shown to have happened with pygmoid races in Africa.

Information provided by Court (pers. comm.) shows Aniwan to be an Australian language also, so that a homogeneous Australian Macro-phylum may now be established, even if the relationship between the Phylic Families is in some cases distant.

Wurm (1970) presents a further classification of Australian languages, based on the most recent information available. Since 1966, the number of Phylic Families has been reduced from twenty-nine to twenty-five.

Sommer (pers. comm.) has just recently shown that the languages of the western side of Cape York Peninsula may not belong to the Pama-Nyungan family as had previously been thought. There also seem to be links between Cape York Peninsula languages and

Arandic which is considered as a Group within the Pama-Nyungan Family. The completion of Sommer's research will undoubtedly throw more light on the problem.

During the last decade knowledge of Australian languages has increased considerably. However, it is felt that any definitive classification of the languages is still some way off. As far as the history of Aboriginal migrations and origins is concerned, there remains a vast amount of analysis and reconstruction to be done before linguistic evidence can offer any solution which may be regarded as anything like definite. In the meantime, Capell's typological theory has not been invalidated by later work based on lexico-statistics and the reconstruction of proto-forms.

A summary of the present state of knowledge concerning the linguistic evidence and Aboriginal origins, therefore, necessarily follows Capell, as follows:

The distribution of some of the reflexes along parts of the marginal zone of the continent, especially in the south and west, is significant, and such occurrences could be regarded as remnants of languages which were pushed out to the margins of the continent by later migrations, unless there was immigration into Australia along the southern and western coasts by one common group, which would seem most unlikely. The evidence suggests that these later migrations could have been migrations of speakers of Common Australian, the present day languages re-classified by O'Grady, Hale, Wurm, and C. F. and F. M. Voegelin. Capell also established the existence of definite regional vocabularies, distinct from Common Australian, in marginal areas of the continent—Dampier Land and east and west Victoria—although these regional vocabularies are permeated with Common Australian to a greater or lesser degree. These also suggest a sub-stratum, pre-dating the forward surge of Common Australian.

The Common Australian theory of Capell, then, suggests the powerful and extensive spread throughout the continent of one of the diverse but genetically related languages which developed in the north, if indeed it did not originate outside Australia. It pushed existing elements out to the margins of the continent, and superimposed itself on all it met until only vestiges of the original languages remained.

Linguists would date the Common Australian surge at between 5,000 and 10,000 BP, if any reliance can be placed on glottochronology.

One puzzle remains. How does one explain the great phonological similarity right across the continent, for example Tiwi, which shows only minimal (8%) Common Australian influence, and Common Australian itself? In other words, how do Common Australian and Original Australian share almost identical phonologies in areas where it can be shown that Common Australian had very little influence? Two possible solutions present themselves. If Common Australian was a migrant from outside Australia it must have originated near to the setting out point of the Original Australians, or Common Australian was a dominant break-away group from the Original Australians, who must themselves have formed a phonologically homogeneous group.

In brief, there are still many problems which await an answer. There is still a vast amount of linguistic evidence to be gathered before any definitive statement about Aboriginal origins can be contemplated, and with only synchronic data and as yet no satisfactory dating technique, it would seem that linguistics is still far from achieving this end. However, with inter-disciplinary co-operation, more leads should be provided which could lead ultimately to a definitive theory of the origins of the Aborigines. No single discipline would seem capable of solving the problem alone.

**References**


Capell, A. 1937. The structure of Australian languages. *Oceania* **8**:27-61.

——1939. Mythology in northern Kimberley, north-west Australia. *Oceania* **9**:382-404.

——1940. The classification of languages in north and north-west Australia. *Oceania* **10**:241-72, 404-33.

——1941. Notes on the Wunambal language. *Oceania* **11**:295-308.

——1942. Languages of Arnhem Land, north Australia. *Oceania* **12**:364-92; **13**:24-50.

——1956. *A new approach to Australian linguistics*. Oceania Linguistic Monographs No. 1. Sydney: University of Sydney.

——1958. Aborigines: languages. In *The Australian encyclopaedia*. Vol. 1, pp. 21-8. Sydney: Angus and Robertson.

——1962a. *Some linguistic types in Australia*. Oceania Linguistic Monographs No. 7. Sydney: University of Sydney.

——1962b. The development of language in Australia. In *Some linguistic types in Australia*, pp. 3-14. Oceania Linguistic Monographs No. 7. Sydney: University of Sydney.

——1963. *Linguistic survey of Australia*. Sydney: Australian Institute of Aboriginal Studies.

——1965. A typology of concept domination. *Lingua* **15**:451-62.

——1967a. Sound systems in Australia. *Phonetica* **16**:85-110.

——1967b. Pronominalisation in Australian languages. In *Papers in Australian linguistics No. 2*. pp. 21-42. Pacific Linguistics, Ser.A, Occasional Papers No. 11. Canberra: Australian National University.

—— and Elkin, A. P. 1937. The languages of the Kimberley division. *Oceania* **8**:216-46.

Curr, E. M. 1886-7. *The Australian race*. 4 vols. Melbourne: John Ferres.

de Charencey, Hyacinthe. 1880. *Recherches sur les dialectes Tasmaniens*. Actes de la Société Philologique 11, Fascicle 1. Alençon: E. de Broisé.

Dixon, R. M. W. 1966a. Mbabaram: a dying Australian language. *Bull.Sch.Oriental Afr. Stud.* **29**:97-121.

——1966b. *Mbabaram phonology*. Hertford: Stephen Austin and Sons Ltd.

Douglas, W. H. 1958. *An introduction to the Western Desert language*. Oceania Linguistic Monographs No. 4. Sydney: University of Sydney.

——1968. *The Aboriginal languages of the south-west of Australia*. Australian Aboriginal Studies No. 14 (Linguistic Ser.No.4). Canberra: Australian Institute of Aboriginal Studies.

Dyen, Isidore. 1956. Language distribution and migration theory. *Language* **32**:611-26.

——1962. The lexico statistical classification of the Malayo Polynesian languages. *Language* **38**:38-46.

Elkin, A. P. 1937. The nature of Australian languages. *Oceania* **8**:127-69.

Greenway, J. 1963. *Bibliography of the Australian Aborigines and the native peoples of Torres Strait to 1959*. Sydney: Angus and Robertson.

Hale, K. L. 1962. Internal relationships in Arandic of central Australia. App. to *Some linguistic types in Australia* (A. Capell), pp. 171-85. Oceania Linguistic Monographs No. 7. Sydney: University of Sydney.

——1964. Classification of northern Paman languages, Cape York Peninsula, Australia: a research report. *Oceanic Linguistics* **3**:248-65.

——1965. Australian languages and transformational grammar. *Linguistics* **16**:32-41.

——1966. The Paman group of the Pama-Nyungan phylic family. *Anthropological Linguistics* **8** (2):162-97.
——1967. Some productive rules in Lardil (Mornington Island) syntax. In *Papers in Australian linguistics No. 2*, pp. 63-72. Pacific Linguistics, Ser.A, Occasional Papers No. 11. Canberra: Australian National University.
Holmer, N. M. 1963. *On the history and structure of the Australian languages*. Australian Essays and Studies **3**. Uppsala: Lundequistska Bokhandeln.
Kroeber, A. L. 1923. Relationship of the Australian languages. *J.Proc.R.Soc.N.S.W.* **57**:101-17.
Laycock, D. C. 1960. Language and society: twenty years after. *Lingua* **9**:16-29.
Macintosh, N.W.G. 1952. Stature in some Aboriginal tribes in south-west Arnhem Land. *Oceania* **22**:208-15.
Mathew, J. 1899. *Eaglehawk and crow*. Melbourne: Melville, Mullen and Slade.
Milligan, J. 1890. *Vocabulary of the dialects of some of the Aboriginal tribes of Tasmania*. Hobart: William Thomas Strutt.
Murdock, G. P. 1964. Genetic classification of the Austronesian languages: a key to oceanic culture history. *Ethnology* **3**:117-26.
Nekes, H. and Worms, E. A. 1953. Australian languages. In *Micro-bibliotheca anthropos*. Vol. 10 (microfilm). Freiburg: Anthropos-Institut.
O'Grady, G. N. 1957. Statistical investigations into an Australian language. *Oceania* **27**:283-312.
——1963. *Nyangumata grammar*. Oceania Linguistic Monographs No. 9. Sydney: University of Sydney.
——1966. Proto-Ngayarda phonology. *Oceanic Linguistics* **5**:71-130.
——, Wurm, S. A., and Hale, K. L. 1966. *Aboriginal languages of Australia (a preliminary classification)*. Map. Department of Linguistics, University of Victoria, British Columbia.
——, Voegelin, C. F., and Voegelin, Florence M. 1966. Languages of the world: Indo-Pacific fascicle six. *Anthropological Linguistics* **8**(2).
Ray, S. H. 1907. *Reports of the Cambridge anthropological expedition to Torres Straits* (ed. A. C. Haddon), Vol. 3, Linguistics. Cambridge: Cambridge University Press.
Schmidt, W. 1919a. *Die Gliederung der australischen Sprachen*. Vienna: Mechitharisten-Buchdruckerei.
——1919b. *Die Personalpronomina in den australischen Sprachen*. Vienna: A. Hölder.
——1952. *Die tasmanischen Sprachen*. Comité International de Linguistes. Publications de la Commission d'Enquête Linguistique **5**. Utrecht: Comité International de Linguistes.
Smythe, W. E. 1948. Elementary grammar of the Gumbáingar language (North Coast, N.S.W.) *Oceania Monogr.* No. 8.
Strehlow, T. G. H. 1944. Aranda phonetics and grammar. *Oceania Monogr.* No. 7.
Tindale, N. B. 1940. Distribution of Australian Aboriginal tribes: a field survey. *Trans.R.Soc.S.Aust.* **64**:140-232.
—— and Birdsell, J. B. 1941. Tasmanoid tribes in north Queensland. *Rec.S.Aust.Mus.* **7**:1-9.
Tryon, D. T. 1968. The Daly River languages: a survey. In *Papers in Australian linguistics No. 3*. pp. 21-46. Pacific Linguistics, Ser.A, Occasional Papers No. 14. Canberra: Australian National University.
Voegelin, C. F., Voegelin, Florence M., Wurm, S., O'Grady, G., and Tokuichiro, M. 1963. Obtaining an index of phonological differentiation from the construction of non-existent minimax systems. *Int.J.Am. Linguistics* **29**:4-29.
Wurm, S. A. 1963. Aboriginal languages: the present state of knowledge. In *Australian*

*Aboriginal studies* (ed. H. Shiels), pp. 127-48. Melbourne: Oxford University Press.

——1964. The present state of New Guinea (non-Melanesian or Papuan) and Australian historical and comparative linguistics. In *Proceedings of the ninth International Congress of Linguistics (1962)*, pp. 575-9. The Hague: Mouton.

——1970. Classifications of Australian languages, including Tasmania. *Current Trends in Linguistics* **8** (Linguistics in Oceania), in press. The Hague: Mouton.

# 24 Art and Aboriginal Prehistory

Robert Edwards

The study of the cave art of the Australian Aboriginal started with the earliest European contacts. Flinders (1814:**2**, 188-9) discovered paintings on Chasm Island while examining the Gulf of Carpentaria in January 1803; King's survey around Cape York in 1821 located ochre paintings on Clacks Island (King 1827:**2**, 26-7); the famed Wandjina paintings of the Kimberley region were recorded by Grey (1841:**1**, 201-7) in 1837. As each section of the continent was opened up the strange new art attracted immediate attention and the first of a long series of descriptive papers soon appeared.

The records of Aboriginal paintings are as comprehensive and voluminous as those for any other aspect of their culture. A wide range of different styles has been identified in this art and linked with specific tribes or localities. An obvious one, recognised by Spencer in 1912, is the elaborate X-ray art of western Arnhem Land with its detailed portrayal of the skeleton and internal organs as well as external features (Spencer 1914: 433-9). Equally distinctive are the large mouthless figures of the Wandjina art (Crawford 1968). Another style has emerged from studies of the small human figures of the cave art of western New South Wales, while in central Australia the Aboriginal of the desert portrayed his totemic ancestors by using simple line, track, and geometric motifs (Strehlow 1964:44-59). The final analysis of the large number of paintings recently discovered in Cape York should establish the existence of a further distinctive type in that area. A more precise distribution of styles throughout the continent will almost certainly emerge as regional studies are pursued to greater depth.

In some places it has been possible to carry out intensive studies of the local art with the aid of Aborigines who still regard it as normal cultural expression (Davidson

*Fig. 24:1 Art sites mentioned in text*

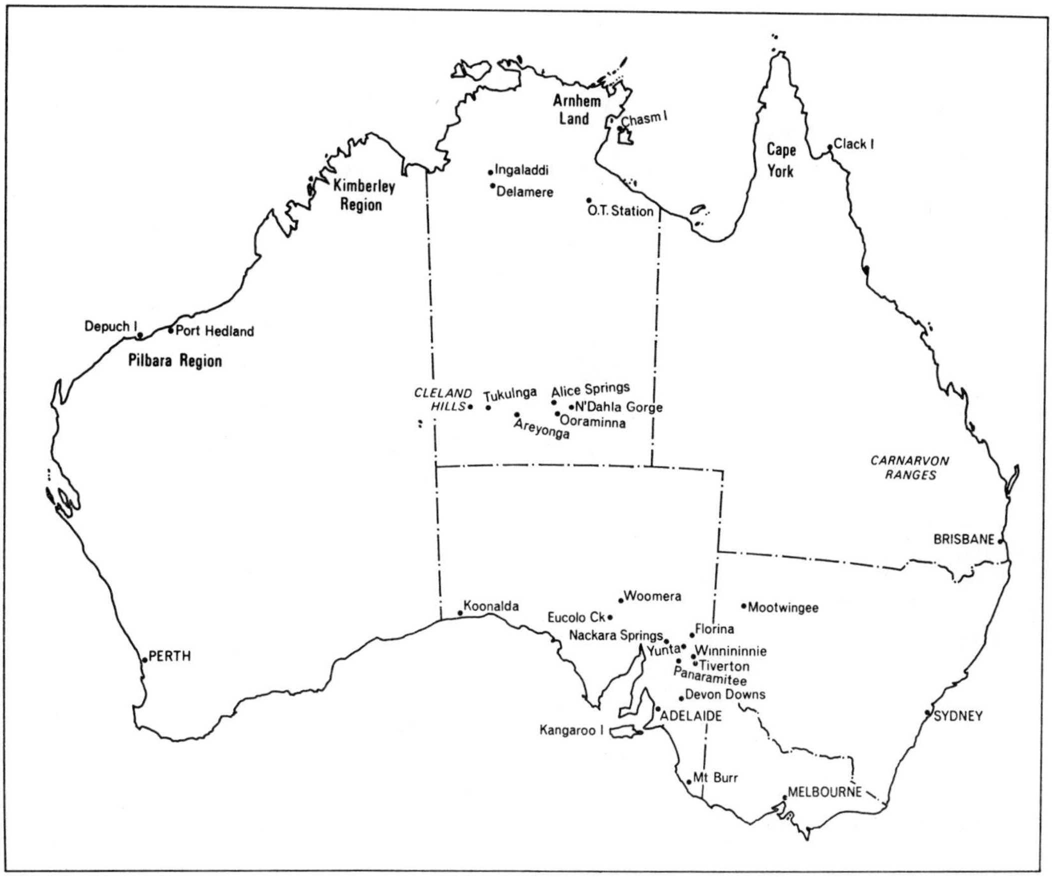


1936:108-20; Macintosh 1952:256-74; Tindale 1959:305-32; Arndt 1962:298-320). In central Australia the Walbiri rituals associated with the well-known Ngama and Ruguri cave painting sites play a significant role in what survives of tribal ceremonial life. Although in the traditional situation the degree of sanctity is known to have varied from group to group, the drawing or redrawing of the figure of a mythical ancestor was usually intended to have some effect, direct or indirect, on their physical or spiritual environment. There were instances when individuals painted simply to pass the time but over a wide area the artistic act, the art itself, and the associated ritual had magical and religious significance in the traditional culture (Berndt 1964:1-10). All this has been studied in the broad ethnographic context of the comparatively recent situation.

Despite the impermanence of this art, which demands continual retouching for its preservation, we have been able to look back some way into its past. In particular the art of central Australia had a code of rigid adherence to designs ordained by totemic ancestors at the time of creation of the Aboriginal world. Heavy penalties were enforced on those who failed to conform (Strehlow 1964:44-59). This rigid discipline and the isolation of the arid centre from external contacts combined to preserve the forms of the art which have probably remained unchanged for many centuries.

The situation in northern Australia is in direct contrast. Here a rich array of elaborate and colourful paintings, many showing undoubted Macassan influence, overlie a much simpler style (McCarthy 1964:39-40). In most regions the Aboriginal was also moved to record in his galleries the first contacts with Europeans and there are many compositions featuring the white man with his guns and domesticated stock. Because of their liability to decay and the lack of any firm evidence of antiquity, cave paintings can only reliably be taken to present the comparatively recent situation. The earlier manifestations of Australian art have to be traced through the rich heritage of rock engravings distributed widely across the continent. The permanence of the engraved designs, cut deeply into rock, affords the only direct access to the artistic expression of the culture of long-past generations.

The study of rock engravings began in much the same way as that of paintings. From the arrival of the First Fleet in 1788 the existence of large outline engravings on the Sydney sandstone has been noted and many detailed records of the designs have been made (Campbell 1899:1-73; McCarthy 1959: 203-15; Sim 1966:1-42). During his explorations of the northwest Australian coast in 1840 Stokes (1846:**2**, 170-2) noted that the Aborigines had recorded human figures, animals, birds, weapons, and implements on the rock surfaces of Depuch Island by removing the hard outer coating and baring the natural colour of the rock. Gradually other sites were found and recorded. The initial Depuch Island finds were but a small extension of a large mainland series (McCarthy 1961:121-48; McCarthy 1962: 1-73; Wright 1968:1-78). Among other important discoveries were sites in the Flinders Ranges in South Australia (Basedow 1914:195-211; Mountford 1929:337-66), in western and northern New South Wales (Black 1943:9-50; McCarthy and Macintosh 1962:249-98; McBryde 1964:201-10), at Mt Cameron West in Tasmania (Meston 1932:1-6) and in the Carnarvon Ranges of Queensland (Goddard 1941:368-72).

The simple engravings, neither so obvious nor so immediately attractive as the colourful paintings, were often overlooked by explorers and early settlers and their study has been slow to develop. As late as 1936 there were only two recorded sites for the Northern Territory (Davidson 1936:56-60) and it has been only in the last decade that the great number and wide distribution of engravings has begun to be realised (Mountford 1960: 145-7; Edwards 1968:665-6).

Although a pattern of regional distribution of rock engraving styles is emerging, this is not so clearly developed as that claimed for paintings. In fact it is becoming increasingly apparent that there are a number of constant features common to many sites in widely separated areas of the continent. These features are: close proximity of engravings to regular sources of water; association with occupation sites; advanced weathering and surface patination; consistent relative frequencies of designs.

The association of engraving sites with water supplies is particularly obvious in some areas, such as the semi-arid Yunta district of South Australia, where every permanent spring and rockhole is surrounded by a group of engravings (Edwards 1964:658-9). Sites in western New South Wales are also associated with permanent water supplies while engravings often occur on rock faces adjacent to claypans, rockholes, and springs in central Australia. In northern Australia there are always plentiful water supplies in the vicinity, but with its sub-tropical climate this evidence is of less significance.

As regular water supplies were essential to the survival of the nomadic Aborigines it is only to be expected that evidence of their visits to such sites is often found, usually in the form of stone implements. However, it is rare for large numbers of implements to be linked with engraving sites. It is true, for example in the Woomera area, that there are engravings at Eucolo Creek (Hall, McGowan, and Guleksen 1951:375-80) not far distant from some of the densest concentrations of artifacts in Australia. Here can be found a full range of industries from large trimmed cores, hammers, and millstones to microliths, pirris, endscrapers, tulas, and ground axes. This is an exceptional situation and in most instances in south and central Australia no such representative collections have been recorded. Normally there is only a sparse scatter of implements, with perhaps only a few types represented and then only in small numbers.

It may be significant that at an increasing number of engraved sites trimmed core implements are being found, identical in type with the so called 'Kartan' industry first identified from a number of sites on Kangaroo Island and the Adelaide Plains (Cooper 1943: 343-69). There are even some situations where large collections of these tools have been found to the exclusion of other more sophisticated industries (Edwards 1964: 655-8). During the recent examination of engraving sites in central Australia numbers of these same implements were found, thus extending their known distribution far beyond anything previously recorded (Edwards 1968:668).

The progressively emerging association of core implements with engravings presents an aspect worthy of further study, as although no firm dating has been established for large trimmed core industries there are pointers to their great antiquity. Several fine examples were recovered from an occupation floor situated 3 m below the land surface at Fulham, near Adelaide (Howchin 1919: 81-4). They are also common on Kangaroo Island, unoccupied at the time of its discovery by Flinders in March 1802 (Cooper 1960:481). It has also been shown that core and flake industries existed in Australia for at least 11,000 years from about 14,000 to about 3,000 B.C. (Mulvaney 1966:89). Doubtless the 'Kartan' is one of the industries belonging to this early phase.

While considering the subject of implements, it is interesting to note that no tools obviously made specifically for the engraving of a rock have been found. There are a number of different techniques involved, principally abrading, pecking, scratching, and hammering. Experiments have shown that engravings made by these methods can be duplicated with any piece of hard rock with a suitable natural shape, edge, or point. The primary conclusion that no predesigned tools were used for the execution of engravings is supported by the fact that the individual marks in the pecking and other techniques are of widely varying sizes and shapes.

Another line of inquiry associated with occupation sites, which could provide further clues for the establishment of the age of rock engravings, is the frequency of fire hearths near engraving sites, many only exposed by erosion of the land surface. Already one hearth at a site on Panaramitee Station in the northeast of South Australia has yielded several detached fragments of engraved rock (Plate XXVIII), an obvious indication that it post-dates the time of engraving.

Positive dating of charcoal in the hearth is planned. There is always a chance that engraved fragments will be found incorporated in hearths at greater depths, and the search for such an occurrence should not be overlooked because of the initially unpromising surface situation.

Another consistent feature of most engraving sites is the advanced weathering and disintegration of engraved rock surfaces. These processes sometimes start with frost action, which allows moisture and dust to penetrate the engraved rock pavements, gradually prising them apart. In some places trees and shrubs take root in the crevices, accelerating the destruction. Almost without exception pavements are splitting and slabs are working loose. Earth movements have caused engraved blocks to fall away from cliff faces, and large boulders have been shattered, while there are innumerable examples where the critical stage has passed and only small engraved sections survive to indicate where presumably extensive galleries once existed. It possibly is indicative of great age that the missing pieces do not lie loosely on the surface but have eroded completely. These situations obtain in central Australia (Plates IX, XXVI), western New South Wales (Plate X), South Australia (Plate XI), Tasmania, Western Australia, Queensland, and northern Australia. No precise time scale can be provided for this destruction, but several geologists are of the opinion that a considerable period is involved.

The surface patination produced by weathering also provides additional evidence of great age. When the engravings were first made, they would have penetrated the existing patination, allowing the designs to stand out in the original colour of the unweathered rock. The damage caused to the hard outer skin and underlying rock would have provided a fresh surface prone to reweathering, initially at a comparatively rapid rate. Many geologists believe this process to to be very slow, and Trendall (1964:88) assumes an extreme view, that a weathered skin 6 mm thick takes well over one million years to form. Whatever the time period involved, and this must vary according to local circumstances, most engraved designs at Australian sites have completely reweathered to match the natural surface colour of the rocks (Plates IX, X, XIX, XXVII).

In an attempt to utilise patination as a criterion for dating, thin sections have been made of rock samples from a series of widely distributed sites. As engravings are found on all types of rock from sandstones and limestones to granites and quartzites, there are many variables to be taken into account. Microscopic examination has shown that the depth of penetration varies according to the type of rock, but that for each type there seems to be a depth beyond which weathering appears to have little or no effect. This is probably due to the development of a case-hardened skin, which prevents the penetration of moisture and so reduces or even eliminates the process of chemical weathering which is largely responsible for surface changes. It is interesting that some rock sections show a greater degree of alteration on the underside than upon the exposed surface. Apparently moisture penetrates through cracks and is trapped along fractured bedding planes, where chemical weathering continues as long as moisture remains. On the exposed surface, however, the sun and wind dry the rock quickly, thus retarding and often stopping the process.

In the very arid conditions of central Australia the lack of moisture inhibits chemical weathering and there is usually only a thin, dark weathered skin covering the light-coloured rocks. Sections cut through exposed fragments show that the weathering of the engraved portions ceased when the shallow, but critical depth was reached. At some sites there are instances where the hard patinated skin is beginning to fret, causing the destruction of the rock surface and many engravings (Plates XI, XXV).

In wetter environments weathering penetrates to a greater depth with disastrous results. For instance in the Sydney-Hawkesbury district, where large concentrations of outline engravings exist on soft sandstone rocks, the salt laden atmosphere and prolonged saturation of the rocks, brought about by the even spread of the 125 cm rainfall, combine to accelerate weathering to the point where engraved sailing ships have been all but removed in under 200 years.

In some areas, the most obvious being the Pilbara region in northwest Australia (Wright 1968:1-78), Depuch Island (McCarthy 1961: 121-48), and a number of sites in the Western Desert of central Australia (Mountford 1955: 345-52), a form of engraving persisted into historical times (Crawford 1964:46). In these places the Aborigines often pounded or bruised the surface of the rock with hammerstones to mark out their designs, and these have not yet re-weathered. At some sites this later pounding technique is found in association with the older, patinated, pecked and abraded engravings, the variation in the colour of the engraved portions giving some general indication of succession.

In the arid Negev of Israel, Anati (1963: 187-9) has used shades of patination to assist in determining the relative dates of various styles of engravings situated on the same surface or in the immediate vicinity of one another. However, no engravings have re-weathered to match the natural dark rock surface. As some of them are associated with the Iron Age, Anati believes it takes a minimum of 2,500 years for a thin, initial surface patination to form in that region.

The misguided habit of rubbing engraved designs with abrasive material for the purpose of photography permanently removes the weathered surface and is leading to an increasing degree of confusion, as a recent origin is being claimed for many old engravings (Plate XII). At present the study of surface weathering can tell us only that some sites are old and that others are recent. Obviously there are too many variables to admit positive dating by this means, but sophistication of the method of examination may result in a more useful bracketing of chronologically related sites.

The interpretation of rock engraving designs has been a matter of great fascination to many people and there seems to be an irresistible temptation to assign a meaning to every design. As there is a proportion of abstract symbols at most sites, this leaves the field open for speculation and the earlier overseas pattern of superficial and often fantastic interpretations (Steward 1936:407) has spread to Australia, where some engravings are attributed to the Assyrians, Egyptians, and lost tribes of Israel. This has tended to distort perspective, and the rare and unique rather than the great number of basic simple designs have attracted wide attention.

The identification of a few engravings of outsized kangaroo, wombat, and bird tracks as those of extinct animals seems to some almost as speculative as the theories of visitors from the Northern Hemisphere; yet as dates for man and his engraved art go back in time, so do the dates for extinct animals come forward. Basedow (1914: 201-2) was convinced that large wombat-like tracks at Yunta Springs in northeast South Australia were those of *Diprotodon*, an extinct species of giant wombat. A series of bird

*Table 24:1 Frequency of occurrence of designs*

| | | Percentages | | | |
|---|---|---|---|---|---|
| **Site** | **Units** | **Tracks** | **Circles** | **Total** | **Other designs** |
| Florina | 985 | 48·3 | 39·9 | 88·2 | 11·8 |
| Panaramitee | 1,003 | 57·6 | 26·7 | 84·3 | 15·7 |
| Tiverton | 2,736 | 72·0 | 16·7 | 88·7 | 11·3 |
| Winnininnie | 3,236 | 71·3 | 16·7 | 88·0 | 12·0 |
| Tukulnga | 3,186 | 71·3 | 24·4 | 95·7 | 4·3 |
| Cleland Hills | 387 | 50·0 | 33·0 | 83·0 | 17·0 |

tracks, each some 45 cm long, at a site near Woomera are thought by Tindale (1951: 381-2) to be those of *Genyornis*. Large kangaroo tracks of similar size found at a site on Tiverton Station could represent the giant kangaroo, *Procoptodon* (Edwards 1965a:229). These engravings may simply be those of giant mythical creatures of the Dreamtime, present only in the minds of the Aborigines who made them. Alternatively the increasing evidence of great antiquity for engravings admits the possibility of a contemporaneous existence for the rock engraving people and giant animals. If this is shown to be correct, it would not be surprising if engraved tracks of extinct creatures are found in prehistoric Australian art.

It may be significant that the tracks of the dingo are rarely found among the many thousands of animal tracks at widely distributed engraving sites. Occasional examples have been noted, but these are usually not highly patinated and appear to be of comparatively recent origin. As the dingo was brought to Australia by the Aborigines, and its earliest presence is so far not attested before about 7,000 years ago, its absence from this presumed ancient art form may be more than coincidence (Campbell, Edwards, and Hossfeld 1966:10).

Recent work has shown a healthy trend towards classification rather than interpretation. The tally and classification of designs and the preparation of percentage tables are providing a useful framework for comparative studies. As Table 24:1 indicates, the figures obtained so far show that there is a surprising consistency of motif on widely separated sites (Plates XIII, XIV). At these galleries animal tracks and circles clearly predominate while a comparatively small proportion consists of human footprints, plan aspects of animals, and linear designs. An even smaller percentage includes abstract symbols or designs which are usually prevalent at specific locations, but which sometimes serve to link distant sites. The large fern-like designs found at N'Dahla Gorge in eastern central Australia (Plate XVIII) occur at other sites including one on Panaramitee Station hundreds of kilometres to the south. It is significant that the relative frequencies are very similar whether the sites being compared are close to one another, or 1,300 km or more apart (Edwards 1966:33-8).

Beyond the simple portrayal of man and animal by individual tracks and elementary plan views, the only identifiable development in artistic expression is tracks which occur in a regular series (Plates XV, XVI). At Panaramitee there are several outstanding examples where a series of tracks vividly portrays the movement of man and animal across the landscape (Mountford and Edwards 1963:140-1). Similar combinations have been found on central Australian sites.

A further distinction is found at some sites, where among the numerous tracks,

*Plate I Upland and sand desert in the Rumbalara Hills, N.T., showing the manner in which ephemeral channels die out in the piedmont fringe (photograph by J. Cavanagh).*

*Plate II Riverine desert: the anastomosing channel tract of the Bulloo River, southwestern Queensland; the floodplain is flanked by sandplain on the east (R) and by a larger claypan on the west (photograph by J. A. Mabbutt).*

*Plate III Koonalda Cave from the groundsurface. The cave system leads off from the bottom of the deep collapse doline (photograph by R. V. S. Wright).*

*Plate IV The Gallus site, Koonalda Cave, excavation in progress. Note the large black flint nodules in the cave walls. This flint served as a quarry for prehistoric Aborigines.*

*Plate V Wright's deep pit (almost 6 m) at Koonalda Cave, showing the upper chalky rubble and the lower water-lain red sediments.*

*Plate VI Detail of the lower red unit, Koonalda Cave, consisting of up to 100 individual graded beds of water-lain sediments. Some artifacts were embedded in the deposit.*

*Plate VII Group of Tasmanian Aborigines around their camp fires on a shell midden. Original field drawing by Piron, made in 1792 or 1793, on the southern shores of the D'Entrecasteaux Channel between Bruny Island and Recherche Bay. From Labillardière (1799-1800, Atlas, no. 4). (The Nan Kivell Collection, National Library of Australia.)*

*Plate VIII Circular depression on the surface of a shell midden at West Point (W.P. 1), northwest Tasmania, 1967 (photograph by Rhys Jones).*

*Plate IX Large engraved pavement fractured by natural weathering, Ooraminna, near Alice Springs, central Australia. Note the uniform patination.*

*Plate X Uniform patination and fractured rock surfaces indicate great age for these human figures at Mootwingee, western N.S.W. Many engraved sections have disintegrated or disappeared completely.*

*Plate XI Surviving fragment of a former extensive engraved pavement at Panaramitee Station, northeast South Australia, viewed from above. The patination itself has weathered in some places.*

*Plate XII Repeated rubbing of engraved designs with abrasive material to assist photography removes the age-determining patination. Eucolo Creek, near Woomera, in northern South Australia, is an example of such vandalism.*

*Plate XIII Representative designs on a rock face at Panaramitee Station.*

*Plate XIV Engravings at Amulda Springs, near Areyonga, central Australia, are similar to those at Panaramitee Station, hundreds of kilometres to the south.*

*Plate XV Simple portrayal of an adult emu walking away from a nest of eggs. Florina Station, South Australia.*

*Plate XVI Emu tracks at Tukulnga rockhole in western central Australia bear a striking resemblance to the series at Florina Station, some 1,300 km distant.*

*Plate XVII Unusual male and female figures at N'Dahla Gorge, central Australia, have large ceremonial headdresses.*

*Plate XVIII Large fern-like designs similar to these engravings at N'Dahla Gorge, central Australia, occur at a number of sites, including several in South Australia.*

*Plate XIX Abraded grooves and pecked engravings are concentrated on the walls of the Ingaladdi rock shelter, Northern Territory. Engraved fragments with similar motifs were excavated in layers older than 5,000 years. They had presumably fallen from the walls.*

*Plate XX The clarity of outline, lack of surface patination, and the sailing ship, all testify to the recent origin of these motifs at Ingaladdi, Northern Territory, where they have obliterated earlier designs.*

*Plate XXI Abraded grooves at Nackara Springs, South Australia, are identical with those found at sites in central and northern Australia, cf. Plates XIX, XXII, XXIV.*

*Plate XXII Abraded grooves at the remote Cleland Hills site in central Australia are similar to those in the north and south.*

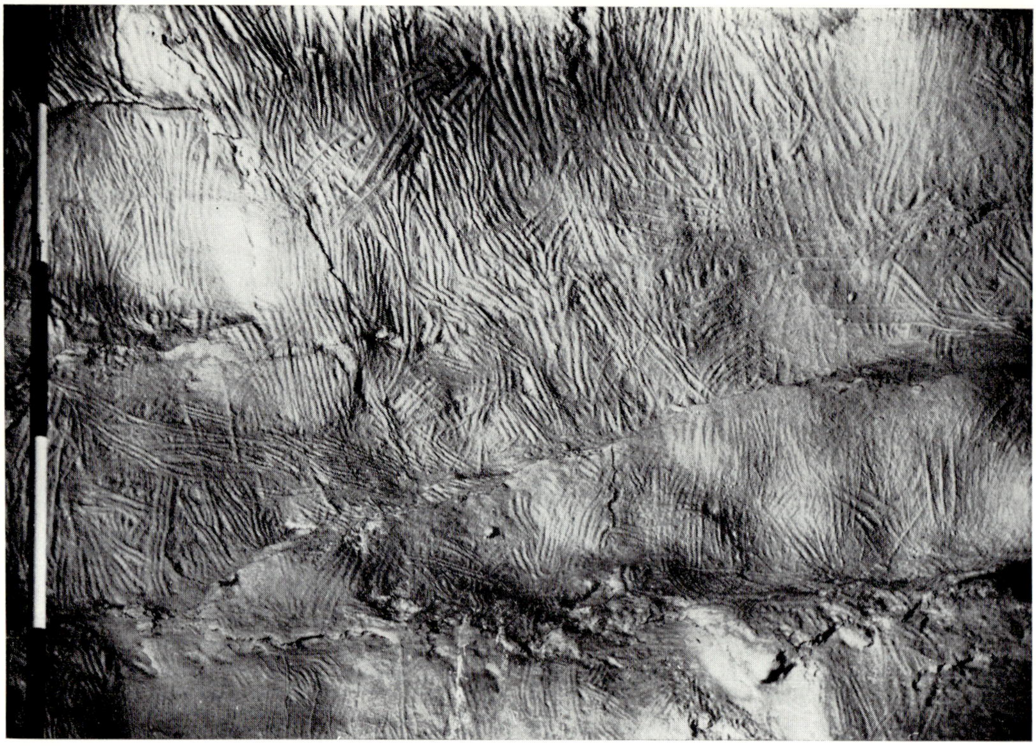

*Plate XXIII These large areas of finger markings on the soft compressible surface of an inner passage of Koonalda Cave, South Australia, are unique in Australian rock art.*

*Plate XXIV Abraded grooves, Koonalda Cave, South Australia. Some of the art on these cave walls must be at least 20,000 years old.*

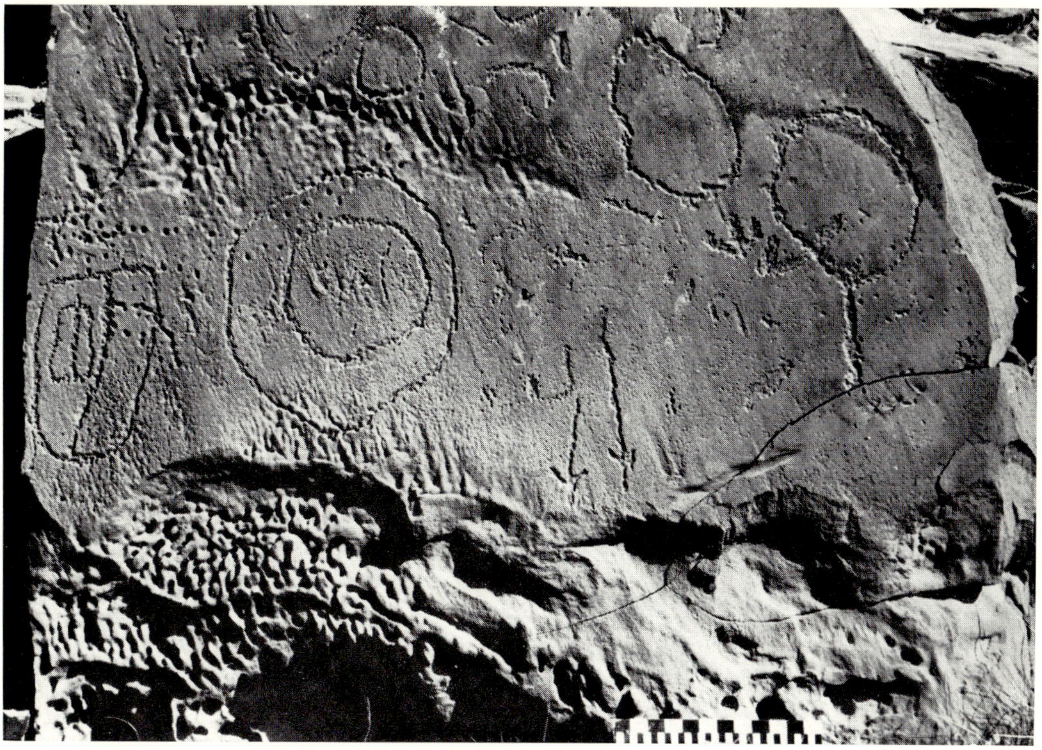

*Plate XXV Surface fretting and dissolution of the sandstone rocks at Ooraminna, near Alice Springs, central Australia, is threatening the survival of many engravings.*

*Plate XXVI Heart-shaped faces at the Cleland Hills site, central Australia, form only a small proportion of the designs. This deeply patinated and eroding sandstone slab has fallen from the rock face.*

(a)

(b)

(c)

(d)

*Plate XXVII Cleland Hills art, central Australia:* (a), *an economy of effort endowed this face with character;* (b), *uniform surface patination indicates antiquity for the comparatively sophisticated face engravings;* (c), *emotion is expressed by this unique human figure;* (d), *some crude facial outlines on broken pavements may provide clues to the development of the more expressive face engravings.*

*Plate XXVIII Excavated hearth at Panaramitee, South Australia. Rocks included fragments with pecked designs of kangaroo tracks.*

*Plate XXIX The large gallery at Mt Cameron West, Tasmania, recently excavated, and in an advanced stage of weathering.*

*Plate XXX Grinding the edge of a diorite axe blank at a quarry near Yuendumu, central Australia.*

*Plate XXXI Stages and techniques of hafting: aspects of reality which usually escape archaeological definition. The axes are deliberately ground; the millstones between the two men are only ground incidentally to their real function.*

circles, and other designs, man and animal are portrayed in profile. A large kangaroo at Mootwingee is duplicated at a recently located site on Delamere Station in the Northern Territory. Small human figures clutching boomerangs at Mootwingee (Plate X) have their counterparts at N'Dahla Gorge, where there are some interesting and unusual figures wearing large ceremonial headdresses (Plate XVII). Despite their apparent sophistication, the profile engravings are not of recent origin, because the age indicators of advanced weathering and heavy patination also are constant. It is only at sites on the west Australian coast (Wright 1968:1-78; McCarthy 1962:1-73) that a proliferation of the human figure has occurred. That this rich array of animated and sometimes grotesque human figures is certainly a later development is indicated by the freshness of the engravings and their corresponding lack of patination.

Many engraved sites are linked by the four constant features dealt with—proximity to regular water supplies, association with human occupation sites, advanced weathering and heavy surface patination, and consistent relative frequency of designs. Such sites have been found generally throughout south and central Australia, in western New South Wales, and in the far north and west. It seems significant that this pattern cuts across the multiple divisions of customs, language, artifacts, and decorative and cave art recorded in these same areas. It might be inferred that these motifs pre-date the time when tribal boundaries became rigid and separate cultural entities developed.

The first positive indication of the real antiquity of rock engravings came as early as 1929, with the discovery of abraded grooves on the back wall of the Devon Downs rock shelter on the lower River Murray at a depth of between 3 and 4 m below ground surface (Hale and Tindale 1930:208-11). However, the rather blurred image of an ancient prehistoric art in Australia came sharply into focus in 1966, when conclusive datings were obtained for engraved material found in an excavation undertaken by Mulvaney on Willeroo Station, 160 km southwest of Katherine in the Northern Territory.

The site is in a small amphitheatre adjacent to the permanent Ingaladdi waterhole. The walls of the sandstone outcrop are completely covered with engravings and paintings (Plate XIX). The weathered and patinated appearance of many engravings suggests antiquity, but others are obviously of recent origin, including one of a ship (Plate XX). Evidently the site had been an important ritual centre for many generations, and the excavation trench was sited adjacent to a heavily engraved part of the wall in the hope that the engravings and paintings would continue below ground level where they could be assigned minimum ages by dating charcoal in the overlying levels. Unfortunately the back wall of the shelter was smooth and bare except for one small group of short abraded grooves.

However, detached rock fragments bearing emu and kangaroo tracks and abraded grooves were excavated later in the main body of the trench. These engraved pieces were recovered between a level with an approximate depth of 1·1 m to 1·2 m, dated (ANU-58) to 4,920 ± 100 BP (2970 B.C.), and a lower level at approximately 1·7 m to 1·8 m, dated (ANU-60) to 6,800 ± 270 BP (4850 B.C.) (Polach *et al.* 1968:187-8). As the engravings were deeply weathered and unlikely to have been executed on loose rock fragments, this age must be considered conservative. The first positive date for engravings from a systematic excavation therefore gives strong support to the circumstantial evidence for the considerable antiquity of other rock engravings, and it is to be hoped that other situations will be found where precise datings can be obtained for a wider range of designs and techniques.

At Koonalda on the Nullarbor Plain, detailed study of simple wall engravings on an inner passage of the deep limestone cave has produced evidence for even greater antiquity of Australian rock art (Edwards and Maynard 1967:11-17; Gallus 1968: 43-9). Charcoal collected between boulders near dense concentrations of markings has been dated to about 20,000 years ago. As the area is in total darkness, this charcoal is thought to have come from crude torches used by prehistoric man, who came to the cave to mine flint (cf. Wright, see ch. 9). There are two main types of markings: hand smears made by running fingers over areas of the wall which are soft and compressible (Plate XXIII), and incised lines abraded or scratched into the harder surfaces (Plate XXIV). It is interesting that the abraded grooves are similar to those found in the Ingaladdi excavation and at other widely distributed sites (Edwards 1965b:17-18) (see Plates XIX, XXI, XXII).

At Koonalda it seems unlikely that any direct association between engravings and charcoal will be firmly established, but their antiquity is supported by the existence of large areas of engravings under 15 m of rock-falls, some of which must date to 20,000 years ago. Although the constant environmental conditions in the cave have reduced the extent of weathering, its effects may be clearly seen. The edges of all the finger markings on the wall, except those few known to be of recent origin, are smooth. At the same time, scrape marks caused by the ancient fall of boulders now wedged below massive rock-falls still retain a fresh appearance.

Koonalda offers unique evidence for Aboriginal art in Australia. Similar markings have been found on the walls of a number of European caves including Altamira, and these are believed by Breuil and Berger-Kirchner (1961:26) to be the earliest signs of man's artistic endeavours in the Upper Palaeolithic, some 30,000 years ago. Possibly the simple wall markings at Koonalda are an isolated and extensive example of the earliest stage of artistic evolution in Australia which is consistent with the long-held view that abraded grooves constituted the first phase of Aboriginal rock art (McCarthy 1964:33).

During field work in the Northern Territory over the past four years, twelve new sites of prehistoric engravings have been located and studied and the position of further sites has been recorded for later investigation. These new discoveries are concentrated in central Australia, where an attempt is being made to link regional groups in southern Australia with those of the north of Western Australia and Arnhem Land. This work has shown that a continuous series of sites extends from south to north and west. There is increasing evidence that other sites extend the range from central Australia into Queensland. The discovery of engravings on O.T. Station, near the Gulf of Carpentaria, is a relevant example. New finds on Delamere and Willeroo Stations in the Northern Territory have extended motif distribution well beyond the established limits of a few years ago, while the frequency of discoveries suggests strongly that further sites will be found in remoter areas.

The rock engravings of Tasmania provide further evidence for an even wider distribution. The engraved designs at several sites bear striking resemblance to those of the mainland and appear to be a southern extension of the same art form. In view of this, it is surprising that no engravings have been reported from southeastern Australia.

The most outstanding Tasmanian site is at Mt Cameron West, on the northwest coast, where deeply engraved circles and simple arrangements of lines are closely massed on a large outcrop of aeolianite on the beach, a little above high tide mark (Plate XXIX). Such a situation is unusual for Australian rock engravings. Charcoal samples overlying some engravings were collected during excavations sponsored by the Australian Institute of Aboriginal Studies, and their age determination may provide a basis for linking the prehistoric art of Tasmania with that of the mainland.

Central Australia is the most rewarding study area at the present time. Recent field investigations were highlighted by the recording of unique finds at Thomas Reservoir in the Cleland Hills, demonstrating that a wealth of important evidence awaits collection. All the major features represented are consistent with other sites studied. There is permanent water; there are occupation sites nearby; the rocks are badly weathered and the engraved surfaces have an even patination; the relative proportions of designs fall within the established range.

Of 387 designs at the site, 50 per cent are animal tracks. Circles account for another 33 per cent, and all other designs 17 per cent (see Table 24:1). Although the combined percentage of tracks and circles is slightly lower than at any of the other sites for which we have figures, it is still high and generally consistent. However, in the 17 per cent of other motifs, slightly more than usual, there are 16 unique designs depicting human faces. Their apparent sophistication is in complete contrast with previously known prehistoric Australian rock engravings. Among them are several heart-shaped faces of owl-like appearance, with eyes formed by concentric circles with an inner pit (Plate XXVI). Others, less stylised, are complete with eyes, nose, and mouth, but lack full facial outlines (Plate XXVIIa). This economy of line is seen particularly in one face, where the circles for the eyes also partly outline the cheerful face (Plate XXVIIb). Perhaps the most striking engraving is a human figure, whose peculiar body suggests movement. The facial expression, unmistakably one of happiness, is testimony to the skill of the prehistoric artist in achieving impressive detail with a comparatively crude technique (Plate XXVIIc). Indeed the portrayal of emotion in some of the Cleland Hills faces is a unique feature in Australian Aboriginal art. Such discoveries indicate the richness of a field as yet only partially explored.

There are several lines of inquiry to be followed. The investigation of sites in an ethnographic context is worthwhile and perhaps the most promising area for such studies is central Australia. Although the Aborigines claim that engravings belong to the Dreaming and are not the product of their own culture, there is a marked affinity between the designs of their present sacred art and those of the engravings. A study of such sites and their present associated mythology should be pursued while there are still tribal elders conversant with traditional sacred lore.

The location of new sites and plotting their distribution in relation to known sites is important. For each new find the listing of individual designs is essential to enable comparisons to be made. Only by this means can a knowledge of design distribution and development be obtained. The continued study of weathering and patination is necessary to enable bracketing of related sites. Perhaps more important is the extended investigation of rock engravings in the purely archaeological context in the search for more positive dating, which will remain the only direct means of placing them in their correct chronological perspective.

The continued systematic study of rock engravings scattered throughout the Australian continent is certain to widen the horizons of our knowledge of Aboriginal prehistory.

**References**


Anati, E. 1963. *Palestine before the Hebrews*. New York: Alfred A. Knopf.

Arndt, W. 1962. The Nargorkun-Narlinji cult. *Oceania* **32**:298-320.

Basedow, H. 1914. Aboriginal rock carvings of great antiquity in South Australia. *Jl R.anthrop.Inst.* **44**:195-211.

Berndt, R. M. (ed.) 1964. *Australian Aboriginal art*. New York: Macmillan.

Black, L. 1943. *Aboriginal art galleries of western New South Wales*. Melbourne: Bread and Cheese Club.

Breuil, H. and Berger-Kirchner, L. 1961. Franco-Cantabrian rock art. In *The art of the stone age* (eds. H. G. Bandi *et al.*), pp. 15-70. Art of the World Ser.No.5. Baden-Baden: Holle, London: Methuen.

Campbell, T. D., Edwards, R., and Hossfeld, P. S. 1966. Archaeological excavations in the south east of South Australia. Canberra: Australian Institute of Aboriginal Studies (unpublished).

Campbell, W. D. 1899. *Aboriginal carvings of Port Jackson and Broken Bay*. Sydney: William Applegate Gullick.

Cooper, H. M. 1943. Large stone implements from South Australia. *Rec.S.Aust.Mus.* **7**: 343-69.

——1960. The archaeology of Kangaroo Island, South Australia. *Rec.S.Aust.Mus.* **13**:481-503.

Crawford, I. M. 1964. The engravings of Depuch Island. In *Report on the Aboriginal engravings and flora and fauna of Depuch Island Western Australia* (eds. W. D. L. Ride *et al.*) Spec.Publ.W.Aust.Mus.No.2:23-63.

——1968. *The art of the Wandjina*. Melbourne: Oxford University Press.

Davidson, D. S. 1936. Aboriginal Australian and Tasmanian rock carvings and paintings. *Mem.Am.phil.Soc.* **5**.

Edwards, R. 1964. Rock engravings and stone implements of Pitcairn Station, north-eastern South Australia. *Rec.S.Aust.Mus.* **14**:643-62.

——1965a. Rock engravings and incised stones: Tiverton Station, north-east South Australia. *Mankind* **6**:223-31.

——1965b. Rock engravings and aboriginal occupation at Nackara Springs in the north-east of South Australia. *Rec.S.Aust. Mus.* **15**:9-27.

——1966. Comparative study of rock engravings in South and central Australia. *Trans.R.Soc.S.Aust.* **90**:33-8.

——1968. Prehistoric rock engravings at Thomas Reservoir, Cleland Hills, western central Australia. *Rec.S.Aust.Mus.* **15**:647-70.

—— and Maynard, L. 1967. Prehistoric art in Koonalda Cave. *Proc.S.Aust.Brch R.geogr. Soc.Aust.* **68**:11-17.

Flinders, M. 1814. *A voyage to Terra Australis*. 2 vols. London: G. and W. Nicol.

Gallus, A. 1968. Parietal art in Koonalda Cave, Nullarbor Plain, South Australia. *Helictite* **6**:43-9.

Goddard, R. H. 1941. Aboriginal rock sculpture and stencilling in the Carnarvon Ranges. *Oceania* **11**:368-72.

Grey, G. 1841. *Journals of two expeditions of discovery in north-west and western Australia*. 2 vols. London: T. and W. Boone.

Hale, H. M. and Tindale, N. B. 1930. Notes on some human remains in the lower Murray valley, South Australia. *Rec.S.Aust. Mus.* **4**:145-218.

Hall, F. J., McGowan, R. G., and Guleksen, G. F. 1951. Aboriginal rock carvings: a locality near Pimba, South Australia. *Rec.S.Aust.Mus.* **9**:375-80.

Howchin, W. 1919. Supplementary notes on the occurrence of aboriginal remains discovered by Captain S. A. White at Fulham . . . with remarks on the geological section. *Trans.R.Soc.S.Aust.* **43**:81-4.

King, P. P. 1827. *Narrative of a survey of the intertropical and western coasts of Australia.* 2 vols. London: John Murray.

McBryde, Isabel. 1964. The linear engravings of the Clarence Valley, northern New South Wales. *Oceania* **34**:201-10.

McCarthy, F. D. 1959. Rock engravings of the Sydney-Hawkesbury district, part 2: some important ritual groups in the County of Cumberland. *Rec.Aust.Mus.* **24**: 203-15.

——1961. The rock engravings of Depuch Island, north-west Australia. *Rec.Aust.Mus.* **25**:121-48.

——1962. The rock engravings at Port Hedland, northwestern Australia. *Pap. Kroeber anthrop.Soc.* **26**:1-73.

——1964. The art of the rock-faces. In *Australian Aboriginal art* (ed. R. M. Berndt), pp. 33-43. New York: Macmillan.

—— and Macintosh, N. W. G. 1962. The archaeology of Mootwingee, western New South Wales. *Rec.Aust.Mus.* **25**:249-98.

Macintosh, N. W. G. 1952. Paintings in Beswick Creek Cave, Northern Territory. *Oceania* **22**:256-74.

Meston, A. L. 1932. Aboriginal rock carvings in Tasmania. *Pap.Proc.R.Soc.Tasm.* 1932: 1-6.

Mountford, C. P. 1929. Aboriginal rock carving in South Australia. *Rep.Aust.Ass. Advanc.Sci.* **19**:337-66.

——1955. An unrecorded method of Aboriginal rock marking. *Rec.S.Aust.Mus.* **11**: 345-52.

——1960. Simple rock engravings in central Australia. *Man* **60**:145-7.

—— and Edwards, R. 1963. Rock engravings of Panaramitee Station, north-eastern South Australia. *Trans.R.Soc.S.Aust.* **86**: 131-46.

Mulvaney, D. J. 1966. The prehistory of the Australian Aborigine. *Scient.Am.* **214** (3): 84-93.

Polach, H. A., Golson, J., Lovering, J. F., and Stipp, J. J. 1968. ANU radiocarbon date list II. *Radiocarbon* **10**:179-99.

Sim, I. M. 1966. *Rock engravings of the Macdonald River district, N.S.W.* Occasional Papers in Aboriginal Studies No. 7 (Prehistory and Material Culture Ser.No.1). Canberra: Australian Institute of Aboriginal Studies.

Spencer, B. 1914. *Native tribes of the Northern Territory of Australia.* London: Macmillan.

Steward, J. H. 1936. Petroglyphs of the United States. *Rep.Smithson.Instn* 1936: 405-25.

Stokes, J. L. 1846. *Discoveries in Australia.* 2 vols. London: T. and W. Boone.

Strehlow, T. G. H. 1964. The art of circle, line, and square. In *Australian Aboriginal art* (ed. R. M. Berndt), pp. 44-59. New York: Macmillan.

Tindale, N. B. 1951. Comments on supposed representations of giant bird tracks at Pimba. *Rec.S.Aust.Mus.* **9**:381-2.

——1959. Totemic beliefs in the Western Desert of Australia. Part I. Women who became the Pleiades. *Rec.S.Aust.Mus.* **13**: 305-32.

Trendall, A. F. 1964. Examination of rocks with aboriginal engravings. App. to *Report on the Aboriginal engravings and flora and fauna of Depuch Island Western Australia* (eds. W. D. L. Ride *et al.*). Spec.Publ.W.Aust. Mus.No.2:83-8.

Wright, B. J. 1968. *Rock art of the Pilbara region, north-west Australia.* Occasional Papers in Aboriginal Studies No. 11 (Prehistory and Material Culture Ser.No. 2). Canberra: Australian Institute of Aboriginal Studies.

# 25 Aboriginal Social Evolution: A Retrospective View

D. J. Mulvaney

This paper opens with thoughts on the evolution of a prehistorian and leads to thoughts on the evolution of a prehistoric society. There are several morals to be drawn, although they are not all proposed explicitly. It is only twelve years since I lectured to the Adelaide ANZAAS Congress of 1958, on the results of the first Fromm Landing excavations. A retrospect of this past twelve years provides a useful platform from which to view trends and prospects. Some participants in this seminar series have expressed impatience at what has not been done; on the contrary, an antediluvian perspective emphasises those vantage points gained. I am unrepentant for my backward glances, despite Rhys Jones's recent warning (1968: 535), that 'We preen ourselves with mutual flattery, content to compare our achievements with those of our predecessors in Australia, rather than with the best of our contemporaries overseas'.

In 1958, Norman B. Tindale chaired my lecture. As far as I can remember, no other professional archaeologist was present. At that time, the oldest *positive* radiocarbon date for toolmakers in Australia came from Cape Martin, South Australia, 8,700 $\pm$ 120 BP; the only *major* excavation reports published since the classic Hale and Tindale report (1930) on Devon Downs and Tartanga, were those by Macintosh (1951) on Tandandjal and McCarthy (1948) on Lapstone Creek. Tindale (1957) published the most elaborate discussion of his five-stage cultural sequence the year before, claiming that his scheme was valid across the continent. His concepts had been assailed in a spirited reply during that year by McCarthy (1958), who also defended his own two-stage sequence (Bondaian and Eloueran), first formulated in his Lapstone Creek report.

Despite their differences of interpretation, however, both Tindale and McCarthy agreed that cultural change was a basic fact of Australian prehistory. This was a major contribution. It seems such a truism today that it is worth noting that in 1958 many older authorities saw matters differently. S. R. Mitchell often said that the Murray Valley evidence was untypical and that a monocultural interpretation was sounder. He had written (1949:5) that 'contrary to the opinion of some ethnologists, the writer believes that the various types of aboriginal stone artifacts represent one industry'. Oblivious to the evidence recorded even within his own book, he concluded (1949:106-8) that stratified sites were few and shallow and that 'there is nothing . . . comparable with the large mounds of occupational debris common in Denmark . . . '.

In 1958 also, bone utilisation was a subject of interest. Tindale (1957) stressed the cultural implications of the bi-pointed bone muduk, while McCarthy (1958:187) offered other explanations. In this year, Mitchell (1958) published a description of bone tools from Victorian middens. He simply noted earlier papers by Tindale and McCarthy on bone points, without comment. As usual, he eschewed theoretical issues, observing (1958:194) that this industry possessed 'few parallels'. It is interesting that he did not consult any historical sources for possible parallels and he sought no ethnographic analogies.

It is therefore not surprising that at Adelaide in 1958, and in my subsequent publication on Fromm Landing (1960), my interests and horizons were more confined than would be anticipated from current interpretive practice. I tied my discussion firmly to the Devon Downs sequence; I discussed the evidence relating to Tindale's cultural synthesis; I examined in detail the occurrence and possible function of stone tools and bone points; while I stressed the rarity of type tools, I discussed their significance and stratigraphic provenance with a boldness which would be challenged in these more statistically conscious times.

Tindale concluded that my evidence supported his cultural interpretation. Indeed, to a large extent it did, but I claimed that the differences were also significant. In my conclusion to the written report (1960:80), I appealed for regionally oriented, objective research—'the time is not ripe for attempts at cultural and chronological syntheses'.

While in autobiographical vein, I should place in context my 'Stone age of Australia' memoir (1961). I began work when still preparing the Fromm Landing report, and much of it was written during 1959. It reflects the same approach, but my disquiet about Tindale's synthesis prompted me to examine this in critical detail. During 1960, however, my own experience broadened and this is mirrored in certain sections of the paper written during that year. Passing references were inserted to my preliminary survey on Mt Moffatt station. Honesty demands the comment that because of my South Australian orientation, I was more excited by the comparative potential of The Tombs site, and I had no expectation either that the Kenniff Cave sequence was more significant or that it represented a new time dimension.

In the context of 1960, the Glen Aire excavations contributed more to my thinking, because I analysed the finds and wrote the report during that year. In retrospect, the report is deficient as an exercise in midden analysis, but I consider that the excavation still fulfilled Collingwood's dictum (1944:83), that an archaeologist 'must first of all decide what he wants to find out, and then decide what kind of digging will show it to him'. I am not equally convinced that all contemporary excavations employing more sophisticated techniques are necessarily as purposeful, or that the excavator could answer adequately Collingwood's question (1944: 85): 'what are you doing this piece of work for?' In a sense, midden refuse was not the problem of the day. The Murray Valley sites had raised a different problem—how to explain the disappearance of various specialised stone tool types in the later occupation of those sites.

At Glen Aire the contrast between a rich bone industry and meagre stone implements of recent date was striking. In discussing this evidence (1962:11-14), I concluded that it demonstrated the reality of local adaptation and changing habits in Aboriginal material culture. As it appeared to support the Murray Valley situation, it strengthened my conviction that a flexible regional interpretation offered the most plausible explanation of later prehistory. This was the essence of my 1961 approach (e.g. 1961:82-3).

The Glen Aire analysis convinced me also of the relevance of using historical evidence and museum ethnographic collections as an adjunct to interpreting the excavated material. This was no sudden revelation, because I had employed such sources previously; indeed Tindale and McCarthy had used them for years. In the past, however, they were sometimes used so selectively and generally that they were divorced from any historical-cultural context. Here is another aspect of prehistoric research which today is taken for granted, but which a few years ago received only token treatment by many writers. Tindale had demonstrated further that field research amongst living Aborigines possessed great relevance, and I wish to discuss those sources at some length, which today are given the terminological status of ethnohistory and ethnoarchaeology.

There is an understandable note of irritation in Tindale's expostulation (1965:162), that 'it is high time that at least a few archaeologists should . . . emerge from their cave holes to study at first hand the data provided by living peoples'. He partly blamed apathetic archaeologists for the sad neglect of ethnographic research in Australia. Writing exactly a century earlier, Sir John Lubbock (cf. 1890:546) was more sanguine, that 'those who have the opportunity of observing stone implements among modern savages will give us more detailed information both as to the exact manner in which they are used, and also about the way in which they are made; that they will collect not only the well-made weapons, but also . . . the humble implements of every-day life'. Yet a century of lost opportunities intervened.

As one of the few archaeologists from the era to which Tindale referred, I accept his charge of neglect but wish to examine it. This emphasis on the value of surviving material culture for prehistoric interpretation poses the problem of the validity of present analogies as a guide to the past. In some cases to be considered later, they are used to reconstruct the nature and basis of society, rather than merely to restore an individual artifact.

When Tindale published his strictures in 1965, archaeologists were distributed thinly and unevenly over the Australian landscape; few had been there before 1961, and those who had were critically short of research funds for distant projects. (Those who dug at Fromm Landing paid their own expenses; the fund available for two seasons' work was less than $500). This is the simple answer to Tindale; but a historical perspective is essential to appreciate fully that lack of interest in detailed ethnography which he correctly deplored, for it also applied to anthropologists.

The blame rests partly a century back, with Lubbock and his social Darwinist colleagues, who enthused about the potential Australian field harvest, but garnered it selectively as an evolutionary storehouse of living fossils. Their concern was with abstract problems and never with individual persons or fully dimensional specific situations; they assumed the innate uniformity of the 'savage' mind and not its complexity of response. Writing in 1874, Pitt-Rivers (see 1906:19) laid it down as the basic principle of ethnographic classification that 'Progress is like a game of dominoes—like fits on to like. In neither case can we tell beforehand what will be the ultimate figure produced by the adhesions; all we know is that the fundamental rule of the game is *sequence*'.

Indeed, this Games Theory of social development may be said to have conditioned field research into both Aboriginal material culture and social structure, with unfortunate consequences. Urging Lorimer Fison to action in Australia, Lewis Henry Morgan affirmed on 20 September 1872 that the record of progress

> is still preserved to a remarkable extent in . . . inventions and discoveries which stand to each other in the ages of savagism, of barbarism and of civilisation in a progressive series; and in . . . domestic and civil institutions . . . in an unfolding series. These afford us two independent lines of investigation by means of which to recover the threads of man's progress. . . .

Writing in the 1860s, Lubbock (see 1890: 429) and Pitt-Rivers (see 1906:53) established the trends in ethnography which, in the words of General Pitt-Rivers, 'establish it as a maxim, that the existing races, in their respective stages of progression, may be taken as the bona fide representatives of the races of antiquity . . . They thus afford us living illustrations . . . [of] the ancient races'. Edward B. Tylor's paper (1893) 'On the Tasmanians as representatives of Palaeolithic man' was an application of this dogma. Baldwin Spencer (1901:8, 12) echoed his teacher, Tylor, when he listed the Tasmanians as 'living representatives of palaeolithic man', and the Australians 'as a relic of the early childhood of mankind'. W. J. Sollas (1911) wrote the notorious extreme. *Ancient hunters and their modern representatives* contained chapters on 'palaeolithic' Tasmanians and 'Mousterian' mainlanders.

Such philosophical notions still attract adherents. Tindale (1950) described the Western Australian kodj axe as 'of palaeolithic type'. Frederick G. Rose (1960:237-8) postulated that the social institutions of Pleistocene Aborigines, including Tasmanians, were gerontocratic and polygamous, 'not significantly different from what we know today'. He argued further, that it appeared legitimate to assume that European Upper Palaeolithic social conditions were comparable to the structure of nineteenth-century Tasmanian society.

In his interest in comparative institutions, Rose followed the lead set by L. H. Morgan. Morgan informed Fison on 8 May 1872 that

> in Australia and Polynesia you are several strata below barbarism into savagism, and nearer to the primitive condition of many than any other investigator. You have in their institutions of consanguinity, marriage and tribal organisation, far reaching and intelligent guides, not only to their present, but also to their past, condition.

Elsewhere I have shown (1958, n.d.) that the anthropological teams—Fison and Howitt, Spencer and Gillen—held similar evolutionary concepts and exerted profound influence upon social theorists of their day. It is particularly relevant that the succeeding anthropological generation reacted strongly against what Malinowski in 1913 (see 1963:5) termed their 'polemical attitude'. Malinowski and Radcliffe-Brown (1930-1) particularly, attacked their assumptions and methodology.

In Australia, Radcliffe-Brown also contended with current excesses of the diffusionist hypothesis school, which scattered disembodied traits across the map. In promoting the functional analysis of society, he attacked the validity of historical reconstruction. Rhys Jones has drawn my attention to Radcliffe-Brown's scornful review (1930) of D. S. Davidson's paper, 'The chronological aspects of certain Australian social institutions as inferred from geographical distribution'. Its publication heralded the nadir of ethnographically or historically oriented anthropological research although Tindale and Donald Thomson perhaps may be cited as notable exceptions to this generalisation. 'If the aim of anthropological investigations is to enable us to arrive at the same sort of understanding of the phenomena of culture that other sciences give us of other phenomena', Radcliffe-Brown wrote (1930:367), 'then such studies as this do not in any way serve that aim'. He concluded (1930:370) that Aboriginal research 'will make little progress until we abandon these attempts at conjectural reconstructions of a past about which we can obtain no direct knowledge in favour of a systematic study of the culture as it exists in the present'. For culture, he understood social structure and not material objects.

This antipathy to the uncritical invocation of history or ethnographic analogy is understandable, but it resulted in the neglect of themes of profound significance for prehistoric research. Grahame Clark (1951:52) considered that in Britain a rift developed between archaeology and ethnography which proved detrimental to both aspects. In Australia, the gulf between anthropology and ethnography widened. An ethnographer was considered by some archaeological and anthropological colleagues as not quite respectable. Perhaps this explains why prehistorians overseas and social anthropologists everywhere evinced so little interest in Australian material culture. Even Australia's foremost prehistorian, V. Gordon Childe, virtually ignored the potential of the ethnographic field. He revealed his hierarchy of values in a discussion of surviving food-gatherers, when he observed (1951:73) that the archaeology of such groups was 'little more reliable than the ethnographic' evidence. Possibly this explains why students of my vintage at Cambridge University received little encouragement or training to undertake what is now fashionable ethnoarchaeological research.

In my own case, I believe that my thinking on ethnographic work was adversely affected by Tindale's rather strained insistence upon the lessons of the 'living culture'. I felt that there was a danger inherent in his approach, of assuming what one is trying to discover while sliding from possibility to certainty. If my 'Stone age of Australia' paper paid rather grudging and limited attention (1961:58-9) to evidence which 'could illuminate the possible content and purpose of many fragmentary prehistoric material remains', it should be read within the context of my doubts (1961:69, 74-5) on the validity of Tindale's inferences.

Just to set the record straight, I wish to quote one of the recommendations from my paper on prehistory, which I submitted from England to the foundation conference of the Australian Institute of Aboriginal Studies, in May 1961. Unfortunately, in my opinion, five pages on 'the broad requirements of a research programme' were deleted from my paper and were not published. This recommendation still expresses my approach to the subject, while indicating that my intentions were honourable.

While there is still time to visit peoples in Australia and Melanesia who preserve traditional technological processes, priority should be given to the study of technology. It should be best achieved through archaeologists and anthropologists working in close co-operation. The one is best equipped to comprehend the process, while the other can appreciate its social significance and act as linguist. Very detailed written and photographic records of all processes involved are required. The camera is only an aid to understanding, it is *not* the substitute for a clear, detailed description.

The year 1961 was climactic in Australian prehistoric studies. The Australian Institute of Aboriginal Studies took shape; prehistorians arrived from overseas to work at the University of Sydney and the Australian National University; the Western Australian Museum set a precedent for other museums by appointing a trained archaeologist. The winds of change blew from new and wider horizons, as an increasing awareness of research overseas disseminated. Many of the new approaches were reflected in the A.I.A.S. conference of typology and nomenclature held early in 1963, not least of which was the emphasis which Jack Golson placed upon American methodological and conceptual techniques. Prehistorians who today turn more naturally to *American Antiquity* than to *Antiquity* would not have done so once; but to judge from the 1961 membership list of the Prehistoric Society, even English influences were peripheral—four institutional and possibly fewer ordinary members.

It is only by comparing 'then and now' that the full growth and sophistication of field methods, laboratory techniques, and conceptual grasp is evident. In my opinion, the papers presented at this series, including his own, are the answer to Rhys Jones's question (1968:535): 'Where is any glimmer of independent archaeological thinking . . . ?'

Times and attitudes change. It is salutory to read that W. H. R. Rivers (1926:122) found difficulty in studying American ethnology, 'because it is unusual in that country to deal to any great extent with general theoretical problems'.

After years of cautioning against premature schematic synthesis, in 1965 I rashly formulated a model for ordering prehistoric artifacts. My concept of two technological stages—correlated tentatively with non-(pre-) hafting and hafting devices—has provoked strong criticism, including comment at this seminar series. I wish to elaborate the ramifications of this hypothesis as the concluding section of this paper (Plates XXX and XXXI).

The evidence should be considered under two aspects. First, the positive data which I assembled from Kenniff Cave and the other excavations which I correlated with it, for a marked change in implement technology and typology. Second, my inference that these changes resulted from the rapid adoption of hafting techniques throughout mainland Australia.

My concern in the 1961 memoir had been to criticise what I considered were inflexible and confused systems of cultural labelling; in 1965 I sought a solution 'at the expense of conventional cultural terminology' (Mulvaney and Joyce 1965:193). Some may consider that the complaint was preferable to the cure. My thesis (Mulvaney and Joyce 1965: 193) was 'that a technological pattern may be emerging, in which regional sequences are found to have many points in common, but many contrasts in local elaboration and adaptation to needs'. I still maintain that one of the basic common elements was the technological change to which I referred; whether it was related to hafting is more questionable, but the suggestion continues to merit serious consideration.

Although I hinted at other far-reaching implications (Mulvaney and Joyce 1965: 207-11), these were not formulated in detail. An element of evolutionary philosophy is discernible, because the division between earlier flake and core industries and the later blade industries presupposed technological improvement and diversification; the hafting concept is a further 'progressive' feature. There is also a trace of diffusionist theory, while the adoption and rapid spread of new techniques implicitly raised the question of innovation in Aboriginal society. It is time to examine some of these issues.

During the past three years nothing has been excavated which confounds my primary thesis, that there were at least two widespread and major technological stages in prehistoric Australia. The occurrence of edge-ground axes in the Oenpelli region (White, C., see ch. 12) and of ground flakes at Laura (Wright, see ch. 11) in flake and core industry complexes does not affect this issue, for the axe is also a core tool which is subjected to a specialised process. My dating of the arrival of the new technological elements should be pushed back perhaps 2,000 years, to about 7,000 BP, but the effectiveness and relative rapidity of response remains evident. Early flake and core industries have been substantiated since 1965 by further work on material from Ingaladdi and in the Oenpelli area. Dr Richard Gould (1968) claims similar evidence at Puntutjarpa shelter near the Warburton Range mission, while R. J. Lampert has comparable material of Pleistocene age at Burrill Lake (see ch. 10).

At all these sites, and others, the early industry is overlain by one in which new features are added to the previous components. In some areas, backed blades are the dominant introduction, while in others, point production proliferated; common to both distributions in some regions are adze-flakes. In addition to these tool types, analysis of cores and waste flakes points to new techniques of manufacture. Compared with my 1965 thesis, the only necessary revisions are the removal of edge-ground axes and grinding techniques from the catalogue of introductions. Term it what you will, therefore, the case for a two-stage, or phase, pattern in mainland prehistory is stronger now than it was in 1965. In my opinion, however, the later centuries of prehistory, at least over much of temperate Australia, may require a third phase designation, because they were characterised by a decline in the variety and quality of stone tool production.

On the other hand, the case for hafting has weakened. I accept that, adopting my criteria, the Arnhem Land Pleistocene axes must be classified as hafted. Fragments of probable ancient axes also occur at Laura, while in this seminar series, Carmel White (ch. 12) has urged a re-evaluation of the evidence from other sites. My difficulty is that posed by all subjective assessments—proof. It is possible to excavate artifacts preserving evidence of hafting, but non-hafting cannot be established by objective criteria, unless microscopic research on comparative use-wear provides it. For example, possibly the leverage provided by the haft may produce a different edge fracture from an implement held in the hand.

All I can appeal to at present is the appearance of the tool—morphology and size being used to suggest its function. Numbers also matter. I do not consider that the concept of a prehistoric, non-hafting period is invalidated by citing occasional ethnographic examples of *hafted* 'non-hafted' stone tools, because the concept could have been applied in 'post-hafting' times.

In reference to the occurrence of ground stone in early deposits being interpreted as evidence for (hafted) axe production, a further clarification is necessary (Plates XXX and XXXI). Grinding on a surface may be deliberate, as is the case in sharpening the edge of an axe blade, or it may be unintentional, as when seeds are ground or pounded between upper and nether millstones and similar devices. Possibly the invention of axes resulted from experience of the latter process or ochre grinding being experimentally adapted to produce shaped forms. But the presence of mortars alone in a deposit cannot be validly used as proof that intentional stone grinding techniques were known. Although I failed to elaborate the point in my Kenniff Cave report, I considered millstones to be incidentally ground artifacts (Mulvaney and Joyce 1965:192).

The significance of millstones for reconstructing botanical and dietary history was grasped by Tindale (1959:49), who inferred that their first use postdated the Pleistocene and was associated with Pirrian times (i.e. my second technological phase). One large mortar from Kenniff Cave and smaller types from Nawamoyn, Malangangerr, and Ingaladdi, in the Northern Territory, together with a specimen from Puntutjarpa, Western Australia, possibly in a 7,000-years-old context (Gould 1968:173), indicate the probability that varieties of grindstone were employed well before 5,000 BP, even into the late Pleistocene (i.e. in my first technological stage).

And there is Tasmania. In A.D. 1800 the Tasmanians were demonstrably 'non-hafters'; back to 8,000 BP there is nothing in the available evidence to suggest they were ever anything else; their stone tools possess affiliations with mainland ones. Possibly they lost the various arts, and the relevant evidence lies drowned in Bass Strait. But if so, they were bereft of the skills long before the mainlanders adopted hafting techniques for points, adzes, and backed blades. It is relevant to add, however, that grindstones were used by the Tasmanians.

Even if Pleistocene men hafted axes, the idea does not appear to have been applied to other elements of material equipment; the crucial adoption and adaptation occurred around 7,000 years ago, on present evidence. All speculations must heed the differences of response: backed blades of diverse kinds perhaps diffused from the west, but points of equal variety may have spread from the north.

It is the magnitude of the change which is significant and the reason why I compared it (Mulvaney and Joyce 1965:210) to the consequences of the arrival of Upper Palaeolithic blade users in Europe. Childe (1951:73) drew attention to the European technological situation before their arrival: over a long period of time there was a gradual standardisation of tools in respect of shape, and some specialisation of tool types. Yet the range was limited. The same might be observed of the scrapers and utilised core tools in Australia during its first phase, although the ground axe must be added to the tool kit in some areas. The tradition was also long-lasting, for it extended from before 20,000 years ago.

Elsewhere, Childe (1942:32) noted that 'modern men appear in the upper palaeolithic enormously better equipped than any group' before them. The new techniques spread rapidly and regional variation was evident; stone implements apart, it is known that this was a time of inventiveness (cf. Childe 1951:75). This is how I interpret the Australian evidence. The new technology was a relatively sudden phenomenon; it diffused rapidly; there was considerable variety of response even with respect to working particular types, such as points. To judge from present distribution maps, the idea-migration-invention-diffusion occurred at more than one place. Were innovations characteristic of other aspects of culture? Unlike in Europe, there is no archaeological evidence as yet, but there is inference. On the mainland, boomerangs, spearthrowers, and shields were common, but they were absent in Tasmania when the Europeans arrived. Either they are further examples of lost arts, or they were adopted in Australia after Tasmania became isolated.

From inferences concerning technological evolution, it is tempting to pass to social evolution. It is a commonplace to list Aboriginal society as one of hunters, fishers, and foragers, thereby stressing its essential economic homogeneity and its similarity with the basis of Old World pre-agricultural societies. Thus Thurnwald (1932:36), who discussed the scope and limitations of such social groups at comparable economic and technical levels, classified them as one 'cultural horizon'. From Morgan to Childe another designation of their state is 'savagery'. Another generally accepted explanation is that hunter-gatherers live in adjustment with their environment, which is envisaged as a relatively stable medium, at least for long periods. Thurnwald (1932:42), Childe (1951:166), and Quimby (1960:383) appear to have reached this conclusion by different paths. All accept that in a given environment the best adapted group will prevail. To Quimby, 'when the technology of a hunting culture is perfected in relation to its habitat, it can be said to be in equilibrium with its environment'.

Such generalisations are acceptable, but today they verge on the platitudinous. In his important essay on culture change, J. H. Steward (1955:17) observed that such 'postulated cultural sequences are so general that they are neither very arguable nor very useful'. No one can dispute the hunter-gatherer state of savagery, or its place in the order of social development from simple to complex, but as a category it is too broad.

I have few constructive suggestions which may fill out the missing details of Aboriginal culture change. A rigorous application of Steward's methods to the Australian scene could repay the effort, but one turns in vain for assistance from his own thoughts on Australia. In his comments on the New World, Steward (1955:18-42) is empirically oriented and seeks coherent assemblies of cultural data, eschewing comparisons of isolated facts. He stresses environmental controls and allows the environment a creative role, rather than one of passive stability. This is surely in keeping with current Australian opinion, as expressed in this seminar by Rhys Jones. When Steward cites Australian examples, however, he is guilty of those faults which he criticises in others. His own dated knowledge chiefly derives from Radcliffe-Brown, and he treats Australia as a unitary example of an 'arid' region adaptation. Recent detailed investigations of the relationships between habitat and economy have directed research along Steward-type avenues.

The fact that there were two major technological stages, the earliest long-enduring and fairly standardised, while the second reached its climax relatively rapidly and evoked varied response, indicates that these hunter-gatherers cannot be treated in any unitary manner. The mechanics of innovation perhaps 7,000 years ago pose crucial but imponderable questions. Prehistorians are accustomed to approach problems such as local invention versus diffusion of ideas or objects by plotting trait distributions. I suspect deductions from such work are at the same general level at which 'savagery' describes the prehistoric economic condition. For example, detailed evaluation of individual traits such as backed blades or points may establish minor regional variations which are obscured on the general distribution pattern. These small differences could reflect functional, ecological, or cultural preferences or adaptations. (As a parallel, consider the archaeological problem of deducing that a boomerang served as a weapon in one region and a musical instrument in another.)

I believe that some relevant lessons may be drawn from Aboriginal tribal society. There are two studies of innovations penetrating tribal society, which evaluate the consequences of their adoption. I commend them as stimulating reading, but cannot expound them here. The first case study was presented by Lauriston Sharp (1952) who examined the complex social and mental behaviour associated with stone axe ownership amongst the Yir Yoront of Cape York, and the socially disruptive consequences of their substitution by steel axes. The other was D. S. Davidson's observation (1935:168-72) that the *idea* of manufacturing pressure flaked spear-heads was adopted by the Northern Territory Wardaman people, situated on the periphery of the physical distribution of prototype Kimberley points.

In both these cases, the social acceptance of the new objects or skills possibly reflects the belief that it would facilitate activities. (In fact, it put the people who use them at a disadvantage, for various reasons.) Steward (1955:38) inferred that any appropriate technique which improved hunting-gathering methods might gain acceptance. Childe's claim (1951:172), that 'one society can borrow an idea . . . only when it fits into the general pattern of the society's culture', echoes Thurnwald's postulate (1932:38), that 'we can expect new technical methods of obtaining food to be adopted by a tribe with the same cultural horizon as the innovators'. In Australia, Meggitt's conclusion on Walbiri society (1962:253) is relevant. He concluded that any 'innovations' were borrowed from comparable neighbouring groups, and 'were likely to have been fully compatible with the existing framework'.

Possibly, therefore, the rapidity with which new ideas spread during the second stage of Australian prehistory is explicable on grounds of their utility and of the similar socio-economic status of their users. Perhaps there were further reasons. In his book *Innovation*, H. G. Barnett (1953:40-3) points to the importance of contact between peoples for transmitting ideas; and the more sizeable the cultural inventory, the more likely are new combinations and adaptations. Increased interchange and a more complex tool kit may have played their part.

Contact during Australia's recorded past largely occurred during regional tribal congregations or through ceremonial exchange patterns. Alternative means of increasing the frequency of meeting between groups could result from direct migration or merely through increased population density. In a recent paper, Tindale (1967) has emphasised the consequences of the eustatic rise in sea level, for occupants of the emerged plains in both Australia and island Southeast Asia. Australia shrank by perhaps 10 per cent, and by 7,000 years ago these changes were almost complete, while Asian islands also dwindled in area. The necessary resettlement on available land presumably would lead to an increase in population density. Tindale (1967:350) advances this as the possible cause of migration to Australia of the Carpentarian racial group.

While I doubt the reality of this racial hypothesis, Tindale's suggestion of eustatic cause and migratory effect merits serious examination. Population pressure amongst island hunter-fishers may have stimulated seaward migration of some bands. Withdrawals of tribal groups from submerged coasts could provoke a chain reaction of tribal adjustment across the continent. These environmentally conditioned movements might be correlated also with linguistic distribution patterns, and they might shed light both on the confined distribution of some language families and groups, and on later shifts in others.

One further problem merits comment. Anthropologists frequently comment on the 'notorious stability' (Sharp 1952:87) and conservatism of tribal society. In this context, Sharp demonstrated the failure of the Yir Yoront to adopt the canoe, despite its useful existence only 45 miles away. Strehlow (1947:5, 170) stressed the 'absolute authority' of Aranda elders, whose methods were 'conservative in the extreme'. Change, he concluded, could have infused only from 'fresh and foreign impulses coming from without; and the age-old isolation of Australia effectively prevented the influx of any new creative ideas which might have brought about a change in the old order'. Even though Meggitt's analysis of Walbiri institutions led him to conclude that the society had no formal governmental hierarchy or political elders, he acknowledged (1962:247-53) its general conservatism and the 'comparatively stable' socio-cultural and physical environment. Barnett (1953:43) agrees that societies whose institutions encourage secrecy and exclusiveness militate against change.

There are difficulties here. Even if contact-period societies appeared as unchanging as the Orient was to European eyes, it was not always so. Dynamic changes *did* occur across Australia several millennia ago, and adaptations continued throughout subsequent prehistory. On the principles expounded above, that utilitarian borrowing occurs between neighbouring groups of hunter-fishers, the Yir Yoront should have adapted their ideologies to embrace the canoe, just as Arnhem Land Aborigines enthusiastically borrowed the dug-out canoe from Macassan trepangers.

How frequently have anthropologists fallen into the fallacy of the 'unchanging land and people'? Perhaps there are hints of a modified approach. As regards tribal control, Hiatt (1965:141-7) has found that anthropological opinion has turned a full circle during a century. Perhaps his analysis has implications for other aspects of tribal life. The rapidity with which European dogs were adapted by Tasmanian hunters is a further relevant indicator that tradition did not die as hard as some anthropologists believed. R. M. and C. H. Berndt (1964:281) reflect this less dogmatic view. They deny 'that there is no scope for individual variation or individual initiative, as some earlier writers suggested: but it does imply that innovation is at a minimum'. If we are intending to use ethnographic analogies to interpret the past, let us ensure that the initial field interpretations are valid.

## References


Barnett, H. G. 1953. *Innovation. The basis of cultural change*. New York: McGraw-Hill.

Berndt, R. M. and Berndt, Catherine H. 1964. *The world of the first Australians*. Sydney: Ure Smith.

Childe, V. G. 1942. *What happened in history*. Harmondsworth: Penguin.

——1951. *Social evolution*. London: Watts.

Clark, G. 1951. Folk-culture and the study of European prehistory. In *Aspects of archaeology in Britain and beyond* (ed. W. F. Grimes), pp. 49-65. London: H. W. Edwards.

Collingwood, R. G. 1944. *An autobiography*. Harmondsworth: Penguin.

Davidson, D. S. 1935. Archaeological problems of northern Australia. *Jl R.anthrop. Inst.* **65**:145-83.

Gould, R. A. 1968. Preliminary report on

excavations at Puntutjarpa rockshelter, near the Warburton Ranges, Western Australia. *Archaeol.& phys.Anthrop.Oceania* **3**:161-81.

Hale, H. M. and Tindale, N. B. 1930. Notes on some human remains in the lower Murray valley, South Australia. *Rec. S.Aust.Mus.* **4**:145-218.

Hiatt, L. R. 1965. *Kinship and conflict*. Canberra: Australian National University Press.

Jones, R. 1968. Editorial. *Mankind* **6**:535-6.

Lubbock, J. Baron Avebury. 1890. *Prehistoric times*, . . . 5th edn. London:Williams and Norgate.

McCarthy, F. D. 1948. The Lapstone Creek excavation: two culture periods revealed in eastern New South Wales. *Rec.Aus.Mus.* **22**:1-34.

——1958. Culture succession in south eastern Australia. *Mankind* **5**:177-90.

Macintosh, N. W. G. 1951. Archaeology of Tandandjal Cave, south-west Arnhem Land. *Oceania* **21**:178-213.

Malinowski, B. 1963. *The family among the Australian Aborigines*. New York: Schocken.

Meggitt, M. J. 1962. *Desert people. A study of the Walbiri Aborigines of central Australia.* Sydney: Angus and Robertson.

Mitchell, S. R. 1949. *Stone-age craftsmen.* Melbourne: Tait Book Co.

——1958. An Aboriginal bone industry. *Mankind* **5**:194-9.

Mulvaney, D. J. 1958. The Australian Aborigines 1606-1929: opinion and fieldwork. *Hist.Stud.Aust.N.Z.* **8**:131-51, 297-314.

——1960. Archaeological excavations at Fromm's Landing on the lower Murray River, South Australia. *Proc.R.Soc.Vict.* **72**:53-85.

——1961. The stone age of Australia. *Proc. prehist.Soc.* **27**:56-107.

——1962. Archaeological excavations on the Aire River, Otway Peninsula, Victoria. *Proc.R.Soc.Vict.* **75**:1-15.

——n.d. The ascent of Aboriginal man: Howitt as anthropologist. In *Come wind, come weather* (ed. M. H. Walker), in press. Melbourne: Melbourne University Press.

—— and Joyce, E. B. 1965. Archaeological and geomorphological investigations on Mt. Moffatt Station, Queensland, Australia. *Proc.prehist.Soc.* **31**:147-212.

Pitt-Rivers, A. L. 1906. *The evolution of culture and other essays.* Oxford: Clarendon Press.

Quimby, G. I. 1960. Habitat, culture, and archaeology. In *Essays in the science of culture* (eds. Gertrude E. Dole and R. L. Carneiro), pp. 380-9. New York: Thomas Y. Crowell.

Radcliffe-Brown, A. R. 1930. The diffusion of culture in Australia. *Oceania* **1**:366-70.

——1930-1. The social organization of Australian tribes. *Oceania* **1**:34-63, 322-41, 426-56.

Rivers, W. H. R. 1926. *Psychology and ethnology.* London: Kegan Paul.

Rose, F. G. G. 1960. *Classification of kin, age structure and marriage amongst the Groote Eylandt Aborigines.* Berlin: Akademie-Verlag.

Sharp, L. 1952. Steel axes for stone age Australians. In *Human problems in technological change* (ed. E. H. Spicer), pp. 69-90. New York: Russell Sage Foundation.

Sollas, W. J. 1911. *Ancient hunters and their modern representatives.* London: Macmillan.

Spencer, Sir W. B. 1901. *Guide to the Australian ethnographical collection in the National Museum of Victoria.* Melbourne: Robt S. Brain.

Steward, J. H. 1955. *Theory of culture change.* Urbana: University of Illinois Press.

Strehlow, T. G. H. 1947. *Aranda traditions.* Melbourne: Melbourne University Press.

Thurnwald, R. 1932. *Economics in primitive communities.* Oxford: Oxford University Press.

Tindale, N. B. 1950. Palaeolithic *Kodj* axe of the Aborigines and its distribution in Australia. *Rec.S.Aust.Mus.* **9**:257-73.

——1957. Culture succession in south eastern Australia from late Pleistocene to the present. *Rec.S.Aust.Mus.* **13**:1-49.

——1959. Ecology of primitive Aboriginal man in Australia. In *Biogeography and ecology in Australia* (eds. A. Keast, R. L. Crocker, and C. S. Christian), pp. 36-51. Monographiae biol. **8**. The Hague: W. Junk.

——1965. Stone implement making among the Nakako, Ngadadjara and Pitjandjara of the Great Western Desert. *Rec.S.Aust. Mus.* **15**:131-64.

——1967. Peopling of the lands southeast of Asia. *Colorado Quarterly* **15**:339-53.

Tylor, E. B. 1893. On the Tasmanians as representatives of Palaeolithic man. *Jl R. anthrop.Inst.* **23**:141-52.

# Index


Abbie, A. A., 316, 317, 321
Acheron Aboriginal Station, 289, 302
Adelaide Plains (sites), 359
Africa, 48, 51, 147, 170, 184, 191, 208
Africans, 330, 331, 332, 339, 341
Agriculture, *see* Domesticated plants
Agriculture, adoption of, 170, 174, 176, 177, 178, 182-5, 190, 192, 208-9, 271
Ainu, 320, 338
Alice Springs, 66, 74, 77
America, 48, 134, 171, 176, 184, 208, 281, 339
Andaman Islands, 235, 317, 350
Animals, *see* Dingo, Domesticated animals, Dog, Fauna, Goat, Pig
Arafura Sea, 4, 5, 143, 147
Aranda tribe, 73, 74, 259, 334, 335, 378
Arandic languages, 348, 350, 352
Archaeological cultures, periods, sequences, 61-2, 118-20, 121-7, 129, 137-9, 144-6, 149, 152-3, 159-60, 191, 192, 281, 322, 359, 368-9, 373-4, 375-6
Archaeological evidence, 41, 61-4, 110, 118-26, 144-6, 159-61, 163-77, 189-91, 208, 242, 246, 273, 274, 276, 278, 281, 359, 360, 363-4, 365-6, 368-70, 373-4
Arid Period, mid-Recent, 21-2, 52, 77, 84, 88, 96
Arnhem Land, 141-55, 184, 185, 188, 192, 196-209, 210, 211, 230, 232, 233, 236, 237, 242, 243, 246, 321, 329, 330, 332, 335, 336, 346, 347, 348, 349, 351, 356, 364, 374
Art, rock, 356-66; *see also* Engravings, rock Paintings, rock
Artifacts, prehistoric, 154, 170, 190; bone, 120, 121, 123, 125, 126, 128, 129, 130, 134, 153, 159, 189, 190, 369, 370; shell, 121, 124, 125, 126, 128, 129, 130, 139, 154, 159, 174, 178; stone, 41, 61, 62, 108, 110-12, 117, 118, 119-29 *passim*, 136, 137-9, 144-6, 148, 149-53, 154, 159, 160, 168-9, 172-4, 178, 189, 190, 191, 208, 240-1, 243-4, 359, 369, 370, 374, 376-7; *see also* Axes, Pleistocene, Material culture, Metal, Millstones, Mortars, Pottery
Aru Islands, 233
Asia, *see* East, Southeast Asia
Atambua (site, Timor), 160, 179
Atherton Tableland, Cairns rainforest, 139, 316, 347
Australia, *see* Central, Northern, Northwestern, Southeastern, Southwestern Australia
Australian Alps, *see* Mountains
Australian Institute of Aboriginal Studies, 107, 109, 133, 253, 365, 372, 373
Australian Museum, *see* Museums
Axes, Pleistocene, 145, 149, 152-3, 208, 374

Baguia (site, Timor), 159
Ba:gundji tribe, 259
Bali, 172

Balranald, 39, 42, 58
Bamboo, 171, 172, 188
Banda Sea, 6
Bantu, 339
Bare Hill (site), 139
Barkly Tableland, 72, 74, 329, 348, 350
Barrabool tribe, 301
Barrau, J., 170, 197, 235, 237
Barrineans (racial type), 319
Basedow, H., 198, 232, 358, 361
Bass Strait, 4, 5, 6, 7, 9, 282, 322
Bass Strait Islands, 283
Bates, Daisy, 106, 107, 108
Bathurst Island, 328, 330
Baucau (sites, Timor), 161, 162, 164
Bêche-de-mer, 188
Bendemeer (site), 119
Ben Lomond tribe, 275, 280
Bentinck Island, 327
Bermingham, A., 110, 119
Berndt, C. H. and R. M., 73, 184, 188, 253, 274, 281, 357, 378
Big River tribe, 275, 279
Bindibu, Bindubi tribe, *see* Pintubi tribe
Binford, L. R., 239, 243
Bird, E. C. F., 2, 121
Birdsell, J. B., 90, 281, 283, 296, 316-21 *passim*, 329, 334, 335, 347
Black, J. M., 197, 198, 230, 231, 232, 235
Blood groups, 188, 321, 327-30, 334, 335, 336, 337, 338, 339, 340, 341
Bomaderry Creek (site), 125, 126, 128
Borneo, 153, 178
Botany Bay, 116, 126
Bowler, J. M., 2, 18, 41, 44, 52, 63, 77, 323
Braidwood, R. J., 183, 184
Breuil, Abbé, 364
Broadbeach burial ground, 323
Broken Hill, 55, 84
Brookfield, H. C., 176, 185, 190
Buandik tribe, 347
Bühler, A., 159, 160, 170, 171, 178
Bui Ceri Uato (site, Timor), 162, 164, 178
Bulmer, R. N. H., 1, 190
Bulmer, S., 1, 176, 189, 190
Bunurong tribe, 301, 302
Burbidge, N. T., 149, 198, 230, 231, 232, 235, 237
Burials, 63, 124, 125, 134, 321, 323; *see also* Cremation
Burkill, I. H., 163, 170, 172, 178, 197, 206, 207, 208, 209, 231, 232, 235
Burning, 85, 91, 161, 177, 243, 254, 256
Burrill Lake (site), 36, 121-8 *passim*, 130, 152, 374
Bushmen, !Kung tribe, 184, 242, 245, 338
Butler, B. E., 19, 29-31, 40, 96
Butzer, K. W., 147, 154, 185

Cadell Fault, 21, 44
Cairns rainforest people, *see* Atherton Tableland
Calaby, J. H., 16, 22, 49, 80, 86, 116, 149, 154
Cambridge University Museum, *see* Museums
Campbell, T. D., 89, 90, 176, 284, 316
Canning Stock Route, 263
Canoes, *see* Watercraft
Capell, A., 163, 279, 334, 345-50, 352
Cape Martin (site), 368
Cape Otway (site), 120
Capertee (site), 118, 119, 120, 126, 127, 152, 322
Cape York, 8, 10, 133-40, 184, 185, 186, 187-8, 192, 196-209, 211, 230, 231, 232, 233, 236, 237, 252, 255, 256, 257, 258, 259, 320, 327, 330, 331, 335, 345, 349, 350, 351, 356
Carbon dating, *see* Dating
Carnarvon Ranges (sites), 358
Carnegie, D. W., 262, 263, 264, 267, 268, 269
Carpentarians (racial type), 316, 320, 334, 335
Carstenz, J., 133, 134
Carter, G. F., 189
Caucasians, Caucasoids, 332, 339
Cavalli-Sforza, L. L., 338, 339
Celebes, 81, 159, 160, 172, 191
Central Australia, 16, 66-78, 83, 89, 196-209, 211, 231, 232, 233, 236, 242, 252, 254, 257, 258, 260, 262-70, 331, 332, 356, 357, 359, 360, 361, 362, 364, 365
Central Australian languages, 346
Ceylon, 191, 338
Chandi Mendut (site, Java), 177
Chang, Kwang-chih, 208, 239
Change, Conservatism, Innovation, cultural, 126-7, 137, 138-9, 144-6, 184-5, 187-8, 191, 192, 242, 243-4, 258, 260, 281, 357-8, 368-9, 376-8
Chasm Island (sites), 356
Childe, V. G., 372, 375, 376, 377
Chillagoe (site), 139
China, 170, 171, 178
Chowilla Dam burials, 323
Clacks Island (sites), 356
Clark, Desmond, 184, 185, 191
Clark, Grahame, 372
Clark, Manning, 280, 284
Cleland, J. B., 89, 198, 231
Cleland Hills (site), 365
Climate, 4-6, 14-23, 27, 32, 35-7, 42, 44, 46-64, 66, 67, 69, 74-8, 84-8, 94-6, 97, 102, 105-6, 116, 133, 143-4, 147-9, 161, 185, 264
Cohuna cranium, 321, 322
Colani, M., 178
Collingwood, R. G., 369, 370
Conservatism, cultural, *see* Change, cultural

Cook, James, 116, 118, 187
Cooper, H. M., 359
Cooper Creek, 70, 72, 73, 77
Coranderrk Aboriginal Station, 289, 291, 297, 298, 302, 308, 309
Costin, A. B., 17, 27
Cotton, C. A., 4, 19, 147
Coutts, P. J. F., 119
Crawford, I. M., 356, 361
Cremation, 63, 318, 323; *see also* Burials
Crocker, R. L., 5, 18, 51, 53, 98, 102
Cumeroogunga Aboriginal Station, 291, 297
Curr, E. M., 106, 107, 252, 259, 307, 344-5
Curracurrang (site), 120, 125, 126, 127, 128, 130, 149, 153, 255, 322
Currarong (site), 123, 125, 127, 128, 130
Curray, J. R., 4, 9, 10, 112, 147

Dampier Land, 349, 352
Darling River, 55, 71, 253, 254, 257
Darlington, P. J., 80
Darwin, 143, 199, 209, 211, 236
Darwin, Charles, 284
Dating, 2, 4, 6, 7, 9, 10, 18, 19, 21, 29, 32, 33, 35, 36, 37, 38, 42, 44, 45, 52, 58, 60, 61-4, 76, 86, 87, 98, 99, 102, 110, 112-13, 119-21, 122, 124-9 *passim*, 135, 137, 138, 139, 144-6, 147, 149, 152, 153, 154, 159, 167-8, 170, 172, 176-7, 178, 189, 190, 191, 242, 281, 318, 320-1, 322, 323, 340-2, 352, 359-66 *passim*, 368, 374, 375, 376, 377
Davidson, D. S., 242, 258, 356, 358, 372, 377
Decay of organic remains, 127, 128, 136, 139, 145, 149, 155
Delamere Station (sites), 363, 364
Deniliquin, 39, 41, 43
Depopulation, *see* Population decline
Depressions, *see* Pits
Depuch Island (sites), 358, 361
Devon Downs (site), 363, 368, 369
DeVore, I., 242
Diamantina River, 70
Dieri tribe, 349
Dingo, 90, 107, 125, 176, 189, 191, 192, 255, 269, 322, 362
Disease, 277, 282, 288, 302, 303, 305-9, 310, 332; *see also* Mortality
Division of labour, 118, 188, 241, 268
Dixon, R. M. W., 347, 351
Djauan tribe, 349
Dog, 160, 163, 175, 176, 177, 185, 187, 189, 190, 191, 378; *see also* Dingo
Domesticated animals, 159, 160, 163, 175-7, 178, 189, 190, 191, 358; *see also* Dog, Goat, Pig
Domesticated plants, 159, 163, 170, 178, 185, 187, 189, 208, 209, 210, 232, 233; *see also* Bamboo, Tamarind
Dravidian languages, 345
Dunes, 18, 41, 50, 55, 59, 69, 72, 74-6, 78
Durras North (site), 121, 122, 123, 125, 126, 127, 128, 255
Dury, G. H., 19, 21, 69, 147
Dyen, I., 350

East Alligator River, 143
East Asia, 339, 341
Ebenezer Aboriginal Station, 298
Echuca, 52, 53, 61
Ecology, human, 245-6, 249, 262-70, 272-85
Edwards, R., 90, 110, 191, 358, 359, 362, 364
Elcho Island (sites), 331
Elkin, A. P., 182, 281
Emiliani, C., 20, 147
Engravings, rock, 358-66; *see also* Paintings, rock
Environments, Aboriginal impact on, 85, 90-1, 242-3; coastal, 114-16, 252, 255-6, 257-8, 259, 281, 284; desert, 66-78, 252, 253-4, 257, 259, 262-70; estuarine, 256, 257; lacustrine, 46-64; monsoonal, 141-4, 147, 149, 198-201, 252; riverine, 252-3, 254-5, 257, 259, 281; semi-arid, 4-6, 46-64, 106-7, 108-9, 252-3; Timor, 161-3
Enzyme groups, 328, 331-4, 335, 336
Ethnobotanical evidence, 196-8, 203
Ethnographic evidence, 106-9, 116, 118, 128, 130, 185, 187-9, 239-47, 250-2, 262-70, 273, 274, 276, 277-8, 279-81, 296-7, 299, 310, 319, 357, 365, 370-3, 378
Eucla, 106, 107, 108
Eucolo Creek (sites), 359
Excavations, 109, 110, 121-6, 129, 134-5, 136-7, 144-6, 159-60, 163-6, 363-4, 365, 368, 369, 370, 373, 374, 375
Eyre, E. J., 252, 259

Fairbridge, R. W., 1, 6, 9, 10, 11, 18, 147, 154
Fauna: estuarine, 123, 125, 126, 144, 154, 256; extinct, 82, 89-90, 90-1, 159, 160, 177, 179, 191, 192, 243-4, 361-2 (on Australian mainland), 154, 322; freshwater, 16, 48, 49, 51, 61, 62, 63, 64, 77-8, 81, 82, 144, 254; marine, 16, 114, 116, 123, 125, 144, 255-6, 257, 278; terrestrial, 16, 21-2, 49, 72, 77-8, 80-91, 116, 125, 144, 159, 160, 163, 177, 178, 189, 190, 191, 243-4, 254-5, 256, 264, 268, 269, 270, 322, 361-3, 365; *see also* Dingo, Domesticated animals
Fiji, 230, 233, 234
Finke River, 71, 74
First Fleet records, 116, 130, 255, 358
Fison, Lorimer, 371
Flannery, K. V., 183, 184, 254

Flinders, Matthew, 356, 359
Flinders Island (Bass Strait), 7, 9, 279
Flinders Ranges (sites), 358
Flint, R. F., 15
Flora, *see* Bamboo, Domesticated plants, Sandalwood, Subsistence, plant, Tamarind, Toxic plants, Vegetation
Flores, 178
Fossil man, 321-2, 323
Framlingham Aboriginal Station, 289, 307
Frank, R., 97, 110
Fromm Landing (site), 368, 369, 370
Fulham (site), 359

Galloway, R. W., 15, 16, 17, 18, 33, 35, 84, 147
Gallus, S. A., 109-13, 364
Gates, R. R., 316
Genetic evidence, 326-42; *see also* Heterogeneity, Physical anthropological evidence
Gentilli, J., 14, 15, 55
Ghanaians, 339
Giles Meteorological Station, 263
Gill, E. D., 6, 7, 11, 16, 18, 22, 33, 52, 86, 154
Gillen, F. J., 74, 258, 371
Gippsland, 291, 296, 297, 298, 300, 301, 308, 309, 310
Glaciation, 16-17, 32, 33, 36, 281
Glen Aire (sites), 369-70
Glover, I. C., 161, 164, 168, 172, 177, 179, 190, 191
Goat, 160, 163, 175, 176-7
Godwin, H., 7, 9, 10
Gogobera tribe, 349
Golgol (stratigraphic unit), 58-9, 61, 62, 63
Golson, Jack, 127, 149, 152, 153, 176, 178, 184, 187, 189, 208, 373
Gorman, C. F., 170, 176, 208
Goulburn Valley, 44, 288, 289, 291, 298, 302, 309, 310
Gould, R. A., 152, 240, 242, 250, 265, 267, 269, 270, 374, 375
Graman (sites), 119
Great Artesian Basin, 69, 71, 72
Great Dividing Range, *see* Mountains
Great Sandy Desert, 4, 5
Greenberg, J., 350
Green Gully skeleton, 321
Green Point midden, 278
Grey, Sir George, 356
Groote Eylandt, 346, 347
Group composition, size, 259, 265-6, 277-8, 280, 281; *see also* Tribes, distribution, organisation, size
Gubabwingu tribe, 349
Gulf of Carpentaria, 2, 4, 5, 10, 143, 147, 209, 211, 327, 336, 337, 350
Guwa Lawa (site, Java), 176
Gymea Bay (site), 120, 127, 255

Haasts Bluff, 263
Haddon, A. C., 187, 188
Haemoglobins, abnormal, 334
Hafting, 107, 138, 149, 152, 191, 322, 373-5
Hale, H. M., 253, 254, 255, 257, 363, 368
Hale, K. L., 348, 350, 352
Half castes, 283, 289, 290, 291, 292-3, 294-5, 297, 299, 303-4
Halmahera, 350
Hawkesbury River, 253; sites, 361
Heterogeneity, homogeneity, Aboriginal genetic, physical, 316-21, 334-6, 341-2
Hiatt, Betty, 89, 240, 242, 250, 273, 277
Hiatt, L. R., 184, 281, 378
Hills, E. S., 41, 51, 55
Hillston, 39, 41, 58
Historical evidence, 116, 128, 130, 133-4, 250-2, 273, 277-8, 279-84 *passim*, 288-313, 370
Hoabinhian (-Bacsonian) culture(s), 176, 208
Hobart, 274, 277
Holmer, N., 350
Hooijer, D. A., 160, 176, 179
Hooton, E. A., 316, 320
Horticulture, *see* Agriculture, adoption of, Domesticated plants
Hossfeld, P. S., 2, 189, 362
Howchin, W., 359
Howells, W. W., 316, 320
Howitt, A. W., 107, 116, 130, 371
Hrdlička, A., 316
Hunter-gatherer studies, 171, 184, 244-6, 281
Huts, 133, 134, 188, 278, 279

Implements, *see* Artifacts, prehistoric
India, 171, 177, 191, 198, 199, 230, 231, 234, 320, 338, 341, 346
Indochina, 178, 232, 234, 235
Indonesia, 159, 163, 172, 177, 185, 188, 192, 196, 232, 235
Industry, *see* Archaeological cultures, periods, sequences, Artifacts, prehistoric
Infanticide, 282, 298, 302, 303
Ingaladdi (site), 149, 320, 322, 363, 374, 375
Innovation, cultural, *see* Change, cultural
Inscription Point (site), 126
Israel, 361

Jajowrong (Jaara) tribe, 301, 303, 308
Japan, 159, 176
Java, 5, 6, 176, 177, 178, 191, 234
Jennings, J. N., 6, 7, 18, 20, 35, 68, 69, 76, 97, 100, 147, 153, 154, 185, 209
Jones, Rhys, 1, 14, 16, 20, 63, 82, 84, 86, 90, 91, 126, 153, 189, 192, 239, 243, 245, 271, 273, 278, 281, 317, 368, 372, 373, 376

Joyce, E. B., 88, 96, 120, 127, 149, 191, 242, 322, 373, 374, 375

Kafiavana (site, New Guinea), 176
Kaiadilt tribe, 330
Kalumpang (site, Celebes), 159
Kangaroo Island, 283; sites, 359
Karawa tribe, 330
Katherine, 155, 199, 209, 211, 236, 363
Ka:wadji tribe, 257, 259
Keast, A., 185
Keilor cranium, 321
Keith, Sir A., 319
Kenniff Cave (site), 88, 120, 149, 152, 153, 322, 369, 373, 375
Kimberleys, 209, 211, 236, 320, 327, 330, 331, 335, 346, 347, 348, 351, 356
Kintore Cave (site), 155
Kiowa (site, New Guinea), 176
Kirk, R. L., 321, 327, 330, 331, 332, 334, 339, 341
Knopwood, R., 277, 280
Koonalda Cave (site), 96-102, 105, 107-13 *passim*, 364
Kow Swamp skeletons, 321, 322, 323
Krefft, G., 85, 88
Kroeber, A. L., 281, 345
Kulin tribe, 347

Labillardière, J.J.H.de, 277, 278, 318
Lachlan River, 39, 44, 58
Lakes, 16, 17-18, 33, 35, 41-2, 46-64, 71, 77, 87, 89, 160, 263-9 *passim*, 349; Condah Aboriginal Station, 289; Eyre, 66, 76, 87, 349; Gassi Issi (site, Timor), 160, 162; George, 17, 33, 35; Menindee, 16, 55, 56, 87; Tyers Aboriginal Station, 290, 291, 299; Victoria, 5-6, 259
Lampert, R. J., 36, 121, 125, 152, 246, 255, 374
Languages, 163, 187, 188, 189, 262, 275, 279-81, 345-53; *see also* Linguistic evidence
Lapps, 341
Lapstone Creek (site), 118, 127, 368
Larapinta tribe, 74
Lardiil tribe, 330
Launceston, 274, 280
Laura (site), 136-40, 149, 152, 322, 374
Lawrence, R., 73, 74, 89, 114, 130, 187, 196, 197, 198, 203, 208, 230, 231, 232, 240, 249, 254
Leakey, L. S. B., 319
Lee, Richard, 171, 242, 245
Lene Hara (site, Timor), 160, 162
Lesson, A., 316
Liang Djenilu (site Timor) 160, 162
Liang Leluat (site, Timor), 160, 162
Lie Siri (site, Timor) 162, 164, 176
Linguistic evidence, 187, 189, 273, 274, 275, 279-81, 321, 345-53, 378; *see also* Languages
Linton, Ralph, 171
Loddon River, valley, 39, 42, 288, 289, 301
Long, J. P. M., 262, 310
Lord Howe Island, 231
Lubbock, Sir J., 284, 370
Lundelius, E. L., 21
Lunettes, 18-19, 41, 42, 46-64, 71

Mabbutt, J. A., 72, 77
McArthur, Margaret, 89, 250, 251
Macassans, 188, 189, 199, 320, 330, 358, 378
McBryde, Isabel, 119, 126, 130, 149, 358
McCarthy, F. D., 89, 118-21, 122, 125, 126, 127, 152, 183, 188, 242, 250, 251, 258, 358, 361, 363, 364, 368, 369, 370
McConnel, Ursula, 198, 256, 258
Macdonnell Ranges, 70, 72, 77, 263
Macintosh, N. W. G., 1, 7, 90, 159, 191, 316, 317, 318, 319, 320, 323, 357, 358, 368
Macknight, C. C., 188, 320
Malangangerr (site), 145, 146, 147, 149, 375
Malay Peninsula, 206, 231, 234, 235
Malays, 330, 334
Malaysia, 196-209, 211, 232, 233, 234, 235, 236
Malinowski, B., 371
Mallee, 55
Maloga Aboriginal Station, 297, 298
Mammoth Cave (site), 86-7
Mansuy, H., 178
Manton Site (New Guinea), 190
Martin, P. S., 90, 95, 189
Massola, A., 126, 258
Material culture, ethnographic, 107-8, 118, 133-4, 187-8, 253, 254, 255-6, 257-9, 268-9, 279, 363, 376; *see also* Artifacts, prehistoric, Watercraft
Mathew, Rev. J., 345
Maung tribe, 349
Maynard, L., 364
Megaw, J. V. S., 116, 118, 120, 125, 126, 127, 130, 149, 153, 255
Meggitt, M. J., 73, 74, 89, 90, 182, 184, 205, 231, 242, 259, 260, 281, 377, 378
Melanesia, 158, 159, 163, 170, 178, 185, 189, 316, 317, 319, 320
Melanesians, 331, 335, 336, 341
Melbourne, 288, 297, 301, 307, 310
Melville Island, 320, 328, 330
Merrilees, D., 14, 16, 55, 82, 84, 86, 87, 90, 191, 271
Mesopotamia, 183
Meston, A. L., 358
Metal, 163, 167-8, 171-2, 177, 178, 188, 241, 263, 268, 377

Middens, shell, 122, 125, 126, 133-6, 145, 146, 153, 154, 255, 278, 369-70
Midland Valley, 274
Millstones, 152-3, 257, 375
Mirning tribe, 106-8, 113
Mitchell, S. R., 369
Moluccas, 178
Monaro, *see* Tablelands, Southern
Mongoloids, 316, 330, 341
Morgan, L. H., 371, 376
Mornington Island, 327
Mortality, 273, 277, 288, 296, 299, 300-1, 303-9, 312; *see also* Disease
Mortars, 240-1, 246, 267, 375
Mossgiel skeleton, 321, 322
Mt Burr (site), 191
Mt Cameron West (site), 278, 358, 365
Mountains, hills, 14, 16, 26-37, 71, 72, 84, 120, 121, 127, 189-90, 253, 263, 265, 267, 269, 358
Mountford, C. P., 358, 361, 362
Movius, H. L., 140
Mulvaney, D. J., 62, 88, 96, 120, 121, 123, 127, 149, 152, 155, 159, 179, 191, 192, 242, 254, 321, 322, 368-78
Munggan tribe, 256
Mungo: skeleton, 63, 323; stratigraphic unit, 58-61, 62, 63
Murdock, G. P., 350
Murramarang (site), 125, 126
Murray River, basin, 39, 42, 44, 48, 53, 55, 64, 96, 253, 254, 257, 259, 288, 289, 291, 297, 298, 300, 309, 320, 369, 370
Murrayians (racial type), 316, 319, 320, 334, 335
Murrumbidgee River, 39, 44, 59
Museums, 118, 121, 127, 130, 321, 323
Musgrave Ranges, 72, 253

Narrinyeri tribe, 347
National Museum of Victoria, *see* Museums
Nawamoyn (site), 145, 146, 147, 149, 153, 242, 375
N'Dahla Gorge (site), 362, 363
Negritos, 316, 317, 318
Neolithic Revolution, *see* Agriculture, adoption of
New Britain, 317
New Caledonia, 233, 317
New England district, 119
New Guinea, 2, 4, 6, 10, 15, 17, 80-2, 83, 88, 144, 147, 149, 153, 172, 174, 176, 178, 184, 185-7, 189-91, 196, 198, 199, 201, 230, 231, 233, 243, 317, 329, 331, 332, 334, 337, 338, 339, 341
New Guinea languages, 187, 345, 346, 350
New Hebrides, 233, 331
New South Wales, 16, 19, 26-37, 38-46, 47-64, 83, 85, 86, 87, 96, 114-30, 209, 211, 232, 236, 237, 246, 318, 319, 320, 323, 358, 360, 363; *see also* Darling River, Murray River, Riverina
New South Wales languages, 346
New Zealand, 20, 233
Ngadadjara tribe, 334
Ngadanja tribe, 256
Ngalia tribe, 73
Ngama (site), 357
Niah Cave (site, Sarawak), 153
Nikiniki (site, Timor), 159, 160, 162, 171
Non Nok Tha (site, Thailand), 171
Noola (site), 149, 322
Northern Australia, 141-55, 183, 236, 319, 332, 358, 359, 360; *see also* Arnhem Land, Cape York, Northern Territory
North East tribe, 275
Northern Territory, 141-55, 185, 197, 199, 230, 231, 236, 319, 320, 330, 332, 346, 349, 350, 358, 363, 364; *see also* Arnhem Land, Central Australia
Northwestern Australia, 83, 85, 361
Nullarbor Plain, 5, 69, 96-102, 105-9, 112-13, 268, 364

Oceania, Pacific Islands, 196, 231, 234
Oceanic languages, 346
Ochre, 139, 149, 174, 279, 356, 375
Oenpelli (sites), 141-55, 322, 374
O'Grady, G.N., 187, 350, 351, 352
Oyster Bay tribe, 275, 277, 279

Pacific Islands, *see* Oceania
Padypadiy (site), 146, 154
Paintings, rock, 136, 139, 164, 356-8, 363; *see also* Engravings, rock
Pakistan, 177
Pama-Nyungan Language family, 351-2
Panaramitee (site), 359, 362
Papua, *see* New Guinea
Papuans, 320
Papuan languages, *see* New Guinea languages
Pedley, L., 198, 230, 231, 232, 237
Pels, Simon, 19, 41, 43, 44
Péron, F. A., 277
Petermann Ranges, 253
Peterson, Nicolas, 154, 174, 184, 239, 240, 242, 282
Philippines, 172, 178, 196, 234, 235
Physical anthropological evidence, 159, 189, 316-23; *see also* Genetic evidence, Heterogeneity
Pig, 159, 160, 164, 170, 175, 176, 177, 185, 187, 189, 190, 191
Pintubi (Bindibu, Bindubi) tribe, 257, 259, 262-70
Pitjandjara Pitjanjadjara tribe, 259, 334, 335
Pits, depressions, 123-5, 278

Pitt-Rivers, A. L., 371
Plants, *see* Bamboo, Domesticated plants, Subsistence, plant, Tamarind, Toxic plants, Vegetation
Plomley, N. J. B., 273, 274, 277, 278, 281, 282, 284
Poisons, animal, fish, 208, 254, 257, 258
Polach, H. A., 33, 63, 98, 110, 145, 171, 363
Polynesia, 176, 317
Population, 73, 106, 107, 170-1, 178, 259-60, 264-5, 272-85, 288-313, 377; decline, depopulation, 273, 274, 277, 282-3, 284, 285, 288, 291, 293, 297-300
Port Dalrymple tribe, 275
Port Davey tribe, 274
Port Jackson, 116, 118, 130
Port Phillip, 7, 33, 257
Pottery, 159, 160, 163, 170, 172, 176, 177-8, 189
Puntutjarpa (site), 152, 374, 375

Quarries, 112, 269, 364
Queensland, 72, 77, 85, 87, 96, 118, 130, 149, 187, 197, 209, 211, 230, 231, 232, 236, 237, 327, 329, 332, 335, 345, 346, 347, 349, 351, 358, 360, 364; *see also* Cape York
Quimby, G. I., 376

Radcliffe-Brown, A. R., 260, 264, 273, 281, 372, 376
Ray, S. H., 345
Reed, C. A., 177, 190
Ride, W. D. L., 84, 85
Risdon Cove, 283
Riverina, Riverine Plain, 38-46, 55, 76, 252, 254, 293, 296; *see also* Darling River, Murray River
Rivers, 19, 40-1, 42-6, 70-1, 76, 77, 143, 253, 254
Rivers, W. H. R., 373
Robinson, G. A., 116, 273, 274, 277, 278, 279, 280
Robinson tribe, 274, 275, 277-9, 280, 281
Rose, F. G., 371
Roth, W. E., 116, 134, 135, 187, 198, 203, 230, 231, 232, 273, 280, 282
Ruguri (site), 357

Sahlins, Marshall D., 184
Sahul Shelf, 20-1
Salinity, 46-64
Sandalwood, 163, 172, 178
Sandy Cape tribe, 277
Sarasin, F., 159, 160, 170, 171
Sassafras (site), 120, 127
Schmid, Elizabeth, 94, 95, 96
Schmidt, Father W., 273, 279, 345
Schmidt language group, 275, 279-80
Sea levels, 1-11, 107, 109, 112-13, 123, 134, 135, 143, 147, 153-4, 282, 377
Seasonal movement, 107, 129-30, 134, 154, 184, 245, 254, 255, 256, 259, 265-6, 277, 279
Sediments, 19, 20-1, 40-5, 94-102, 122, 123, 136, 145-6, 164-7
Seelands (site), 119, 126, 149, 322
Seiçal River (sites, Timor), 164
Serum protein groups, 328, 330-1, 335, 336, 337, 338, 340, 341
Service, E. R., 171, 242
Sharp, Lauriston, 377, 378
Shell middens, *see* Middens, shell
Shepard, F. P., 9, 10, 112, 154
Shoalhaven River, valley, 29, 33, 114, 126, 129
Sim, I. M., 358
Simmons, R. T., 327, 329, 334
Simpson, G. G., 80
Simpson Desert, 74
Sleisbeck (site), 149, 152, 155, 349
Smyth R. Brough, 252, 308
Snowy Mountains, *see* Mountains
Soils, 19-20, 28-33, 41, 42, 96, 98-9, 102
Solheim, W. G., 171, 172
Sollas, W. J., 371
Solomon Islands, 350
South Australia, 50, 51, 53, 86, 96, 105-13, 197, 198, 230, 231, 232, 264, 318, 319, 334, 347, 358, 359, 360, 361, 363, 364; *see also* Central Australia, Nullarbor Plain
Southeast Asia, 81, 149, 158, 159, 171, 172, 176, 178, 192, 198, 199, 206, 208, 209, 377
Southeastern Australia, 2, 26-37, 47-64, 86, 88, 114-30, 209, 211, 252, 255; *see also* New South Wales, Victoria
Southeastern Australian languages, 346
Southern Tablelands, *see* Tablelands, Southern
Southwestern Australia, 51, 85, 87-8, 106
Specht, R. L., 143, 144, 182, 196, 197, 198, 199, 201, 203, 207, 208, 209, 211, 230, 231, 232, 235, 237
Spencer, W. B., 258, 356, 371
Stanner, W. E. H., 134
Steward, J. H., 245, 361, 376, 377
Stirton, R. A., 87
Stirling, E. C., 198, 231
Stony Creek tribe, 275
Strehlow, T. G. H., 356, 358
Subsistence, animal, 48, 74, 88-90, 107, 116, 123, 125, 126, 130, 154, 170, 175-7, 253, 254, 255, 256, 257-8, 268; bird, 121, 125, 126, 128, 154, 170, 187, 253, 254, 255, 256; fish, 48, 74, 82, 114, 116, 118, 121, 123, 125, 126, 128, 130, 134, 154, 187 254 256, 257-8; insect, 82-3, 130; plant, 74, 107, 118, 121, 134, 153, 154, 170, 184, 187, 196-237, 240-1, 254, 255, 256, 267-8, 375; *see also* Domesticated plants, Tamarind; reptile, 107, 170, 253, 256, 268, 270
Sundaland, 6
Sydney, 250, 255, 257, 259, 358, 361

Tablelands, Southern, 29-36, 116, 130, 299; *see also* Atherton Tableland, Barkly Tableland
Talaut (Indonesia), 81
Talgai cranium, 321, 322
Tamarind, 188, 199
Tandandjal (site), 368
Tartanga (site), 368
Tasmania, 3, 4, 7, 9, 14, 15, 16, 17, 19, 35, 36, 52, 84, 86, 89, 90, 147, 153, 230, 231, 271-85, 317-19, 320, 344, 358, 360, 364; Aborigines, 153, 271-85, 317-19, 320, 371, 375, 378; art, 364-5; artifacts, technology, 121, 153, 375; cultural sequence, 281; languages, 273, 275, 279-80, 344, 345, 347, 349, 350, 351; material culture, 278, 279, 375, 376; physical anthropology, 317-19, 320; tribes, distribution, organisation, size, 273-80; *see* Ben Lomond, Big River, North East, Oyster Bay, Port Dalrymple, Port Davey, Sandy Cape, Stony Creek, West Point tribes
Taungerong tribe, 302
Taylor, W. G., 245, 246
Technology, *see* Archaeological cultures, periods, sequences, Artifacts, prehistoric
Tectonics, 2, 7, 44, 45
Tedford, R. H., 16, 55, 84, 86, 87
Tennant Creek, 66, 199
Thailand, 171-2, 232, 234, 235
The Tombs (site), 149
Thompson, P. W., 116, 125, 175
Thomson, D. F., 116, 143, 184, 187, 188, 192, 253, 255, 256, 257, 258, 259, 262, 264, 266, 267, 268, 269, 372
Thorne, A. G., 63, 188, 318, 320
Threlkeld, Rev. L., 116
Thurnwald, R., 376, 377
Timor, 4, 5, 6, 7, 80, 81, 158-79, 190, 191, 350
Tindale, N. B., 55, 83, 91, 149, 191, 253, 254, 255, 257, 259, 271, 283, 296, 316, 347, 350, 368-78
Tiverton Station (site), 362
Tiwi (language), 352
Torres Strait, 4, 5, 10, 126, 185, 186, 187, 237, 329
Torres Strait Islands, 185, 186, 187-8, 237
Torres Strait Islands languages, 187
Toxic plants, 207-8; *see also* Poisons
Trans-Fly (New Guinea), 185, 187
Tribes, distribution, organisation, size, 273-81; *see also* Group composition, size; names, 73-4, 253, 274; *see also* Aranda, Ba:gundji, Barrabool, Bindibu, Bindubi, Buandik, Bunurong, Dieri, Djauan, Gogobera, Gubabwingu, Jajowrong (Jaara), Kaiadilt, Karawa, Ka:wadji, Kulin, Larapinta, Lardiil, Maung, Mirning, Munggan, Narrinyeri, Ngadadjara, Ngadanja, Ngalia, Pintubi, Pitjandjara, Pitjanjadjara, Taungerong, Walbiri, Waljbiri, Wardaman, Warramanga, Wik-Munkan, Woiwurrung, Yalpari, Yintingga, Yir Yoront; *see also* Tasmania, tribes
Turnbull, C. M., 183, 192
Twidale, C. R., 2, 55, 70, 96
Tyimede (sites), 142, 145-6, 149, 152, 154
Tylor, E. B., 284, 371

Uai Bobo (sites, Timor), 162, 164-76
Uai Ha Ie (site, Timor), 162, 164
Ulnam (site, Timor), 160, 162

van Andel, T. H., 4, 6, 7, 20, 143, 147
van Heekeren, H. R., 159, 172, 176, 177, 178, 191
van Stein Callenfels, P. V., 172
Veddoids, 338
Veevers, J. J., 20, 161
Vegetation, 15-16, 27-8, 32, 33, 35-6, 42, 72-3, 106, 116, 133, 144, 161-2, 185, 198-9, 201, 209, 263-4; *see also* Bamboo, Domesticated plants, Sandalwood, Subsistence, plant, Tamarind, Toxic plants
Venezuela, 331
Verhoeven, Father Th., 159, 160, 161, 178, 179
Victoria, 16, 19, 21, 50, 51, 52, 55, 85, 86, 88, 90, 231, 288-313, 321, 347, 349, 352; *see also* Murray River, Southeastern Australia
Voegelin, C. F. and F. M., 350, 351, 352
von Koenigswald, G. H. R., 177

Wakefield, N. A., 22, 84, 85, 86, 88
Walbiri, Waljbiri tribe, 74, 231, 259, 357, 377, 378
Walker, J. B., 273, 274, 277, 279, 280, 281
Walker, P. H., 17, 19, 29, 33, 36, 96
Walker tribe, 275, 279-80
Wallacea, Wallace's Line, 81, 149, 159
Walls of China, 58, 61, 63
Warburton Range, 263, 267
Wardaman tribe, 377
Warner, W. L., 188, 242
Warramanga tribe, 258
Watanabe, Hitoshi, 245
Watercraft, canoes, 188, 254, 257, 258, 279, 378
Water supplies, 68-71, 106, 107, 252, 254, 256, 264, 266-7, 358-9, 363
Wattamolla (site), 126, 129
Weipa (sites), 133-6, 139, 140
Wells, *see* Water supplies
West, J., 273, 277, 280, 282, 284
Western Australia, 15, 18, 52, 70, 73, 74, 76, 85-6, 90, 152, 262, 264, 322, 330, 331, 360, 363, 364; *see also* Nullarbor Plain, Western Desert
Western Desert, 262-70, 327, 329, 331, 335, 361
Western Desert languages, 262, 348, 350
Western District, 288, 289, 291, 298, 301, 302, 309, 310
West Indies, 208
West Point (site), 278, 318-19
West Point tribe, 277

White, Carmel, 86, 141, 144, 154, 172, 178, 185, 191, 208, 239, 242, 243, 321, 374
White, J. P., 116, 153, 159, 174, 176, 178, 189, 190, 243
Wik-Munkan tribe, 198
Willandra Creek, 39, 42, 59, 64
Willems, W. J. A., 159, 160, 178
Willeroo Station (sites), 363, 364
Willey, G. R., 183, 184, 239
Williams, F. E., 185
Wilson Promontory, 6, 7, 9, 119
Wimmera, 42, 55, 64, 289, 291, 298, 309
Woiwurrung tribe, 301, 302
Wollumboola Lake (site), 125, 126, 129
Wood-Jones, F., 316, 318
Woomera (sites), 359, 362
Worslcy, P. M., 143, 182
Wright, R. V. S., 98, 120, 149, 152, 358, 361, 363, 364, 374
Wurm, S. A., 350, 351, 352

Yalpari tribe, 74
Yintingga tribe, 255
Yir Yoront tribe, 377, 378

Zambia, 191
Zanci (stratigraphic unit), 58-61, 62, 63
Zeuner, F. E., 177

*Designed by the A N U Architecture/Design Section*
*Text set in Monotype Baskerville, 10 point, 1 point leaded, and printed on Burnie English Finish 85 gsm at The Griffin Press, Adelaide*

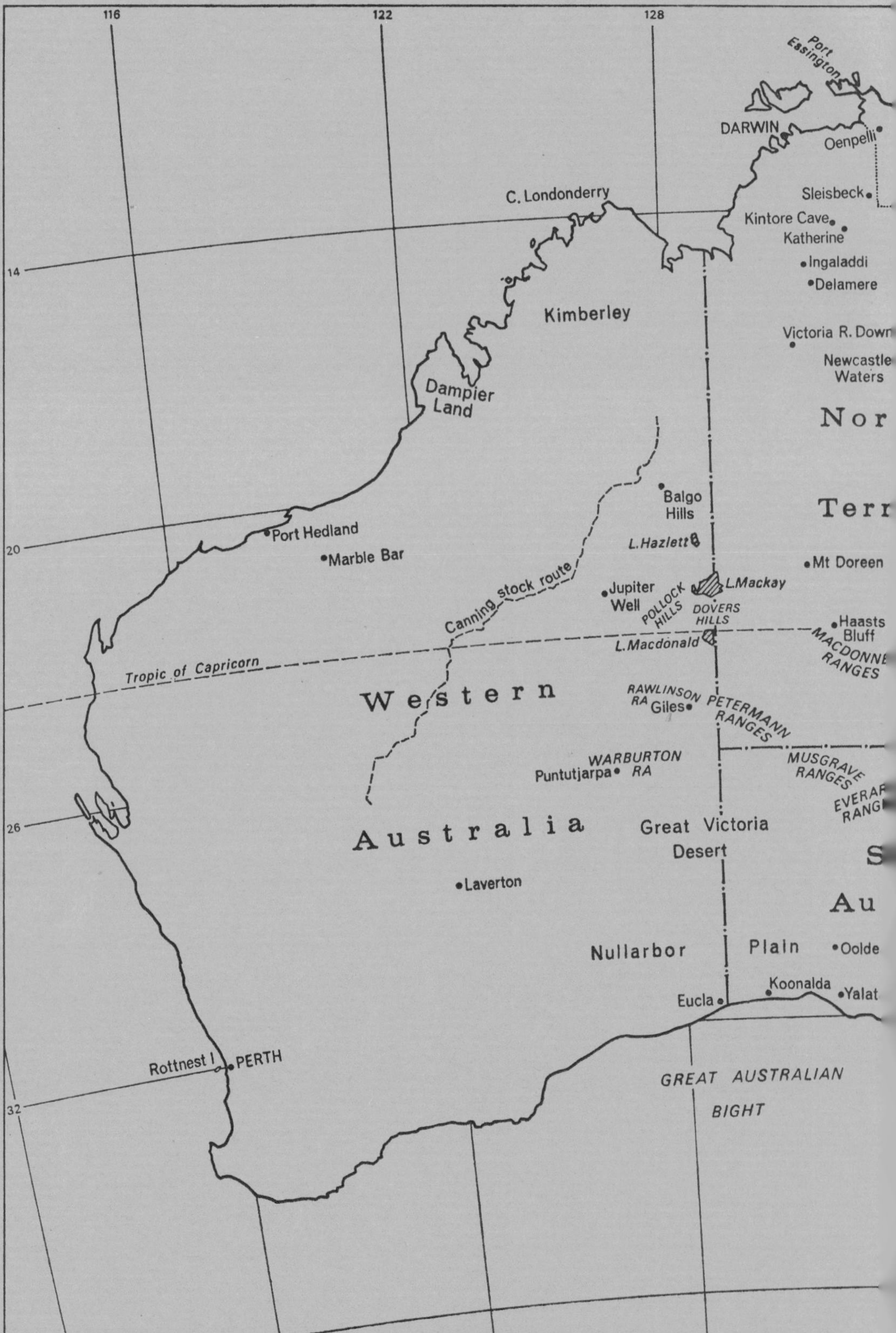
116
122
128
Port Essington
DARWIN
Oenpelli
C. Londonderry
Sleisbeck
Kintore Cave
Katherine
Ingaladdi
Delamere
14
Kimberley
Victoria R. Down
Newcastle Waters
Dampier Land
Nor
Balgo Hills
Terr
Port Hedland
Marble Bar
L. Hazlett
20
Mt Doreen
Canning stock route
Jupiter Well
L. Mackay
POLLOCK HILLS
DOVERS HILLS
Haasts Bluff
L. Macdonald
MACDONNE RANGES
Tropic of Capricorn
Western
RAWLINSON RA
Giles
PETERMANN RANGES
WARBURTON RA
Puntutjarpa
MUSGRAVE RANGES
EVERAR RANG
26
Australia
Great Victoria Desert
S
Laverton
Au
Nullarbor
Plain
Oolde
Koonalda
Yalat
Eucla
Rottnest I
PERTH
GREAT AUSTRALIAN BIGHT
32
38
110
116
122
128